1977 Yearbook of Science and the Future

Encyclopædia Britannica, Inc.
Chicago Toronto London
Geneva Sydney Tokyo
Manila Seoul

1977 Yearbook of Science and the Future

Library of Congress Catalog Card Number: 69-12349
International Standard Book Number: 0-85229-319-4

THE UNIVERSITY OF CHICAGO
The Yearbook of Science and the Future is published with the editorial advice of the faculties of the University of Chicago.

Editorial Advisory Board

About Science

A year in which there are no dominant, overriding concerns often is marked by a wide diversity of enterprise. The past twelve months have been such a period for science. Projects begun in previous years continued, but at a less intense pace. These included the manned space flight program, much reduced from the Moon-landing era, and efforts to protect the environment and to develop new sources of energy.

The *1977 Yearbook of Science and the Future* reflects the past year's diversity. The 16 feature articles deal with subjects ranging from complex investigations of subatomic particles and early stellar evolution to the proper design of sailboats and iceboats for maximum speed.

An innovation in this volume is the grouping of features that are related in subject matter. The first group, containing four entries, tends to belie this, but the rather heterogeneous assortment of articles presented is deemed to be of general interest. "Colonies in Space" analyzes the feasibility and potential benefits of building for human habitation enclosed spinning cylinders 240,000 miles into space. New techniques of scientific exhibition by the Milwaukee Public Museum, including the use of appropriate sounds, smells, and materials, are described in "The Museum as Environment." The evolution in the design of sailboats for speed and efficiency, from the use of intuitive/comparative methods to the application of principles of aerodynamics, is the subject of "Dynamics of Sailboats and Iceboats." And last in this group is "The Structure of Human Memory," which describes the revolutionary change that recent research has wrought in understanding the composition and workings of the remembering process.

Astronomy and astrophysics are the subjects of the second group of features. "The Formation of Stars" presents new information that astronomers have acquired about the earliest stages of stellar evolution. "Radio Telescopes" tells of the many discoveries that could not have been made without these instruments. One discovery in which radio telescopes were of great value was that molecules, including those essential for life on Earth, exist in interstellar space. "Molecules in Space" discusses the rapid expansion in the kinds of known interstellar molecules that has occurred in the last eight years.

The next group of features brings the reader back to Earth, treating man-made and natural environmental factors. "The Climate of Urban Areas" describes changes in temperature and precipitation that are generated by large cities and advises that such effects be taken into account in future planning. The causes and disastrous results of natural flooding are portrayed in "River Floods."

From the readily observable sphere of weather and high waters the Yearbook next turns to the submicroscopic world of the physicist and chemist. "Particle Physics: A Realm of Charm, Strangeness, and Symmetry" discusses the proliferation in the number of known subatomic particles, and also the likelihood of finding an ultimate fundamental building block of matter. The second article in this group, "Laser-Induced Isotope Separation," treats the use of lasers to separate isotopes of chemical elements from one another. The process is particularly significant in attempts to increase the supply of fissionable uranium.

The biological sciences are represented by several entries. "The Jojoba: Cinderella Crop for the '70s?" discusses the scrubby North American desert bush that contains in its seeds oil with properties strikingly similar to those of sperm-whale oil, a desirable industrial lubricant. One of the most complex systems of living organisms, displaying a high degree of interdependency, is portrayed in "The Coral Reef Community." And plants and animals that are poisonous to man are discussed in "Biotoxins: The Poisons of Life." The use of such toxins to make compounds with therapeutic properties is an interesting aspect of this subject.

Two final articles stand by themselves. "Number Theory: The Fibonacci Sequence" tells of an ordered set of numbers that is manifested in nature in such ways as the genealogy of male honeybees and the "divine proportion" of human aesthetics. Why this occurs is a question that continues to challenge scientists. "Superconductivity in Electronics" describes the development of a new "ultrasensitive, ultraaccurate, and ultrafast" technology, the application of superconducting materials to electronic instruments and circuits.

Full-color photographs and drawings abundantly illustrate the articles. Distinguished authorities have been selected to write these features, as well as the Year in Review entries, which treat the advances that have occurred in individual disciplines within science and technology. Also included is a section on scientists of the year, covering obituaries and major awards to individuals.

And the Future

This annual volume is not only devoted to current research and advances in science and technology but, in the broad sense, the intent is to convey to our readers the nature of the world of the future—to the extent that it is possible to do so. For this reason, and because the audience addressed basically consists of *Britannica* readers, it seems appropriate to describe here the results of a poll of U.S. contributors to the new, 15th edition of the *Encyclopædia Britannica*.

Ten questions were asked of this rather illustrious and select group of authorities. Not every question turned on science and technology but all required prediction of the future and, hence, the answers are of interest and concern here. Nearly 60 percent of the 1,681 individuals polled responded, an extraordinarily high figure which certainly indicates substantial interest and concern on the part of *Britannica*'s contributors. The results, indicated in parentheses, were as follows, eliminating the numbers of "undecided" in each instance:

BRITANNICA CONTRIBUTORS' POLL—1975

		Yes	No
1.	As the nation approaches its 200th birthday, do you think that democracy as we know it will survive?	(586)	(338)
2.	Will lasting peace eventually come to the Middle East?	(487)	(405)
3.	Do you think that the United States will experience continued inflation (691) or recession (272) during the next five years, or will a depression (147) occur during this period?		
4.	Will the energy crisis ultimately be resolved by the development of alternative energy sources? If yes, what source do you consider the most likely? (Answers included: nuclear, 398; solar, 346; coal, 104; winds and tides, 48; and geothermal, 26.)	(778)	(129)
5.	Do you expect the quality of life in this nation for the next generation to be better (213), worse (410), or about the same (293) as in your lifetime?		
6.	Will a cure be found for cancer in this century?	(553)	(266)
7.	Is widespread famine inevitable in the less developed, third-world countries?	(603)	(323)
8.	Will there be another Ice Age on this planet?	(572)	(194)
9.	Does the increasing popularity of such pseudoscientific books as *The Ancient Astronauts* and *The Bermuda Triangle* indicate a rising wave of anti-intellectualism in this country?	(368)	(518)
10.	Do you think that Toynbee's theory of the inevitable decline of civilizations is applicable to our society today?	(504)	(385)

Space precludes the interesting commentaries provided by the respondents; even as you and I, they were markedly reluctant to provide a simple yes or no to such queries without some qualification. Representative of many, perhaps, was the humorous note by Derek Gregory, contributor of the *Britannica* article "Batteries and Fuel Cells," who stated: "If we knew the answers to half of these questions, we would be rich, geniuses, and foolish to tell anybody else!"

THE EDITOR

Contents

Colonies in Space

by Gerard K. O'Neill

In an age when raw materials and land are at a premium, concerned scientists and engineers are examining the feasibility of man's migration to a new home in space.

The unique location and zero-gravity industrial facilities of a completed colony in space offer exciting possibilities for the construction of such specialized structures as orbiting solar power generating stations. In one concept of the future, as depicted on the overleaf, a linked pair of large rotating vessels has been built at a gravitationally stable region 240,000 miles from the Earth. In the foreground at the upper left, a docking complex, jutting from the sunward end of one of the colonies, provides arrival and departure terminals for rocket-powered space transports from the Earth and Moon and for articulated intercolonial shuttles that derive their linear velocity by peeling away from the rotating docking complex much as one might be flung from a spinning carousel if one let go. Pacing the colony in its movement through space is a recently completed power-generating station, scheduled to be towed into a geosynchronous orbit 23,000 miles above the Earth. Solar energy falling on eight large panels of photovoltaic cells will be converted into microwaves by generators housed in the twin conical pods at the ends of the collector bank and then broadcast to a permanently installed antenna grid on the Earth's surface for reconversion into high-voltage electricity.

One sunny Saturday morning, Jim Johnson and his wife, Ann, step out of their Italian-style villa, mount their bicycles, and pedal to a nearby lake for a swim. At noon they picnic with some friends in a forested park and later enjoy a soccer game. Jim's sister, who lives in a community 50 miles away, has invited them to her house for the remainder of the weekend, so that evening they board the local shuttle and are whisked to their destination at a speed of 400 miles per hour. As always, the Johnsons are captivated by the brief but spectacular view through the shuttle's observation ports—the Earth and Moon, suspended like pendant jewels, standing in striking contrast against the coal-black backdrop of outer space.

The year is 1990 and the Johnsons have been living in space for almost two years. This Saturday evening they are traveling between their home and the second of an orbiting pair of self-sustaining colonies, 240,000 miles from the planet of their birth.

The contemplation of such a future for man is old and has been a theme of scientific speculation and of fiction for several decades. In this regard, it is reasonable to ask why the future world of the Johnsons should be considered remarkable. In 1969, after space exploration and related research provided a data base for serious calculations, Princeton University physicist Gerard O'Neill realized the need for a quantitative reexamination of the concept of space colonization. Largely through his efforts, scientists, engineers, lawyers, and social scientists from universities, the aerospace industry, and the U.S. National Aeronautics and Space Administration met on numerous occasions to consider the matter. By late 1975 some startling conclusions had emerged: (1) man *could* colonize space using only existing materials and technological knowledge; (2) if work was begun soon, the biosphere of the Earth could be free of heavy industrial activity in less than a century; and (3), given the resources of the space environment, the problems of famine, pollution, energy procurement, and living space could be solved even if, in the distant future, the human population were to grow to 20,000 times its present size. Many people who studied the problem suggested that the U.S., possibly in cooperation with other countries, should begin a program aimed toward the establishment of a colony in space; they were motivated in part by the likelihood that such a migration of people would encourage self-sufficiency, independence, small-scale government, and cultural diversity. Thus, to the individuals and organizations in a position to transform dreams to realities, Jim and Ann Johnson represent far more than characters in a science-fiction story. To them, Jim and Ann are the very real children of the 1970s.

***GERARD K. O'NEILL** is Professor of Physics at Princeton University.*

The pictorial translation of scientific concepts was accomplished by Ben Kozak and John Youssi.

Feasibility considerations

Among the surprises to develop during early calculations on the feasibility of space colonization was the realization that space, contrary to its name, is abundant in many of the necessities of normal human life and that it possesses unique qualities of definite advantage to colonists. Sunlight is available in a continuous, unfiltered stream to power turbogenerators 24 hours per day, to grow crops throughout the year, and to furnish light and warmth during days whose length can be set at will. Unlike the Earth's delicate biosphere, the vastness of space, equivalent to a heat sink perpetually at a temperature near absolute zero, can absorb all the waste heat that an industrial society might produce. Without gravity, atmosphere, or weather, the movement of materials between points in space could be far less expensive a process than on Earth. With energy needed only to change speed and direction, rather than to maintain them, large shipments could be floated between sites much as barges float on the rivers of Earth.

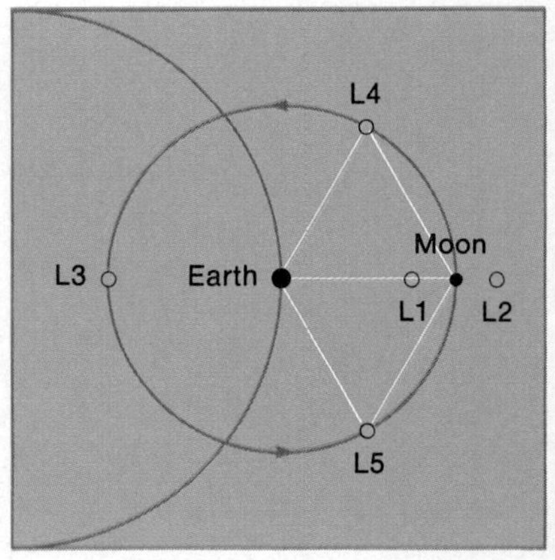

Lagrangian points or regions of gravitational stability produced by the interaction of two large planetary masses within certain mass limits, such as the Earth and Moon. Located equidistant from both masses at the Earth–Moon distance of 240,000 miles, the regions designated L4 and L5 appear particularly attractive as construction sites for the first space colonies.

The possibility of meteoroid impact damage was also considered. Model studies based on spacecraft data and lunar seismic recordings have indicated that most meteoroids are relatively soft conglomerates of dust possibly bound by frozen gases and that, for a space community 1,000 square kilometers in cross section, the mean interval for a 100-gram meteoroid strike causing easily repairable local damage would be three years. For major collisions involving a meteoroid one metric ton in weight, the mean interval would be an astronomical one million years. On the other hand, solar radiation and cosmic rays do pose serious problems. A large colony containment vessel, shielded by its own construction materials of aluminum, plastic, soil, and a sizable volume of atmosphere, would provide some protection against these hazards. However, more efficient protection could be afforded by surrounding the containment vessel with a stationary shield, built of the slag generated by the colony's construction operations.

The gravitational interaction of the Earth-Moon system has provided possible future colonists with several advantageous locations in space where certain problems of construction and stabilization would be minimized. In 1772 the mathematician Joseph-Louis Lagrange demonstrated that for every pair of large planetary masses within certain mass limits there existed in space a set of five points, called libration or Lagrangian points, at which a small body will remain approximately at rest relative to the two large masses. Of these points only two, denoted L4 and L5, are stable and for the Earth-Moon system they lie along the Moon's orbit—L4 precedes the Moon by 60° and L5 follows it by 60°. The Earth, Moon, and either point form the three vertices of an equilateral triangle whose sides are about 240,000 mi (390,000 km) long, the Earth–Moon separation distance. The Earth–Moon interaction is further complicated by the gravitational effect of the Sun; hence, the regions of stability actually become orbits about L4 and L5 rather than fixed points. An object pushed or pulled out of those orbits will tend to return. Such a gravitational lock would maintain the completed colony indefinitely without the need for periodic corrections.

Another surprise to emerge from early feasibility studies was the discovery that a closed, pressurized vessel containing a surface area measured in the hundreds of square miles could be built in space with currently available construction materials. An initial modest project,

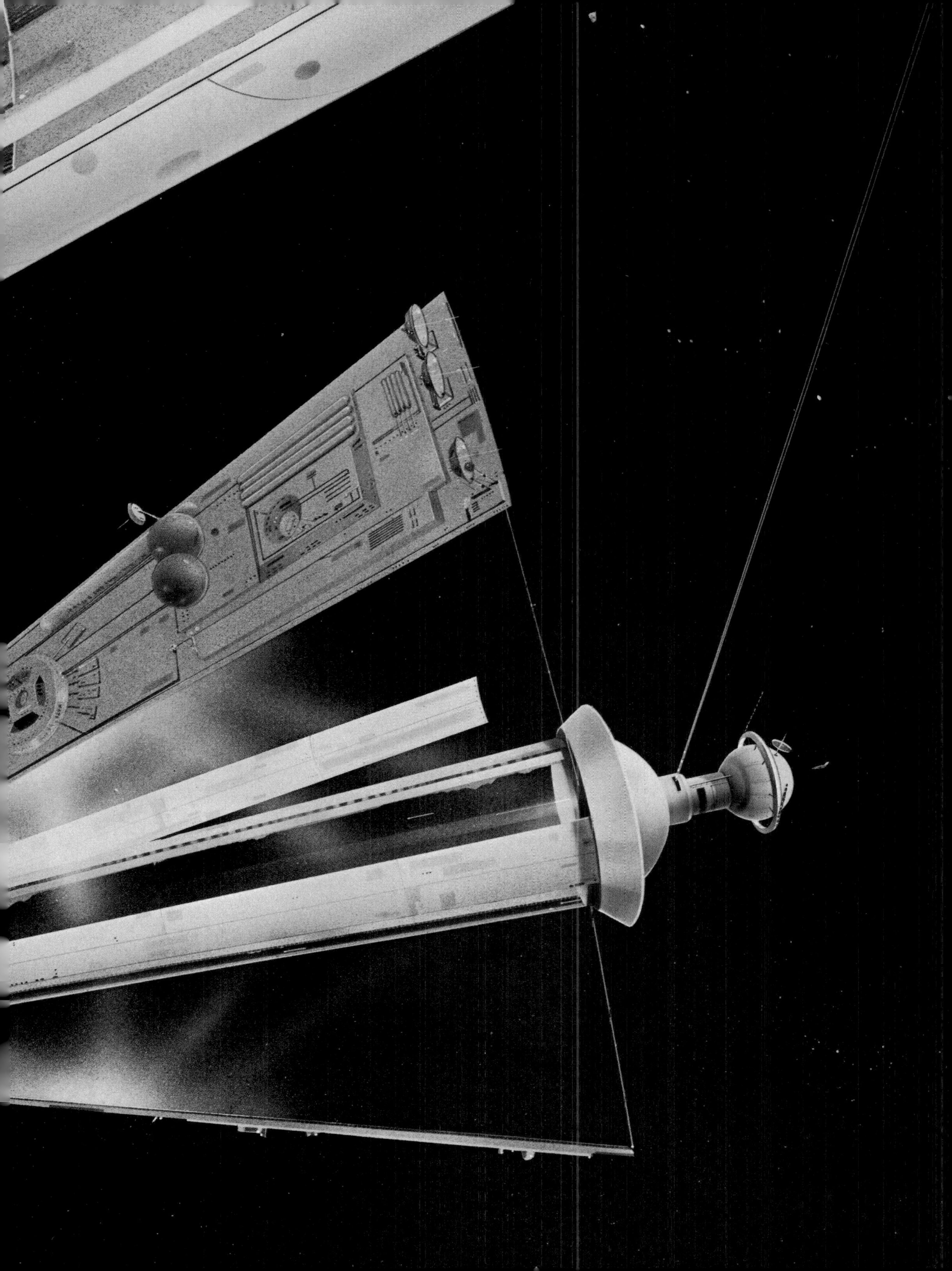

Space colonists of the mid-21st century may find themselves surrounded by the immensity and lavish conveniences of a paired system of rotating pressure vessels each 15–20 miles in length, as pictured on pages 12 and 13. To maximize the efficiency of the design, the long axis of each cylinder is pointed toward the Sun. Restrained by tethers, three external rectangular mirrors open outwardly above window strips to provide light and warmth to the land regions. An arrangement of external pods with environments tailored for specialized manufacturing, agriculture, and scientific research surrounds the outer rim of a parabolic solar-energy collector. The focus of the collector is a narrow band of closely spaced boiler tubes embedded in the surface of the main cylinder; these transfer absorbed energy in the form of a hot fluid to turbogenerators that provide power for the colony. Projecting from the spaceward end cap, the colony's waste-heat radiator emits a red glow. At the sunward end is a docking complex for handling traffic to and from the sister colony as well as passengers and supplies from the Earth and Moon.

housing some 10,000 people on about 750,000 square meters (approximately 0.3 square mile), would require an expenditure not many times greater than that spent on the Apollo space program—about $40 billion in 1975 dollars. For economic feasibility, the work force and production machinery of the completed first colony would be required to contribute in large measure to the construction of the second. This leapfrogging technique could be applied in turn to third- and fourth-generation colonies.

Mining the Moon

The fabrication in space of a habitat for humankind requires vast amounts of materials: that for a closed vessel or system of vessels of sufficient strength to contain an atmosphere; additional matter of sufficient mass and thickness to provide a shield against radiation and meteoroids; more for the structures and equipment within the vessel; and an initial supply of soil, nutrients, and water from which vegetation can grow. The estimates of construction materials required range to hundreds of thousands of metric tons for a population of 10,000, with additional amounts for radiation shielding. Even if materials were abundant on Earth, the energy expenditure required to transport them into space against the force of gravity would be enormous.

Fortunately, the Earth has a large natural satellite whose accessibility by spacecraft has been amply demonstrated. Apollo soil samples have shown the average lunar surface to be composed of 40% oxygen by weight, 20% silicon, and 20–30% metals. The Moon is a source of many metals, including aluminum and titanium, of silicates for glass, and of oxides for precious oxygen. Of the elements most needed by a colony under construction, only carbon, nitrogen, and hydrogen are lacking in the lunar surface. The required masses of these elements are relatively small, however, and Earth can afford to provide them. For example, of the total mass of required water, only 11%—the weight of the constituent hydrogen in each molecule of H_2O—must be shipped from Earth. Likewise, soil from Earth is not necessary for agriculture in space; if provided with nitrates and trace elements from Earth, and with water to be derived as already noted, lunar soil will adequately support plant growth. For the initial colony it has been estimated that as much as 98% by weight of all needed materials can be obtained from the Moon and can be shipped to a construction point such as L5 for one-twentieth the energy cost of a comparable shipment from Earth.

Utilizing a transportation system derived from space-shuttle technology, a mining crew of perhaps 200 with 10,000 metric tons of equipment would set down on a preselected site on the lunar surface. A power plant and mining machinery would be included, but much of the equipment would comprise the means for transporting ore from the Moon to the construction site in space. To this end, several novel devices resembling catapults have been proposed which would take advantage of the absence of frictional air resistance on the Moon and the low lunar escape velocity, about one-quarter of Earth's 11.2 km/sec. One such mechanism, called a transport linear accelerator or mass driver, employs a closed aluminum track some 14 km in length and a system of small bucketlike vehicles each weighing about 5 kilograms and containing superconducting coils. Levitated and guided along the track by synchronized magnetic pulses, each bucket would accelerate a 10-kg payload of compacted lunar soil to the escape velocity. The bucket would then be decelerated and recirculated, while the payload would continue in free flight, gradually slowing in the gravitational field of the Moon. Recent calculations suggest that the optimum place at which to catch the catapulted payloads may be L2, an unstable Lagrangian point directly behind the Moon as viewed from Earth, where the energy cost of moving materials to a construction site like L5 would not be high.

The first colony

While mining proceeds on the lunar surface, it will be necessary to establish a temporary but hospitable environment for the work force at the construction site of the first colony. This would probably take the form of a prefabricated space station where technicians, engineers, and laborers can live with their families, produce much of their own food, smelt lunar ore, and fabricate the metal, ceramics, and glass that will become the pressure vessels of the colony. A pilot colony of 10,000 and a fairly luxurious vessel system of 500,000 metric tons has been estimated to require a construction crew of 2,000, some of whom with their families may choose to remain aboard the completed project as part of the colony.

Given sufficient limiting conditions, the geometry of a space community roughly will define itself. To satisfy the requirements of normal gravity, a normal day-night cycle, natural sunlight, an appearance resembling Earth, and the most efficient use of materials and solar power, one possible design appears to be a pair of spinning cylinders, spherically rounded at the ends like medicine capsules. For Model 1, one concept of the initial colony that has been analyzed in some detail, each cylinder would be only as large as deemed workable, about 1 km long and 100 m in radius. Its outer walls would be divided into six longitudinal regions of alternating land and window. The three land regions plus the end caps, built up of lunar rock and soil and comprising a total surface area of 750,000 sq m, would be devoted to living space, parks, forests, lakes, and a population of animals. The window regions, comparable in size to the land strips, would allow natural sunlight to enter the cylinder. Constructed of transparent glass, they would be subdivided into panels and reinforced by the structural members of the cylinder. Three external rectangular mirrors, hinged near one of the cylinder ends to open out above the window strips, would reflect the image and light of the Sun into the land regions in a manner comparable to that on Earth. As the mirrors were moved, the angle at which sunlight strikes the land regions would change,

An advanced transport linear accelerator may take a form similar to the design shown on pages 16 and 17. Serving both as the means for transporting ore from the Moon to the site of colony construction in space and as living quarters for the miners and transport crew, the device employs a flexible, closed aluminum track supported at intervals by mobile power and support towers. Rectangular bucketlike vehicles with a core of superconducting coil windings ride the track, levitated and propelled by magnetic pulses synchronized over the length of the device. Each tower is a self-contained unit, providing its share of the power input to the accelerator from a parabolic solar collector, and is capable of self-propelled locomotion when the mining camp is relocated or when fine adjustments are needed in the trajectory of the launched ore. In modified form, such a device might be used to push an entire asteroid toward the Earth by launching bits of it in the opposite direction.

and closing them completely would allow night to fall. Such adjustments would control the length of day and the intensity of the incoming light, which in turn would affect the average temperature and produce Earth-like seasonal variations.

Calculations for the structural framework of the cylinders have been made. One version has been based on the use of steel cables, or longerons, that run the length of the cylinder and sustain the atmospheric pressure on the two hemispherical end caps, together with circumferential bands that carry the forces of the atmospheric pressure and the spin-induced artificial gravity on the cylinder walls. Strength can be further improved by the extensive use of titanium in the overall structure. To allow for the window regions, both longerons and bands can be bunched to form a coarse mesh with openings about 50 cm square that will provide greater than 80% transparency and be nearly invisible against the incoming sunlight.

It is essential to the energy efficiency of this design that the long axis of each cylinder point continually toward the Sun. To simulate gravity on the internal surface of the cylinder, it is also necessary that the cylinder rotate about that axis. To prevent the axis from gyroscopically precessing away from solar orientation, it has been suggested that two cylinders, oriented with parallel long axes and rotating in opposite directions, be coupled with a tension cable at one pair of ends and a compression tower at the other. The tension cable would act to prevent out-of-parallel outward movement of the ends and the compression tower would resist inward movement. Together they would form a system that has zero net angular momentum along the long axes of the cylinders and is able therefore to remain pointed toward the Sun without the need for corrective thrusters.

Electric motors powered by solar energy would apply static torque through the structure of the compression tower to start the cylinders spinning simultaneously. To provide a radial acceleration equal to the gravitational acceleration on the surface of the Earth—9.8 m/sec^2—the necessary rate of revolution can be computed easily using the formula $a = v^2/r$, in which a is the radial acceleration, r is the cylinder radius, and v is the linear velocity of any point on the cylindrical surface. For $r = 100$ m, a possible radius for Model 1, $v = 31.4$ m/sec or 1,880 m/min. For each revolution of the cylinder, a point on its surface must travel the circumference, $2\pi r$ or 630 m. Thus, in one minute the cylinder must complete 1,880/630, or about three revolutions. For a pair of cylinders with a total mass of 500,000 metric tons, a spin-up time of about one week has been calculated to require a constant power input of 150 kilowatts, which is only a fraction of the estimated power-generating capacity of the colony.

Power for spin-up and for all the daily necessities of the colony can be derived from simple paraboloidal mirrors, boiler tubes, and closed-cycle turbogenerators. Considering that the solar power falling on the end cap of a Model 1 cylinder is on the order of thousands of megawatts, a system with 33% thermal efficiency can provide for 10,000 people more than ten times the power per person currently demanded (ten kilowatts) in highly industrialized regions of the world. Waste heat could be sent into space by infrared radiators without any fear of adversely affecting the environment.

Though the raising of crops and food animals could well be practiced within the main cylinders at the expense of living space, suggestions have been offered that, to enhance the quality of life in the colonies, the cylinders themselves should contain only structures and facilities of a personal and social nature; *i.e.*, residences, schools, stores, medical facilities, entertainment centers, parks, and outdoor recreational sites. Food production would be carried on outside the main cylinders in groups of ancillary pods that could be tailored in geometry and in solar orientation to suit the product being grown. Wheat, for example, will thrive in hot continuous sunlight; on the other hand, fruits and vegetables respond well only to a rhythmic day-night cycle. Calculations based on an ideal 12-month growing season and on experiments using double-planting and multiple-cropping techniques have indicated that a colony in space should be able both to house and to support an ecological limit of 143 people per 10,000 sq m of land region with an adequate diet of 3,000 calories, 52 grams of usable protein, and 4.3 lb (1.9 kg) of total food per person per day. This would impose an agricultural requirement of 440 lb of standing biomass to maintain each person. It also dictates the maximum population for a colony of any given usable area; for example, for a Model 1 system of 750,000 sq m, where agriculture must be carried out within the main cylinders, the maximum population would be 750,000 sq m × 143/10,000 sq m, or about 10,700.

Other external pods might be devoted to factories, workshops, and research laboratories. On Earth, heavy industry expends much of its budgeted energy simply in fighting gravity. The weightlessness of space would permit superior manufacturing techniques and facilitate automated construction, not only of conventional items but also of other and larger colonies in space. Under conditions of zero gravity and almost absolute vacuum, the processing and production of exotic metal and metal-ceramic alloys, high purity glasses, large single crystals of semiconductors, and other materials almost impossible to make under terrestrial conditions would be both practical and commercially rewarding for the colony.

Alternatives to Model 1

It should be emphasized that Model 1 is not the only plan for a space colony under consideration; it is simply one version to which substantial analysis has been devoted. Scientific and industrial planners are not yet at the point of systematic comparison of design alternatives and cost minimization, and many designs at the present stage of evaluation appear feasible. Regardless of the variety of geometries and functional details, however, many of the problems and solutions of space colonization that were discussed above still apply.

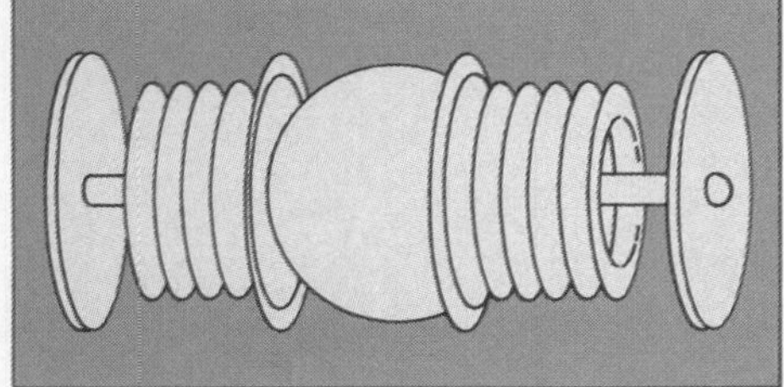

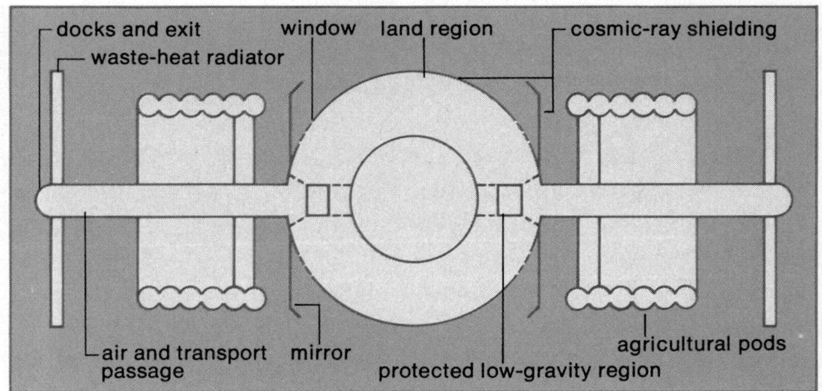

An alternative to the cylinder could be a sphere some 250 m in radius, corresponding to a total surface area of $4\pi r^2$ or 785,000 sq m. It would rotate about twice per minute inside a stationary cosmic-ray and meteoroid shield built from an accumulation of slag produced by the smelting of lunar ore in space. Within the sphere the usable land area would run from its equator, nearly 1,600 m in circumference, to the 45° line of latitude in each hemisphere. The topography of the land might resemble in some respects the hills of San Francisco, and the population density would also be similar. Again, there could be substantial streams and ponds and a lush vegetation of grass, flowers, and trees. Sunlight, entering at an angle of 45° by way of mirrors, could be adjusted to a normal diurnal cycle in the residential sections and optimized in the agricultural sections for particular crops.

Another proposed form is that of a huge rotating torus or doughnut-shaped wheel, more than 1.5 km across and connected by a framework of struts or tubes to a central hub. The torus half that faces outward from the center would comprise the habitable land region; the inner half would be transparent, fitted with louvers that controlled sunlight being received from large reflectors positioned around the hub. Pointing the axis of rotation toward the Sun would maximize energy efficiency by ensuring a continuous exposure to reflected sunlight of the transparent section of the hull.

The quality of life

With food in abundance, clean electrical energy, controlled climates, and pleasant weather, living conditions in the colonies should be much more satisfying than in most regions on Earth. Besides increasing living space, locating industry and agriculture outside the main vessel will produce at least two additional benefits: their hazards and disruptive qualities will be isolated from home and family life, and the temperatures, illumination levels, and humidities best suited to agriculture can be chosen without affecting the quality of life in the living areas. For the short distances of the vessel interiors—up to 30 km for some fourth-generation designs—bicycles and low-speed, pollution-free electric vehicles will be ideal. Conventional gliding may be a popular recreational activity. Analysis of the thermal scales that should exist in the atmosphere of a large vessel indicated that gliders would find sufficient atmospheric instability for lift.

Depending on the size of the colony, most Earth sports and recreations— as well as some unique new ones—would be possible in such communities. One highly interesting aspect of spin-induced artificial gravity is that it decreases linearly as one approaches the rotational axis. If the interior landscaping were to include either freestanding mountains or, as has been suggested for large cylindrical and spherical colonies, a mountain profile up an inside surface toward the rotational axis, a climber would find the going easier as his altitude increased and would appreciate that the air did not thin. In carefully enclosed areas near the axis of rotation, man's long-standing dream of flying solely by muscle power could come true, and an entirely new set of interests built around the ornithopter might be born. This region also could house a low-gravity gymnasium and pool.

As colonies multiplied and space technology increased, colony sites would eventually expand beyond the Lagrangian points to more distant orbits. Because each colony would have a negligible intrinsic gravitational field and probably be about the same distance from the Sun, travel would consume little energy. Hence, the recreational space vehicles of the colonial age would need to be hardly more than pressure vessels with little complexity beyond basic life-support equipment and the comfortable furnishings typical of mobile homes of the 1970s. With this ease of travel, visiting, tourism, and permanent exchanges of people would soon become commonplace.

Even with such interaction, however, the self-sufficiency of space communities would have a strong effect on government and society. A large community, eager to preserve its language and culture, might choose to isolate itself. Each colony would be free to develop its own life style, free to experiment with clothing, aesthetic design, music, art, literature, law, government, and personal relationships. The options for creativity would be almost unlimited.

Envisioned as a more elaborate variation of a spherical space colony, the vessel depicted on the opposite page comprises two concentric spheres tethered together at their equators and a symmetrical arrangement of agricultural pods, transport passages, and waste-heat radiators. Rotation about the long axis provides Earth-normal gravity at the equator of the outer sphere and a large fraction of it in the inner sphere, depending on its radius. A solid band of window circles each pole, and flat, circular mirrors centered at the poles direct solar illumination into the larger sphere. Nonrotating cosmic-ray shielding, obtained from the slag generated in the smelting of ore, backs each mirror and surrounds the main sphere; it is also used to form conical shield extensions at the outer edges of the mirrors to fully protect land regions in the sphere from cosmic rays that might otherwise enter through the windows. A zero-gravity region extends along the length of the vessel's rotational axis, facilitating materials handling in the transport passages; in the space between spheres, this region is bounded by light nylon netting to provide a gymnasium and recreational area for the colonists.

Power for the Earth

Concern for the accelerating consumption of finite supplies of coal and petroleum has generated a renewed interest in alternative sources of energy, among them the harnessing of solar radiation. Filtered by the atmosphere and unavailable for half of every 24-hour day, however, the Sun's energy as received at the Earth's surface can be only inefficiently collected and utilized. The benefits of capturing much more of this enormous solar flux in the continuous day of space led engineer Peter Glaser and others in the 1960s to develop a new concept called the satellite solar power station (SSPS). The power collector, in the form of a satellite positioned in a synchronous 24-hour orbit, 22,300 mi above a fixed point on the Earth's surface, would receive solar energy in space and convert it to a highly directed microwave beam of an intensity sufficiently low to meet environmental

The quality of life that could be expected in a large, autonomous, self-sustaining colony of the 21st century is shown on pages 22 and 23. Parks, woodland, flowing water, and mountainous country extend for miles in a panoramic vista. Following the vessel's axis of rotation, a lightly tethered structure houses low-gravity laboratories and medical facilities, recreational areas, observation decks, and a launch platform for gliders and ornithopters; in addition, it provides a convenient avenue for people and materials moving to and from the docking complex.

safety regulations. The beam would be continuously locked onto an antenna grid on the Earth and converted into high-voltage current suitable for distribution over conventional power lines.

Studies have indicated that the technology to build the SSPS is available. Recent microwave experiments have already demonstrated overall broadcast efficiencies of 54% at power levels of many kilowatts. A number of critics have observed that low-intensity microwaves beamed to an enclosed antenna would be preferable to nuclear power or strip-mining for coal. In addition it can be shown that, for equal quantities of power production, the amount of land needed for microwave antennas and electrical converters would be as little as 5% of that required for completely Earth-based solar-energy reception. To date the major stumbling block has been economic: the inability to design both a power satellite with very low mass and a system for placing the satellite components in synchronous orbit at a cost competitive with other energy sources.

It has been proposed by some proponents of space colonization that these economic difficulties may well be avoided if the SSPS were constructed in space using the zero-gravity industrial facilities and lunar mines of the completed space colony. As with the building of the colony itself, the cost for the transport of lunar ore to a construction site like L5 would be a small fraction of the cost for lifting those same materials from Earth. After fabrication of the power plant, the cost of its movement from L5 into geosynchronous orbit should also be low.

The construction of power stations for positioning over many parts of the world could become a lucrative business for an established colony. Several independent estimates have indicated that it could soon generate sufficient cash flow to finance the entire colonization project. Assuming that productivity in space will be similar to that of heavy industry on Earth, that interest rates on capital will be 10%, over and above inflation, and that the construction of the first space colony will require six to ten years, it appears that, within 15 years from the beginning of major investment, the total power input from space using colony-built SSPSs could exceed the peak capacity of the trans-Alaska pipeline, or two million barrels of oil per day, and that within a few more years the further growth of SSPS systems would be limited only by saturation of the world market. Although these estimates are based on low power costs at power-distribution points on Earth (initially $0.015/kw-hr and decreasing thereafter), the benefit-to-cost ratio for the entire program, calculated from economists' standard discount formulas, appears to be much higher than one.

Beyond Model 1

It has been shown that the ability of an initial colony like Model 1 to fabricate most of its structural needs from low-cost lunar materials using energy-saving zero-gravity building techniques, to supply a terrestrial demand for exotic space-manufactured goods, and to construct power stations for an Earth sorely in need of energy sources should allow it to finance itself. It is also reasonable to expect that an additional economic payoff in the form of greatly reduced construction and transport costs would come from the development of a second-generation colony, a Model 2 perhaps larger than the first. Following this path of expansion, there may come a day when space colonies would look beyond the Earth–Moon system for essentially all of their material needs. The asteroid belt, almost 300 million km beyond the orbit of the Earth, comprises thousands of small metal-rich bodies a few hundred kilometers or less in diameter. Composed mostly of iron and nickel, one cubic mile (4.2 cu km) of asteroidal material could provide ten times the current annual world steel consumption. The asteroids also contain significant amounts of nitrogen for fertilizer and of hydrocarbons, the chemical basis of plastics and many fibers and fabrics as well as an extraterrestrial source of hydrogen for water.

Transport of materials from the asteroid belt can be accomplished in a manner analogous to ocean freight on Earth; *i.e.*, in very large quantities, with very low fuel costs and minimum crews. In space, the optimal technique may be to eliminate the freighter hull altogether. The mass driver discussed earlier not only can hurl small loads in one direction by using the Moon as a reaction mass but by using these loads as reaction mass it can also push relatively large bodies in the opposite direction. With such a means of propulsion it should be possible to move an entire million-ton asteroid toward a construction site simply by launching bits of it in the opposite direction.

It is interesting to imagine a possible scenario of the 21st century: An abundant supply of asteroidal materials has given rise to completely self-sufficient colonies, immense in scale and lavish in conveniences. By 2025 to 2050 the number of large colonies has multiplied rapidly, and emigration into space becomes an attractive and economically practical option for a substantial fraction of the Earth's population. This trend continues to accelerate so rapidly that in 30 years Earth's population falls to 1.2 billion, a figure corresponding to the year 1910. Due to the ease of zero-gravity materials handling, most manufacturing takes to space as well, and the Earth slowly begins to return to the relatively pastoral state that existed before the coming of the industrial age. Just into the 22nd century, new colony construction thins out the population of the colonies themselves to a rather stable density of 1.43 persons per 10,000 sq m, about 1/100 of their ecological limit. By then the total land area among the colonies is more than three times that of the Earth.

Aside from providing material and economic benefits, there seems a good chance that opening the door into space can do much to improve the human condition in other respects. Relieved from the need to contest for the diminishing resources of his planet, man can anticipate a more peaceful and famine-free future than would otherwise be his lot. And humankind will be guaranteed to endure for generation upon generation to come.

The Museum as Environment

by Kenneth Starr and Shirley Marine

Proving that "museum" need not be synonymous with "dull" and "dusty," the Milwaukee Public Museum has created exhibits that involve the visitor in a vivid learning experience.

Courtesy, Milwaukee Public Museum

"Education," said Augustus G. Pohlman, M.D., in a lecture delivered before the St. Louis Natural History Museum Association in 1917, "is founded on one of the most important human characteristics—curiosity. The human being is naturally curious about things which have gone before, about things as they are, and about things as they will be." Among all educational institutions, museums of natural history and history have the satisfaction of that curiosity as their chief raison d'être. But museum directors are learning that it is not enough merely to display the relics of distant times and places. The museumgoer must be helped to see them in context; he must become, if only for a little while, a part of the living environment that gives them meaning.

One of the major innovators in the effort to involve the museumgoer, rather than treating him as a disengaged spectator, is a middle-sized museum in the Midwestern city of Milwaukee, Wisconsin. Pioneering in museum exhibition techniques is nothing new to the Milwaukee Public Museum. As early as 1889, five years after the museum opened, the noted naturalist and taxidermist Carl Akeley created for it the first diorama, a three-dimensional exhibit showing mounted muskrats in a simulation of the animals' native habitat. Today, in a building completed in 1963, the museum has three floors of exhibits that use not just sight but other senses as well to bring to life what would otherwise be dusty remnants of bygone times and alien cultures.

KENNETH STARR *and* ***SHIRLEY MARINE*** *are, respectively, Director and Public Affairs Officer of the Milwaukee Public Museum.*

(Overleaf) Photograph by Bill Arsenault—EB Inc.

(Bottom right) Bill Arsenault—EB Inc.; others courtesy, Milwaukee Public Museum

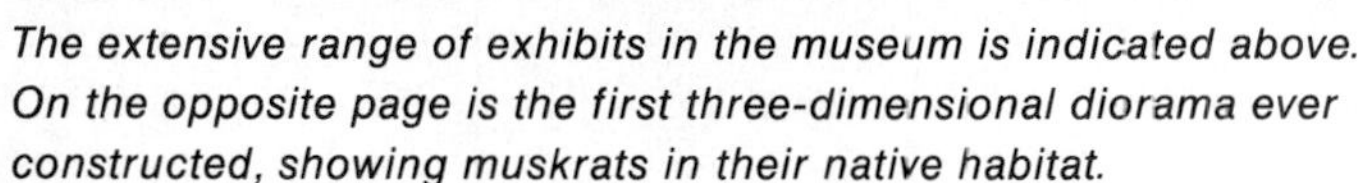

The extensive range of exhibits in the museum is indicated above. On the opposite page is the first three-dimensional diorama ever constructed, showing muskrats in their native habitat.

Good-bye to glass cases

The first break with tradition made in planning these exhibits was the development of overall themes for floors or wings. The exhibits on the first floor follow a historical sequence from the early days of Earth through the rise of Old World civilization, the age of trade and exploration, and New World Wisconsin to turn-of-the-century Milwaukee. Supplementary exhibits illustrate ongoing geologic processes and the development of religious concepts and means of communication. On the second floor a "Wonders of Life" wing depicts classifications of life forms, their interrelationships, and the environments in which they live. Also on the second floor and continuing to the third, exhibits with the theme "The World and Its Peoples" provide an environmental and cultural tour of North America, the Arctic, Asia, the oceans, Oceania, South America, Middle America, and Africa.

The second break with tradition was in the manner of presentation. Not only specimens but also ideas and concepts are exhibited. In atmospheric settings, the visitor is exposed to visual and sometimes auditory and olfactory sensations designed to make the habitat simulation a living experience.

In an American Plains Indians exhibit 40 feet in diameter, shown without the distortion or distraction of glass, eight bison appear poised in flight from the goading spears and yells of Crow archers and spearmen. The visitor hears the taped sounds of the thundering hooves, while a visitor-activated rattle on a rattlesnake in the exhibit adds an unexpected bit of authenticity. Adjacent exhibits, combining miniature dioramas and actual artifacts, explain how the lives of the bison and the Plains Indians were interwoven. Through this, the relationship of both with other living forms and with their environment becomes apparent.

Courtesy, Milwaukee Public Museum

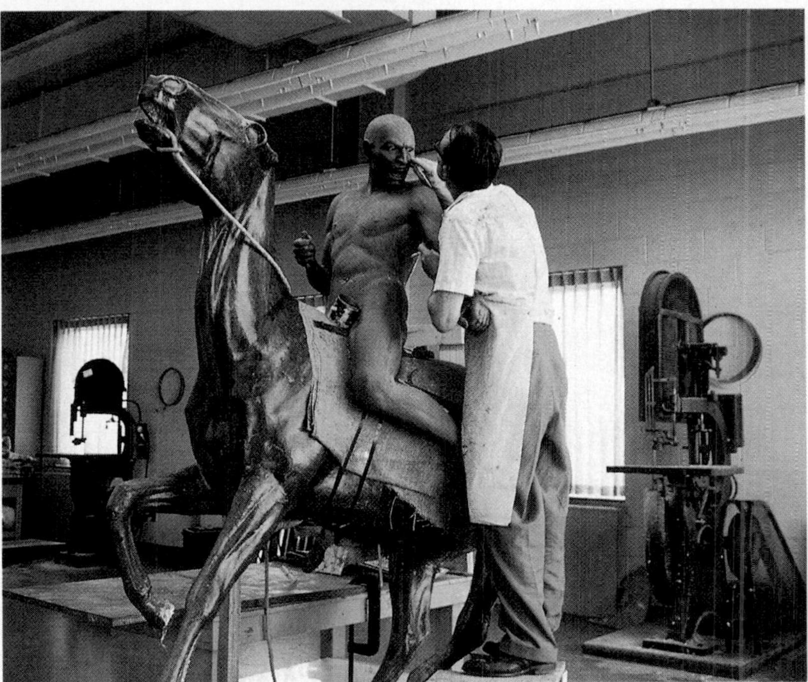

Huge, life-size diorama (above) is a highlight of the American Indian exhibits. Museum worker constructs the Crow hunter (left) in position to kill the charging bison (opposite page).

(Top) Bill Arsenault—EB Inc.; (bottom) courtesy, Milwaukee Public Museum

(Bottom right) Bill Arsenault—EB Inc.; others courtesy, Milwaukee Public Museum

The building and decorating of the temple of Amon (above and at right) reveal the scale of the architecture and the construction equipment of the ancient Egyptians. The religious atmosphere of India is conveyed by the temple setting of the statue of Siva (opposite page), lighted to emphasize the Hindu god's divinity.

Courtesy, Milwaukee Public Museum

Courtesy, Milwaukee Public Museum

North American Indian exhibits include a section of a Hopi kiva (above) and many highly dramatic painted wooden masks (opposite page).

A similar variety of techniques are used to present the life cycle of the Hopi Indian in the American Southwest. The exhibits are housed in a simulated two-story pueblo. Epoxy flooring gives the impression of hard-packed, sandy ground beneath the visitor's feet. On entering, the visitor sees four rooms with manikins engaged in the everyday activities of pueblo life 100 years ago. The father, seated crosslegged on a fur rug, is weaving. Mother and daughter are cooking at a hearth. A woman is baking piki bread and another woman is grinding grain. Another area is devoted to the religious ceremonies that delineate Hopi life. A simulated section of a kiva or clubhouse of the Hopi Antelope Society is shown here, with an altar and a sand painting on the floor. A ladder, such as that used by Hopi men to enter the subterranean chamber, extends outside the kiva and forms part of the exterior of the exhibit. Smaller exhibits combine actual kachina masks and dolls with miniature dioramas showing how they are used in ceremonies. Additional miniature dioramas portray the Drab Flute Ceremony and an overall cross section of a typical Hopi pueblo.

Continuing on, the visitor enters an igloo, strolls past Japanese shops and through a Japanese garden to a house where a tea ceremony is being performed, stands in the bazaar of Old Delhi amid the sounds of taxi horns and the cries of merchants hawking their wares, walks through a Melanesian ceremonial house with skulls in the doorway and initiation masks hung on poles, peers into an opened burial of pre-Columbian Peru, looks through the windows of a Spanish colonial home in Ecuador, hears and sees the bargaining in a Guatemalan market, and visits the African continent with all its geophysical, biological, and cultural diversity.

The exhibits in the African wing are the most spectacular in the museum. Illustrating the Islamic culture of northern Africa, the Moorish rooms, with their inlaid woodwork, brass, tooled leather, tile, and glass, are viewed through windows from a tiled courtyard. Another aspect of northern African life is revealed by a desert tent of the nomadic Tuaregs. Moving from this area, past a case exhibit on ancient black cultures, the visitor unexpectedly comes upon two great open dioramas, among the largest such displays in the world.

The first shows a bamboo forest and the forest's edge in East Africa, complete with taped sounds of animals, seemingly coming from all sides. Included are a trumpeting elephant, Colobus monkeys, antelope, rhinoceros, and a pride of lions. This diorama has special significance for museum employees. The male lion was the mascot of the 1928 Cudahy-Massey-Museum African Expedition, which acquired many of the animals in the wing. Rescued from a fire as a cub, Sim was brought back to the museum and became the staff's pet, living on the roof of the old museum building a block south of the modern structure. Eventually he became too large for those quarters and was moved to the county zoo, but in death Sim returned to the museum to be mounted in a masterful pose with a lioness and cubs. The elephant, a resident of the museum for half a century, presented a problem

Courtesy, Milwaukee Public Museum

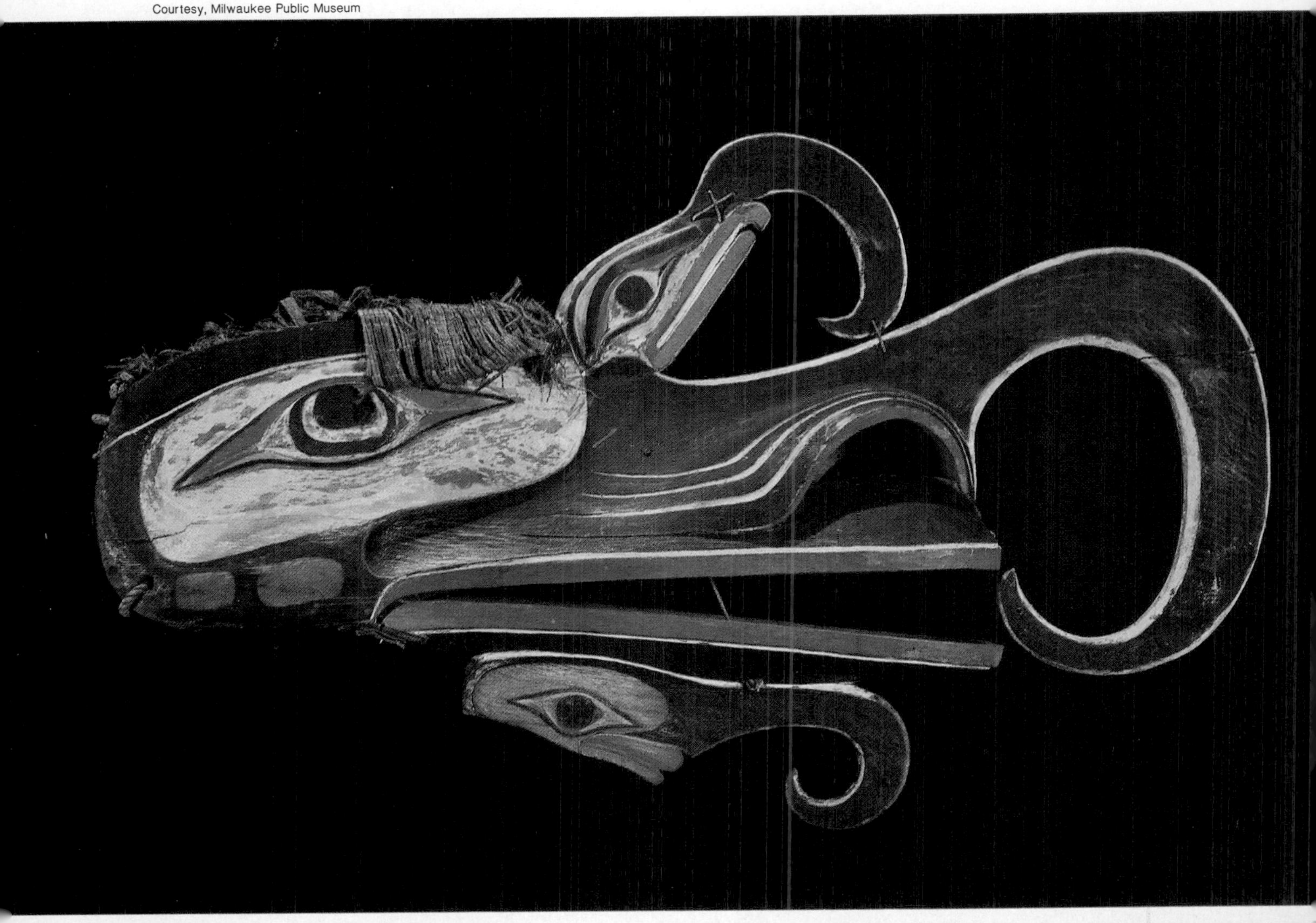

Courtesy, Milwaukee Public Museum

Large diorama (above) portrays a successful African lion hunt. Attention to accuracy of detail is also shown by the wooden replica of a tribesman's head (right).

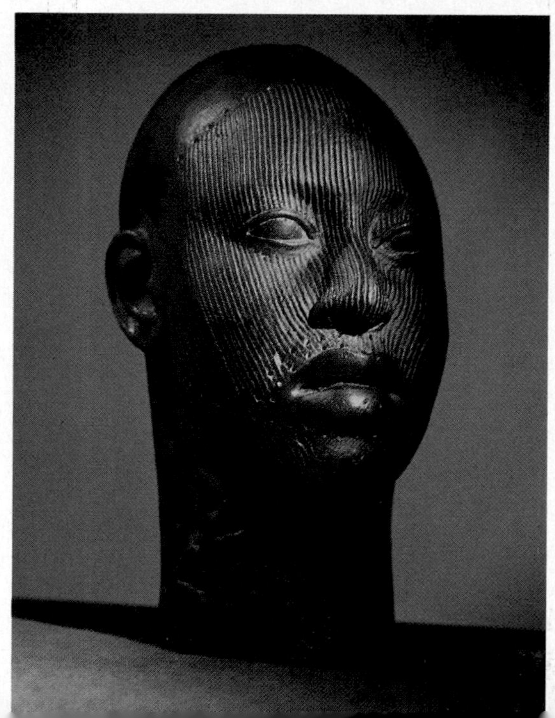

when the move was made to the new building. In the end it had to be cut into several large sections and then reassembled.

The second large diorama depicts a savanna waterhole and a Masai lion hunt. The design is a triumph over perspective and space. The exhibit area, shaped like a curved shell, has been transformed into a typical savanna environment. On one side of a walk-through path are baboon, Grant's gazelle, crowned crane, cattle egret, marabou stork, warthog, weaverbird, Grant's zebra, topi, Thomson's gazelle, fringe-eared oryx, eland, and aardvark. An acacia tree, *Adenium* bush, and a termite nest add to the authenticity. On the other side of the path is a traditional lion hunt of the Masai, frozen in the moment of truth.

Behind the scenes

The Milwaukee Public Museum began with a donation by the Wisconsin Natural History Society, which was organized in 1857 specifically to maintain a natural history museum for the German-English Academy. In 1881 the society presented its natural history collection of 19,000 specimens to the city of Milwaukee, with the proviso that the city establish and maintain a free public museum. Enabling legislation for the establishment of a "Public Museum of the City of Milwaukee" was passed in 1882, and the museum opened on May 24, 1884. At that time it had two staff members, a budget of $6,000, and 6,000 square feet of exhibit space in the former Exposition Building.

By 1899, when the museum moved to new quarters in a building it was to share with the public library, its budget had increased to $21,033, its staff to seven, and its exhibit space to 46,000 square feet. With the move in 1963 to its present quarters, a structure built in the Civic Center for the museum alone, it increased its exhibit space to 150,000 square feet. The budget in 1976 totaled more than $2.5 million, and the staff fluctuates between 175 and 200, depending upon seasonal, part-time, and contractual help.

In addition to providing vastly increased exhibit space, the move to modern facilities made possible an expansion of the areas devoted to public services and to the many behind-the-scenes activities that constitute a vitally important part of a museum's work. The scientific staff occupies offices and laboratories on the fourth, fifth, and sixth floors of the central core of the building, and each of the seven departments within the Scientific Division (Anthropology, Botany, Geology, History, Invertebrate Zoology, Vertebrate Zoology, and Taxidermy) has a storage area in its basement. Also in the central core are the administrative offices, the museum library, public facilities including a lecture hall, cafeteria, and lounges, classrooms where university classes in anthropology and museology are held, exhibition and graphics and paint and carpentry shops where the exhibits are prepared, and a print shop and complete photographic studio. On the roof is a greenhouse operated by the museum's Greenhouse Garden Club. This volunteer group maintains the live plants in the exhibit halls and assists the museum staff on horticultural matters.

Courtesy, Milwaukee Public Museum

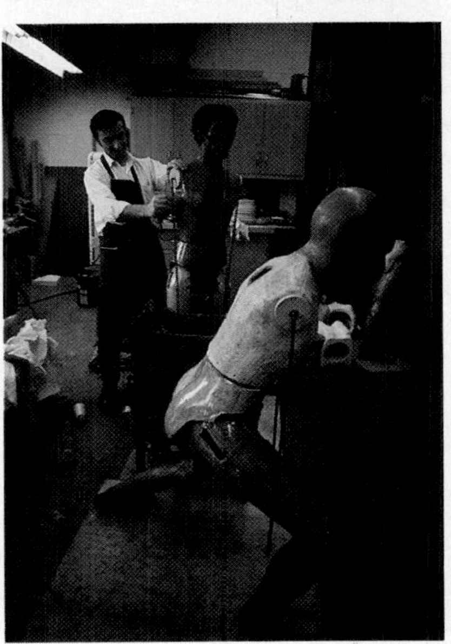

Painting backgrounds of dioramas and making life-size figures for exhibits are among the many activities that take place in the museum workshops.

Courtesy, Milwaukee Public Museum

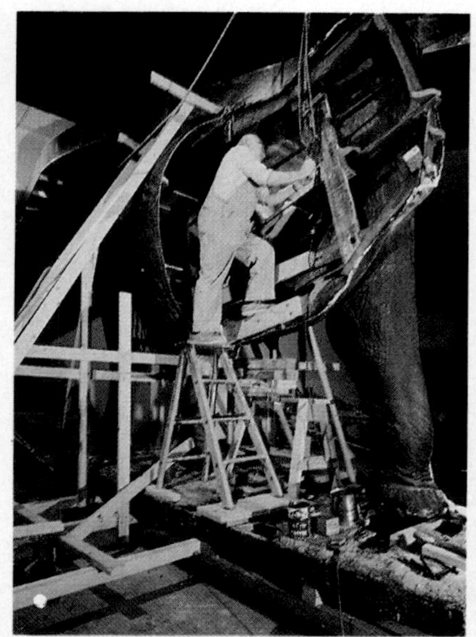

Diverse talents and equipment are required to construct the museum's exhibits (above and right). On the opposite page an artist (bottom) paints a background for a desert scene (top).

Courtesy, Milwaukee Public Museum

The three exhibit floors extend outward from the central core, and there also are a Youth Center, an Audiovisual Center, and the TV control room. A volunteer group of teenagers, operating out of the TV control room, give puppet theater productions in the Nickelodeon in the Streets of Old Milwaukee exhibit and outdoors on the piazza overlooking the Civic Center. Exhibition is not the only area in which the museum has experimented since its early days. It was one of the pioneers in working with schoolchildren, presenting public lectures, and making specimens available for loan as teaching tools. The Education Division cooperates with local school systems in presenting curriculum-coordinated programs based on museum exhibits, as well as providing a variety of programs for both children and adults independent of the school system.

With the recent donation of a new press, the museum has extended its publications program. Publications on the museum's extensive American Indian, pre-Columbian, African, gun, and typewriter collections are in constant demand by scholars, students, and collectors. Another service provided by the museum is the identification of specimens brought in by citizens. Staff members cooperate with other agencies and museums through advisory services and through participation on various commissions and committees. The museum also sustains field expeditions to maintain its collections. Three major expeditions were made in 1974, namely to Syria, the Amazon region, and Nepal. The latter was principally paleontologic in purpose but also involved climbing on Makalu, the fifth highest mountain in the world. An expedition to the Arctic in 1975 resulted in a major find of fossilized remains of bony fishes, soft-shelled turtles, crocodilians, and primitive terrestrial mammals in rocks about 50 million years old, showing relationships with northwestern Europe. Continued research in Syria, Nepal, and the Arctic and new fieldwork in Pakistan were undertaken in 1976.

Another major historical event in the life of the museum occurred in 1976. The museum was taken over on April 1 by Milwaukee County, thereby broadening both the tax base and the audience served.

Looking toward the future

With almost a century of leadership in the museum profession behind it, the Milwaukee Public Museum continues to reevaluate its programming and to search for new and better ways to serve its community. One major project nearing completion during the year was a new exhibition area, scheduled to be opened in 1976. Funded by a two-year, $150,000 grant from the National Endowment for the Humanities, the exhibits will make extensive use of the innovative techniques for which the museum has become known to illustrate the theme "The Urban Habitat: The City and Beyond" and will mark the Bicentennial.

In line with the system used to create all the exhibitions in the new building, the new area—called the Environmental Hall in its planning stages—was developed by a committee representing the Scientific,

Courtesy, Milwaukee Public Museum

Open-air market in Guatemala, a diorama conveying a great deal of information on customs, dress, products, and architecture with minimal material in a relatively small area.

Exhibition, and Education divisions. This group determined the overall approach, and the detailed plans were made by a subcommittee of two scientists and an artist-designer. For purposes of presentation, the overall theme was broken down into nine components, each with an objective, a definition of the techniques to be used to achieve that objective, and an outline of the exhibit contents.

To deal with biological cycles, human survival strategies of altering the environment, and the beginnings of city development in the earliest settlements, a lighting-narrative presentation was developed for a life-size diorama of a prehistoric rock shelter. This exhibit will set the stage for a later statement on the false security of humankind as it ignores the dynamics of nature and builds increasingly complex and complete environments. Subsequent exhibits being planned will illustrate man's movements across the world and his need to invent technology and social practices to keep population from outstripping available resources, the emergence of true cities, the growth of industrialization epitomized by the great industrial fairs of the 19th century, the fruits of technology 100 years later, and what the future might hold. All of these exhibits will include a variety of visual, audio, and other sensory techniques.

Closer to the immediate interests of the museum's own community is another project currently under way, a European Village being constructed on the first floor. Set up in approximately 17 geocultural units, the Village will illustrate the cultures of over 30 ethnic groups that emigrated from Europe to the Wisconsin area. Arts and crafts will be utilized, but the main thrust of the exhibition and the educational

Courtesy, Milwaukee Public Museum

Interior of a Spanish house, revealing great attention to accuracy of detail in such exhibits.

programming scheduled around it will be to depict the life-styles and the values of the different groups and the contributions they have made to American culture. Through the International Institute of Milwaukee County, the committee in charge is working with ethnic groups in the Milwaukee community to develop the exhibits and programming. The European Village is scheduled to open in 1978.

To round out the educational experience offered by the museum and further touch every segment of the community, the administration plans to supplement its natural and human history exhibits with a science center, proposed for construction in an area immediately north of the present building. In a conceptual proposal prepared to stimulate funding, the intent of the science center is stated as: to integrate natural science, social science, art, and physical science-technology for a comprehensive insight into life.

And to assure that the museum is providing the most up-to-date information, existing exhibitions are constantly being revised to incorporate the findings of current research. Special temporary exhibits are scheduled to supplement the permanent exhibition program. New exhibition techniques are being applied and experimental teaching techniques are being considered as the Milwaukee Public Museum—like all museums—is increasingly called upon to supplement more traditional, formal schooling. For, to return to Dr. Pohlman, mankind's natural curiosity, directed into proper channels, "results in greater happiness for the individual, in a better understanding of his environment and his relation to it, both individually and collectively, and finally, in a better environment."

Dynamics of Sailboats and Iceboats

by Olin J. Stephens II

Achieving maximum speed and performance from a sailing ship is a challenge that has occupied boat designers from the time of the ancient Egyptians to the present day.

George Silk from TIME-LIFE Picture Agency, © Time Inc.

Douglas Baglin from ALPHA

Cadet crewmen practice sail furling on square-rigged West German naval training vessel "Gorch Fock." The 100-foot schooner "America" (above right), first winner of the America's Cup in 1851, defeats the "Shenandoah" in an ocean race.

But the principal failing occurred in the sailing
And the Bellman, perplexed and distressed,
Said he had *hoped, at least, when the wind blew due East*
That the ship would not *travel due West.*

—Lewis Carroll

They no longer carry the world's freight and passenger traffic across the seas, but sailing ships have by no means become extinct. On the contrary, each year ever increasing numbers of them can be seen on the world's lakes, oceans, and rivers. They are sailed in a leisurely manner for pleasure and raced intensely for sport. Sizes range from 100-foot oceangoing yachts to craft of less than 10 feet, and the facilities for a passenger's comfort vary similarly, from the luxurious to the nonexistent.

The design of sailing ships to achieve maximum performance is one of man's oldest engineering challenges. Nations vied for the military and commercial advantages that could be gained by having the fastest and most maneuverable vessels. Today it is the sportsman who studies the dynamics of sailboats and iceboats, trying to determine the design that will make his boat the fastest on the water.

Early ships

The design of sailing vessels and some understanding of the wind's propulsive force go far back into history. The first sailing ships seem to have been used by the Egyptians as early as 3000 B.C. These people learned about sail forces as they affected steering, and placed the single mast well forward for sailing down the Nile before a constant southerly breeze. For use on the Mediterranean the mast was located

__OLIN J. STEPHENS II__ is President of Sparkman & Stephens Inc., designers and builders of racing yachts. He has also sailed in many races.

nearly amidships. Such vessels engaged in trade between Egypt, Crete, and probably Mycenae during the second millennium B.C. They were followed by the ships of the Homeric Greeks, which became more specialized and larger as the Mediterranean horizons widened. In China sailing rafts can be documented as early as 472 B.C.

The early vessels, although built with masts and carrying sails, all used the sail as auxiliary power to assist oarsmen. Oars continued in use well into the 18th century until men learned how to sail into the wind (windward sailing), which was a long, slow process. The Gokstad ship from Scandinavia was fitted with a mast as well as a long projecting keel, and it has been suggested that in the hands of the Vikings (9th to 11th centuries) such vessels must have been sailed rather close to the wind. It is written that one Rand the Strong was put to death in the 11th century by the king of Norway because he had the "wind at will." The Chinese may well have led in the development of windward sailing by the early (8th or 9th century) combination of a battened matsail (one in which coarse material stiffened with bamboo laths is used) with a deeply submerged rudder hung on a sharply raked sternpost so as to provide lateral support against the pressure of the wind on the sail.

Sailing-ship performance improved slowly through the centuries. An eye accustomed to seeing the yachts of today finds only small differences between the vessel of the early Atlantic explorers and the later cargo ships that traded with the Americas, although some of the latter were much larger. It is said that the English fleet under Sir Francis Drake which defeated the Spanish Armada in the 16th century was made up of smaller and more maneuverable ships than the huge Spanish vessels, which were presumably patterned after the galleons that carried Spain's American treasure. The English Navy, thus, won something of a technological as well as a strategic victory.

From Drake's time to that of Lord Nelson in the first years of the 19th century progress occurred in many small ways. The smaller vessels, the coasters one can see in Dutch and Venetian paintings of the time, had less freeboard (distance between water and deck at side) than did earlier boats, and the towering stern structures that were previously seen had disappeared. The drafting of ship's lines on paper began in the 17th century, but most ship's lines were taken from models. Drawings of the hull lines, which delineate various plane intersections through the hull surface—vertically, transversely, horizontally, and diagonally—made possible the development of hydrostatics, by which the quantity and distribution of displacement were calculated and stability thereby determined. Yet advances in ship design were uncertain and came slowly. Lord Nelson's "Victory" was 45 years old at Trafalgar and had been the flagship of 13 admirals before him.

During the 19th century, many theories of ship resistance were proposed. There was a rapid trend toward finer forms and greater length, which brought about major improvements in sailing ship performance. These eventually culminated in the clipper ships. The U.S.

George Silk from TIME-LIFE Picture Agency, © Time Inc.

Ted Hood tests 3-foot model of his 12-meter yacht "Nefertiti" in the ship-model towing tank at the Massachusetts Institute of Technology. In the vector diagrams on the opposite page, figure 1 shows in horizontal projection the wind and water forces acting on a boat as it sails to windward. All terms are described in the accompanying text except for F_h, which is the heeling or lateral force; V_{mg}, the speed made good to windward (in direction of true wind); β, the angle between V_a and the vessel's head; γ, the angle between the true wind and the vessel's course; and λ, the leeway angle. Figures 2a and 2b show that an iceboat can go faster when on a course at a slight angle to a following wind (2b) than when running directly before the wind (see text). In figure 2c a sailboat can attain a V_{mg} of 4.25 knots in a true wind of 8.5 knots by tacking downwind at an angle of 45° between V_t and V_s. In a situation where the aerodynamic lift and drag of a boat's rig are equal (figure 3a), the best course is at right angles to V_a. In figure 3b, $\tan^{-1}$ D/L(hull) + $\tan^{-1}$ D/L(rig) equals the angle between the apparent wind and the course, designated by $\beta + \lambda$.

shipbuilders in New York and New England led in their development, largely due to the incentive of the California trade, while the British followed with slightly more conservative but equally fast vessels built to carry Chinese tea and Australian wool.

Model testing

Rational ship design took a long step forward in 1870–71 when a British civil engineer, William Froude, advocated the determination of a ship's speed and power by means of testing a model. Model theory had been explored by Sir Isaac Newton, who had failed to consider the effect of friction, but Froude believed that he could determine this element of resistance by measuring the drag of long slender planks. The idea of model tests was not new; Benjamin Franklin, among others, had tested models of canal boats in 1768. Prior to Froude, however, the extrapolation of model resistance to a full-scale ship had been only a theoretical exercise.

Froude suggested that frictional resistance constants obtained by his tests of planks could be applied to both models and full-sized ships in such a way that after subtracting the frictional component from the measured model drag, the remainder, which he called residuary, could be expanded by known laws; the total ship resistance would then be the sum of this expanded residuary element and the frictional resistance recalculated for the full-scale condition. The British Navy financed the construction of a towing tank for Froude and tested the results, establishing the validity of the method by towing "HMS Greyhound," an early destroyer. Residuary resistance, which is the principal component of drag at higher speeds, results largely from pressure differences on the hull surface that generate waves.

The period immediately following Froude's work marked the decline of the sailing ship. For some 60 years, while steamships carried more and more of the world's trade, there was only sporadic interest in model testing of sailing vessel hulls and none at all in predicting their performance by applying sail forces to tested hulls. In 1928, however, a professor of aerodynamics at Massachusetts Institute of Technology, E. P. Warner, and a student, Herman Whiton, who was an enthusiastic sailor, placed instruments on a sailing yacht and measured pressures on the sails. This appears to be the first systematic study of sail characteristics, and it greatly interested K. S. M. Davidson of the Stevens Institute of Technology. Soon afterward Davidson began to test small yacht models in the institute's swimming pool.

The calculation of frictional resistance on which Froude's classic method of model tests depends is inexact at best. Friction on a bulky form is not necessarily the same as that on a plank of the same area, and the accepted law of comparison for friction (Reynolds number, proportional to length times velocity) is very different from that for wave making, the Froude number (the square root of length times the gravitational constant). The model, therefore, can never be tested at a speed that will satisfy both conditions. To compare wave-making

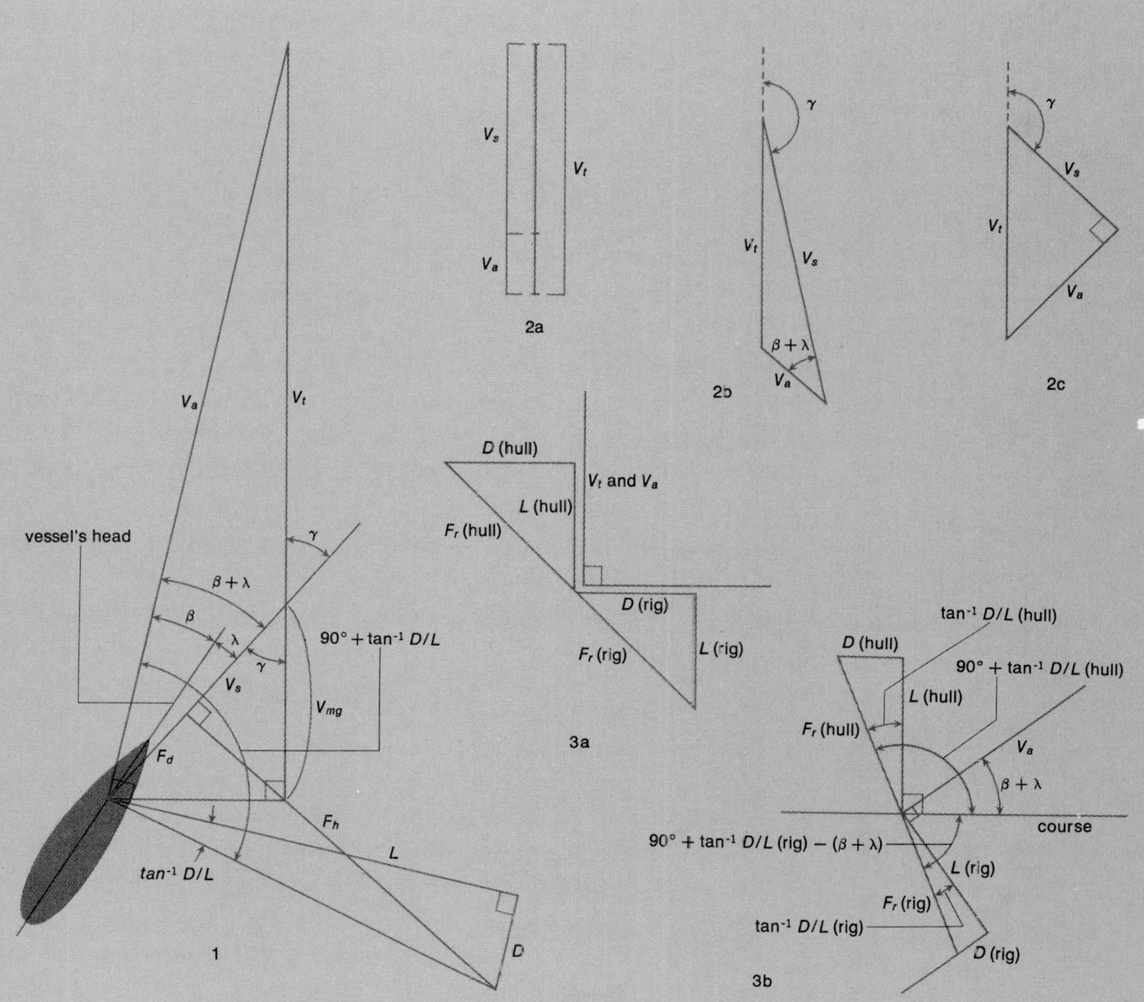

characteristics the small model is tested at a much lower speed than the full-scale vessel, but this is exactly the wrong way to achieve a comparison for friction. To this problem must be added a further difficulty, one which results from the unstable flow conditions over an uncertain area of transition that lies between normal or turbulent flow over the model surface and laminar flow, which occurs at very low Reynolds numbers.

Ship models and full-scale ships or yachts were normally large enough so that the test velocity could be kept out of this transition range. Davidson, who wanted to test yachts, found that for reasons of cost he would have to work with small models (about five feet long). This put his tests in the transition range so that uncertainties were magnified. His study of related work in aerodynamics and fluid friction in pipes suggested that he could induce turbulence and thus obtain a reasonably predictable condition. He did this by roughening the

The 12-meter yacht "Mariner" (opposite page, top) is tested as the possible U.S. entrant in the 1974 America's Cup races. "Courageous," the winner of the America's Cup for the U.S. in 1974, is subjected to a test of the paint (bottom left) on its hull, which is seen under construction (bottom center). Another boat that sought to represent the U.S. in the America's Cup, "Valiant," undergoes a test of tension on its mast (bottom right).

leading edge of the model hull and keel. Despite persistent questions and growing concern about the accuracy of predictions, Davidson's five-foot models provided yacht designers with a valuable tool.

At first, Davidson's models were tested like power-driven vessels, for upright resistance only. But it soon became evident, because of the disappointing windward performance of yachts that had tested very well, that the rig forces had to be taken into account. Davidson, therefore, proceeded theoretically to put the rig into the model, causing his reports to yacht designers to become predictions of full-scale speed. The rig forces, expressed as sail driving and heeling coefficients, were developed through full-scale sailing tests of a 34-foot yacht, the "Gimcrack." Its speed, course, and heel angle along with the relative wind strength were carefully measured under various conditions. Then, after determining the hull characteristics in the tank, appropriate coefficients relating the wind strength, the sail area, and the principal forces were worked out.

In order to simulate the rig forces, the yacht models were tested with applied longitudinal and transverse forces as determined from the "Gimcrack" coefficients. As the testing continued, it became clear that the keel had two primary functions in yacht performance, to provide lateral resistance and stability.

Davidson's early work took place during 1931–35. In 1936 the last J-class boat to defend the America's Cup was designed after a series of tests at his tank in Hoboken, New Jersey, and in 1937 its outstanding performance was convincing proof of the value of this type of testing. Since that time, the vast majority of successful racing yachts have been tested by similar methods. New questions recently have come to the fore, however, and the accuracy of small models as applied to certain hull configurations, particularly those with excessively full sterns, is subject to question.

Vector diagram analysis

Today, propulsion by sail applies primarily to pleasure craft and iceboats. Methods of yacht design have moved a long way from the intuitive/comparative toward the analytical/quantitative. Vector diagrams have been especially useful in achieving this progress, providing illustrations of the forces that move a sailing vessel or iceboat. Figure 1 illustrates in horizontal projection the typical wind and water forces acting on sails and hull under windward conditions.

Combined sail forces can be expressed in aerodynamic terms as lift (L), which is taken as that component of force acting at a right angle to the wind, and drag (D), the component acting in the direction of the wind. The resultant force (F_r), which represents the single direct acting force on the sails, can be diagrammed as an angle to the wind equal to 90° plus the angle represented by the tangent of D/L.

The wind direction against which the force angle is measured is the wind as felt by the boat and sails, that is, the apparent wind (V_a). In direction and strength the apparent wind depends on two quantities

Photos, Stanley Rosenfeld

Photos, Stanley Rosenfeld

Sail is taut as ocean-racing boat tacks into the wind (above). At the right the America's Cup winner, "Courageous," demonstrates the proper narrowness of the slot between the sails as it tacks into the wind.

that may be represented by vectors, the direction and velocity of the true wind (V_t) and the course and speed of the yacht (V_s). The force available for propulsion (F_d) is the projection in the direction of the yacht's course of F_r against V_s.

In trigonometric terms, the angle between F_r and V_a is defined as $90° + \tan^{-1} D/L$. If the yacht's course and F_r were at right angles, the yacht would be pressed sideways without advancing. This condition would exist when the angle between V_s and V_a was equal to the angle $\tan^{-1} D/L$, meaning that $90° - \tan^{-1} D/L + \beta + \lambda = 90°$. Thus, in order to move the yacht forward along its course, $\beta + \lambda$ must be large enough and its complement, the angle defined by drag/lift, sufficiently small to show a positive vector F_d. In order to make progress to windward, the sails must be efficient enough and the course assumed must be far enough away from the apparent wind to produce this forward force; at the same time, the ship's course must be close enough to the actual or true wind, V_t, so that it is forward of abeam and the movement along the course has a component toward the wind. The angle designated γ must be as much less than 90° as possible.

In assuming some typical values for the diagram, if the lift-to-

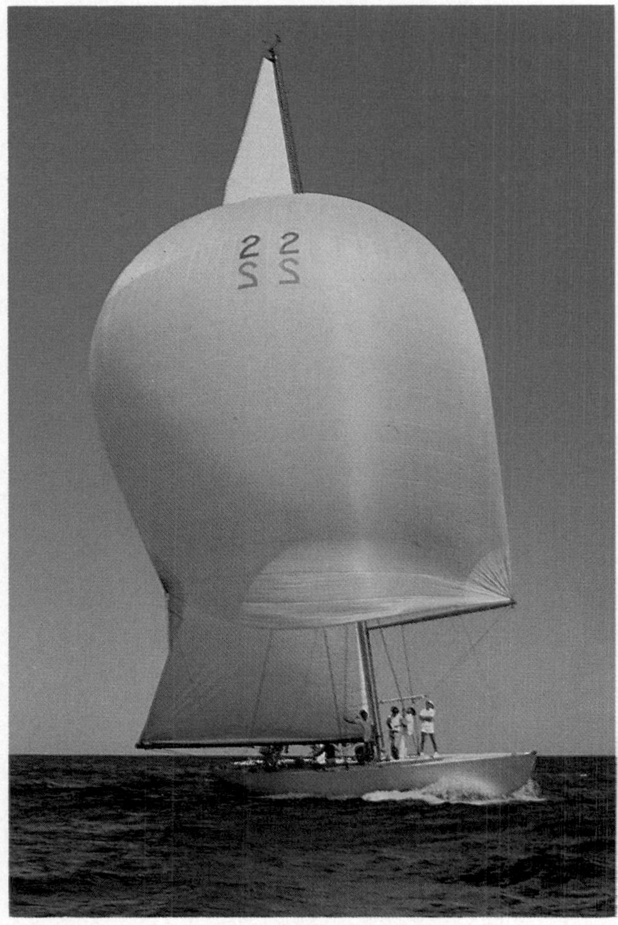

Spinnakers are tested on "Intrepid" during trials for the America's Cup race. Gold spinnaker at the right is larger along the base than the white one. The tests were not completed before the Cup races and are expected to be resumed for future competition.

drag ratio is 5:1, then D/L defines an angle of 11.3°. These values may seem to describe a very inefficient airfoil. This is because the air drag of hull, spars, and rigging is included. If the boat is sailing at 6 knots on a course 40° to a true wind of 10 knots, V_a equals 15.1 knots, V_{mg} equals 4.6 knots, and $\beta + \lambda$ equals 25.2°.

These general terms can be used to describe and evaluate the performance of any sailing vessel, be it an Egyptian barge, a modern racing yacht, or an iceboat. For these various boats, however, the proportions of the vector diagrams would differ greatly as would the speed of the vessel in relation to the velocity of the wind. An iceboat, for example, can frequently outsail the wind. Its resistance is so low that a designer of sailing yachts has to think hard to explain how its speed is limited. The answer would seem to be that the primary factor is wind drag, which increases proportionally to the square of the apparent wind velocity. Some further drag must be induced by the slight sideways velocity of the runners, leeway that is very small in comparison with the sailing yacht.

A vector diagram for an iceboat (figure 2a) reveals that as the speed of the vessel grows in proportion to the wind speed the apparent wind

(Top left) Thomas Zimmermann from ALPHA, (right) David Moore from Black Star, (bottom) Fred Ward from Black Star is drawn forward; that is, when V_s is large with respect to V_t, the angle $\beta + \lambda$ will be small as compared with γ. It is also notable that when the iceboat runs straight before the wind, making the direction V_s the same as V_t, it can go no faster than the wind. In such a case it would make perhaps 26 knots in a 30-knot wind. If the course is shifted slightly, as in figure 2b, however, making an angle between V_t and V_s of 15°, in a 30-knot wind the speed made good in the direction of the wind would be 36.7 knots. This is a typical case; often the values are much higher. Sailing yachts also can tack downwind to advantage in light weather (figure 2c).

Sails are trimmed according to the angle of the wind upon them. A boat going into the wind (top left) must have its sails trimmed more tightly than one moving at a right angle to it (top right). Sails are most fully extended when a boat is running in front of the wind (bottom).

If the aerodynamic lift and drag of a vessel's rig are equal, the resultant wind force lies at an angle of 135° from V_a (figure 3a). Therefore if the best lift the hull can produce is again not more than the drag, then the best course that can be held would be 45° above F_r or at right angles to V_a. Figure 3b shows that the true wind angle to course must be greater still, thus eliminating the possibility of windward performance. It is apparent that the sum of the applicable aerodynamic and hydrodynamic drag/lift angles equals $\beta + \lambda$. Over the years, as both hulls and rigs have been improved for windward sailing, both angles have been gradually reduced.

Aerodynamic principles

The vector diagrams have shown relative hull and sail forces represented on a horizontal plane. In this way, it can be seen how the wind blowing in a certain direction can propel a vessel in another direction and how progress can be made into the wind by tacking (turning the ship's bow into the wind and shifting the sails so as to fall off on the other side by the same angle as before). The principles are simple, and progress has been slow but steady. Improvements in sailboat design have also been greatly assisted by knowledge gained from work on aircraft. Rigs in particular have been improved, but yacht hulls and keels have also been influenced by the application to boats of aerodynamic research and practice.

A fundamental problem of sailing vessel design, which applies also to iceboats, is the heeling effect of the sails. In round figures, the side force on the rig of a modern yacht that would cause it to heel over is approximately two pounds for each square foot of sail in an apparent wind of about 15 knots. The forward force would be somewhat less than a half pound in the same wind. These values are not far from the aerodynamic drag-to-lift ratio of 1:5.

The side force acts at a level determined by the height of the sails, while the resisting force exerted on the hull by the water acts below the waterline. Being equal and opposite, they exert a powerful force to heel the boat over. This is resisted by the righting moment, which results from the width of the support; this width is related to the moment of inertia of the waterline plane and the position of the center of gravity in the sailing boat, and to the spread of the runners in an iceboat. In a modern yacht, the righting moment will usually increase

Photos, Stanley Rosenfeld

almost in proportion to the heel angle until the rail is in the water, after which it may be reduced but normally remains positive at least until the yacht's rig is in the water. This gives the normal yacht complete freedom from the possibility of capsizing.

In the case of an iceboat, once the weather runner lifts off the ice, the righting moment begins to lessen and a quick capsize can easily follow unless the helmsman is alert. Catamarans are much like iceboats in this respect.

The designer of a sailing yacht must always recall that the forward component of the wind power is closely related to the side force so that lack of stability results in limiting the driving force and, consequently, the speed of the yacht. Thus, stability in itself is an important factor in achieving speed, and although weight is normally frowned upon in yacht racing, it is not always detrimental to speed. Very light weight is always beneficial when a yacht is sailing on downwind courses in which the wind forces are aligned generally along the boat's course rather than across it, but more weight can be positively beneficial in certain boats when sailing to windward.

Instrument panel on the open deck (above) was an innovation, later disallowed, that was tested on "Mariner" during the 1974 America's Cup trials. On the opposite page, men work rigging (top left) on "Tenacious" during an ocean race in the Admiral's Cup series. Buoy (top right) helps demarcate course in America's Cup competition. Oceangoing yachts race in Buzzards Bay off the coast of Massachusetts (bottom).

Boat design

The preceding discussion has described one typical method of looking at sailing dynamics. The principal forces are sensitive to speed. For example, the wind forces that are picked up and transmitted by the sails will vary according to the square of the wind velocity as it strikes the sails. Hull forces follow no such general rule and are either determined in the testing tank or by comparison with a similar model that has been tested. Yacht designers use a wide range of comparative figures in their efforts to find the best combination.

As recognized by Froude, length is a critical factor in the speed of any vessel. If a 16-foot boat is sailing at 4 knots, this is commonly called a speed-to-length ratio of one. To achieve the same ratio, a 100-foot boat must sail at 10 knots. At higher speed, resistance increases on a progressively steeper curve that ultimately sets a top limit that is sometimes called hull speed. This may be at a speed-to-length ratio of 1.4 or even close to 2 in exceptional cases, but whenever similar large and small boats sail at the same relative speed the larger will be faster in approximate proportion to the square root of the speed-to-length ratio. Like most explanations, this is oversimplified. There are other influences, some favoring the smaller and others the larger boat, but on an overall basis the larger will be the faster and to the approximate extent of Froude's law.

Currently, a good deal of effort goes into the study of actual boat speeds in the course of designing and building a racing yacht. Boat performance is commonly measured in terms of a size index known as the boat's "rating." On the basis of this rating a time allowance is derived for application to a boat's actual elapsed time in a race, with the objective of compensating for speed differences due to size, hull proportions, and rig. It is evident that every boat owner and designer

George & Judy Manna from Photo Researchers

George Silk from TIME-LIFE Picture Agency, © Time Inc.

Iceboat of the DN60 class is sailed in Massachusetts (above). The name of the craft is derived from the Detroit News, *where it was designed, and the fact that it has 60 square feet of sail. At the right, sailors in Wisconsin have maneuvered their iceboats into the "hiking" position, in which the windward runner is lifted off the ice. Unless the boat is quickly turned closer to the wind, it will capsize; expert sailors have learned to sustain hikes for about 30 seconds.*

must attempt to obtain the highest possible speed for a given rating. Consequently, much ingenuity is used in both the formulation of rating rules and the effort to find the best combination of proportions and features.

The observation of individual boats and race results have both been used to put together the formula that constitutes a rating rule; then, using similar information, the designer will try to find the shape and proportions that will give the best result. Performance is never mirrored exactly in the rule's provisions, and various methods are used to hunt for the right variation on previous designs.

Among the new approaches is the very careful analysis of the performance of existing boats. Racing boats are fully equipped with instruments to measure boat and wind speed as well as course, wind direction, and heel angle. Records of performance can be kept on tape and later plotted. Comparative results yield clues to features that give better and worse performance and can pinpoint characteristics having a more or less favorable effect on performance than that created by the rating rule. One boat may accelerate after a tack much more rapidly than another, and the study of recorded performance data can help the helmsman to steer the best course for quick recovery of speed as well as to find the best angle to the wind for optimum speed either upwind or down. By knowing a boat's speed and heel angle in varying winds, a designer is able to judge the correct sail area and weight of ballast. Faults can be found and often corrected when the predicted performance is not realized. Performance analysis may

Herb Dallinger from ALPHA

P. Gilbert, Jr., from Stock, Boston

Hiking out to the edge of a boat to try to prevent capsizing takes place on a catamaran (left) and a single-hulled craft (above). Once it has been lifted partially off the water, a catamaran is more likely to capsize than is a single-hulled boat.

also either confirm tank-test predictions or suggest where such predictions fall down.

An interesting application of performance analysis was recently prepared by David Pedrick, whose study compares the potential performance of a modern ocean racing yacht and the record-setting clipper ship "Flying Cloud" on a voyage from New York around Cape Horn to San Francisco. "Flying Cloud's" dimensions are recorded as: length 229 feet; beam 40 feet 8 inches; and depth 21 feet 6 inches. She left the Boston yard of her builder, Donald McKay, for New York. Sailing from there on June 2, 1851, "Flying Cloud" reached San Francisco on Aug. 31, 1851, a passage of 89 days.

A modern challenge to "Flying Cloud" is the yacht "Kialoa III." With an overall length of 79 feet, about one-third that of "Flying Cloud," its speed potential could be taken as about 58% that of "Flying Cloud." Based on tank-test results checked against records of its performance, "Kialoa's" speed in winds of various strengths and directions is plotted in figure 4.

The likely conditions for a record-breaking attempt against "Flying Cloud" are based on a departure from New York City in January under conditions predicted by the Pilot Chart of the North Atlantic for January 1974, number 16. Numbers on the Beaufort scale, which indicates wind strength, were converted to their equivalents in knots, while the prevailing wind directions were plotted against an assumed course so that the boat speed can be obtained from figure 4. With allowances made for gales and calms, the average speed for 2,700 miles through

Stanley Rosenfeld

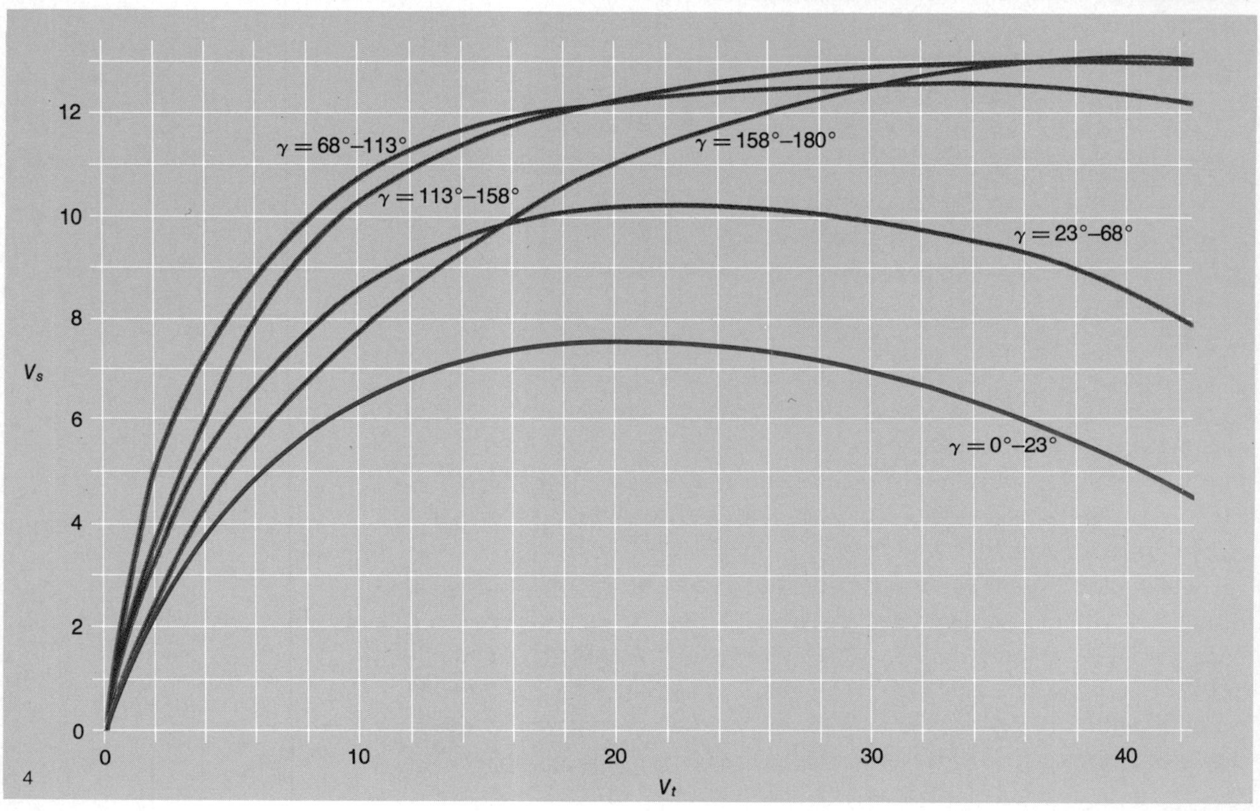

4

the North Atlantic is 9.93 knots. Similar estimates for the balance of the journey, with an additional allowance of 15% for sail changes and other "slows," give an estimated speed for "Kialoa III" of 8.06 knots. Over the course distance of 14,400 miles this speed yields an elapsed time of 74 days 11 hours, or 15 days less than the record passage of "Flying Cloud."

It must be clear that this is a paper exercise and as such is very different from any single actual voyage with its calms, gales, damage to sails and hull, and all the hazards of the sea. On the other hand, as a study, it correlates interestingly with the record performance of the two leading yachts, "Great Britain II" and "Kriter II," in a recent race from London to Sydney, Australia. In this race the two boats, each similar to "Kialoa III," broke the clipper ship, "Patriarch's," 69-day sailing record between those two ports by nearly two days.

Records of performance and detailed study of the results also featured the 1974 campaign to defend the America's Cup. All of the 12-meter yachts involved had sensing equipment, and records were generally kept. One yacht, "Courageous," the successful defender, used an onboard computer; on demand it could display such derived values as the speed made good either upwind or downwind, and the apparent wind direction and strength on a new course. Also stored in the computer was a complete polar diagram calculated to give the boat's speed under any conditions in order to compare these theoretical figures with the on-the-spot performance. The importance of crew work was highlighted by the fact that only when racing against another boat did actual performance equal that predicted.

Similar, though somewhat less sophisticated, equipment is used on many of the offshore racers, which comprise the greatest number of large sailing boats. In long races the courses to be sailed are often based on polar diagrams and a statistically based assumption on the weather to come.

It has often been said that two vessels may be built identically and yet behave quite differently. To a sailboat designer, this is only an indication that subtle differences can have important effects. Continued development in the study of yacht dynamics will gradually eliminate the element of mystery in these differences.

FOR ADDITIONAL READING

K. S. M. Davidson, "Some Experimental Studies of the Sailing Yacht," *Transactions of the Society of Naval Architects and Marine Engineers* (1936).

Czeslaw A. Marchaj, *Sailing Theory and Practice* (Dodd, Mead & Co., 1964).

Joseph Needham, *Science and Civilisation in China,* vol. 4, part 3 sec. 29 (Cambridge University Press, 1971).

Charles Singer *et al.* (eds.), *A History of Technology,* vol. i chap. 28 (Oxford University Press, 1954).

Hal McKusick from FPG

The 79-foot yacht "Kialoa III" (opposite page, top) was tested to determine whether it could sail from New York City around Cape Horn to San Francisco faster than the 19th-century clipper ship "Flying Cloud." In figure 4 various hypothetical performances of "Kialoa III," based on tank tests and past records, are determined by plotting the direction and force of the true wind (V_t) against the relative speed and course of the vessel (V_s). The figures $\gamma = 68°–113°$, etc., indicate the range of the angles between the vessel's course and the true wind. The bottom curve, representing angles from 0°–23°, is an example of speed made good to windward (V_{mg}). The tests indicated that "Kialoa III" should be able to sail the course distance in 74 days 11 hours, 15 days less than required by "Flying Cloud."

The Structure of Human Memory

by W. K. Estes

Often taken for granted, the abilities of the human brain to remember and to forget are quite remarkable. Both seem to depend upon a system of encoded collections of interrelated information.

One of the most striking trends in psychology during the past decade has been the revolutionary change in the status of research on memory. Previously the term memory was virtually synonymous with merely remembering. Rote memorizing was regarded as a dreary business compared with, for instance, arithmetical reasoning or problem solving, and the abilities underlying memory were regarded as special capacities that bore little relationship to intellectual performance. By contrast, the current view of memory, as seen in the light of research in cognitive psychology, is that of an incredibly complex and ever changing assemblage of capacities, skills, and strategies which evolves during one's lifetime like a tropical forest after the rains.

One of the most basic learning functions of every living creature is the accumulation of information that enables it to find its way in its environment. In higher forms, beginning at least with homing pigeons and migrating waterfowl, this learning must result in a complex memory structure amounting to a mental map of spatial relations in the environment. For man, however, the capacity to remember takes on much more significance. It is almost literally the case that memory makes man human. Aside from a few innate dispositions characteristic of the species, an individual learns to function in society because he or she is influenced during every waking moment by a store of memories, acquired at a very early age, concerning customs, folkways, and laws. Considered in this light, the problems of understanding memory may seem too vast and complex to be solved even by means of the scientific methods that are unraveling the secrets of the living cell.

A first thought might be to look into the brain, presumably the place where memory resides. Inspired by the dramatic elucidation of the genetic code, one might expect that some "mnemonic code" will be the next to yield. It should be recalled, however, that the first stretch of

__W. K. ESTES__ is a Professor of Psychology at the Rockefeller University, New York, New York.

Illustrations by John Craig

U.S.NAVY
robin

the road to an understanding of the structure of DNA was paved by the work of Mendel and his successors, who uncovered the empirical rules of genetic inheritance. Likewise, before an intelligent search for the location of memory in the cell junctures or macromolecules of the brain can begin, the basic laws of memory must be characterized.

The immediate priorities for the study of memory thus begin to emerge in clear outline. The functioning of the individual must be examined, not just as a biological organism but as an information-processing system. Ways must be found to measure the information that enters the memory system as a consequence of a learning experience, and of following the transformations that convert incoming sensory information to a form that can be stored in the brain. How stored information can be located on demand and how it is converted to a usable form must be understood. Current progress toward these goals reflects a convergence of two very different lines of effort. One is the patient and painstaking development, for nearly a century, of experimental techniques for the analysis of memory in the laboratory. The other is the sudden and salutary infusion of ideas and methods from the new computer science of information processing.

Memory in the laboratory

Memory and forgetting have been studied over intervals of time ranging from seconds to days and, in a few instances, even to years. Included at the short end of the time scale are studies of the kind of memory involved in remembering an automobile license number for the time needed to write it down. In a typical experiment, a human subject is seated before a closed-circuit television screen controlled by a small laboratory computer. Every few seconds a starting signal appears on the screen and is followed by a random sequence of digits or letters, usually on the order of eight to ten characters. This is followed perhaps by some distracting material, such as simple arithmetic problems, and then by a signal to begin recall.

Simple as it seems, this experimental task has yielded some surprising findings. It has demonstrated, for example, that memory for a sequence of letters or digits over the first few seconds of a retention interval may be increased by as much as 100–150% if the subject pronounces each of the characters aloud as it appears, compared with a condition in which pronunciation is omitted or actively prevented. In addition, it has been found that similarity in sound among the members of the sequence has an important influence on recall. As the members of the sequence are made more similar in sound—for example, BTVC rather than BRFM—an individual's memory for content (*i.e.*, for identity of the items that were presented, regardless of order) is often improved, but his memory for the order of presentation is drastically impaired.

On a slightly longer time scale, basic aspects of an individual's memory for verbal material are frequently studied with a test known as the free-recall task. Usually by means of a tape recording, the subject is presented with a random series of perhaps a dozen or more common words and then is asked to write down all the words that can be recalled as they occur to him. Again, though apparently simple, the task yields some interesting results. Even highly verbal adults find it exceedingly difficult to remember as many as 15 or 20 words within only a few seconds of presentation of the words. It is possible, nevertheless, to develop a good deal of skill at such recall. One insight into the development of such skill may be obtained by looking at the order in which an individual recalls the words. Typically it is found that the words are not recalled in the order presented but tend to be grouped so that words of similar meaning occur together. Evidently the individual does not simply register the words that he hears in a passive fashion but, rather, imposes some organization on them.

A third kind of experiment enables psychologists to study memory for sentences. The subject is presented with a short sequence of sentences by means of slides or filmstrips, then after a retention interval is given a test sentence and is asked whether the sentence was or was not included in the previously displayed sequence. One interesting group of findings has arisen from studies of the speed of response on this recognition test. An individual's speed of saying "yes" or "no" in response to a test sentence has been observed to vary systematically with changes in structure between the sentence originally presented and one presented on the recognition test. For example, subject response time increased when a sentence was first presented in the active voice and then asked as a test sentence in the passive voice—*e.g.*, "The boy fed the dog" later asked as "The dog was fed by the boy." Thus, this experiment provides a means of measuring the time required for the mental operations that are required to compare a test sentence with one previously experienced.

Using the same form of recognition test, investigators have come across the curious fact that under some circumstances, individuals often "recognize" sentences that they have never seen before. For instance, if the original list of sentences includes both "The boy pulled the trigger" and "The sparrow fell to the ground," some individuals on a later test will believe that the sentence "The boy shot the sparrow" was on the list. This kind of finding indicates that the human memory system does not typically store sentences by recording them in the manner of a tape recorder. Instead it dissects or parses the sentence and stores semantic and syntactic information about it in a form that later may enable one either to reconstruct the sentence or to answer questions bearing on the information that it conveys.

Varieties of memory

The effective study of any subject depends on dividing it into parts in a rational manner. One primary categorization of memory, that of short-term versus long-term memory, is based on the observation that a great part of the information impinging on one's senses remains available to recall or recognition for a few seconds, then is lost without a

Figure 1. Representations of an association between memory traces in terms of early versions of association theory (upper left) and one that accounts for the role of context (bottom and right). *See text, page 66.*

trace. The small part that survives this critical period is likely to endure indefinitely unless subjected to specific interference from material learned later. However, before approaching the central question of what determines survival within the memory, a second categorization must be introduced.

Perhaps the most fundamental classification that has resulted from research on memory is University of Toronto psychologist Endel Tulving's distinction between episodic memory and semantic or categorical memory. The first type comprises memories for particular experiences, such as a view of the Grand Canyon at sunset or a bout of seasickness. In instances of this sort, one's impression is that memory constitutes an image, albeit somewhat faded and fragmented, of the original experience, and research essentially confirms this impression. An individual's collection of episodic memories is of importance in laying the foundation for his ability to recognize the almost innumerable scenes, objects, and people that he has encountered in the past. On the other hand, these memories are relatively unorganized; the constituent elements cannot be called up readily and systematically on command as can, for example, the words in one's vocabulary.

Semantic memory, including memory not only for words but for concepts, rules, and abstract ideas, is quite different from episodic memory. When called upon to use the word "red," one does not remember the particular circumstances in which this word was first encountered or previously used, but rather one recalls the meaning of the word. In order to solve a problem, to read a newspaper, or even to carry on a conversation, one must be able to summon up many semantic or categorical memories at extremely high speed. It does not suffice to recognize a printed word as a familiar object. One must recall a meaning and relate it to what has gone before in the sentence —all in a tiny fraction of a second. This kind of performance is possible only because semantic or categorical memories are organized into a system that permits efficient retrieval.

The basis of memory structure: traces and associations

With some of the varieties of memory now properly categorized, the question of memory survival can be approached meaningfully. What determines the difference between short-term and long-term memories? All varieties of memory start from sensory experience, the influx of energies exciting neural structures in the sense organs and ultimately the brain. In current theories, it is generally assumed that short-term memory reflects merely a persisting state of excitation of the neural structures activated by a sensory experience, with the consequence that the structures are readily reactivated if, but only if, the sensory pattern soon recurs. If the sequel to the experience is motivationally significant to the individual, then the memory trace is consolidated or fixed into a long-term memory and provides a basis for recognition when a sensory pattern encountered on some future occasion matches the stored memory trace.

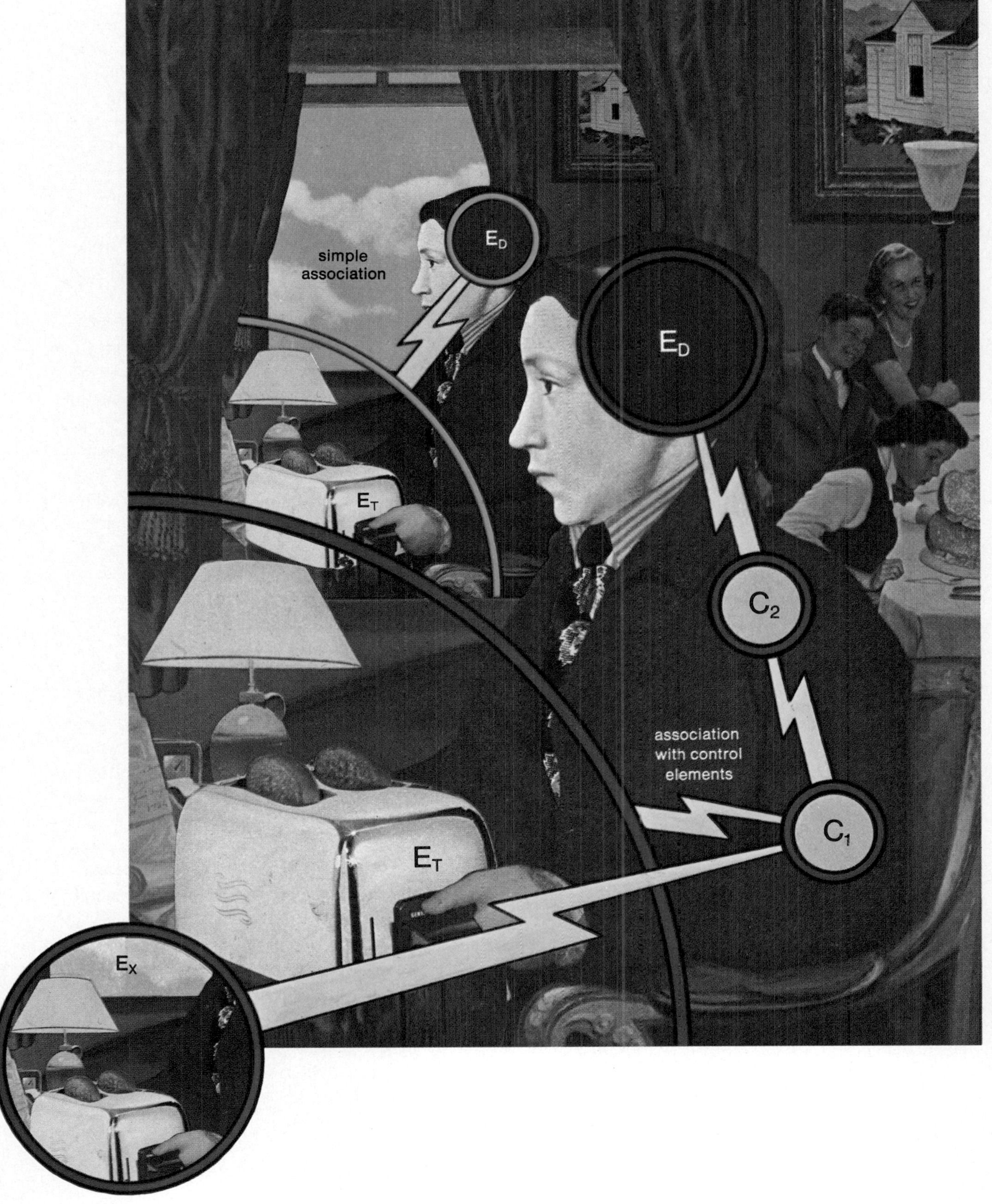

E_D
simple association
E_T
E_D
C_2
association with control elements
C_1
E_T
E_X

Figure 2. A memory representation of the word CAB is shown on the opposite page. The sensory pattern from the printed word (lower left) is filtered through successive levels of control elements, which operate like the "and" gate of a computer circuit, allowing information to flow to the next level of control elements only if the proper messages arrive simultaneously over the input routes leading to it.

Recognition of sensory patterns is only a small part of memory, and the concept of a trace alone does not account for man's ability to recognize the meanings of words, to recall sequences of events, or to answer factual questions. These problems were recognized centuries ago by the association philosophers, who augmented the concept of a memory trace with that of an association or connection. It was proposed that traces of contiguous experiences become associated in memory, so that the later recurrence of one activates the other by association; for example, the recollection that a flash of lightning is followed by a clap of thunder. The concept of an association between memory traces, however, has been shown to be oversimplified because it fails to take adequate account of the role of context or to lend itself readily to the building of complex memory structures.

To illustrate the problem of context, suppose one wishes to develop a memory structure that would account for a person's ability to remember, say, that all the lights in his house went out when he pressed down the handle on his toaster. A representation in terms of the earliest versions of association theory would be as in figure 1 (upper left), in which E_T denotes the experience of pressing the toaster handle, E_D the experience of seeing the house go dark, and a path between them denotes an association in memory. On a later occasion, reactivation of the first element of the chain would lead to reactivation of the second, so that upon trying the defective toaster again, the individual whose memory is represented would expect the lights to go out. But this representation does not take account of the fact that the individual would only expect the lights to go out on a later occasion if he tried to make toast in a similar context; *i.e.*, with the toaster still unrepaired, with the toaster plugged into an electrical outlet in his house, with the lights on, and so on. A more adequate representation, which takes account of the role of context, is shown in figure 1 (bottom and right). In this case, E_T, the original event, and E_X, which denotes the context in which the original event occurred, are both connected to a common control element, C_1, which in turn is connected to E_D by another control element, C_2. It is assumed that C_1 will be activated on a future occasion only if the stimulating circumstances reactivate both E_T and E_X, in which case a neural message is transmitted to C_2, thence activating E_D and leading to the person's recall or expectation of his sudden plunge into darkness.

The conception of association by way of control elements provides a simple means of representing still more complex memories which could not be handled at all in terms of classical association theory. Consider, for example, one's knowledge that a certain combination of marks on paper represents the word CAB. The memory representation is illustrated in figure 2. The sensory pattern from the printed word is filtered through successive levels of control elements, the lowest level representing elementary constituents of letters, the next level representing letters, and so on. A control element is seen to operate like the "and" gate of an electronic circuit, allowing information to pass only if

CAB
A B C D E
BRFM
BTVC
CAB

messages arrive simultaneously over the input lines leading to it. This interpretation reflects both the extensive researches of Eleanor J. Gibson on children's reading and the experimental and mathematical analyses of letter and word recognition in adults by David L. LaBerge, David E. Rumelhart, William K. Estes, and others. Applied to the teaching of reading, it suggests that concern should be devoted first to teaching children to attend to the attributes that differentiate letters, rather than to the attributes of whole words.

The process of combining lower-order into higher-order units by way of control elements appears to be the basic mechanism whereby the associative memory system encodes information and maintains it in organized hierarchical structures that permit efficient access when the time comes for retrieval from memory. An understanding of this process may well be essential to any attempt either to comprehend the basis of memory structure or to investigate its neurological substrate. Yet, a fine analysis in terms of such associative linkages is not ordinarily necessary in dealing with practical problems. Just as in the study of genetics, where the problem of inheritance may be approached either in terms of the redistribution of genes or in terms of the transmission of traits, studies of the psychology of memory may utilize either the concepts of associations and control elements or concepts of information storage and retrieval. In discussing more complex problems of long-term memory, it will be convenient to proceed largely in terms of the latter vocabulary.

Principles of organization

Perhaps the most fundamental and pervasive principle of memory organization is that of coding. Semantic or categorical memories represent highly encoded forms of information in which much of the detail of original experiences has been sacrificed and replaced by a relatively small number of representative attributes. When listening to a spoken message, for example, one may retain a short-term episodic memory trace of the actual sounds sufficiently rich in detail that, if presented with a tape recording within a few seconds, one would be able to say whether it represented precisely what had been heard. But this detailed episodic memory is soon lost. During the same few seconds, however, the perceptual system analyzes this representation, classifying the input during successive instants in terms of simple linguistic properties and then retaining in memory only a record of the pattern of consonant and vowel sounds. Further, these sequences of consonants and vowels are recognized as properties of higher order units—*i.e.*, of syllables or words—that are already contained in the memory system, and it is only these higher order units that are rehearsed and entered into long-term memory. Thus, on a test after the passage of hours or days, the individual will be able to recall the content of the spoken message but not the details of its sound pattern.

Coding in terms of linguistic properties provides a foundation for recognizing an enormous variety of words even in infinitely varying contexts. Yet it is not enough to be able to recognize printed or spoken words; language also requires one to produce words when needed that are not present to the senses. When one desires water, it may be necessary to produce the word even though there is no printed version available to activate the memory system. Once again, the evolution of the human memory system appears to have arrived at a solution by means of an organization in terms of attributes.

The meaning of a word is represented in one's memory structure by the set of attributes of the contexts in which the word is used. In the case of "water," one of these attributes corresponds to the sensation of thirst. Access to a word in a situation requiring recall is gained by way of its attributes. Similarity in meaning between words is a function of the number of attributes they have in common. Consequently, the mention or the recall of a single attribute as a cue will typically bring to mind a number of related words. The sensation of thirst would tend to reactivate the control elements not only for "water" but perhaps also "coffee" and "beer." Among the other attributes of a particular situational context, some might belong only to water and thus provide a basis for finer discrimination.

The organization of memory in terms of attributes has enormous advantages over one that depends solely on the recording of individual episodes. By noting the relevant attributes of a situation that calls for recognition or recall of a particular word, one may quickly gain access to the appropriate item without having to search a record in memory of the hundreds of thousands of occasions on which the word has been encountered in the past. But while problems of memory search are greatly simplified by a system in which attention to one attribute of a situation at a time brings to mind small subsets of relevant words from the larger vocabulary, it would seem inconvenient to have a number of different words clamoring for attention at every successive point in the generation of a stream of speech.

The secret of reducing this competition seems to be found in hierarchical organization. Every user of language can arrange the words he knows concerning any subject in categories of higher and lower order. Hence, a robin is a bird and a bird is an animal. Further, it has been noted in free-recall experiments that individuals tend to recall words in clusters according to categories. This observation has suggested that words are also hierarchically arranged in one's memory structure. The word "robin" then would represent just one node in a treelike hierarchy of associations. The idea seems to be supported by reaction-time experiments, which have shown, for example, that one is faster in verifying the truth of the sentence "A robin is a bird" than the sentence "A robin is an animal," presumably because robin and bird are closer together in the hierarchy than robin and animal. However, this simple relation between reaction time and distance is not always found to hold; other factors, in particular the frequency of occurrence of particular attributes, prove in some instances to be more important determinants of reaction time than distance in the hierarchy.

The results of experiments investigating the organization of human memory have suggested that words tend to be hierarchically arranged in the memory structure according to categories of higher and lower order. For instance, as depicted on the opposite page, the words "robin" and "whale" each represent one node in a branching structure of associations. The basic organization is in terms of sets of attributes, and the individual who possesses an efficient memory has become skilled at consulting these attributes in a systematic order.

The trend of the evidence indicates that the hierarchical organization is not a fixed property of one's memory structure but, rather, is generated by the individual at the time of recall. Thus, while the basic organization is always in terms of sets of attributes, the individual who is efficient at recall has formed habits of consulting attributes in a systematic order. If asked to decide whether a whale is a fish, a child might have to search his memory for all of the experiences in which people had spoken of whales and fish and try to remember what had been said about the relationship. An older individual, but one still unskilled in recall, might have to search through the names of all of the fish in his memory to determine if "whale" was among them. The still more skilled individual would first bring to mind the attributes defining "fish" and then note if all of these hold for "whale," thus completing the task with the necessity for very little search of memory.

The retrieval of memories

Large differences in performance on tests of memory have been apparent from the earliest attempts to measure intelligence. However, the idea that the differences in performance necessarily reflect differences in innate ability or biological capacity to memorize has not held up. Consider, for example, recent cross-cultural studies of mental functioning (such as those conducted by Michael Cole and his associates at the Rockefeller University) that provide comparisons of performance on similar memory tasks by individuals with very different cultural backgrounds. When given familiar tests of the kind that are used in conjunction with standard intelligence scales, children who have grown up in an African hamlet with little schooling perform less efficiently than U.S. schoolchildren. However, patient and imaginative research has shown how to allow for the different habits of information coding and retrieval developed by children in diverse cultural settings. New test methods resulting from this work show the recall of African children to be comparable to that of U.S. schoolchildren when engaged in similar tasks. Using a different approach, British investigators, including Elizabeth K. Warrington and Lawrence Weiskrantz, have shown that even individuals suffering from amnesia as a consequence of brain damage improve their performance on memory tests if they are supplied with some of the attributes of words to be remembered or if they are encouraged to group the material for recall on the basis of auditory or semantic categories and then to attend to the relevant attributes at the time of recall. Hence, in both of these examples, the originally observed differences in performance do not reflect differences in the biological capacity to record information; instead, at least in great part, they reflect differences in the way in which the information is organized, the skills and strategies developed for retrieving information and bringing it to bear on problems.

As a consequence of much investigation, psychologists have begun to see that the difference between transient and enduring memories lies primarily in what the individual does during and immediately

typical attributes

A1—spontaneous movement
A2—quick response to stimulus
A3—warm-blooded
A4—feathers
A5—suckles young
A6—hair, profuse or scant
A7—cold-blooded
A8—fins, scales, and gills
A9—green and gray feathers
A10—eats mostly seeds and rice
A11—brown back and red breast
A12—eats fruit, insects, and worms
A13—lives on land
A14—basically carnivorous
A15—runs on four legs
A16—lives in water
A17—superficially resembles fish in body shape
A18—feeds on plankton, small fish,or other aquatic life

following perception of the material to be remembered. If he attends to given material in isolation, does not rehearse it except perhaps for sheer repetition, and does not expect a need to recall it in the future, then ordinarily the memory quickly becomes inaccessible. Excellent examples in this class are the numbers an individual deals with during a series of calculations or the specific sequences of words that he sees while reading for pleasure. Conditions that make for more enduring memory include attention to the circumstances under which an event occurs, attempts to relate the given event or material to broader categories or to other material already in memory, and definite expectation that later recall will be required.

But merely intending to remember something accomplishes little. When an individual skilled in memory management is perceiving or experiencing the material or activity he intends to remember, he thinks of the circumstances under which he is likely to use it. Consequently, a representation of the material to be remembered is stored in the memory system together with retrieval cues or attributes of the situation in which it may need to be recalled. On a later occasion, when the retrieval cue occurs in a test situation, the cue tends to activate the original memory and make it available for recall.

The problem of capacity

One of the most natural questions regarding memory is that of how much information a memory can hold. It has been deferred until this point because it must be considered relative to the major classifications of long- and short-term memory and episodic and categorical or semantic memory.

Individual differences apart, there is no doubt that everyone is subject to severe limitations on the capacity of short-term memory. The memory-span test of the Binet intelligence scales has shown that normal adults are able to recall only about seven random digits immediately after hearing them and few are able to go beyond eight or nine. These limitations, however, do not appear to be a matter of informational capacity, comparable to the channel capacity of a telephone line, for the amount of information held in short-term memory can be considerably amplified by appropriate grouping or chunking of the material. For example, it has been predicted from mathematical models of memory and confirmed by experiments that recall is most efficient if the items are presented in groups of three. Further, grouped items may be recoded into a single unit that represents the entire group, as the name of a telephone exchange represents three digits, and thus reduces the total number of items to be remembered. The limitations appear to be a consequence, rather, of the dynamics of the system. The memory traces generated by new sensory experiences rapidly decay, or at least become inaccessible, unless maintained by rehearsal until they can be incorporated into long-term memory structures. The rate at which rehearsal can be accomplished limits the number of items that can be maintained.

With regard to long-term memory, matters are more complicated. If only recognition is tested, it appears that capacity for storing episodic memories is virtually unlimited. Experimenters, including Roger N. Shepard and Lionel Standing, have presented subjects with hundreds or even thousands of pictures of complex scenes and then obtained virtually perfect performance on tests of recognition. But when recall is required, the retrievability of episodic memories suffers interference from both earlier learning experiences (proactive inhibition) and later ones (retroactive inhibition). In fact, investigations of interference effects in memory for verbal materials conducted during the 1950s led to the upsetting suggestion that effective long-term memory capacity may actually decrease as a function of practice, and that much of the forgetting that an individual undergoes in everyday life is a consequence of interference from previously learned material.

Fortunately, continuing research has clarified matters considerably. The proactive effects in episodic memory may actually have nothing to do with informational capacity but, rather, may reflect difficulties of retrieval that ensue when events which have many common attributes occur close together in time. Thus, it might be expected that proactive inhibition could be decreased simply by increasing the time intervals between occasions on which related material is practiced, and this expectation has been borne out in experiments.

For semantic or categorical memories, there seems to be no evidence of proactive inhibition at all. On the contrary, increases in an individual's supply of semantic information on any topic make it easier to acquire more. Retrievability of the information may, however, be a problem, one familiar to all who keep stuffing discarded possessions into cluttered attics and find the items increasingly difficult to recover if later needed. Many minds, unfortunately, suffer much the same problem. The secret of expanding the informational capacity of long-term memory appears to lie in habits of noting the relationship between the attributes of newly encountered material and material already in memory as well as in updating habits of orderly retrieval. The new emphasis on continuing education in the society of the 1970s is likely to pay off only to the extent that the fruits of research on memory retrieval can be applied to show people how to manage their continually expanding supplies of information.

FOR ADDITIONAL READING

J. R. Anderson and G. H. Bower, *Human Associative Memory* (V. H. Winston, 1973).

G. H. Bower, "Cognitive Psychology: An Interpretation," in *Handbook of Learning and Cognitive Processes*, vol. 1 edited by W. K. Estes (Erlbaum Associates, 1975).

Charles N. Cofer (ed.), *The Structure of Human Memory* (Freeman, 1976).

Bennet B. Murdock, Jr., *Human Memory: Theory and Data* (Wiley, 1974).

The Formation of Stars

by Stephen E. Strom

Sensitive infrared detectors and radio telescopes have given astronomers new insights concerning the birth and early evolution of stars.

Over the past decade astronomers have charted the life history of stars backward in time virtually to the moment of stellar conception. The new tools of infrared and millimeter-line radio astronomy have provided important insights into the conditions that prevail during the earliest stages of stellar evolution. In turn, the conclusions reached from these studies have major implications for understanding the development of planetary systems. In order to provide the framework for a discussion of recent discoveries, it is perhaps most useful to review current theoretical beliefs that are concerned with the birth and early evolution of stars.

Theory of stellar evolution

In our Galaxy, a star's life begins in cold, dark clouds of interstellar gas and dust. Primarily comprised of molecular hydrogen, these dark clouds pervade the plane of the Galaxy.

Shielded by microscopically tiny dust particles from the heating effects of ultraviolet photons and by their internal magnetic fields from heating by energetic interstellar cosmic rays, these clouds cool to temperatures between 5° and 20° K. At these low temperatures the internal pressure of the cloud is small. The entire cloud may then begin to collapse if its gravity exceeds the opposing internal pressure forces. During this process, initially denser regions within the cloud also begin to collapse on a time scale shorter than that of the overall cloud collapse. These denser regions are the likely centers of formation for stars and clusters of stars. The details of the initial cloud collapse and fragmentation are, however, not well understood, either theoretically or observationally. Astronomers are not sure, for example, whether external pressure plays a significant role in providing the final "push" that forces the clouds to collapse. Furthermore, they do not yet know whether entire clouds or only smaller subregions of a cloud are active in the star-forming process.

The next stages in stellar formation are somewhat better understood. Computer simulations of spherical, protostellar clouds of stellar masses, collapsing under self-gravity, have been developed by Richard Larson and by C. C. Hayashi and his collaborators. Although these models differ in detail, their essential features are similar. At first, these clouds develop a high-density core region, which begins to collapse much more rapidly than the remaining protostellar cloud, on a time scale determined by the core density. During the entire collapse stage, this core region accretes infalling material from the surrounding cloud. The core mass thus increases continuously during these early phases, and its radius decreases simultaneously. As the ratio of mass-to-radius increases, the core temperature increases. This heating reflects the conversion of gravitational energy to thermal energy. Core heating dissociates hydrogen molecules and later ionizes the resulting hydrogen atoms. Additional heating of the core takes place at the "boundary" between the core and the infalling cloud as low-density cloud material crashes into the high-density core region at

***STEPHEN E. STROM** is an Astronomer at the Kitt Peak National Observatory in Tucson, Arizona.*

(Overleaf) Painting by Paul R. Alexander

Gaseous nebula in the constellation Serpens. Such clouds of gas and dust are the birthplaces of stars.

supersonic speeds. When the core reaches a stellar-like mass and radius, its temperature and density achieve sufficiently high values so that the increased internal pressure slows, but does not halt, the further contraction.

At this point further evolution of the core takes place at a rate determined by the ratio of the available gravitational energy to the total amount of radiation emitted per unit time (the core luminosity). During the contraction phase, the core density and temperature continue to increase. Energy is transmitted from the core center to its boundary, at first by convective motions of the gas and later by radiative processes. Finally, the density and temperature in the central regions of the core become sufficiently high to initiate conversion of hydrogen to helium in a series of nuclear fusion reactions. With the development of an internal energy source, contraction of the core ceases. The starlike core has at this stage nearly reached a long-term equilibrium state in which the internal structure is such that its luminosity and radius remain nearly constant. Thus, a "main-sequence," hydrogen-burning star is born.

The stage at which the starlike core becomes accessible to observation may depend on the initial mass of the collapsing protostellar

On a Hertzsprung-Russell Diagram (opposite page) stars are plotted according to their spectral types and surface temperatures (horizontal axis) and their absolute magnitudes or luminosities relative to the Sun (vertical). The well-defined belt is the main sequence, along which lie most of the stars. The locations of various spectral types of stars are indicated: (top left) Rigel, a very massive luminous star of type B8; (top center) Aldebaran, a red giant of type K; (top right) Antares, a red giant of type M; (bottom left) Sirius B, a small and dense white dwarf of type A; (bottom center) the Sun, a main sequence star of type G; and (bottom right) Wolf 359, a red dwarf of type M.

cloud. For stars of high mass, the "envelope" of infalling gas and dust is thought to persist well into the early main-sequence lifetime of these stars. The tiny dust grains present in the protostellar cloud are efficient absorbers of visible light and act to obscure the underlying newborn star at these wavelengths. In the case of stars of lower mass, envelope gas and dust are almost totally accreted by the core at a sufficiently early stage to render these low-mass, pre-main-sequence stars visible during the latter portion of their contraction phases.

The time scale required for stellar-like cores to reach the main sequence depends on the mass of the protostellar clouds. Typically, a star of five solar masses (5 $M_{\odot}$) will reach the main sequence less than one million years after the onset of gravitational collapse; for a star similar in mass to the Sun, more than ten million years are required.

The foregoing summary of events can be predicted from models in which the most plausible estimates of physical conditions have been made. Many simplifications have been introduced, however, in order to simulate numerically the collapse and early evolutionary phases. There are also many uncertainties surrounding the physical processes that are important during a star's pre-main-sequence lifetime. For example, the effects of rotation and magnetic fields have been ignored. A proper understanding of the mechanisms by which binary stars or planetary systems are formed awaits a theoretical approach that incorporates rotation explicitly in the collapse calculations. Moreover, the manner in which a protostar is able to stabilize itself against the disruptive effects of rapid rotation involves an understanding of the evolution of magnetic fields during prestellar phases.

Nevertheless, the spherical collapse models seem sufficiently realistic to provide the essential framework for observational studies of the early phases of stellar evolution. A discussion of these studies follows.

Discovery of young stellar objects

In order to check their theoretical picture of early stellar evolution, astronomers had to find a sample of young stellar objects (YSOs) for observational study. The most logical regions to search for YSOs were the dark interstellar clouds. Before the late 1960s, searches of stellar birthplaces could only be made at optical wavelengths. Three major classes of YSOs were isolated from such searches. In the early 1940s, Alfred H. Joy of Mt. Wilson and Palomar Observatories (renamed the Hale Observatories in 1970) discovered a class of faint stars having highly unusual properties: irregular light variability coupled with a complex spectrum seemingly composed of an ordinary absorption spectrum similar to that observed for stars of 1 $M_{\odot}$ and smaller, and a variable emission-line spectrum characteristic of a moderate-temperature gas. These objects are called T Tauri stars. Later, G. Haro and George Herbig announced the discovery of even more unusual objects associated with interstellar clouds, the "Herbig-Haro" objects. These objects appear as semistellar patches of nebulosity characterized by a highly unusual emission-line spectrum. The co-discoverers believed at

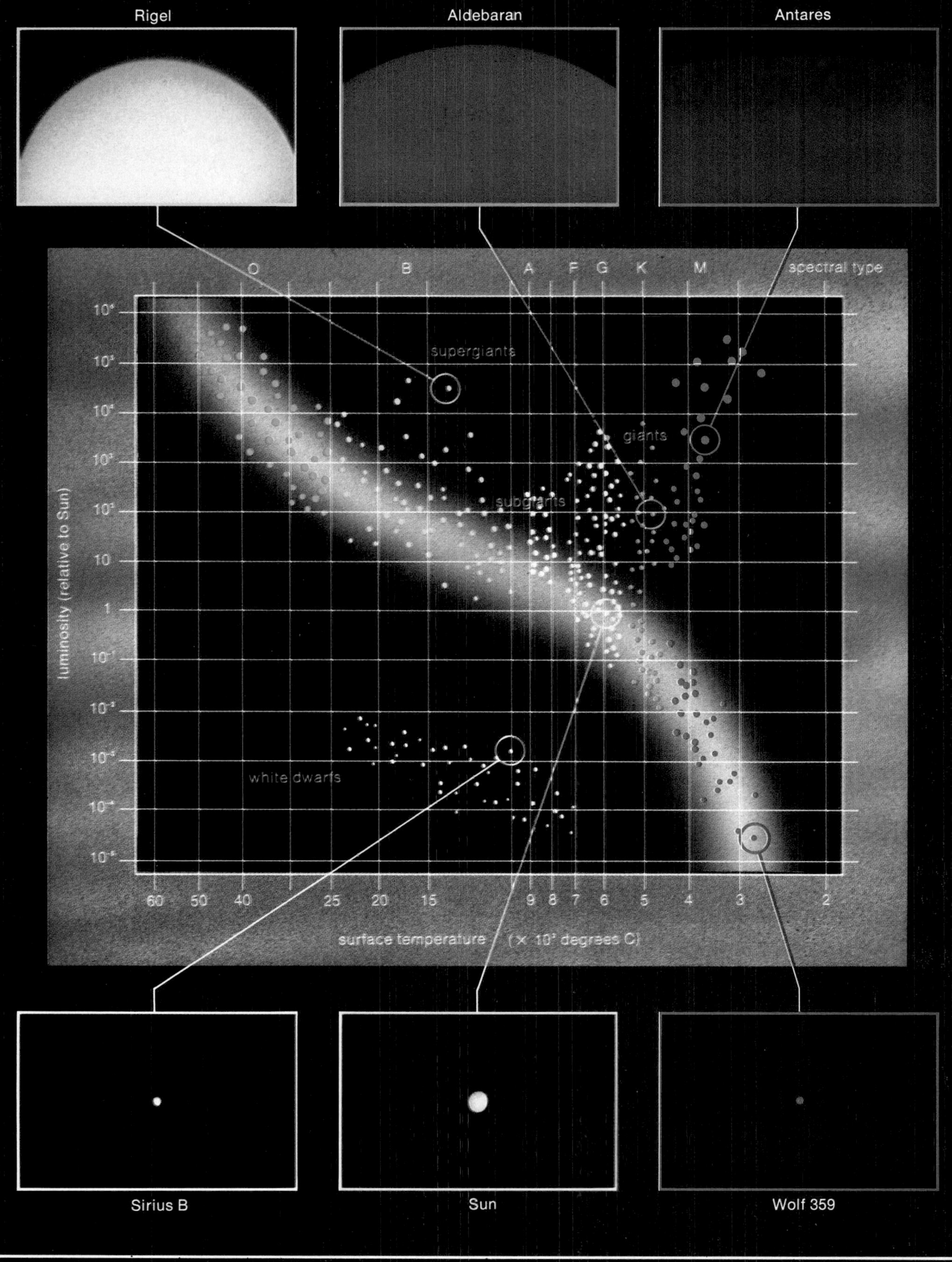

Adapted from Patrick Moore, "The Atlas of the Universe" (1970), Mitchell Beazley Ltd. & Rand McNally & Co.

Photos, courtesy, U.S. Naval Observatory

In the spiral galaxy Messier 51 the bright blue spiral arms outline regions of recent star formation. The next generation of stars produced in the Galaxy will emerge in a few million years from the thin lanes of interstellar dust directly adjacent to the arms.

the time that the objects were visible manifestations of stars in formation. Finally, Herbig discovered a class of stars, the Herbig emission stars, having absorption-line spectra characteristic of stars of mass greater than ~2 $M_{\odot}$. The Herbig emission stars shared with the T Tauri stars the properties of unusual emission-line spectra and irregular brightness changes.

Because interstellar dust efficiently absorbs light at visible wavelengths, optical searches are unable to probe far into the dusty, dark cloud regions or into the "cocoons" of gas and dust presumed to surround newly formed stars. At longer, infrared wavelengths (1 micron [0.000001 meter] or greater), however, the dust is more transparent. Consequently, with the advent of sensitive infrared detectors, astronomers have been able to initiate systematic searches of dark cloud regions at those wavelengths. In comparing an infrared (wavelength of 2.2 microns) and an optical picture of a dark cloud complex it becomes apparent that the optical photograph provides a very deceptive view of the extent of star formation in these interstellar clouds. The infrared stars observed at 2.2 microns and not visible in an optical picture probably represent only the brightest members of a cluster of many thousands of stars previously hidden from view. Dust dims the visible light from many infrared stars by factors of more than a billion times greater than the dimming that is suffered at infrared wavelengths.

Almost every infrared search of dark cloud complexes reveals YSOs obscured by circumstellar cocoons or by dark cloud material.

The blue spiral arms of the galaxy Messier 101 comprise huge complexes of newly formed stars.

Most infrared stars are probably already-formed stars. A few, however, may actually be heavily obscured objects of high luminosity, large radius, and low temperature, which radiate most of their energy in the infrared. Such objects may be protostars, in the stage just subsequent to collapse. The Becklin-Neugebauer (B-N) object discovered in the huge star-forming complex in Orion is thought by many astronomers to be such an object. Undetectable at optical wavelengths, the B-N object has a luminosity estimated from infrared measurements as more than 100,000 times that of the Sun.

Radio astronomers have also joined the search for YSOs. The development of sensitive receivers that can observe in the millimeter wavelength range has resulted in the discovery of a large number of emission lines produced by a wide variety of molecules present in interstellar clouds. Studies of this relatively uncharted region of the electromagnetic spectrum have already presented astronomers with exciting results. These findings may eventually lead to an identification of the factors that determine the dominant chemistry in the interstellar medium.

Observations of emission-line transitions in the millimeter wavelengths also provide astronomers with new tools for probing the physical conditions in dark cloud complexes. In order to be observed, a molecule must be excited to an energy level above the zero energy or ground state. The rate at which such excitations take place depends on the density and temperature of the dark cloud gas. The observed intensity of the emission line is proportional to the excitation rate and,

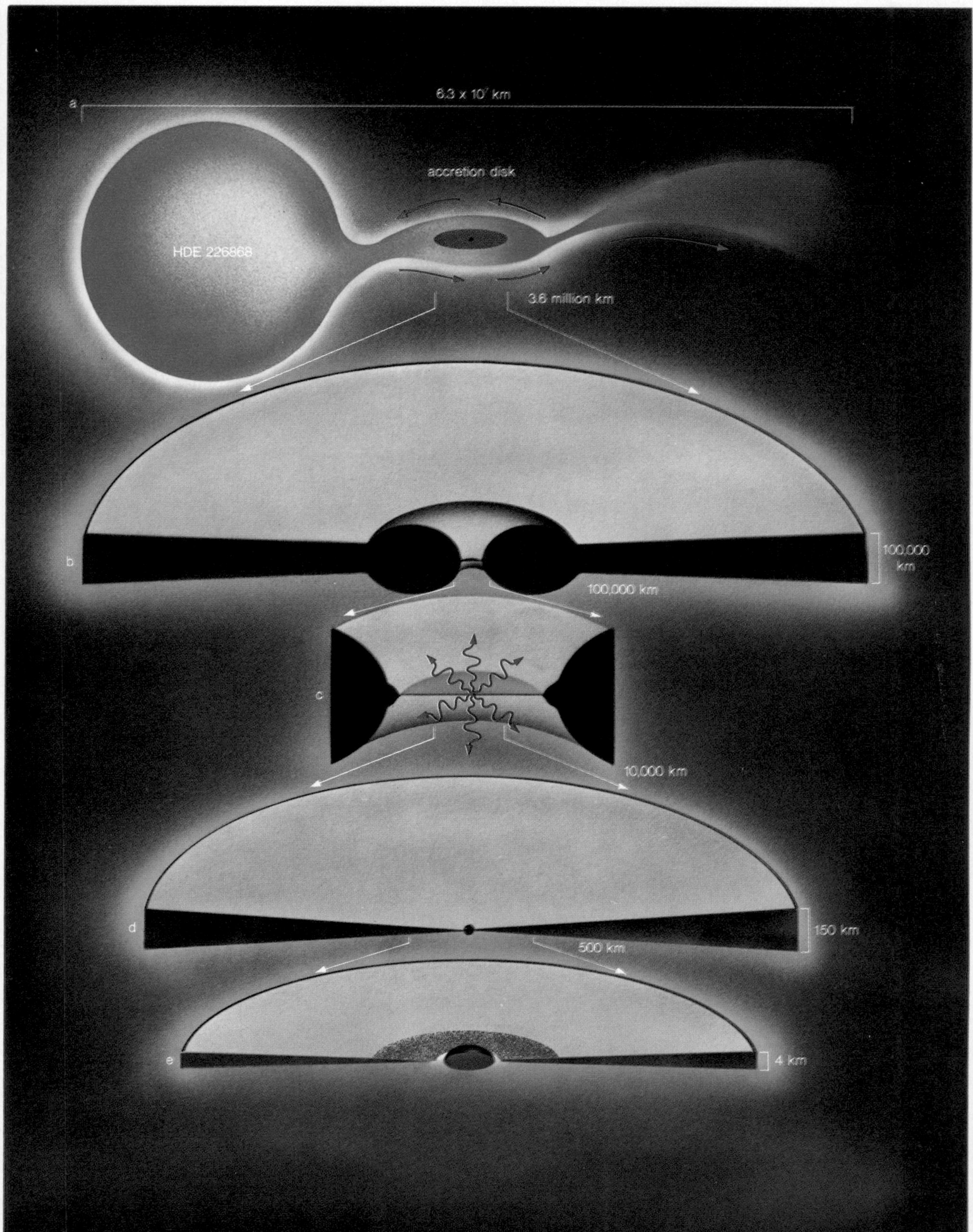

a
6.3 x 10^7 km
accretion disk
HDE 226868
3.6 million km
b
100,000 km
100,000 km
c
10,000 km
d
150 km
500 km
e
4 km

therefore, to the gas temperature and density. Moreover, the observed width (in wavelength space) of a molecular emission line conveys information regarding the distribution of velocities among the excited molecules. Thus, the factors that determine whether the cloud will collapse—the gas, temperature, density, and velocity—can be studied at millimeter wavelengths.

With the increased spatial resolution and sensitivity that will be possible when larger millimeter-wavelength antennas are built, it may be possible to identify dark cloud regions in the process of collapse. Remnants of the collapse process, cocoons of gas and dust surrounding newly formed stars, have probably already been seen at millimeter wavelengths. Maps of regions surrounding several Herbig-Haro objects indicate that the newly formed stars associated with those objects are located near the center of relatively high-density clouds of interstellar material.

Finally, strong maser emission lines of water and the hydroxyl (OH) radical have been discovered in many star-forming complexes. Although theoretical interpretations of these maser sources differ, they are believed to be produced in regions of very high density and perhaps are stimulated to emit by embedded protostars. In some cases luminous infrared sources, possibly protostars, are found in close association with these sources. Astronomers believe that the infrared sources and their associated masers may signal the final stages of formation of a YSO.

Early stages of stellar evolution

Having listed the various denizens of dark clouds discovered over the past few decades, the next step is to ask what these objects reveal about the early stages of stellar evolution. The objects discovered at optical wavelengths, although not the youngest of the YSOs, nevertheless are most amenable to detailed study of their physical characteristics. From optical and infrared observations of Herbig-Haro objects, T Tauri stars, and Herbig emission stars, astronomers have constructed a relatively coherent picture of early stellar evolution.

After the initial collapse phase, a YSO remains embedded deep within the large, dark cloud complex and also within its own smaller, denser "cocoon" of dust and gas. At an early age, perhaps as young as 10,000 years, the YSO develops a strong stellar "wind." This wind is analogous to the stream of ionized gas observed to be escaping from the surface of our own Sun. The strength of the wind of a typical young stellar object is, however, a millionfold greater than that of the Sun in terms of the amount of material leaving the star's surface. The wind is sufficiently strong to halt the accretion of gas and dust from the protostellar cloud.

The presence of a wind of this strength is not a feature expected from the simple spherical collapse picture presented earlier. It is fortunate for scientists that nature is, in fact, kinder than model predictions because stellar winds are sufficiently strong to dissipate, on a time

Model for a black hole in Cygnus X-1 helps explain visible and X-ray observations. Gas is drawn from the supergiant primary star HDE 226868 (a) by the gravitational attraction of a black hole. The gas nearest the black hole is whipped around it into a tight circular orbit, forming a thin accretion disk. A close-up of the accretion disk (b) shows that the gravitational pressure of the black hole compresses it and makes it thin, while at the same time thermal pressures in the gas react against the compression and try to thicken the disk. In the central bulge (c) the pressures, caused by heat from X-rays emitted near the black hole, are strong enough to thicken the disk. Thermal pressures in the core of the disk (d) are even higher than in the bulge, but gravity is so strong that it prevents this disk from thickening. X-rays observed from the Earth are generated only in the innermost 200 km (125 mi) of the core (e), at the center of which is the black hole itself.

Diffraction grating aids astronomers by breaking up light into spectral bands. Light from stars is gathered by a telescope and then dispersed by the fine parallel grooves of the grating into different wavelengths.

scale of 100,000 years, the cocoon of gas and dust surrounding the newly formed stars. At first, the cocoon "breaks" in only a few places. Light from the star leaks out through these breaks, illuminating patches of dark cloud material. These illuminated patches are probably the Herbig-Haro objects. Thus, by observing radiation scattered toward the Earth by the dark cloud material, astronomers are able to study the properties of stars that are perhaps no older than 100,000 years. Eventually the cocoon cloud is dissipated, and if the star is close enough to the boundary of the dark cloud it becomes visible either as a T Tauri star, if it is a low-mass ($\leq 3\ M_{\odot}$) object, or as a Herbig emission star, if it has higher mass.

The ionized matter in the wind is probably responsible for the unusual emission-line spectra characteristic of the Herbig-Haro objects, the T Tauri stars, and the Herbig emission stars. Because the wind is somewhat variable in strength the total amount of ionized material being emitted varies as well, thereby accounting for the observed irregular light variations characteristic of YSOs. Gradually, as the YSO approaches the main sequence, the wind dies down and the object increasingly resembles a normal star.

The source of the stellar wind is not well understood. The most likely explanation is that the wind represents the mechanism by which the envelopes of stars such as our Sun decrease their rotation rates. Some of the T Tauri stars, which will become main-sequence stars of near-solar mass, are still rotating at speeds of 50–100 kilometers per second, while the Sun rotates at only 2 kilometers per second. The removal to large distances of rapidly rotating gases from the T Tauri star's surface by a wind coupled to the rotating surface layers effectively slows down the rapid spinning of these stars.

The winds, if present at ages as early as 10,000 years from stellar "conception," will significantly affect the formation and evolution of primordial solar nebulas. First, any protoplanets must already have condensed from the gas and dust in these nebulas; the wind may well clear out low-density gas and dust to distances many times the dimensions of our solar system. Second, the atmospheres, at least of the planets near the forming star, will probably be removed by collision with the wind. Astronomers believe that further study of Herbig-Haro objects and T Tauri stars may yield critical information concerning the early history of the solar system.

Current research problems

Because detailed knowledge of early stellar evolution is primarily restricted to those objects for which optical data are available, astronomers have yet to test the validity of their theoretical insights into the early collapse phases. Promising candidates for further study of the early evolutionary phases include the YSOs embedded in dark clouds, objects analogous to the B-N object, and the water and OH maser sources. The next few years should witness significant progress in understanding these early phases.

A fuller understanding of the star formation process seems intimately intertwined with solving the problem of dark cloud evolution. How did these complexes of gas and dust form? Are they random events, or does some set of external forces determine their creation? Astronomers have just begun to address these important questions. Photographs of spiral galaxies similar to our own suggest that star formation is restricted primarily to the bright spiral arms that dominate the visible light from these systems. Recently formed, hot, massive stars with short main-sequence lifetimes, in fact, form the outlines of these spiral patterns. None of these young objects is found outside the spiral arms, thus ruling out the possibility that star formation takes place randomly across the disks of these galaxies.

Of great significance is the association of interstellar dust clouds with the spiral arms. These clouds delineate narrow lanes directly adjacent to the bright arms. The next generation of bright, massive stars will probably be formed in these dust lanes. Thus, the formation of dark clouds and, later, stars must be triggered by a mechanism that operates on a galactic scale. W. Roberts, C. C. Lin, and Frank Shu argued that the processes of cloud and star formation are initiated by the interaction of galactic gas with a galaxy-wide gravitational perturbation in the disk stars. This perturbation is called a "density wave." As the gas and dust rotating about the galactic center encounter the density wave, they are compressed. The compressed gas and dust form interstellar clouds, some parts of which collapse and form new stars. If these views are correct, scientists not only have isolated the driving force for present-day star formation in spiral galaxies but may also have developed an essential clue to understanding the evolutionary history of these systems.

Another challenge faced by astronomers engaged in studying the star formation process is to achieve a proper understanding of how the first generations of stars in our universe were formed. Astronomers believe that the universe during its early history did not contain in significant quantities those chemical elements that collect to form dust grains. Consequently, there were, in all probability, no dark cloud complexes similar to those found at present in our Galaxy. As a result, the events leading to collapse of stars from primordial gaseous material were probably significantly different from those now operative. In fact, the precise interplay of physical events that led to the creation of stars under those conditions is not at all well understood. Yet the manner of star formation and its efficiency at this early epoch probably was the most significant factor in determining the form of our Galaxy and of others similar to it. Much research, both theoretical and observational, therefore remains to be done before a full answer can be given to the question: where did stars come from?

Radio Telescopes

by George W. Swenson, Jr.

Radio waves from sources as distant as the known boundaries of the universe continually bombard the Earth. This rich repository of information about the cosmos is tapped by modern astronomers.

Operating principles of prime-focus parabolic radio telescope (opposite, top) and an instrument employing a Cassegrain system of reflectors (bottom).

With his vision restricted to the visible portion of the electromagnetic spectrum and his body to the ground on which he stood, ancient man perceived but feebly the existence of a vast and distant universe beyond his own planet. The first significant attempt to overcome these limitations took place in 1609 when Galileo pointed his newly invented optical telescope toward the Moon and, in effect, approached 30 times nearer this satellite than any previous observer. For more than three centuries thereafter, astronomical knowledge continued to be won almost solely by observing the visible light emitted by or reflected from celestial objects. By the beginning of the 20th century, scientists were aware that light waves differ from radio waves mainly in their length; consequently, any object that emitted light waves could also be expected to emit electromagnetic waves in other parts of the spectrum. Furthermore, optical astronomers knew that some celestial bodies emitted energy in the near-infrared part of the spectrum, where wavelengths were just too long to be detected by eye, and in the near-ultraviolet where wavelengths were just too short. Photographic plates could be made sensitive to these wavelengths, but astronomers were frustrated in investigating wavelength ranges very far from the visible region of the spectrum. They discovered that the Earth's atmosphere is nearly opaque to both infrared and ultraviolet waves but that it is transparent to a spectral region called the optical window, a range nearly identical to the visible spectrum.

By the early 1930s it was suspected that the Earth's atmosphere is also fairly transparent to a certain region of the radio spectrum, a radio window from about 5 millimeters to 30 meters, but no one had ever detected wavelengths in this range entering from beyond the atmosphere. About this time Karl Jansky, an electrical engineer with the Bell Telephone Laboratories, undertook an investigation of radio static, the familiar hissing that had always interfered with shortwave communications. Using an experimental antenna that resembled the wing skeleton of a biplane, he determined that 15-meter radio noise was emanating from a definite point in the sky that moved from day to day—the nucleus of our Galaxy in the direction of the constellation Sagittarius. Jansky's results, published late in 1932, remained largely ignored by the scientific community for several years, until they attracted the attention of Grote Reber, a U.S. electrical engineer and amateur astronomer. Reber's perceptive analysis of the instrumental problem of radio astronomy led him to build, in the garden of his home in Illinois, what may properly be called the world's first radio telescope. It consisted of a paraboloidal reflector about 9½ meters in diameter and a low-noise receiver tuned to a wavelength of about 2 meters. With this apparatus, in the late 1930s and early 1940s, he mapped the Milky Way and observed the Sun.

World War II interrupted the development of the astronomical aspects of the budding science, but the development of ultrahigh-frequency receiving techniques for military purposes and the training of many physical scientists in practical electronics stimulated rapid

GEORGE W. SWENSON, JR., *is Professor of Astronomy and Electrical Engineering at the University of Illinois, Urbana.*

(Overleaf) Painting by Paul R. Alexander. Illustrations by Dennis Magdich

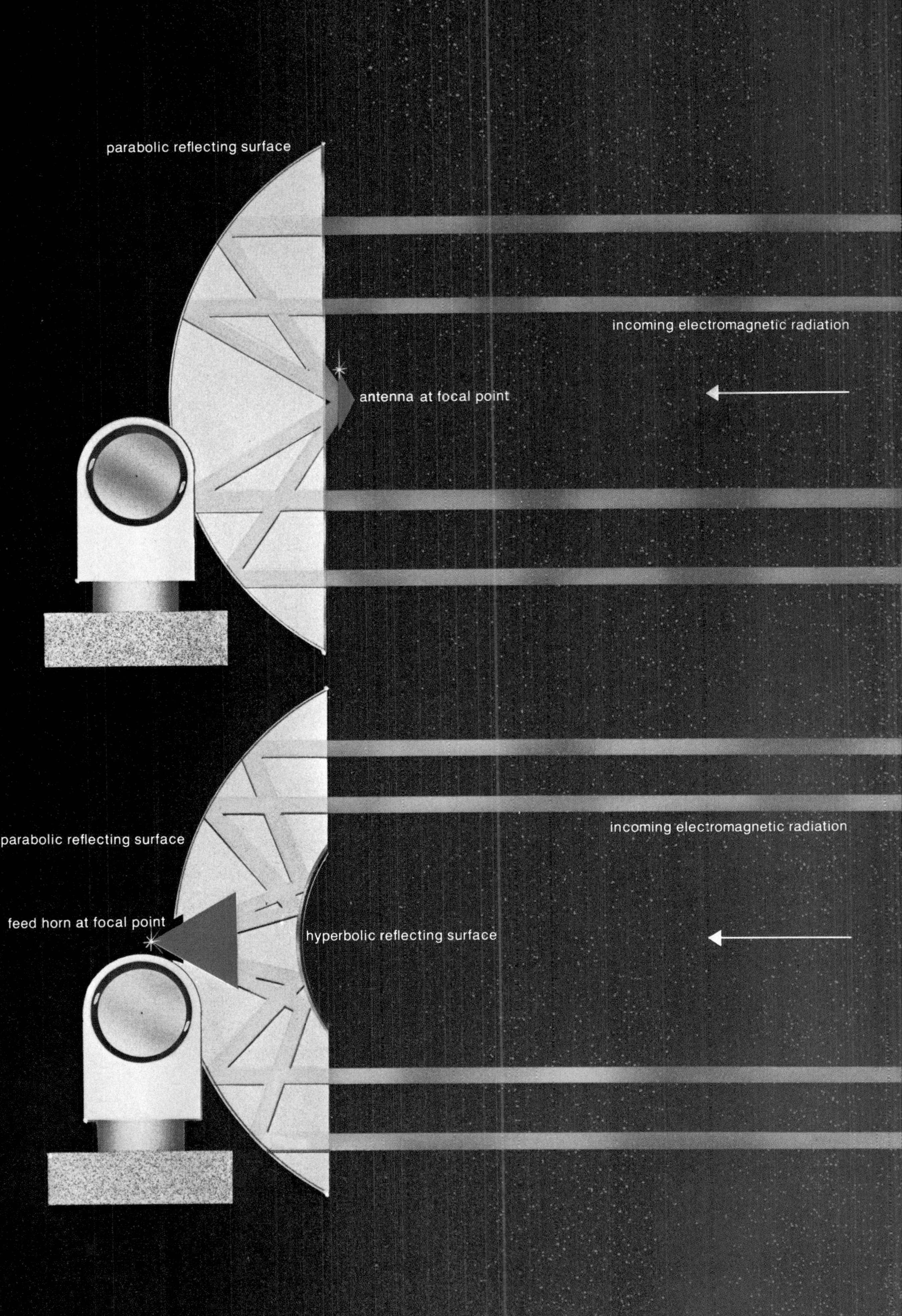
parabolic reflecting surface
incoming electromagnetic radiation
antenna at focal point
incoming electromagnetic radiation
parabolic reflecting surface
feed horn at focal point
hyperbolic reflecting surface

Lying above the optically visible surface of the Sun are regions of gas and magnetic fields in which intense radio signals are generated. Two kinds of radio emission, known as the quiet Sun component and the slowly varying component, are long-lived signals associated with persistent or relatively permanent solar characteristics. Short-lived radio bursts are associated with temporarily active regions of the Sun, especially with the site of solar flares, and can be assigned to a variety of specific causes (opposite page). Type I bursts (not depicted) are the detailed emissions of radio-noise storms that invariably accompany large sunspot groups. Types II, III, and V originate in disturbances moving outward from an optically visible flare. Types II and III are believed to be generated by oscillating plasmas (atoms dissociated into electrons and positive ions) traveling at velocities about 0.003 and 0.2 times the velocity of light. Type V and IV bursts consist of synchrotron radiation generated by plasma electrons moving near the velocity of light through curved paths in the intense magnetic fields (white loops) associated with solar disturbances.

postwar expansion of radio astronomy. During the war years some observations of solar emissions were made by radar engineers, and the young Dutch astronomer H. C. van de Hulst predicted that it might be possible to detect a spectral line at a wavelength of 21 centimeters from clouds of monatomic neutral hydrogen in the Galaxy. With the resumption of peacetime research, the exploitation of available electronic talent and quantities of surplus radar equipment made possible rapid progress in astronomical research. British and Australian scientists discovered that, in addition to the Sun and the Milky Way, numerous sources of radio waves exist in the sky, and that some of them are of such small angular diameters as to suggest that they are very distant from us. Finally, in 1951, H. I. Ewen and E. M. Purcell of Harvard University detected the hydrogen-line radiation predicted by van de Hulst. Continued observations at 21 centimeters over the whole sky soon produced a view of the Galaxy as an exceedingly flat layer of matter with spiral arms, a shape in general agreement with the positions of optically visible components and with the photographs of many of the external galaxies.

Equipment limitations

Since these first discoveries in the science, radio astronomy has been dominated by several instrumental limitations for which there have been as yet no completely satisfactory solutions. One problem is that of angular resolution, which for a given instrument may be defined as the smallest angular separation of two infinitesimal sources that may still be distinguished from one another, or alternatively as the smallest feature that may be distinguished in full detail. The earliest radio telescopes were merely radio antennas of modest size, whose angular resolutions or beams were so broad that they gave only very fuzzy or out-of-focus pictures of the distribution of sources. To yield good angular resolution or a narrow beam, an antenna must be very large compared with the wavelength that is received. In the case of the familiar paraboloidal-reflector antenna, the beam width is approximately 60° divided by the diameter of the reflector in wavelengths. Thus, to distinguish between two small sources 1° apart in the sky, one would need a paraboloidal antenna at least 60 wavelengths in diameter. Resolution at 1-meter wavelength of sources that are 1° apart would require a 60-meter antenna; resolution at 10-centimeter wavelength would require a 6-meter antenna.

Actually, astronomers need much finer detail than this in their measurements. Suppose that one would accept an angular resolution of one second of arc. This is the angle subtended by a ten-cent piece at a distance of one mile. For a wavelength of ten centimeters this would require a paraboloidal reflector about 22 kilometers in diameter, whose surface was constructed with a precision of one centimeter. The apparatus at the focal point of the paraboloid would have to be suspended about ten kilometers above the bottom of the reflector, again with a precision of about one centimeter. Additionally, the

Adapted from H. P. Palmer, *et al.* (eds.) "Radio Astronomy Today" (1963); Harvard University Press and Manchester University Press

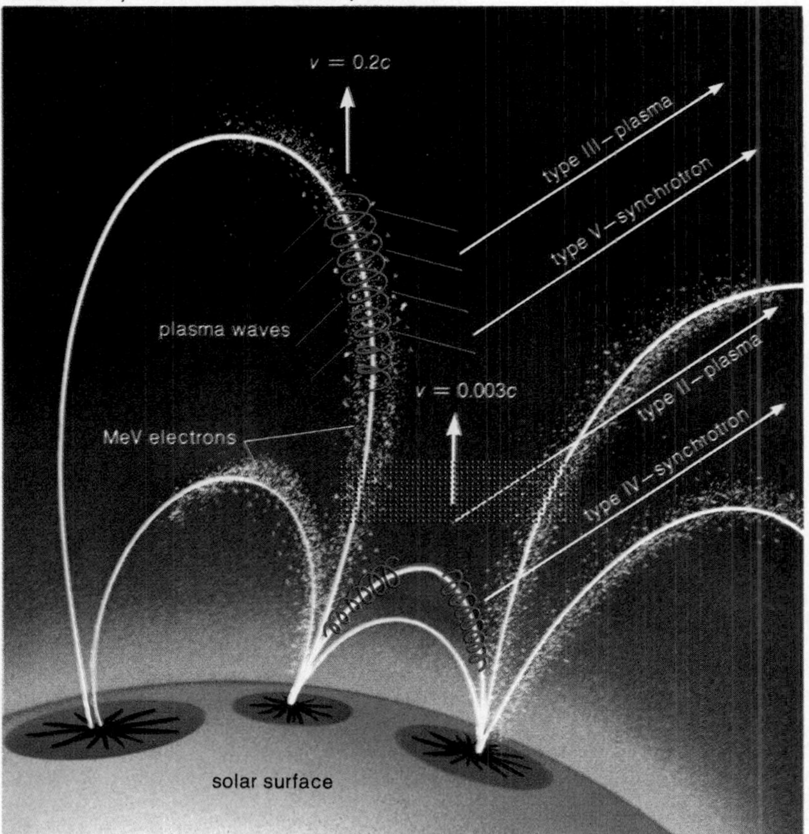

telescope preferably should have the capability of being pointed in any direction and of maintaining its precise shape during the movement. Clearly this proposal would be both structurally and financially impractical. The radio astronomer, therefore, must be satisfied with more modest paraboloidal reflectors, the largest of which at present are about 100 meters in diameter.

Another problem is that of sensitivity. The signals received by a radio telescope are extremely weak. Typical sources presently being studied at decimeter or centimeter wavelengths deliver only about 10^{-19} watts of power to a typical radio telescope. It has been estimated that all the energy collected by all radio telescopes since Jansky's time would only suffice to lift one grain of sand a few centimeters.

The output that the radio telescope delivers to its recording device represents not only the power from the cosmic source but also the radio noise generated in the electronics of the telescope and radiated from the ground near the telescope. This noise tends to obscure cosmic signals and constitutes a real nuisance, for it can be many thousands of times stronger than the desired signal. Obviously it is necessary to employ the best low-noise receiving techniques; masers and parametric amplifiers had their first applications in radio astronomy. Receiver stability, signal-processing techniques, and the theory of weak-signal detection have achieved their highest stages of development in this branch of the electronic art. Nevertheless, the radio

Courtesy, National Radio Astronomy Observatory

Pioneering radio astronomer Grote Reber and his home-built, prime-focus parabolic radio telescope.

astronomer is continually plagued by insufficient signal power and must always seek to increase it. As in the struggle with the problem of angular resolution, the most direct solution is to increase the collecting area of the antenna. Immediately the various problems of large antennas, including cost and precision, are encountered, and compromises must be sought.

Parabolic reflectors

The parabolic reflector is the most widely used form of radio telescope. The principle is also utilized in optical telescopes and in such focused light sources as searchlights and flashlights. The radio waves from a cosmic source, that is, a source at a very great distance, may be considered to travel as parallel rays. If these rays are parallel with the axis of the paraboloid they will all be reflected from the surface at such an angle as to pass through its focal point. A feed antenna at the focal point converts the concentrated electromagnetic field into an electric voltage, which can be amplified by a radio receiver. Thus, the parabolic reflector collects the radiated energy passing through a substantial area and concentrates it for recording. Rays traveling in a direction not parallel with the axis of the paraboloid will not be reflected to a single point, and few of them will intersect the focal point. Hence, the telescope has little sensitivity to off-axis sources. This effect is what makes possible the telescope's ability to discriminate among various sources.

In order to reflect the incoming waves efficiently, the surface of the paraboloid must be a good conductor of electric currents and must approximate the mathematical paraboloid to within about one-tenth of a wavelength. To be effective at a wavelength of 15 centimeters, for example, the surface must have few points which deviate from the ideal surface by more than 1 centimeter. Furthermore, if the surface is made of wire mesh, the mesh openings should be 1 centimeter or less in diameter.

The telescope must be aimed at the source to be observed. If the reflector is a paraboloid of revolution, mechanical motion of the entire reflector is required. Because the aiming must be precise, the motion must be controlled to a few seconds of arc in many cases, even with substantial wind forces on the structure. Driving machinery and measuring and control apparatus have been developed to high degrees of precision and represent substantial parts of the costs of large parabolic telescopes.

At the present time the largest fully steerable paraboloidal radio telescope is the 100-meter instrument of the Max Planck Institut für Radioastronomie near Bonn, West Germany. For such a heavy instrument, deformation from the ideal paraboloid is a serious problem. Before it was built, theoretical studies showed that a 100-meter reflector could be so constructed that, when shifted in spatial orientation, gravitational forces would cause its surface to deform continuously from one paraboloidal shape to another. This preservation of parabolic integrity, together with its aiming precision, permits the Bonn telescope to be used for wavelengths as short as one centimeter, where it produces a beam width of about 22 seconds of arc. This represents the best angular resolution presently obtainable with a single dish, although it is only slightly better than the maximum visual acuity of the unaided human eye.

Courtesy, Max Planck Institut für Radioastronomie

The 100-meter, altazimuth-mount radio telescope of the Max Planck Institut für Radioastronomie near Bonn, West Germany.

Other forms of single reflectors

The large reflector telescope can be realized in other geometrical forms in addition to the paraboloid of revolution with the feed antenna at the prime focus. One of these is the Cassegrain system long used in optical reflecting telescopes. Here a reflector in the form of a hyperboloid of revolution is placed ahead of the main parabolic reflector to reflect the radio rays back into a feed horn at the vertex of the main reflector. This system provides greater accessibility to the electronic equipment at the feed point and tends to receive less radio noise from the surrounding environment than does the prime-focus parabolic telescope. A further evolution of this principle involves a slight departure from true parabolic shape for the main reflector and a compensating departure from hyperbolic shape for the subreflector. This shaped-reflector system can be made more efficient than the prime-focus parabolic; while the latter can extract approximately 50% of the cosmic radio power falling on its surface, the shaped-reflector telescope can extract up to 80%. This principle will undoubtedly be used in the majority of new radio telescopes.

Other large, single-reflector telescopes have been built, utilizing a variety of configurations to achieve large collecting area at relatively modest cost. An example is the 133-meter-wide parabolic cylinder of the University of Illinois Vermilion River Observatory. To reduce costs, this instrument was built of earth, the cheapest suitable material, utilizing a natural north-south ravine. A wire mesh laid on the parabolic surface reflected cosmic radio waves to a row of 276 antennas along the focal line atop the towers. Adjustment of the electrical phase relationships among the antennas permitted steering the telescope beam in the north-south direction, but east-west motion is provided only by the Earth's rotation. This telescope was representative of a number of large instruments, employing a variety of operating principles, which were built to make surveys of the sky during the late 1950s and early 1960s. This phase of radio astronomy, which provided catalogs of many thousands of cosmic radio sources, is largely over and the specialized instruments that did this work are, for the most part, no longer actively employed.

One method of achieving a large collecting area at relatively low cost—although with some sacrifice of versatility—is through the use of a spherical reflector. The best-known telescope of this type is the 1,000-foot-diameter instrument at Arecibo, Puerto Rico. The spherical reflector surface is anchored in a natural depression in the ground. Incoming radio waves are reflected from the surface to a region

Courtesy, Bell Laboratories

Electrical engineer Karl Jansky and the experimental antenna with which he traced the celestial source of radio noise at 15 meters wavelength.

below the center of the sphere where they can be intercepted. Because the reflected rays do not all pass through the same point, the reflector does not have a focal point; hence, it is necessary to provide a number of antennas, called a collector array, either arranged along a line or distributed over an area. In addition, because the various rays from a distant source travel different distances to the collector array, sophisticated electronic equipment must be provided to add the contributions of the various rays properly.

The spherical telescope can be steered by moving the entire collector array along an arc centered on the center of the sphere. Unfortunately its steerability and consequently its tracking capability are limited by the fact that, at extreme angles, much of the incoming radiation can never reach the array. Hence, the portion of the collecting area contributing to the system at any one time is usually substantially less than the projected area of the telescope. An important advantage of the spherical telescope, however, is that a number of collecting arrays can be employed simultaneously to observe in different directions and, if desired, at different wavelengths.

Single-reflector antennas are useful for a wide variety of astronomical programs. In the larger sizes their large collecting areas and relatively narrow beam widths make them especially suitable for radio spectroscopy, the study of molecules in interstellar space by means of the emissions at discrete, characteristic wavelengths which typify individual molecular species. About 40 different molecules currently are known to exist in clouds of gas and dust between the stars of the Galaxy, ranging in complexity from hydrogen and hydroxyl, with two atoms each, to dimethyl ether and ethyl alcohol, with nine. Studying the concentrations of these molecules yields information concerning the processes involved in the birth and death of stars and in large-scale motions of the Galaxy. Large single-reflector telescopes are more sensitive to the radiations from such large, diffuse sources (nebulas) as interstellar gas clouds than are telescopes capable of

Courtesy, University of Illinois Vermilion River Observatory

The 133-meter-wide parabolic cylinder of the University of Illinois Vermilion River Observatory.

much better angular resolutions. For this reason single-reflector telescopes are still under active development, especially those designed to receive wavelengths between 0.5 millimeter and 1 centimeter. Such telescopes require extreme precision of their reflecting surfaces and pointing them requires equally great precision.

Tracking apparatus

One of the most basic requirements of a telescope is that it be able to follow a source across the sky as the Earth rotates. This permits the faint signals from the source to be accumulated for many minutes or even hours, in order that they may be discerned above the noise generated in the receiver and in the local environment of the telescope. This ability to track a source is particularly important for molecular spectroscopy of diffuse sources, which are often imbedded in the very noisy background radiation from the Galaxy. The motion of the telescope that permits tracking is generally achieved by mounting the reflector on either an equatorial or an altazimuth mount.

The equatorial or polar mount is the traditional way of accomplishing sidereal motion, the motion of a star or other fixed point in the sky, both for optical and for radio telescopes. An equatorial mount has an axis parallel with the axis of the Earth. As the Earth turns from west to east, the telescope turns about its equatorial axis to follow a source in the sky. Another axis, the declination axis, at right angles to the equatorial axis, provides the ability to aim at any point in the sky. The advantage of the equatorial mount is that sidereal motion can thereby be accomplished by motion about a single axis, and only one accurate tracking motor is required.

An altazimuth telescope has one vertical axis and one horizontal axis. In order to follow a source in sidereal motion, it is necessary to turn the telescope about two axes simultaneously. This, in turn, requires a computer to compute the required rates of motion about the axes. Until precise, inexpensive digital computers and shaft-angle

Courtesy, G. W. Swenson, Jr.

An example of the aperture-synthesis system employing multiple-antenna arrays is the linear configuration of five 20-meter paraboloids at Stanford University.

encoders became widely available, this was a serious disadvantage of the altazimuth system. The equatorial system has the disadvantage that to track a source from horizon to horizon requires that the reflector be mounted at some distance from the declination axis, an expensive procedure. In addition, those sources which never set, the circumpolar sources, cannot be tracked for 24 hours as they can with an altazimuth telescope. For these reasons, all large radio telescopes of the future probably will have altazimuth mounts. The 100-meter telescope at Bonn is of the altazimuth type. The supporting tower is mounted on railroad wheels and rotates about a vertical axis. The horizontal axis is carried by the tower and is driven by huge spur gears.

Radio interferometers

Much of the present research in radio astronomy is concerned with the detailed examination of cataloged cosmic sources. One method is the study of the spectral emission and absorption lines characteristic of molecules in cosmic gas clouds; this usually involves large single-reflector telescopes capable of tracking a source for long periods of time and having large collecting areas to extract the greatest possible power from the faint signals from the source. The other requirement in the detailed examination of a source is accurate mapping of its structural details. In the case of a very distant source, such as a galaxy external to our own Milky Way or one of the enigmatic quasars, this examination requires very high angular resolution, much better than the approximately one minute of arc available with single reflectors.

In order to overcome the structural and monetary limitations on the

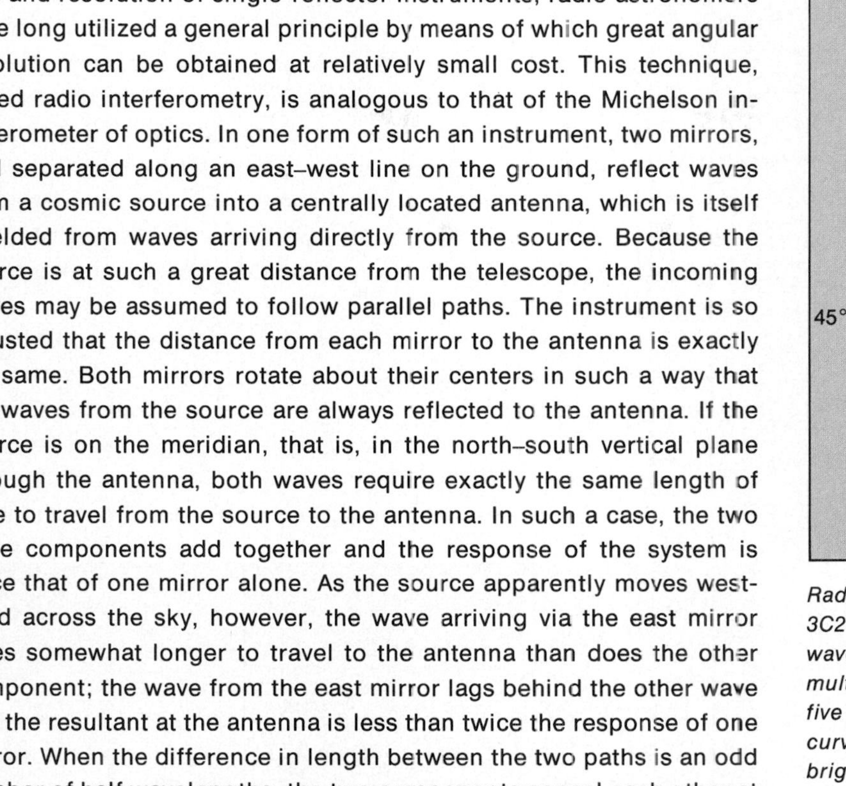

size and resolution of single-reflector instruments, radio astronomers have long utilized a general principle by means of which great angular resolution can be obtained at relatively small cost. This technique, called radio interferometry, is analogous to that of the Michelson interferometer of optics. In one form of such an instrument, two mirrors, well separated along an east–west line on the ground, reflect waves from a cosmic source into a centrally located antenna, which is itself shielded from waves arriving directly from the source. Because the source is at such a great distance from the telescope, the incoming waves may be assumed to follow parallel paths. The instrument is so adjusted that the distance from each mirror to the antenna is exactly the same. Both mirrors rotate about their centers in such a way that the waves from the source are always reflected to the antenna. If the source is on the meridian, that is, in the north–south vertical plane through the antenna, both waves require exactly the same length of time to travel from the source to the antenna. In such a case, the two wave components add together and the response of the system is twice that of one mirror alone. As the source apparently moves westward across the sky, however, the wave arriving via the east mirror takes somewhat longer to travel to the antenna than does the other component; the wave from the east mirror lags behind the other wave and the resultant at the antenna is less than twice the response of one mirror. When the difference in length between the two paths is an odd number of half wavelengths, the two components cancel each other at the antenna and the source, in effect, becomes invisible. The result of the westward motion of the source, therefore, is a steady oscillation in output from the antenna.

The foregoing description is based upon the assumption of a source of point (infinitesimal) diameter, emitting radio waves at a single wavelength. Such cosmic sources as galaxies, planets, nebulas, and the Sun generally emit radiation over a wide range of wavelengths and, of course, these have appreciable angular dimensions. Scientists have shown that the response of an interferometer to such radiations is an oscillation in the output of the receiver, similar in many respects to the response of the simplified system we have described.

In order to understand the application of the interferometer to a more general problem, one may consider a source consisting of two points of equal strength, separated by a small angle. First, let the angular separation be sufficiently small so that the interferometer output due to one source adds directly to the output due to the other source. Then the indicated output will be nearly twice the output from a single source. Next, let the angular separation be increased until the outputs due to the two points nearly cancel each other. The resultant output for this case will be substantially less than twice the output due to either point source alone. Finally, let the angular separation be such that the path difference between the two sources is one-half wavelength. Then the two outputs directly cancel and the resultant is zero. The interferometer, in a sense, is blind to this particular source.

Courtesy, Stanford Radio Astronomy Institute

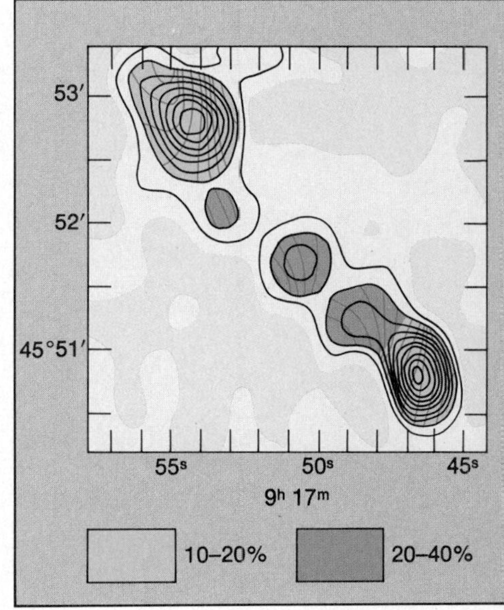

Radio map of the giant galaxy 3C219 observed at three centimeters wavelength by Stanford University's multiple-antenna array reveals five source components. The black curves represent total-intensity brightness contours, and the blue curves indicate the position angle of the plane of radiation polarization. Shaded areas within the contours are regions in which significant polarized radiation flux has been detected, with the darker shade corresponding to a higher percentage of polarization.

From a consideration of the three cases above, it is clear that the response of the interferometer depends upon the size and shape of the source. For the simple, two-component source assumed here, the separation of the source components can be determined from the output response of the interferometer. The angular separation to which the instrument is blind is directly related to the difference between the two path lengths, which, in turn, depends upon the separation, in wavelengths, of the two mirrors. By moving the two mirrors, one can find the separation that produces the blind response, and hence the angular separation of the two sources.

Aperture-synthesis techniques

It is interesting to consider the case of a number of pairs of mirrors—in effect, a number of interferometers—all reflecting waves from the same cosmic source into the same receiving antenna. If each of the mirrors is placed tangent to a parabolic curve whose focus is at the

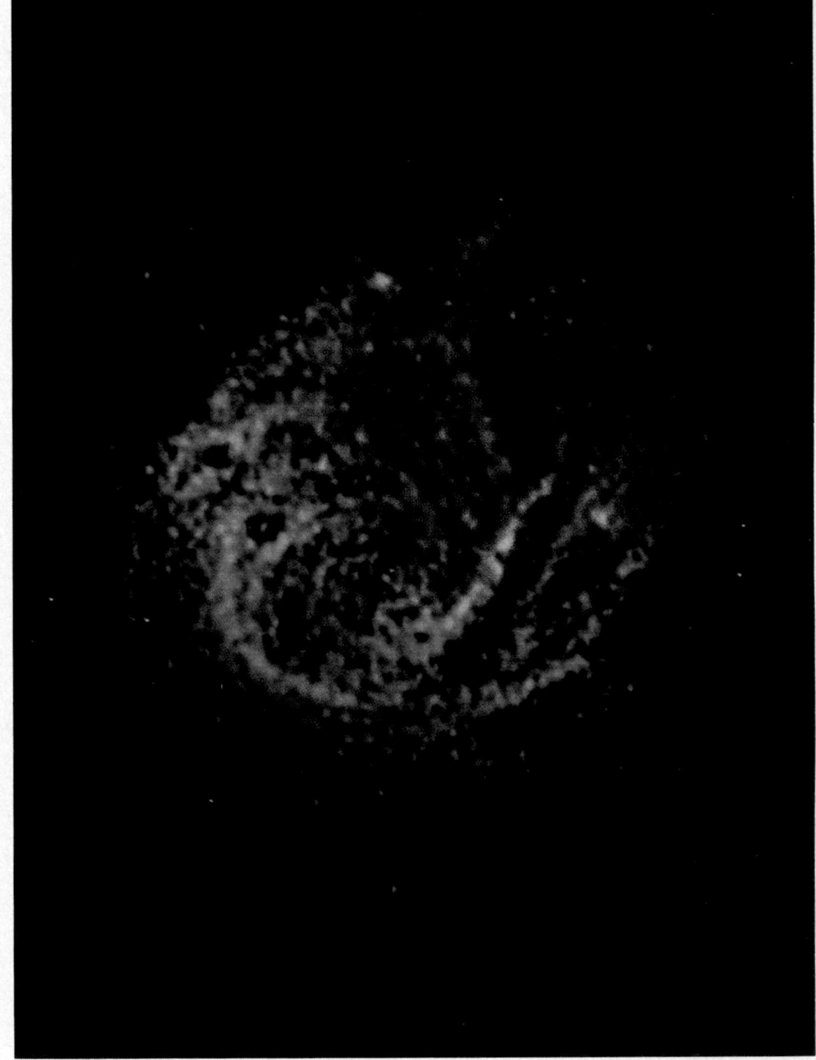

Radio image of the external galaxy M101 obtained with the Westerbork aperture-synthesis radio telescope at the 21-centimeter wavelength of neutral hydrogen and computer-processed by the U.S. National Radio Astronomy Observatory. Color hues in the image are a function of the velocity of hydrogen gas with respect to motion through space of the galaxy as a whole, with the color green representing zero relative velocity. Hydrogen rotating toward the Earth is depicted in colors comprising the blue end of the spectrum, whereas hydrogen rotating away from the Earth is shown in colors comprising the red end. In addition, intensity of the color at any given location is a function of the relative amount of gas moving at that velocity. The image is considered to be a four-dimensional representation because it simultaneously conveys not only general location of a region of gas in terms of horizontal and vertical coordinates but also information on gas velocity and concentration.

antenna, the path length from the source to the antenna is the same, regardless of which mirror is involved. For a source on the meridian, the electric field at the antenna can be considered to be the superposition of the contributions of the several interferometers. Alternatively, the entire device could be considered a parabolic reflector. In either case the physical result is the same; thus, there must exist a close relationship between the results obtained by scanning a source with a number of interferometers of different baselines and those obtained by a parabolic reflector of conventional design.

It should be mentioned that it is unnecessary to place the mirrors tangent to the parabola. Similar results can be obtained by placing them near the ground, properly oriented to reflect the incoming rays into the antenna, and so adjusted in height that the various ray paths from source to antenna differ only by integral numbers of wavelengths. This scheme is seldom used because, for any particular installation, it operates satisfactorily over a rather narrow band of wavelengths. The idea does serve, however, to emphasize the equivalency of an assembly of interferometers and a continuous-aperture antenna.

In astronomical practice, interferometers consisting of mirrors are seldom used, although the earliest such experiment was conducted on a seaside cliff, using the surface of the sea as a mirror. It is much more satisfactory to replace the mirrors with antenna elements and to interconnect them with electrical cables. By connecting a sufficient number of interferometers one may "synthesize" a radio telescope of very large dimensions and, hence, of extremely great resolution. Furthermore, it is unnecessary that all interferometers exist simultaneously. If the characteristics of the source under study do not change within the period of observation, a picture of its brightness distribution may be synthesized with only two antennas, one fixed and one movable over the ground. The source is observed for one day with a certain separation between antennas; then one is moved to a different position and the source observed for another day. The process is repeated until all necessary spacings have been used. Next, the data are processed in an electronic digital computer that delivers a map of the source. The fineness of detail and other properties of the map are equivalent to those that would be produced by a single-reflector telescope of diameter equal to the largest separation of the interferometer antennas. Thus, two antennas and a one-kilometer railway connecting them are equivalent to a single-reflector telescope one kilometer in diameter. Clearly the former system is structurally and economically feasible; the latter is not. Several important two- or three-element interferometers are in operation in various parts of the world.

The penalty one pays for the relative economy of a two-element interferometer is the long period of observing required to synthesize a source map. This time can be reduced substantially by using several antennas, forming interferometer pairs simultaneously between all possible combinations of two antennas. For example, four antennas can form six simultaneous pairs; the number of antenna movements

Courtesy, National Radio Astronomy Observatory

The 140-foot (43-meter) instrument at Green Bank, West Virginia, is the largest equatorially mounted radio telescope ever built.

Barrett Gallagher

The 1,000-foot (300-meter) spherical-reflector radio telescope at Arecibo, Puerto Rico. The triangular framework suspended above the reflector carries arc-shaped tracks, along which the antenna collector array can be moved to steer the telescope toward any part of the sky within 20° of its zenith.

and the total observing time are one-sixth that required for a two-element system.

Several large aperture-synthesis systems of the multiple-antenna type are in operation, including a 1.6-kilometer array of twelve 25-meter paraboloids at Westerbork, The Netherlands, and a five-kilometer array of eight 19-meter paraboloids at Cambridge, England. The array of five 20-meter paraboloids at Stanford University, like many of the world's synthesis systems, is on a straight, east–west baseline. An array on a straight line can produce two-dimensional maps of sources at high declinations, that is, of sources that lie well out of the plane of the Earth's Equator. Sources near the Equator, however, cannot be synthesized by a linear array. For this reason, in some arrays antennas can be placed on intersecting baselines, permitting satisfactory observation of low-declination sources. The most elaborate of such arrays is the Very Large Array (VLA), currently under construction by the National Radio Astronomy Observatory in a high mountain basin in New Mexico. Twenty-seven 25-meter shaped-reflector antennas will occupy stations along a Y-shaped configuration of railroad track, each arm of which is 21 kilometers long. At a wavelength of 1.5 centimeters, this instrument can synthesize source maps with detail as small as 0.1 second of arc, each map requiring one 12-hour observing period. This project is by far the most ambitious ever undertaken for radio astronomy, and its projected cost of $80 million makes it the most expensive ground-based telescope ever designed.

From a project of A. G. Willis, R. G. Strom, and A. S. Wilson of the Leiden Observatory

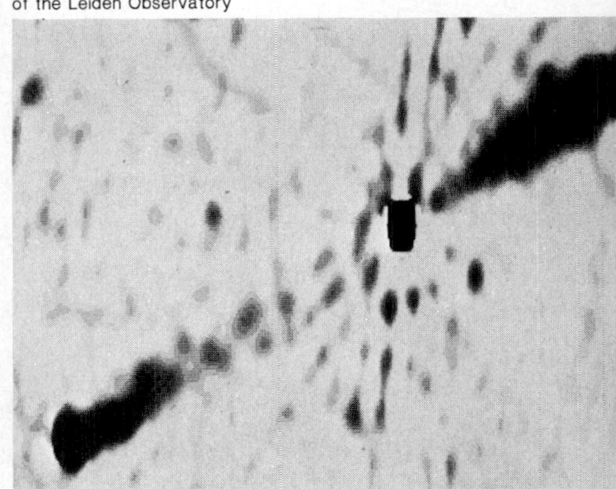

The aperture-synthesis radio telescope at Westerbork, The Netherlands (left), is an interferometer composed of twelve 25-meter antennas located along an east–west baseline 1,600 meters in length. In conjunction with a computer, the instrument was used to generate a radiograph (above) of the largest known object in the universe, a radio source measuring more than 18 light-years across. The object comprises two radio-emitting clouds located on opposite sides of the optically visible galaxy 3C236, the location of which is marked near the center of the radiograph by a dark rectangular spot.

Even the VLA, however, does not represent the ultimate radio telescope. It is known that very distant galaxies and quasars have details as small as 0.0001 second of arc, equivalent to a ten-cent piece viewed from a distance of 10,000 mi. To achieve such extraordinary angular resolution at radio wavelength requires baselines on the order of the diameter of the Earth. For about ten years radio astronomers have been operating transcontinental and transoceanic interferometers with baselines thousands of kilometers long, without direct electrical connections between the antennas. The cosmic impulses are recorded on magnetic tapes at each telescope and the tapes are then brought together at a specialized computer facility where they are combined to yield the output of the interferometer. The process is a technically sophisticated one and its undertaking engenders several difficulties. For example, the clocks at the two stations must agree within about a millionth of a second and must keep steady time to within a ten-millionth of a second per day. Many other technical problems have been solved in this effort. Some partially successful attempts have been made to observe with four and five telescopes in long-baseline synthesis arrays, yielding models of galaxies and quasars with detail as fine as thousandths of a second of arc. With further development this technique promises to shed new light on the enigma of the quasars, and upon many other large-scale physical processes in the depths of space, some of which may well involve hitherto unsuspected fundamental physical processes.

Molecules in Space

by Alan H. Barrett

From 1968 through 1975 the kinds of molecules detected in interstellar space rose from 4 to 39. New instruments and techniques of observation brought about this remarkable achievement.

Radio telescope at the Kitt Peak (Ariz.) National Observatory used in the discovery of approximately 80% of the interstellar molecules.

In 1968 astronomers knew of only four kinds of molecules in the vast regions between the stars. All of them were diatomic molecules, that is, composed of two atoms. By the end of 1975, however, the number of molecules detected in the interstellar gas had risen dramatically to 39, most of them polyatomic (having more than two atoms). This sudden growth in knowledge of the interstellar medium is an example of a phenomenon quite common in science: the application of new experimental techniques often leads to great advances in an otherwise slow-moving area of research. In the case of molecular studies of interstellar gas, the new techniques that brought about the new knowledge were the development of improved instrumentation for observations at millimeter wavelengths and the ability to fly ultraviolet spectrometers and telescopes in space. Thus, technological advances that allowed astronomical observations at opposing ends of the electromagnetic spectrum, microwave and ultraviolet, contributed significantly to the study of molecules in space.

Molecular emission at radio frequencies originates from the dense, dark dust clouds that frequently obscure large areas of astronomical photographs. These clouds cannot be studied at visual wavelengths because of absorption of the visible light waves by the dust. However, radio and, in some cases, infrared wavelengths are not subject to extensive absorption by the dust grains and therefore permit one to study the interiors of the dark dust clouds. From such studies astronomers learned about the physical and chemical composition of the clouds. They found that the clouds can be as cold as 5°–10° K (−268° to −263° C) and have gas densities as much as a thousand times more than those previously estimated. Furthermore, it seems quite clear that such clouds, in some instances, are the sites of star formation. (*See also* Feature Article: THE FORMATION OF STARS.)

The beginning: diatomic molecules

The first interstellar molecules were detected and subsequently identified in the late 1930s and early 1940s. Weak, narrow absorption lines were detected in the visual spectra of some stars and were attributed to interstellar methenyl (CH), ionized methenyl (CH^+), and the cyanogen radical (CN). The experimental and theoretical investigations that followed led to the conclusion that gas densities in space are so low, typically one to ten atoms per cubic centimeter, that only diatomic molecules could be formed in detectable abundances. Although methenyl and cyanogen appeared to be widely distributed throughout the galaxy, attempts to detect other diatomic hydrides, such as the hydroxyl radical (OH), were fruitless.

A quarter-century elapsed before another molecular constituent of the interstellar medium was discovered. During that time the study of astronomical objects by radio techniques had become commonplace, and microwave spectroscopy in the laboratory had provided detailed knowledge of the microwave properties of many molecules in a wavelength range of about 1 mm to 20 cm. Thus the stage was set for the

ALAN H. BARRETT *is Professor of Physics at the Massachusetts Institute of Technology.*

(Overleaf) Painting by Paul R. Alexander

Graphic Films, Inc.

beginning of radio spectroscopy of the interstellar gas. In 1963 the hydroxyl radical (OH) was discovered by detecting absorption lines at the 18-centimeter wavelength from the strong radio source in Cassiopeia. The hydroxyl radicals, located in cold clouds between the Earth and the radio source, revealed their presence by absorbing some of the radio emission of the source at precisely the frequencies known from laboratory studies to be characteristic of hydroxyl radicals. The intensity of the absorption indicated that the abundance of the hydroxyls was about one ten-millionth of the abundance of the hydrogen atoms in the clouds. Similar densities had been deduced for the methenyl and cyanogen molecules.

Many studies of interstellar hydroxyl radicals were undertaken following their discovery in 1963, several of which led to puzzling results that are not yet entirely understood. For example, by the late 1960s astronomers had found small regions in the sky that were radiating extremely intense hydroxyl emission. The regions compare in size to our solar system, which is very small by astronomical standards, and they radiate an intensity in the hydroxyl lines greater than if their temperature was one thousand billion degrees. Astronomers now believe that nature has somehow conspired to turn these regions into giant amplifiers, thus causing radio signals that are small when they enter the regions to be made much more intense before they leave. Such amplifiers had been created in the laboratory. They are called masers. (Maser stands for microwave amplification by stimulated emission of radiation.) It came as a surprise, however, that masers can be created in space by natural processes. Theoretical astrophysicists still have not reached agreement on just what physical mechanisms are responsible for the existence of cosmic masers or where these unusual objects fit in the scheme of stellar evolution.

Polyatomic molecules

Five years elapsed between the detection of interstellar hydroxyl radicals and the next discovery of a molecular constituent of the interstellar gas. In 1968 a group at the University of California at Berkeley detected both ammonia (NH_3) and water (H_2O) molecules at wavelengths near 1.3 centimeters. The leader of the Berkeley group was C. H. Townes, Nobel laureate for his invention of the laboratory maser and an early advocate of the radio detection of interstellar molecules. These two discoveries were particularly important because they laid to rest the idea that only diatomic molecules existed in interstellar gas. Thus radio astronomers felt justified in looking for more complex molecules, which they did with startling success.

The initial detection of ammonia was in the gas clouds in Sagittarius, which are believed to surround and form a part of the center of our Galaxy. Ammonia was later detected in numerous other gas clouds in the spiral arms of the Galaxy. The microwave spectrum of ammonia consists of several transitions between energy levels of the ammonia molecule over a relatively small range of frequencies, thus enabling one to probe the gas clouds over a range of excitations of the ammonia. This led to determinations of the temperature within the clouds, a quantity generally indeterminate from a single transition.

The discovery of interstellar water was unexpected because the microwave transition in the water molecule is not observable under the general conditions of interstellar space. Yet emission from the water molecules is very intense. This is another example of molecular amplification, another cosmic maser. The water emission is found to occur in many of the same parts of the sky as that from the hydroxyl masers. These are predominantly "H II regions," those regions of space immediately adjacent to a group of very hot stars in which the stellar ultraviolet radiation has produced a large volume in which most material is ionized. But not all the material is ionized because many molecules, which are electrically neutral, are found to be present in H II regions. It appears that the molecules exist in dust clouds embedded in, or near, H II regions. It is certain that the connection between H II regions, dense dust and gas clouds, and regions of stellar formation is an intimate one that is only beginning to be scrutinized in detail.

Encouraged by the detection of simple polyatomic molecules such as ammonia and water in the interstellar medium, radio astronomers immediately set out to search for other, more complex, molecules. In 1969 the first polyatomic organic compound of the interstellar gas was detected, formaldehyde (H_2CO). This was the first molecule found that contained two atoms other than hydrogen, and it served to stimulate the search for organic molecules of ever increasing complexity.

Courtesy, Hale Observatories

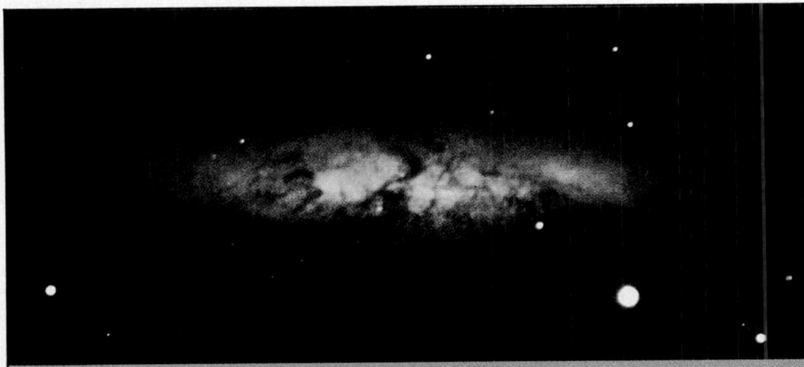

The galaxy M82 is a strong radio source in which several molecules have been found. Below are the molecules that had been detected as of early 1976.

Interstellar Molecules

Name	Chemical symbol	Year of discovery	Wavelength
methenyl radical	CH	1937	4300 Å*
cyanogen radical	CN	1940	3875 Å
ionized methenyl	CH^+	1941	4232 Å
hydroxyl radical	OH	1963	18 cm
ammonia	NH_3	1968	1.3 cm
water	H_2O	1968	1.3 cm
formaldehyde	H_2CO	1969	6.2 cm
carbon monoxide	CO	1970	2.6 mm
hydrogen cyanide	HCN	1970	3.4 mm
cyanoacetylene	HC_3N	1970	3.3 cm
hydrogen	H_2	1970	1100 Å
methyl alcohol	CH_3OH	1970	36 cm
formic acid	CHO_2H	1970	18 cm
ionized formyl radical	HCO^+	1970	3.4 mm
formamide	NH_2CHO	1971	6.5 cm
carbon monosulfide	CS	1971	2.0 mm
silicon monoxide	SiO	1971	2.3 mm
carbonyl sulfide	OCS	1971	2.7 mm
acetonitrile	CH_3CN	1971	2.7 mm
isocyanic acid	HNCO	1971	3.4 mm
methylacetylene	CH_3C_2H	1971	3.5 mm
acetaldehyde	CH_3CHO	1971	28 cm
thioformaldehyde	H_2CS	1971	9.5 cm
hydrogen isocyanide	HNC (?)	1971	3.3 mm
hydrogen sulfide	H_2S	1972	1.8 mm
methylenimine	H_2CNH	1972	5.7 cm
sulfur monoxide	SO	1973	3.0 mm
ionized amino radical	N_2H^+	1974	3.2 mm
ethynyl radical	C_2H	1974	3.4 mm
methylamine	CH_3NH_2	1974	3.6 cm
dimethyl ether	$(CH_3)_2O$	1974	9.6 mm
ethyl alcohol	CH_3CH_2OH	1974	3.3 mm
sulfur dioxide	SO_2	1975	3.6 mm
silicon sulfide	SiS	1975	3.3 mm
acrylonitrile	H_2C_2HCN	1975	22 cm
methyl formate	HCO_2CH_3	1975	18 cm
nitrogen sulfide radical	NS	1975	2.6 mm
cyanamide	NH_2CN	1975	3.7 mm
cyanodiacetylene	HC_5N	1975	3.0 cm

*1 Å = 10^{-8} cm.

M42, the Great Nebula in Orion, is a region in which many interstellar molecules have been detected. It is an H II region, one in which the ultraviolet radiation from an adjacent group of very hot stars has caused widespread ionization. On the opposite page are "ball and stick" models and emission line spectra of five interstellar molecules that form the "building blocks" of life. The sources of the spectra are also indicated.

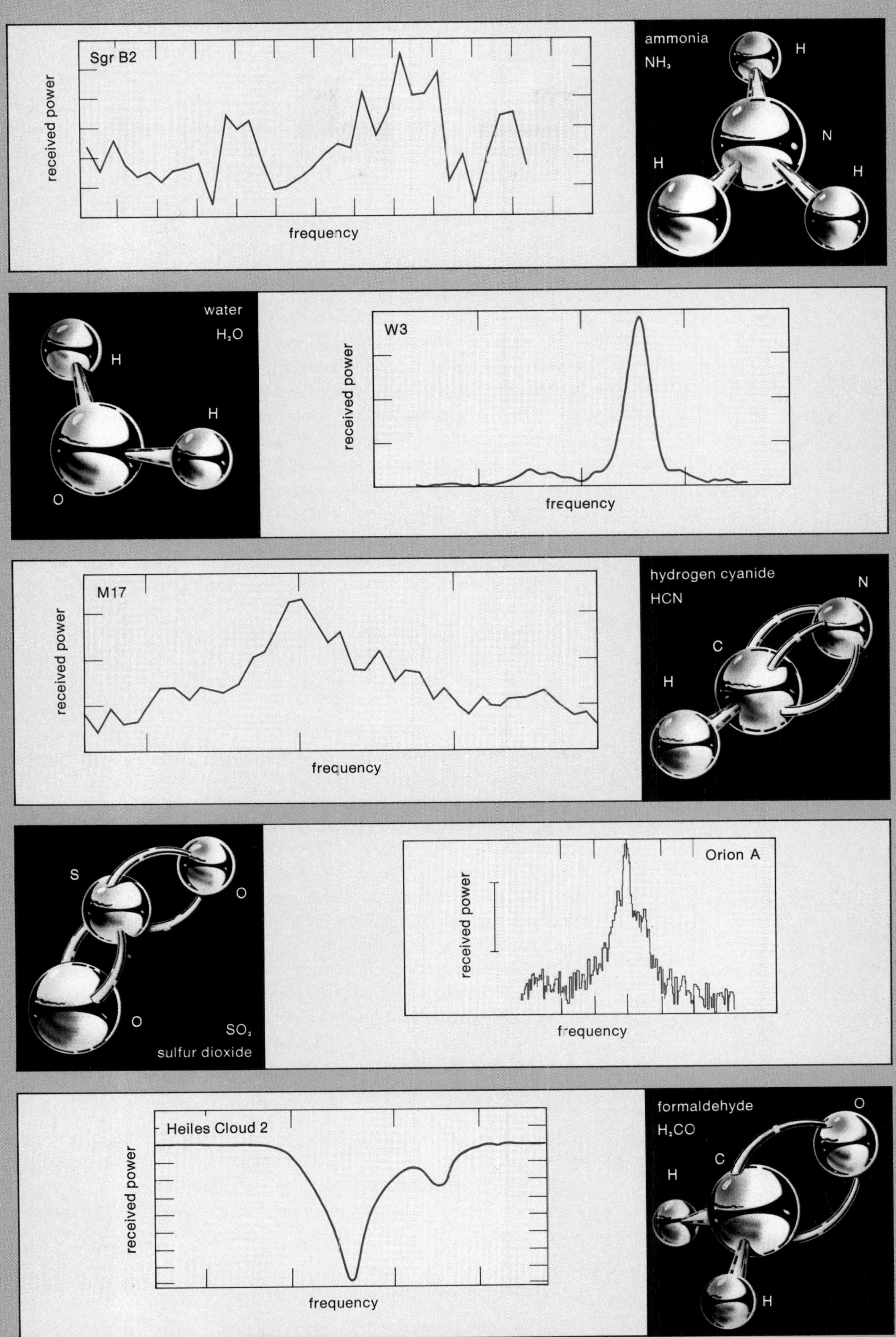

Sgr B2
received power
frequency
ammonia
NH_3
H
N
H
H
water
H_2O
H
H
O
W3
received power
frequency
M17
received power
frequency
hydrogen cyanide
HCN
N
C
H
S
O
O
SO_2
sulfur dioxide
Orion A
received power
frequency
Heiles Cloud 2
received power
frequency
formaldehyde
H_2CO
O
C
H
H

Formaldehyde was found to be widely distributed throughout the Galaxy, demonstrating that it is a common ingredient of interstellar gas. Astronomers discovered it to be strongly correlated with the existence of interstellar dust in the dark clouds. The observations show the interstellar abundance of formaldehyde to be about one-tenth to one-thirtieth that of the hydroxyl radicals, yet it had been expected to be approximately 3,000 times less abundant. Clearly, astrophysicists' ideas about cosmic abundances and the formation of molecules were due for major revisions.

Formaldehyde was detected by its absorption of microwave emissions from galactic radio sources, as was the case for the hydroxyl radical. But a new surprise was in store when one looked for formaldehyde in dark clouds. Such clouds are not sources of radio emission themselves, and so astronomers expected to detect the direct emission from the formaldehyde molecules. The formaldehyde signal, however, still showed up as absorption. In this case, the molecules were absorbing the ubiquitous microwave background radiation believed to be the remnant of the "big bang," the cataclysmic event at the beginning of the universe. The intensity of this radiation corresponds to a temperature of 3° K (−270° C). Therefore, if the formaldehyde could absorb this radiation, the temperature that describes the rotation of the formaldehyde molecules must be less than 3° K. This can be thought of as the inverse of the maser process, where molecules can be made to amplify the radiation. With formaldehyde the molecules are somehow made to have enhanced absorption. No other interstellar molecule detected has this property.

Seventeen molecules—two years

Once it was realized that polyatomic molecules containing two or more heavy atoms could form in the interstellar environment, radio astronomers immediately undertook searches for additional constituents. The results were startling. Seven new molecules were discovered in 1970 and ten in 1971. Many of these discoveries came about not only because of revised ideas about interstellar chemistry but also because microwave technology had advanced to the point that radio astronomers had available sensitive instruments that could receive signals at millimeter wavelengths. These wavelengths were particularly attractive to the spectroscopist because many of the molecules of astrophysical interest have their lowest transitions between rotational energy states in the millimeter wavelength range. For example, carbon monoxide (CO) has its lowest transition at 2.6 millimeters and is unobservable at all longer wavelengths.

The new molecules discovered in 1970 by radio methods were carbon monoxide, hydrogen cyanide (HCN), cyanoacetylene (HC_3N), hydrogen (H_2), methyl alcohol (CH_3OH), formic acid (CHO_2H), and the ionized formyl radical (HCO^+). The identification of HCO^+ was not firmly established until 1975 because no previous laboratory spectroscopy had been performed on the formyl radical. Microwave transitions of the cyanogen radical, previously seen at optical wavelengths, were also detected in 1970.

All microwave transitions detected at millimeter wavelengths have been found in emission spectra. Each molecule has one of a series of possible rotational energies. A spectral line is generated when a molecule makes a transition to a state of less rotational energy and radiates away the excess energy. The radiation is detected by radio astronomers as an increase in the emission from space at a wavelength characteristic of that molecular species.

The mere detection of an emission line allows conclusions to be drawn about the total gas density in the emitting region. For molecules to be seen in emission the overall state of rotational excitation must be continually maintained by collisions with other atoms or molecules in the cloud so that as one molecule drops to a lower state another is raised to a higher state by a collision. Otherwise, the molecules would decay to a lower energy, radiate their energy away once and for all, and then remain forever in the lowest rotational energy configuration. For molecules that emit millimeter wavelengths, the energy would be radiated over a period ranging from about one day to one year, after which all radiation would cease. The final step in the argument is that the collision rate depends on the total density in the cloud; therefore, knowing a minimum collision rate sets a lower limit on the gas density.

Using the line of reasoning outlined, astrophysicists established that many of the dark dust clouds of our Galaxy have total gas densities of 1,000–1,000,000 molecules per cubic centimeter. This can be contrasted with previous estimates, which generally ranged from 100 to 1,000 molecules per cubic centimeter, but it must be remembered there was no way to probe such clouds before radio spectroscopy.

Although the radio study of molecules provided evidence for densities of 1,000–1,000,000 molecules per cubic centimeter, the complex organic molecules themselves make up only approximately one ten-millionth of that density. What then is the rest? The balance is pure hydrogen in either atomic or molecular form, though certainly molecular in the dense clouds. Why then was hydrogen not the first, or certainly among the first, molecule detected? For astronomers, molecular hydrogen is elusive indeed. It has no microwave transitions, and its energy-state transitions at wavelengths of visual light are not readily excited in the interstellar medium. It has very strong transitions in the ultraviolet portion of the spectrum, but these are not detectable from the Earth's surface. The Earth's atmosphere is very effective at blocking out the ultraviolet light from the Sun and other stars. Therefore, molecular hydrogen had to be detected above the Earth's atmosphere. The experimental obstacles were overcome by G. R. Carruthers of the U.S. Naval Research Laboratory (Washington, D.C.) in 1970 when he launched a successful rocket flight of an ultraviolet spectrometer. Eight transitions of molecular hydrogen were detected in the direction of the star Epsilon Persei.

The molecules discovered in 1971 were carbon monosulfide (CS), formamide (NH_2CHO), silicon monoxide (SiO), carbonyl sulfide (OCS), acetonitrile (CH_3CN), isocyanic acid (HNCO), methylacetylene (Ch_3C_2H), acetaldehyde (CH_3CHO), and thioformaldehyde (H_2CS). Also detected was a line at 3.3-millimeter wavelength, the origin of which remains uncertain although a suggestion that it is caused by the radical HNC has been put forward. In many cases, research on these molecules has not progressed much beyond the discovery phase, primarily because the number of radio telescopes capable of operation at millimeter wavelengths is extremely limited.

The molecule silicon monoxide has unusual microwave properties in its astrophysical environment. Astronomers detected emission in separate transition wavelengths of 7, 3.5, and 2.3 millimeters; these represent a combination of rotation and vibration of the molecule. (In addition to their rotational energy, molecules exist in one of a series of vibrational energy states, representing the energy of motion of the atoms as they vibrate about their equilibrium position in the molecule.) Interstellar molecules are expected to be in their lowest state of vibrational energy because their radiative lifetimes in vibrational states are only a few seconds. Thus when one transition at 3.5 millimeters wavelength was detected, the identification as being due to silicon monoxide was treated with skepticism because it implied that the molecule was excited to its second vibrational state. The detection of subsequent transitions, however, removed any doubt that the radiating molecule was indeed silicon monoxide.

Silicon monoxide is another example of an interstellar maser, the first found at millimeter wavelengths. The emission was detected from several small regions that are also sources of infrared emission, perhaps from a cool central star embedded in the cloud of gas and dust. It seems likely that the infrared radiation is responsible for exciting and continually replenishing the silicon monoxide molecules in their higher vibrational energy states. Subsequent investigations were expected to clarify or modify this picture.

The pace of discoveries of new molecules in interstellar gas has slowed slightly since 1971, the year that ten were discovered. From 1972 through 1975 fifteen new species were detected, of which five are inorganic and ten are organic molecules. Two of the latter, dimethyl ether [$(CH_3)_2O$] and ethyl alcohol (CH_3CH_2OH), are composed of nine atoms each and are the most complex molecules found in the interstellar medium through 1975. The heaviest molecule to be discovered was cyanodiacetylene (HC_5N), found in late 1975.

Isotopes

All interstellar molecules detected through 1975 consist of the atoms hydrogen (H), carbon (C), oxygen (O), nitrogen (N), sulfur (S), and silicon (Si). Every atom, however, exists in several forms, called isotopes, all of which have the same chemical properties but different masses. For example, there is a carbon atom of mass twelve (^{12}C) and another of mass thirteen (^{13}C), and both readily combine with oxygen to form carbon monoxide. Similarly, one finds three stable isotopes of oxygen, ^{16}O, ^{17}O, and ^{18}O. Thus, there are six forms of carbon monoxide that are chemically equivalent but physically slightly different because of their different masses ($^{12}C^{16}O$, $^{13}C^{18}O$, etc.). Because each isotopic species has a different mass its rotational energies are slightly different from those of another species and the microwave transitions are, therefore, at different wavelengths. For this reason a radio astronomer can study $^{12}C^{16}O$ and $^{13}C^{16}O$ without the emission from one interfering with that of the other.

The ability to observe isotopic species of the same molecule has been a valuable research tool for two reasons. First, the cosmic abundance of individual isotopes provides direct information concerning the nuclear synthesis of the elements by nuclear processes in the interiors of stars, where all heavy elements are made. Second, one isotopic species of a molecule is usually considerably more abundant than another; in a very dense cloud of interstellar material, therefore, emission from the abundant species may be able to escape from the cloud only if it originates near the edges, whereas emission from a less abundant species may escape from anywhere in the cloud. For example, many clouds are found to contain so much carbon monoxide that the cloud appears opaque when viewed at the wavelength of the $^{12}C^{16}O$ transition, but when viewed at the wavelength of the $^{13}C^{16}O$ transition one "sees" through the entire cloud because ^{13}C is about ninety times less abundant than ^{12}C. Radio astronomers have studied molecules composed of the following isotopes: hydrogen, 1 and 2; carbon, 12 and 13; nitrogen, 14 and 15; oxygen, 16, 17, and 18; and sulfur, 32 and 34.

A new era of exploration of the interstellar medium began in August 1972 with the launching of the Copernicus satellite. The instrumentation on the satellite provided high-resolution ultraviolet spectra so that many ionized atoms and some neutral molecules could be studied in a spectral region unobservable from the ground because of atmospheric absorption. From the standpoint of this review, the important contributions of the Copernicus observations are those related to molecular hydrogen (H_2) and its isotopic species (HD) of mass one and mass two (deuterium). Astronomers have found that the interstellar hydrogen atoms exist in molecular form with abundances either greater than 10% of the total gas density or less than one millionth of the total. There are almost no regions of space where intermediate values are found. This is believed indicative of the destruction of hydrogen molecules by ultraviolet light except in regions where dust and gas molecules block the ultraviolet radiation. There, hydrogen molecules readily form from hydrogen atoms.

The study of deuterium by both radio astronomers and the Copernicus experimenters promises to provide important information on the formation of the elements and on the average density of matter in the universe. Deuterated hydrogen cyanide (DCN) and water (HDO) have

Photos, courtesy, Roy S. Lewis, University of Chicago

Aggregate of crystals of magnetite (Fe_3O_4) was separated from a type of meteorite known as a carbonaceous chondrite. The aggregate may have condensed out of the nebula from which was formed the Sun and the planets. The diameter of the aggregate, seen here in a scanning electron micrograph, is about ten micrometers (0.01 mm).

been detected by microwave methods, and deuterated hydrogen molecules (HD) and atomic deuterium (D) have been detected by the satellite observations. Interpretation of the molecular results is not as straightforward as the atomic deuterium observations, which give an abundance of deuterium of about one atom to every 100,000 regular hydrogen atoms.

Probing the clouds

Microwave studies of interstellar molecules have begun to reveal a large amount of detailed information about the physical and chemical conditions within interstellar clouds. Not surprisingly, there is a wide range of cloud sizes, densities, and masses. The largest appear to be giant molecular clouds surrounding H II regions. A typical dimension is 10 parsecs (1 parsec = 3.26 light-years); total gas densities are 1,000–10,000 molecules per cubic centimeter, mostly hydrogen; and total mass is equal to 100,000 times the mass of the Sun. These dimensions and properties greatly exceed those of the H II region, and many astronomers now believe that such giant clouds spawn H II regions. Condensations within the cloud become gravitationally unstable and collapse. This leads to increased density and temperature, which eventually give rise to the onset of thermonuclear reactions; that is, one or more stars have been created. The ultraviolet radiation from the young stars ionizes the surrounding material, and it becomes an H II region. The molecules found in such giant clouds are carbon monoxide, formaldehyde, hydroxyl radicals, hydrogen cyanide, carbon monosulfide, and ammonia, and studies of these have provided almost all information about these clouds. The two most studied clouds of this type are those in Orion and Sagittarius.

Condensations within the giant clouds are the first step toward stellar formation. The density builds up in the condensations as they contract, and molecules appear to form quite readily. Almost all the interstellar molecules, even the most complex, are found in these regions. The temperature will also rise as contraction takes place, causing these regions to emit infrared energy. Studies of the molecular emission show these regions to be typically 0.5 parsecs in size, to have a gas density as high as 1,000,000 molecules per cubic centimeter, temperatures of 50°–100° K (approximately −220° to −170° C), and a mass several hundred times that of the Sun. This latter figure is interesting for it shows that there is enough mass to form a cluster of stars. Thus the condensation may break up into a number of smaller condensations, each of which may be an embryonic star. Many astrophysicists believe that the intense maser emission given off by hydroxyl radicals and water molecules comes from such objects because the masers are always found to be multiple sources, each with a size less than our solar system and with densities that are many times greater than those of the larger condensations.

The more general interstellar cloud, intermediate in size between the giant clouds and the condensations, has dimensions of the order of one parsec, densities of 1,000 molecules per cubic centimeter, and a total mass 100 times that of the Sun. Such clouds are found to contain the more common molecules, as found in the giant clouds. In many cases these clouds do not seem to differ from the giant clouds except for their size.

Stack of circular plates of magnetite about nine micrometers high was separated from the same chondrite as was the aggregate on the opposite page. Magnetite has not been found in this plate form anywhere on the Earth.

Prebiotic molecules and the future

Though the list of known interstellar molecules is impressive, especially when the progress in the last ten years is noted, many other molecules have been sought in interstellar space but not found. For example, none of the many organic molecules with a basic ring structure has been detected in the interstellar gas. The molecules that have been found are complex by astrophysical standards but are quite primitive biologically. Scientists therefore have focused on detecting the prebiotic molecules, which include ring compounds. Examples that had escaped detection through 1975 were pyrimidine and pyridine. Laboratory experiments have shown that a mixture of water, ammonia, and methane (CH_4) when supplied with additional energy, such as an electric discharge, heat, ultraviolet light, or ionizing particle radiation, gives rise to a wide range of complex prebiotic molecules. Chemical reactions produced adenine, glycine, alanine, and various other amino acids and sugars. These are the ingredients needed to make proteins and nucleic acids, which are necessary compounds for the evolution of life.

The failure to detect some of the fundamental biological molecules may be due to insufficient sensitivity of the equipment. If so, advances in technology will resolve that problem. But the failure may also be due to the fact that such molecules do not exist in the interstellar environment in any appreciable quantity. There is one note of certainty, however; the search will go on.

The Climate of Urban Areas

by T. R. Oke

A large metropolitan area generates a climate of its own that differs distinctly from that of the nearby countryside. An urban climate can affect areas as far as 125 miles downwind.

Photograph, courtesy, A. G. Davenport

Pollutants in the urban atmosphere create smog over Athens, Greece. They include the following particulates, as seen in photomicrographs: (top, left to right) cement, coffee grounds, and coal fly ash; (bottom, left to right) automobile emission particles collected below an exhaust pipe, ragweed pollen, and fly ash from an industrial incinerator burning sawdust.

So many people throughout the world now live in urban places that the atmospheric environment of cities seems quite normal, even "natural," to them. Simple observation, however, reveals that many features of the air have been changed as a result of urbanization of the landscape. These changes, which become a distinct urban climate, are examples of inadvertent modification of the environment.

Some of the features a careful observer will note include the increased frequency and intensity of red-sky coloration when the Sun is near the horizon; snow in the suburbs but rain in the city if temperatures are near the freezing point; excessively high wind speeds and erratic wind directions in the vicinity of tall buildings; stressfully warm summer evenings when the city air seems to fail to cool; the relative lack of dew on downtown surfaces; a longer frost-free period in the city as compared with the surrounding countryside; and the earlier budding and blooming of flowers and trees in the city. Less obvious but known to the trained observer are the changes in the biological structure of the city, including the disappearance of pollutant-sensitive lichens, flowers, shrubs, and trees; the color adaptation of certain moths to provide better camouflage on pollution-darkened surfaces; and the attraction of certain birds to urban habitats because of greater warmth. The implications are also economic, involving greater costs for building maintenance and summer air conditioning but reduced winter heating bills.

The scientific proof of the climatic effects of urbanization comes from meteorological records and specially designed experiments. Before considering some of these results, however, it is necessary to define the basis of urban/rural comparisons. Ideally, to show the influence of a city there should be pre-urban and urban observations. Only rarely are these available, however, and so it is common to compare urban center readings with those at surrounding rural stations. Care is necessary in the choice of such pairings to ensure that extraneous influences do not disturb the data (for example, the effects of topography). It is also important to know where the urban station is located, at ground level or on a roof. A useful two-layer breakdown of the city atmosphere has been devised. Between the ground and some approximate building height is the zone of the urban "canopy," or building layer, which includes the buildings and the air contained between them. Above this is a deeper layer of influence called the urban boundary layer, the upper limit of which may be as little as 100–200 meters by night and as much as 1.5 kilometers (0.9 mile) by day.

In general, observers have found that on an annual basis for a large temperate-latitude city solar radiation received at the city surface is 15–20% less than in the surrounding countryside, air temperatures are about 1° C higher, relative humidities are about 6% less, winds are 20–30% less, and both cloudiness and precipitation are 5–10% greater. These differences refer to observations in the building layer. Such statistics, however, conceal much more interesting and even more dramatic effects to be found from place to place within a particular city

__T. R. OKE__ is Associate Professor of Geography at the University of British Columbia.

(Overleaf) A model of central Vancouver on a scale of 1 : 4,000 simulates the wind field in order to test wind loading on the tower in the foreground.

M. P. Kahl from Bruce Coleman Inc.

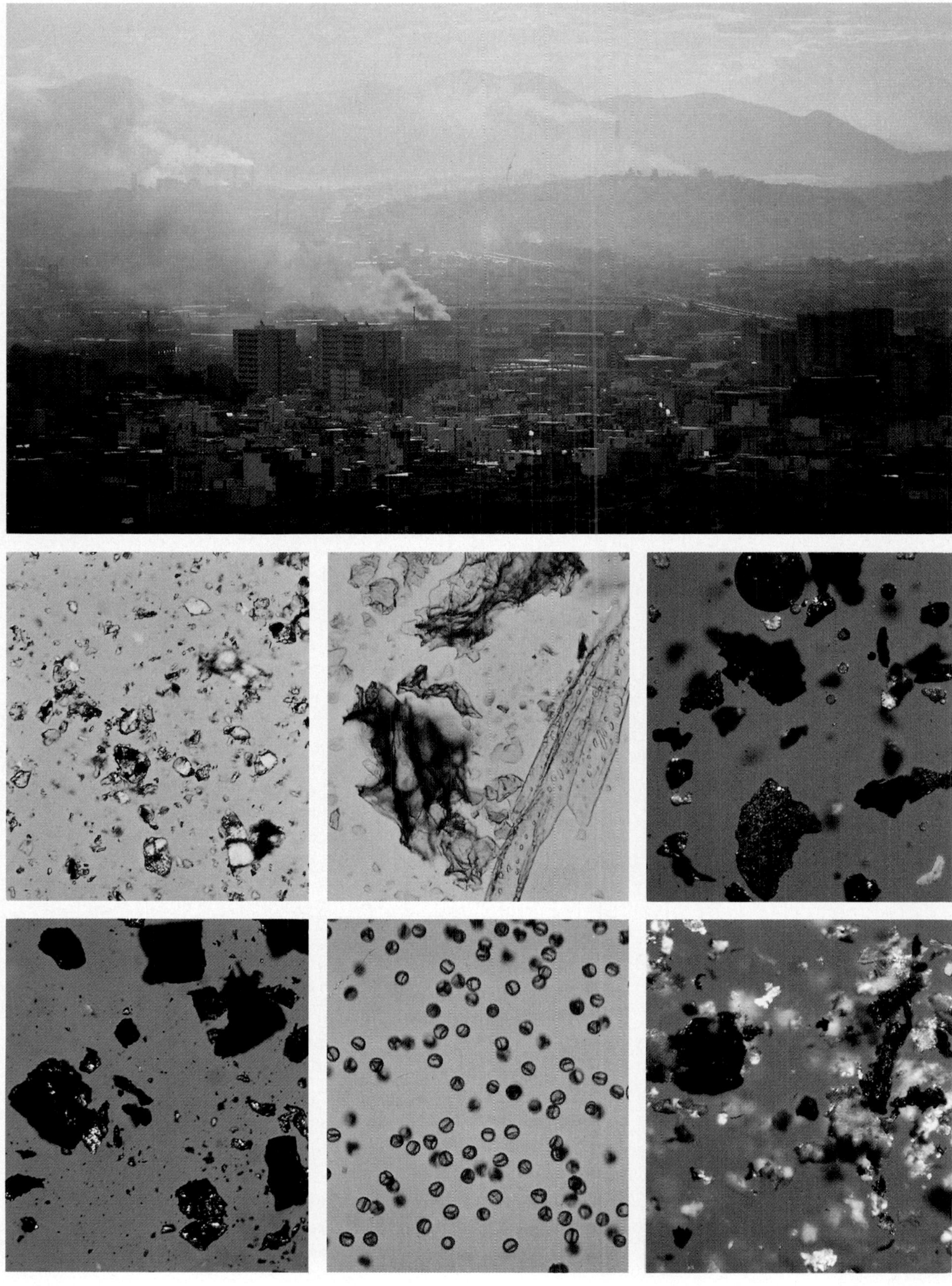

Illustrations by Dennis Magdich

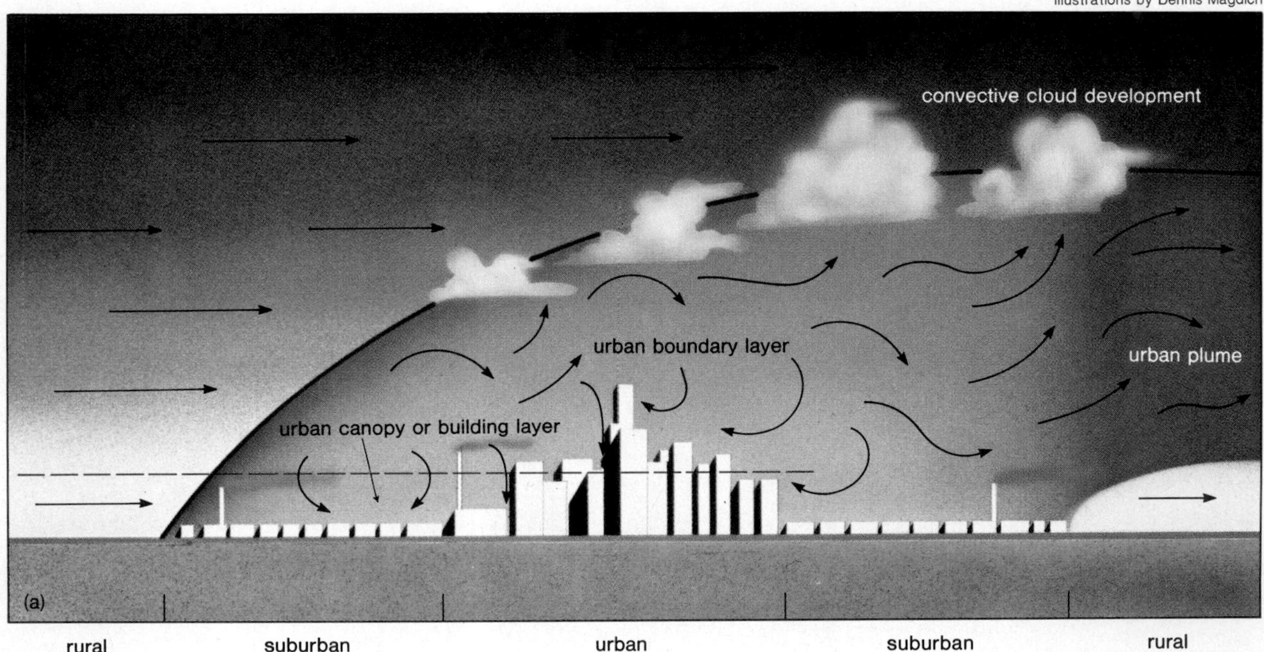

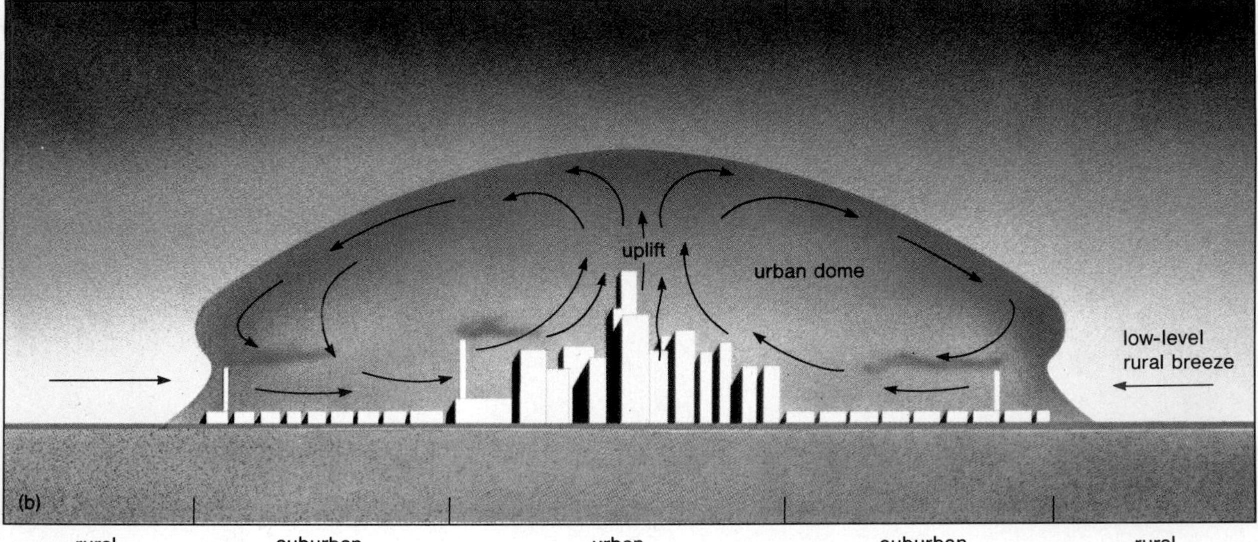

The air circulation above a city with a steady regional wind (a) differs substantially from one over a city with calm regional winds (b).

and on occasions when the weather is most conducive to the development of small-scale climate differences. For example, the 1° C higher annual temperature appears insignificant when it is realized that differences of 5° C around an individual house are quite common.

Physical basis of climate

Before looking at the causes of urban climates it might be helpful to review the mechanisms that govern any climate. The commonly recognized climatological variables such as temperature and humidity are really the end result of a much more basic set of processes that control the cycling of energy (mainly heat in the form of radiation,

Adapted from T. R. Oke and C. East, "Boundary-Layer Meteorology," vol. 1, pp. 411–437, 1971

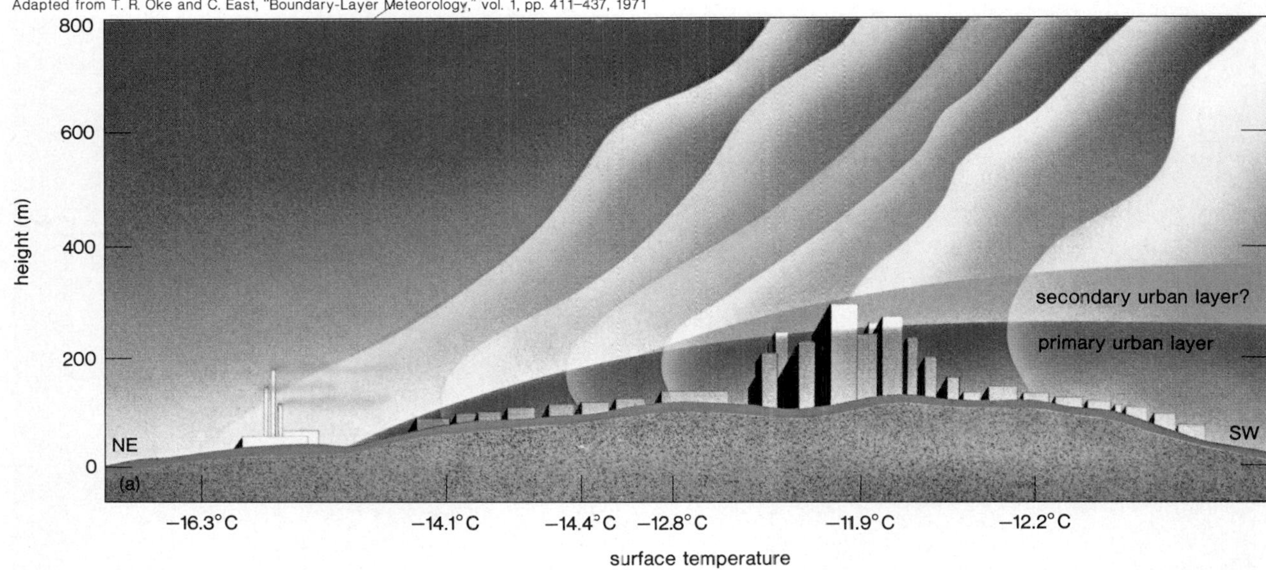

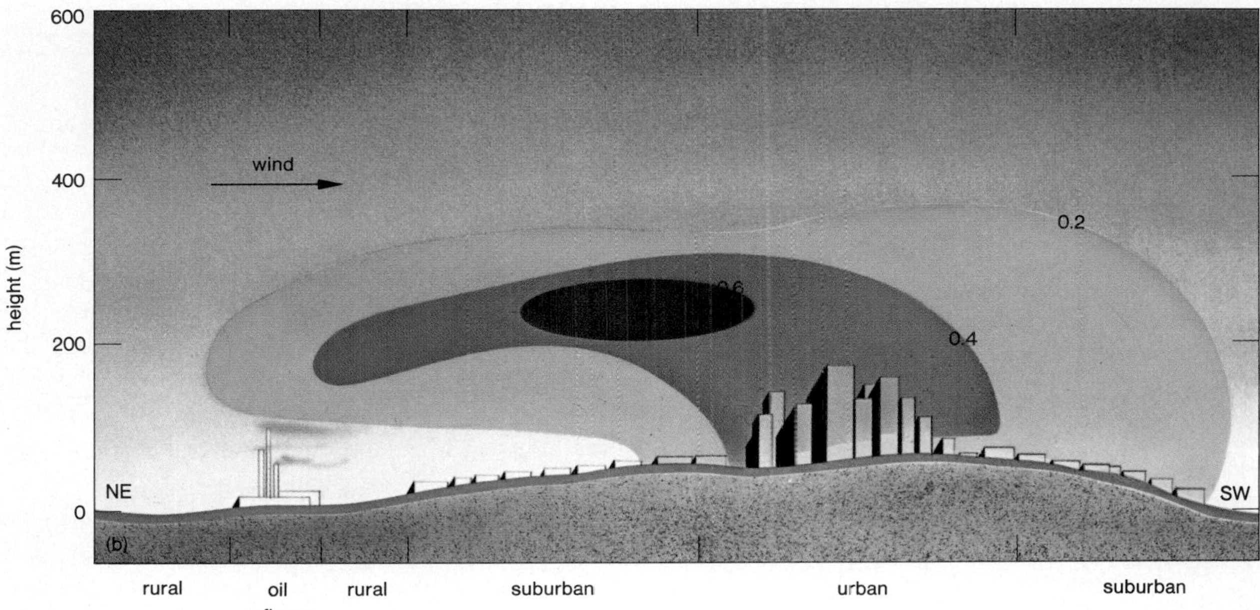

Vertical distribution of potential temperature (a) and sulfur dioxide in parts per million (b) are shown for Montreal on the sunny morning of March 7, 1968. A slight wind at ground level was tenfold higher at 200–300 meters.

sensible heat, or latent heat) and mass (mainly water in any of its solid, liquid, or vapor phases) through a system. The inputs to the system that force the rate of activity are the energy availability (usually the Sun's radiant energy, the reception of which varies with latitude, season, time of day, and the state of the atmosphere) and the water availability (usually precipitation). Both are linked by the continual circulation of the atmosphere and its associated weather systems, which also transport heat and water to and from a site.

Within a region where the supply of radiant energy and precipitation and the effects of weather systems can be considered to be approximately the same, the development of distinct small-scale climates is

Tom Meyers

Courtesy, J. E. Cermak, Colorado State University

Asphalt surfaces of a shopping center in Sacramento, California (left), conduct heat poorly and thus generate hot, dry microclimates. A scale model of the Coalplex Industrial Center near Johannesburg, South Africa, is tested in a wind tunnel at right. The investigation centers on the concentration of acetylene gas at the plant in the foreground that results from the release of acetylene at the plant in the far background. A ground-based stable stratification of the atmosphere is simulated.

the result of the unique mix of surface and atmospheric properties possessed by a specific site. These properties control the direction and the amount of energy and mass channeled through the system. The important properties are radiative (controlling the reflection, absorption, or emission of radiation), thermodynamic (controlling the rate of heat flow, storage, and the resultant temperature change), aerodynamic (controlling the channeling of air flow and the production of turbulent motion), and hydraulic (controlling the water flow or retention and the resultant change in moisture content).

The four properties described above determine the way in which the energy is used: to warm the underlying material, to warm the overlying air, or to evaporate water into the air. If water is freely available as with an open water surface or a freely transpiring plant cover, it may be expected that evaporation will be the most important energy use. Converting heat into this latent form does not heat the environment and leaves less sensible heat to do so; such surfaces, therefore, usually have relatively cool and humid microclimates. If water is restricted, as with a desert or a road, all the absorbed energy is used to heat the underlying medium or the air, the relative proportions depending on the thermal character of the surface and the level of convective activity. Both dry sand (for a desert) and asphalt (for a road) have a poor ability to conduct heat, and so, although the near surface layer becomes very hot, deeper layers are less affected. The majority of the energy is convected into the overlying air. Such surfaces, therefore, have hot, dry microclimates.

In most situations the daily course of events is closely tied to the radiation regime. Short-wave radiation (wavelength 0.3–4 microns) from the Sun largely passes through the atmosphere (depending on the amount of absorbing particles of water vapor and dust present) and is either absorbed at the ground or is reflected. Both the atmosphere and the ground radiate long waves (4–100 microns), but because the ground is warmer than the atmosphere the surface always experiences a net loss in these wavelengths. The combined result of the short- and long-wave exchanges at the surface is called the net radiation balance. By day, because the short-wave gain exceeds the long-wave loss, there is a net radiative surplus; by night, in the absence of the Sun's rays, the net long-wave loss is the only exchange and the surface has a net radiative deficit. As mentioned previously, the daytime surplus is expended to heat the medium and the air above or to evaporate water. The nocturnal deficit is compensated for by heat flows from the medium and the air, and perhaps by release of latent heat as a result of condensation of water on the surface as dew.

To reiterate, this crucial apportioning of the daytime energy surplus and the nighttime energy deficit is controlled by the surface radiative, thermodynamic, aerodynamic, and hydraulic properties. Thus, if one assumes that in the same region the initial input of solar energy and the general weather patterns are similar, the reasons for climate change can be sought in changes in the properties of the surface and atmosphere. To understand why an urban surface climate is different from a rural one, therefore, requires an investigation of the nature of the changes in these properties wrought by urbanization.

Changes in climatic properties

Many of the important changes in climatic properties caused by urbanization are evident from a comparison of rural and urban regions. For convenience these properties may be classified under four headings: pollutants, materials, geometry, and hydrology.

Unfortunately, the urban atmosphere is all too commonly characterized by air pollution. Climatically, pollutants are important because of their influence on the transmission of radiation, and because of their role as condensation nuclei around which cloud droplets may form.

Short-wave radiation is scattered, reflected, and absorbed by many pollutants. This results both in attenuation of the beam and changes in the spectral and directional properties of the radiation. Thus, the total amount of radiation received at the surface is reduced compared with the total received in a clean rural atmosphere (by up to 30%); ultraviolet wavelengths are particularly cut (up to 90%); and the direction of the radiation is more diffuse. The spectral filtering of light explains the red coloration of the Sun, because the shortest (blue) wavelengths are preferentially removed, leaving the light richer in the longer (red) portion of the visible spectrum. The absorbed portion contributes to warming of the atmosphere. Certain of the pollutants, especially water vapor, particulates, carbon dioxide, ozone, and oxides of nitrogen,

From T. R. Oke and C. East, "Boundary-Layer Meteorology," vol. 1, pp. 411–437, 1971

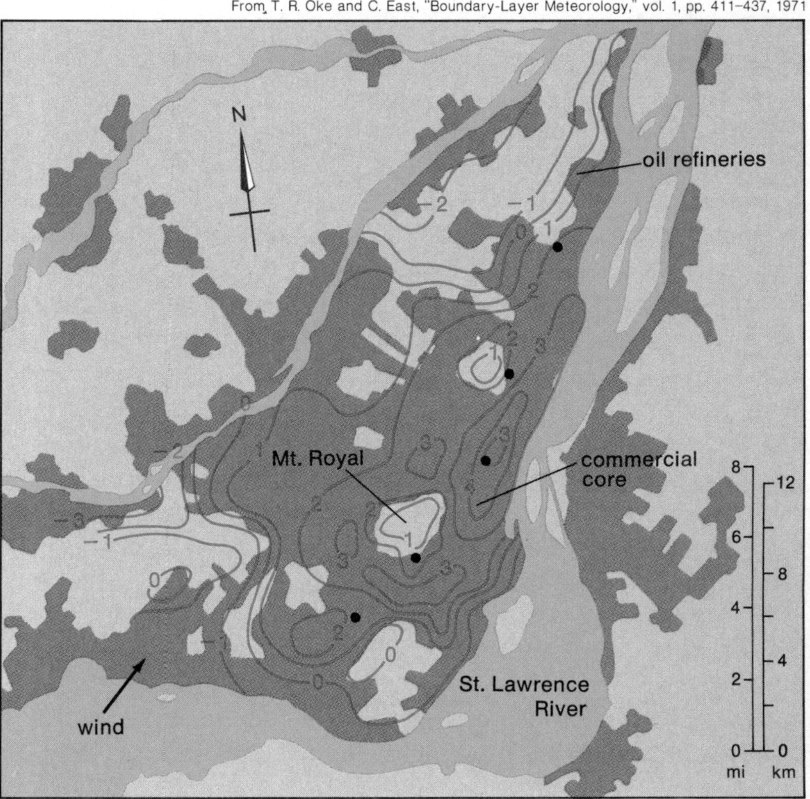

Heat island temperature distribution in Montreal at 9 P.M. on March 18, 1970, under conditions of almost cloudless skies and a light southwest wind. The shading indicates built-up areas, and the isotherms are in degrees Celsius.

also have an effect on the exchanges of long-wave radiation. This is possibly most important in reducing the loss of the radiation emitted from the city surface to the atmosphere and space, thus tending to reduce the ability of the surface to cool. Important exchanges of long-wave radiation also occur within the atmosphere, including the cooling of the top of the pollution layer as it emits up toward the cleaner and much colder upper atmosphere. In summary, the radiative effects of pollutants are mainly thought to be the reduction of incoming short waves and the reduction of the loss of long waves at the surface.

Some of the pollutants are hygroscopic (attract water vapor) and therefore form the soluble nucleus of a droplet. In addition, large amounts of vapor are provided by combustion, and if uplift (and therefore cooling) is provided cloud development becomes possible and precipitation may result.

The release of large amounts of sensible heat (for space heating, or by means of industrial and traffic operations) has no counterpart in the rural situation and is therefore an additional energy source for the city. This energy can be climatically significant, especially in cities with cold winter climates, where the heat available from urban activities may exceed that available as net radiation.

Urban construction results in the removal or covering of much of the natural surface materials. Vegetation and soil are often replaced by bricks, concrete, asphalt, steel, and glass. These new urban materials have very different properties than do soil and vegetation, especially with regard to the disposition of heat and water. For example, their reflectivity for short-wave radiation is often different, but probably most important is that they conduct and store heat much more efficiently. Thus after a sunny day one can physiologically sense the heat being radiated away from a building, and it is warm to the touch.

Unlike natural covers the urban "canopy" is commonly composed of regularly structured blocks with organized interconnecting channels (streets). Aerodynamically, these blocks are tall, sharp-edged, inflexible, and impermeable to airflow. The air must flow over or around these considerable obstacles, and is channeled along streets parallel to the wind direction. The result with strong winds is enhanced turbulent mixing, especially above and in locations around buildings. On the other hand, with weak winds stagnation may occur at street level as the air above the buildings is unable to overcome the frictional retardation and penetrate downward.

An urban "canopy" consists of regularly structured blocks (buildings) with organized interconnecting channels (streets), as seen in this aerial photograph of Manhattan. Air flow is channeled along streets parallel to the wind direction. When winds are strong, deflection and "jetting" occur around buildings.

Charles Moore from Black Star

The city-block structure also provides an effective radiative trap. Compared with an ideal horizontal rural surface, a street is more likely to absorb short-wave radiation simply because the rays are likely to undergo multiple reflection between the walls and street. This effectively reduces the reflectivity of the urban system. In a deep street "canyon" the absorption takes place well above street level, and the floor may be in shade for extended periods. This trap is equally effective in retarding the upward loss of long-wave radiation. A horizontal rural surface can radiate to the complete sky hemisphere, but this most important channel for cooling is partially cut off in suburban areas and very substantially reduced in the downtown canyons.

To a large extent urbanization removes natural water sources and waterproofs the surface. Much of the vegetation is removed, ponds are drained, and the soil is built upon. The urban building materials are largely impermeable and do not, therefore, allow water storage. Following precipitation the water runs off the surface (rather than infiltrating), and is efficiently channeled out of the system by the network of building and road gutters and underground sewers. Thus, except for garden sprinkling, parks, and street trees, the availability of surface water is more restricted in the city, and evaporation may be expected to be reduced compared with that in the country.

Air temperature

Urbanization probably has its most striking effects upon air temperature. A map of the air-temperature distribution (based on a dense network of observations within the building layer) usually reveals a pattern of isotherms running parallel to the city circumference with increasing warmth toward the city center. Because of its obvious topographic analogue (isotherms being considered equivalent to height contours) this structure is dubbed the urban heat "island." The comparison is quite apt. The urban-rural boundary appears as a "cliff" as temperatures rise suddenly upon entering the city; much of the rest of the urban area appears as a "plateau" on the island with a much more

From T. R. Oke, "Atmospheric Environment," vol. 7, pp. 769–779, 1973

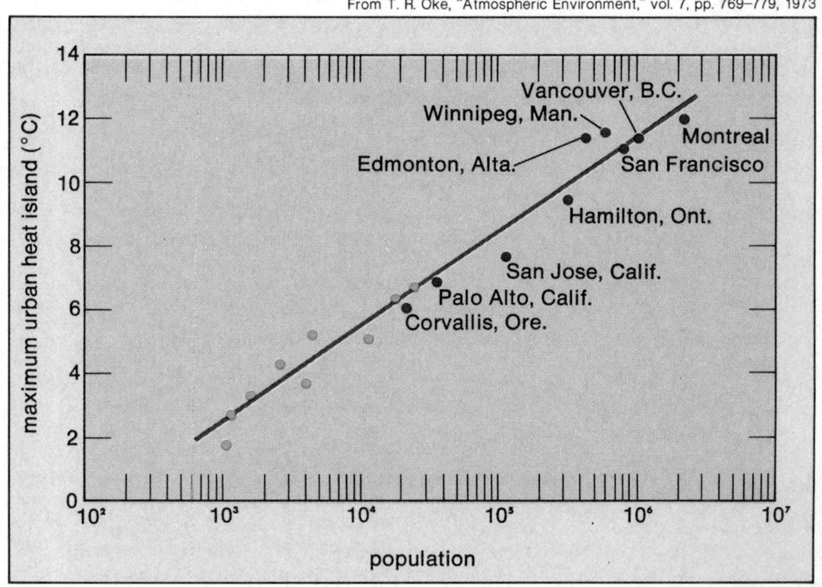

Relationship between maximum urban heat island and population for several cities in North America. The nine unnamed locations are towns in the province of Quebec.

gentle increase toward the center; the city core then appears as a final sharp increase to a central "peak." It is probably not a coincidence that urban building height contours have the same outline.

The above general features are exhibited in all urban areas, but the detailed structure in any particular city depends upon its own set of geographic controls (especially topography, the presence of water bodies, and the urban land use pattern). In Montreal, for example, the "cliff" and "plateau" regions are well illustrated, but there is no single "peak" because the middle of the urban area is dominated by Mt. Royal Park (height 200 meters), which gives a cool core to the island. Isolated cells of warm air surround the cool park, each associated with dense community shopping areas. The overall heat-island intensity (difference between rural surroundings and the city park) in this example is about 7° C, but in other cases is as large as 12° C.

In the absence of major short-term changes in the weather (passage of a front) the intensity of the heat island is greatest three to five hours after sunset and is at a minimum around midday. In fact, some cities show a negative heat island (city cooler than countryside) at midday in the central deep-canyon area. This is thought to be due to shading of the surface. The fact that the heat island is at a maximum after sunset indicates an important point—a heat island can be considered to be a result of the inability of the city to cool relative to the countryside, rather than a result of extra urban heating.

The exact causes of the heat island have yet to be elucidated, but the following mechanisms are thought to be important. Because of the thermal properties of the urban materials (outlined earlier) the urban area is an efficient energy storer, especially by day and in summer. The nocturnal release of this heat is available to warm the air. Equally in winter the nocturnal release of anthropogenic heat (that generated by man's activities) may play a similar role as it seeps out of the buildings. The principal modes of removing this heat from the building layer are via long-wave radiation and turbulent transport and mixing. As outlined earlier, both mechanisms are reduced in effectiveness because of the urban geometry.

It is also interesting to note that the heat-island intensity becomes smaller as cloudiness and wind speed increase. Clouds, like buildings, are effective traps for outgoing long-wave radiation, and therefore with overcast skies urban-rural differences in radiative cooling are small. Equally, the relative protection from turbulent heat loss provided by stagnation in the urban canopy is lost as stronger winds are able to scour in among the buildings.

The heat island is almost ubiquitous. It has been studied most intensively in the temperate latitudes, but it has also been observed in the tropics (Nairobi, Singapore) and in the high latitudes (Inuvik, Northwest Territories, Canada; Fairbanks, Alaska). It is not restricted to large cities, having been found in villages of less than 1,000 inhabitants and even down to the scale of a small group of buildings. All heat islands are unique, being the result of their own combination of geographic and surface properties, but some generalizations are now being established. For example, it might be expected that larger cities will have greater heat islands. Experiments have shown that this is so for the case of the maximum heat-island intensity (the largest differences measured in each city).

The urban heat island also extends up into the urban boundary layer. If there is a steady regional wind, the layer of modified (warmed) air expands with distance from the upwind urban-rural boundary; also, a new rural boundary layer re-forms at the downwind edge, leaving the modified urban air aloft. This structure is called the urban "plume" because it is in some ways similar to a pollution plume leaving a single stack. In this case an entire city is the stack exit, and the contents of the plume are the heat, water vapor, particulates, and other gases that are ejected from the urban area. If the regional airflow is calm, meteorologists believe that an enclosed "dome" forms over the city rather than a plume.

An experiment was performed to show the height distribution of air temperatures as measured from a helicopter following a winter night. The wind was carrying air from the rural area across the urban area of Montreal. The rural air was stable from the surface to at least 600 meters. At progressively greater distances into the city, however, stability in the lowest layers was eroded to greater and greater heights. This erosion defined the vertical extent of the heat island, and on the occasion of the experiment it reached approximately 300 meters over the city center and the downwind suburbs. By day, because of surface heating by short-wave radiation, the depth of the urban boundary layer is greater. How much of the thermal modification of this layer is due to input of heat from the building layer and how much is the result of internal radiative and thermodynamic effects are questions that cannot yet be answered with certainty.

Adapted from S. A. Changnon, Jr., F. A. Huff and R. G. Semonin, "Bulletin of the American Meteorological Society," vol. 52, pp. 958–967, 1971

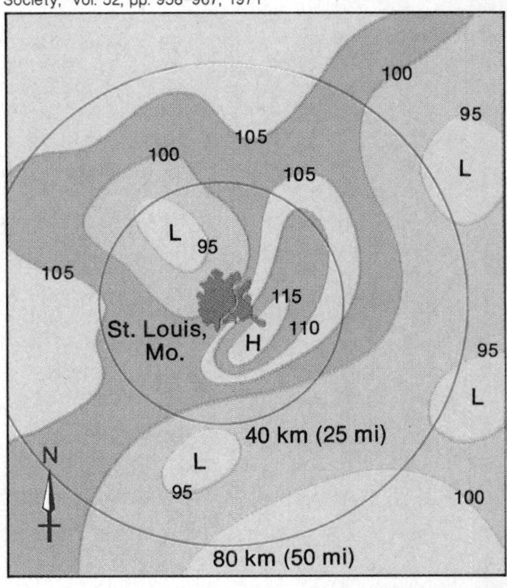

The urban effect on summer rainfall is shown for the St. Louis, Missouri, area for the period 1949–68. The contour lines represent relative values of annual rainfall; for example, places on a 110 line receive 10% more rain than those on a 100 line. The area of high (H) ratios is downwind (east) of St. Louis; it experiences about 15% more rain each year than the central city.

Courtesy, Barnes Engineering Company

Infrared photograph of an automobile with its engine running (above), made with a Barnes thermograph, shows the emission of heat from the vehicle. The hottest areas are red, and the coolest are blue. Cars are a significant source of heat production in a city. Scale model of Denver, Colorado, in a wind tunnel (opposite page, top) reveals how automobile exhausts, visualized by releasing smoke, are channeled to flow down an open area over a freeway. The buildings act as barriers to the exhaust flow. Contrasting views of New York City (opposite page, bottom) show Manhattan on smoggy and clear days.

Atmospheric moisture

The effect of the city on atmospheric moisture is rather uncertain, except that urban relative humidities are lower. This does not necessarily mean, however, that the air is drier in the sense that it has less moisture. Relative humidity expresses the amount of vapor in the air as a percentage of the maximum amount possible at the air temperature. Thus, because urban temperatures are warmer and the air can hold more vapor before becoming saturated, relative humidities are lower even for the same vapor content as rural air.

From available studies it appears that actual urban moisture content is usually lower during the day, especially in summer, and this is attributed to less evaporation in the city. At night some cities are more humid than their surroundings. This may be due to the fact that the city appears to allow weak evaporation throughout the night, with little dewfall, whereas the characteristic heavy dewfall in rural areas depletes atmospheric water. The situation may also be aided by stagnation of moist air within the urban canopy with weak winds. The addition of vapor from combustion to the urban boundary layer may also make the urban heat and pollution plume a moisture plume.

Precipitation

The three important constituents of the urban plume (heat, water vapor, and pollutants) form the basic requirements for convective cloud development and precipitation. The heat island is associated with instability and, therefore, with enhanced uplift, and this is necessary to cool the air and allow condensation. Urban water vapor from combustion is an increased source for cloud droplets, and pollutants form the necessary nuclei around which condensation can take place. Because the process is not instantaneous the results of the addition of these materials (in the form of clouds or precipitation) are not evident until some distance downwind.

Comprehensive work on urban precipitation effects was being conducted in St. Louis as part of Project Metromex (*Metro*politan *M*eteorological *E*xperiment). The research was being undertaken by a group of government, research, and university institutions; they operated one of the most detailed networks of precipitation stations in the world, which was further supplemented by aircraft, radar, balloons, and other means of probing the atmosphere. The St. Louis study clearly demonstrated that the urban-industrial complex influences convective storm behavior, producing up to 20–30% increases in precipitation and increasing the intensity and frequency of severe weather (thunder and hailstorms), especially during the summer. The zone of major effect exists downwind and within a distance of 40 kilometers (25 miles) from the city center. Metromex investigations established that the urban influences of importance to cloud development reach up to the cloud base, and that the nuclei can be ingested into the cloud and washed out again within 16–32 kilometers from the release point. The Metromex airflow studies also revealed that the urban

Courtesy, J. E. Cermak, Colorado State University

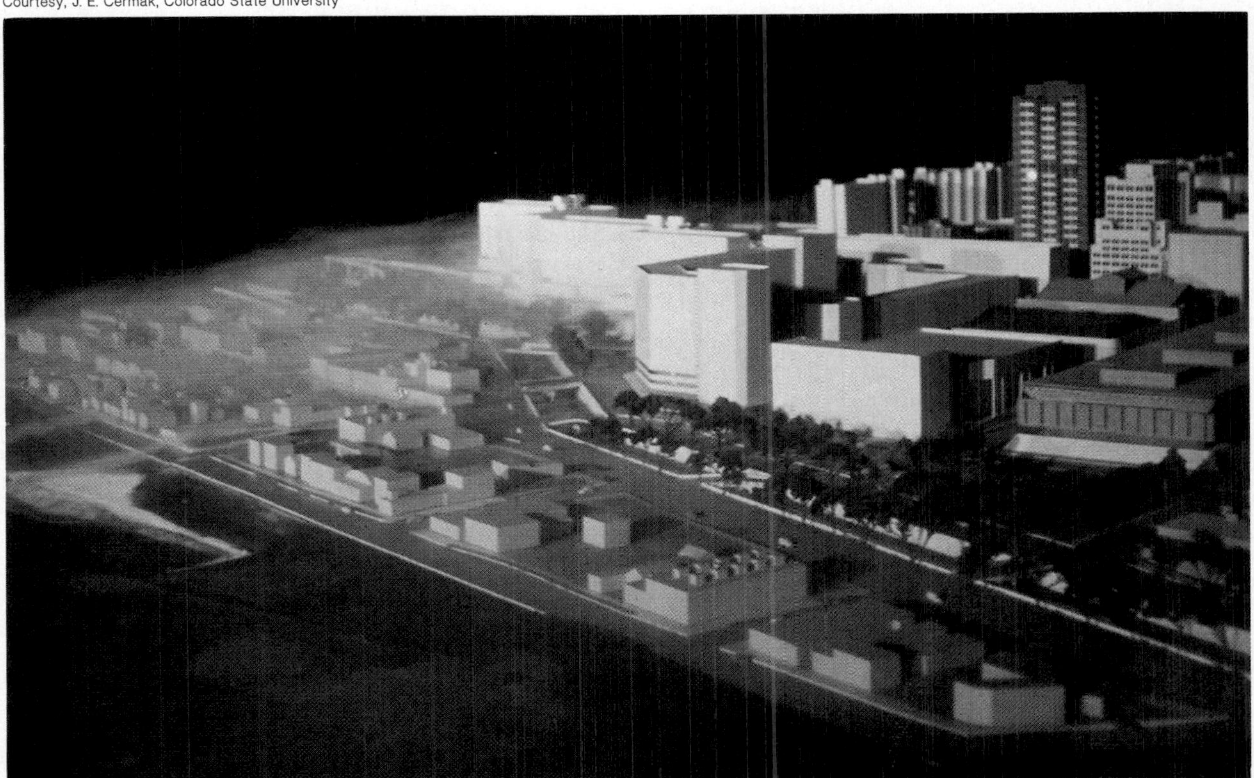

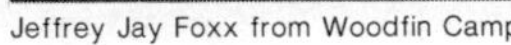
Jeffrey Jay Foxx from Woodfin Camp

Charles Moore from Black Star

area causes air to converge over and downwind of the city, thereby providing a means for concentrating the nuclei and uplifting them to the cloud-base level. The next requirement of the project is to ascertain if the in-cloud processes are also affected by the urban influence.

Airflow

Within the urban building layer air movement is very complex, as every building produces its own flow patterns. Under strong wind conditions, near tall buildings, the ground-level wind climate can become very unpleasant as air is deflected down the windward sides of buildings, "jets" around corners or through passageways, and forms turbulent eddies in the lee of obstacles. Probably the only method to predict the wind structure following new construction is by use of a scale model in a wind tunnel. On average, however, it is to be expected that buildings shield the ground from the air flow over the city and that wind speeds consequently are reduced in cities.

In the boundary layer above the buildings, during moderate winds, the increased roughness causes both greater turbulence than occurs over the countryside and changes in the wind direction. Both are important in regard to air pollution, because the former controls mixing and the latter determines the air trajectory. Under near-calm regional windflow the city sets up its own circulation due to the heat island. This closed cellular pattern is clearly not conducive to the dispersion of air pollutants.

Air pollution

The urban boundary layer is also the urban mixing depth, h, in which the combination of thermal instability from the heat island and the mechanical turbulence due to the rough urban surface leads to fairly homogeneous mixing. In an experiment dealing with sulfur dioxide emitted from both a major oil refinery complex and the whole urban area in Montreal, the upper limit of the boundary layer was found to inhibit mixing to higher levels. Also, the oil refinery plumes initially emitted well above the urban surface were being diffused both upward and downward. The depth h is equivalent to the actual building height in the city center on such mornings, and on some occasions it is even below roof level.

The concentration of pollutants in the air is a direct function of the rate of pollutant emission and an inverse function of ventilation, which is the product of the wind speed and the mixing depth. Obviously, therefore, the potential for urban air pollution problems is greatest for a given rate of emissions when wind speed and mixing depth are both small. This occurs most commonly at night and especially if the large-scale wind pattern is dominated by the presence of a high atmospheric pressure system.

The fate of the urban plume and its contents in the downwind region is only just beginning to be studied. It is clear, however, that the city's influence extends to a considerable distance, perhaps as far as 200

Courtesy, Barnes Engineering Company

Thermogram taken with Barnes thermograph shows heat emission from buildings at the lower tip of Manhattan. Red areas are the hottest.

kilometers (125 miles). Cities, therefore, contribute to regional pollution over a wide area, and if adjacent urban areas are aligned with the wind, their plumes may combine to form a "megalopolitan plume" of giant proportions. This has been detected on satellite photographs, where the industrial haze of the Eastern United States has been tracked to beyond Bermuda.

Future outlook

Scientists remain a long way from understanding the atmospheric environment in which most of the world's population lives. Because the trend toward urbanization is projected to continue, this gap in knowledge must at least be narrowed. Special emphasis will have to be placed on work in the tropics and polar regions (both of which are experiencing rapid expansion) so that cities there will not repeat the mistakes of those in temperate latitudes.

Cities and city-regions must incorporate urban climate planning in their design. Before this can be effectively done, however, there remains the large task of investigating the basic causative processes underlying urban climates. Only then will it be possible to build meaningful numerical and scale models capable of predicting the climatic result of alternative land-use strategies. Ultimately it is to be hoped that this will lead to the rational design of pleasant and healthy cities and to the construction of buildings that are efficient in their use of energy. In some respects failure cannot be tolerated because increasing population and urbanization are already producing regional-scale climatic effects, and it is not beyond possibility that these are the harbingers of larger, even global-scale, climatic changes triggered by human mismanagement.

River Floods

by G. H. Dury

The damage caused by floods has increased steadily as more and more flood-prone land becomes urbanized. Acceptance of risk appears to be a necessary consequence of such development.

Kurz & Allison, 1890

G. R. Roberts, Nelson, New Zealand

Turbulent flow conditions in the Fox River, New Zealand, at bankfull stage. (Opposite) Diagrammatic cross section of a river reach, showing the relation of stage and the river's bed and banks.

G. H. ***DURY*** *is Professor of Geography and Geology at the University of Wisconsin, Madison.*

(Overleaf) The great Johnstown, Pennsylvania, flood of 1889.

The Mississippi River in 1927 inundated an area of approximately 67,000 square kilometers (nearly 17 million acres). Its 1973 flood, although less extensive at 40,500 square kilometers (11 million acres), was agriculturally disastrous because water remained on farmland during what should have been the spring planting season. The Yangtze River of central China in 1931 drowned 140,000 people and inundated 87,000 square kilometers (22 million acres). Even more extensive flooding took place in 1954, when water from the Yangtze spread over 110,000 square kilometers (27 million acres) of farmland, 10% of the nation's total.

The greatest single area liable to river floods lies, however, in the lower valley of the Huang Ho (Yellow River) in northern China, where 140,000 square kilometers (35 million acres) are chronically endangered. Floods on the Huang Ho in 1887 killed two million people, perhaps half of them directly by drowning and the remainder by starvation as farmers were unable to deal with a blanket of flood-laid sediment or with standing floodwater that took as long as two years to vanish from some districts. An even worse disaster occurred in 1931, when the Huang Ho rose in the most calamitous flood in written history: the eventual death roll was estimated at nearly four million. Although such totals are much less than those recorded for major pandemics and exceptional famines, they far surpass the numbers of deaths for the bloodiest recorded battle, namely the First Battle of the

Somme in 1916, in which about one million died, and for the most lethal earthquake, that in Shensi Province, China, in 1556, in which 830,000 perished.

River floods can be regarded either as recurrent or as accidental. Recurrent floods, such as those already described, are the high peak discharges that can be expected as part of the natural order, whereas accidental floods result mainly from the failure of some kind of dam, whether natural or artificial. The sections of the Indus and Ganges rivers that flow through mountainous areas seem especially liable to accidental blockage by landslides and thus to sudden flood flows when such a dam fails. The Birehiganga at Gohna Tal (Lake) in northern India was blocked by a landslide in 1893; rupturing about a year later, the blockade released an initial discharge of perhaps 150,000 cubic meters per second. During the winter of 1840–41 a landslide from the flank of Nanga Parbat fell into the upper Indus. Water accumulating in a lake that reached a maximum depth of 300 meters and a length of 65 kilometers overtopped the barrier the following June, causing it to fail abruptly, and almost instantaneously releasing an estimated two cubic kilometers of flood flow. A Sikh army, encamped near the channel 400 kilometers downstream from the dam, was overwhelmed by a mud-charged wave with a 25-meter-high front.

The best-documented case of prehistoric floods is that of former Lake Bonneville in the Great Basin of the western United States. Originally without an outlet, its level controlled by evaporation, this lake

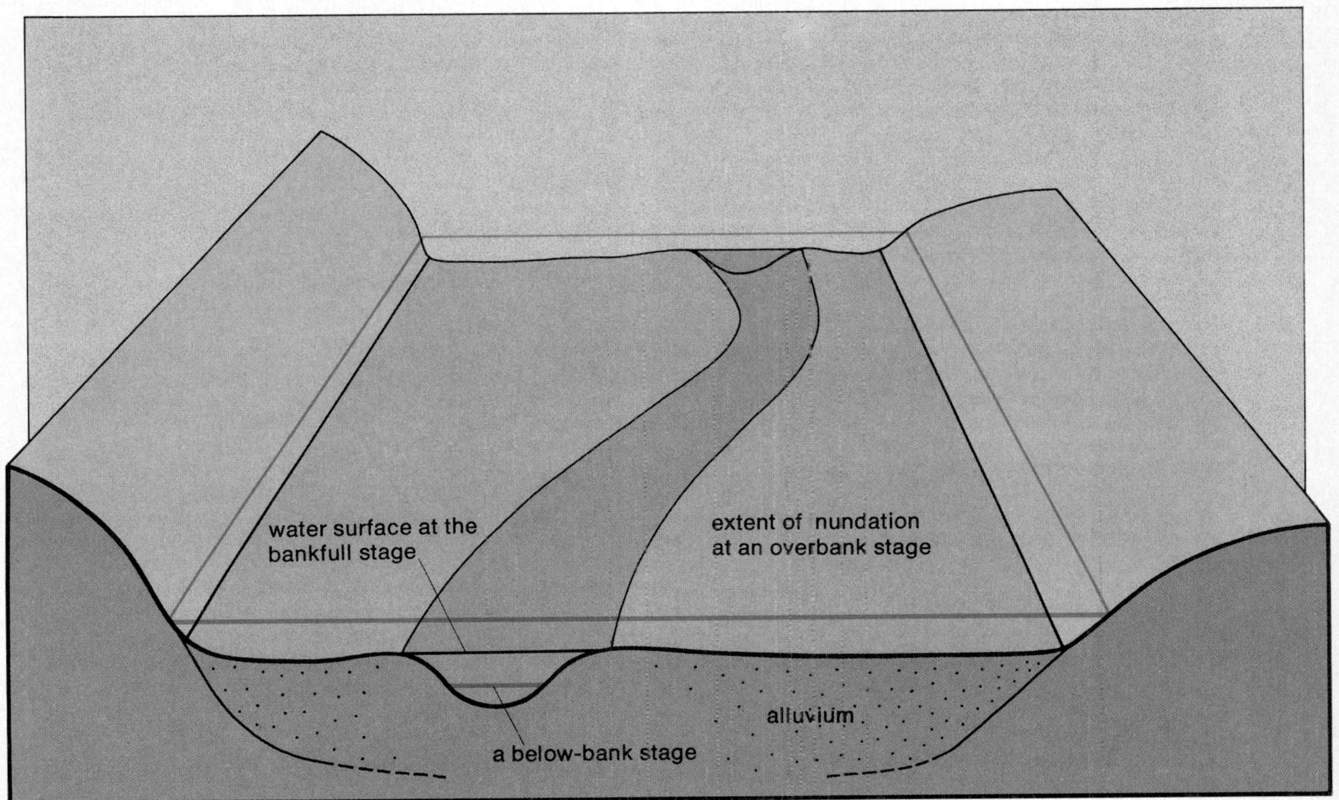

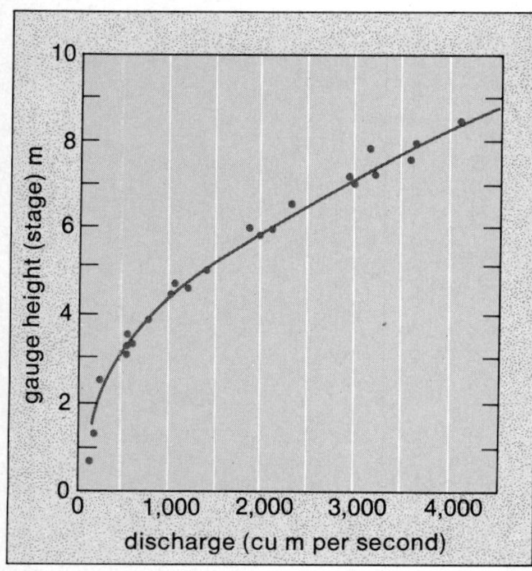

(Above) Rating curve for the Red River of the South at Shreveport, Louisiana, showing the relation of river stage and water discharge. (Opposite) The curve at lower left shows the probability (as a percentage of time) that a river's discharge will exceed or be less than a particular value of water discharge. The arrow indicates the discharge of the most probable annual flood, which is equaled or exceeded 0.5 percent of the time. The histogram at lower right provides the same information in different form, namely as the recurrence or return interval in years. At top is shown the probability plots for peak flow of the Dnepr River at three cities in the Soviet Union (see text).

began a steady rise after a tributary stream captured the Bear River. Perhaps about 30,000 years ago the water overtopped the lowest sill, Red Rock Pass, in the surrounding high ground, rapidly eroding a notch large enough to deliver, at maximum flow, an estimated two billion cubic meters per second. Dams formed by glaciers during the Pleistocene Epoch in the northern U.S. mountains released, on failing, vast floods that shaped the channeled scablands (elevated rocky terrain crossed by dry stream channels). The ice-dammed Lake Agassiz, in what is now the basin of the Red River of the North, spilled through an outlet now followed by the Souris River of North Dakota: the dimensions of the spillway suggest a flow of some 28,500 cubic meters per second. Glacial meltwater also supplies the contemporary glacier bursts of Iceland, although volcanic heat is often, if not invariably, a factor there: the 1934 burst of Grimsvotn produced a peak discharge of 50,000 cubic meters per second.

The most destructive episodes concerning man-made dams have been the failure of the South Fork Dam at Johnstown, Pennsylvania, in 1889, when more than 2,000 lives were lost, and the overtopping of the Vaiont Dam in Italy in 1963. Huge waves, set off by a rockslide, poured over the crest of the Vaiont. Although the structure held, the resulting wall of water was 70 meters high well downstream from the dam; more than 2,600 people died in the flood.

Flood magnitude and frequency

Recurrent floods must be taken into account by land-use planners and flood-control engineers. Such floods occur because stream discharge varies from one part of a year to another, because annual peak discharges vary from year to year, and because rivers do not fashion channels large enough to accommodate the infrequent large flows. In the simplest imaginable case, a meandering river works and reworks the alluvium in its valley bottom, developing a flood-prone belt of low ground adjacent to the channel. This belt, the floodplain, may be essentially horizontal in cross section, or may slope gently away from the channel. There may or may not be natural levees, constructed by the river, along the channel margins.

For most of the time, rivers normally flow at below-bank stages. At the bankfull stage the channel is completely filled with water, whereas at overbank stages a floodplain is inundated. The greatest known discharge at the bankfull stage, determined by modest extrapolation from available gauging records, is that of the Amazon at Óbidos, Brazil, about 300 kilometers upstream from the head of the delta: its value of approximately 283,000 cubic meters per second provides a basis of comparison for the accidental floods discussed above.

For purposes of flood assessment and flood forecasting the depth of flow, termed stage height, is commonly compared with an arbitrary zero point at the gauging station. Reference to stage as opposed to discharge has the advantage that stage can be directly measured, whereas discharge is usually calculated from stage by means of a

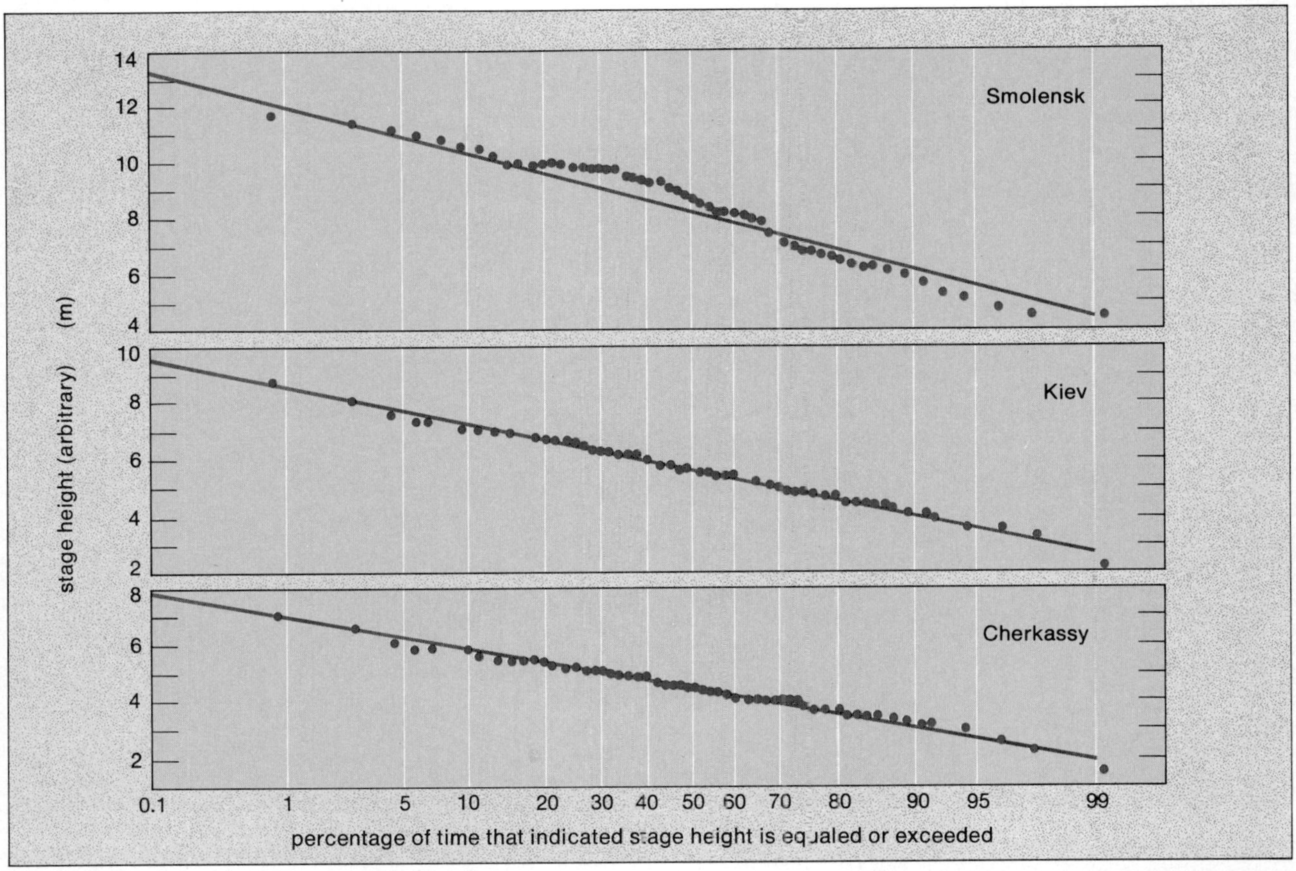

Smolensk
Kiev
Cherkassy
stage height (arbitrary) (m)
percentage of time that indicated stage height is equaled or exceeded

percentage of time discharge is less than shown by graph
discharge (cu m per second)
percentage of time discharge is more than shown by graph

observed distribution
normal distribution
percent
recurrence interval (years)

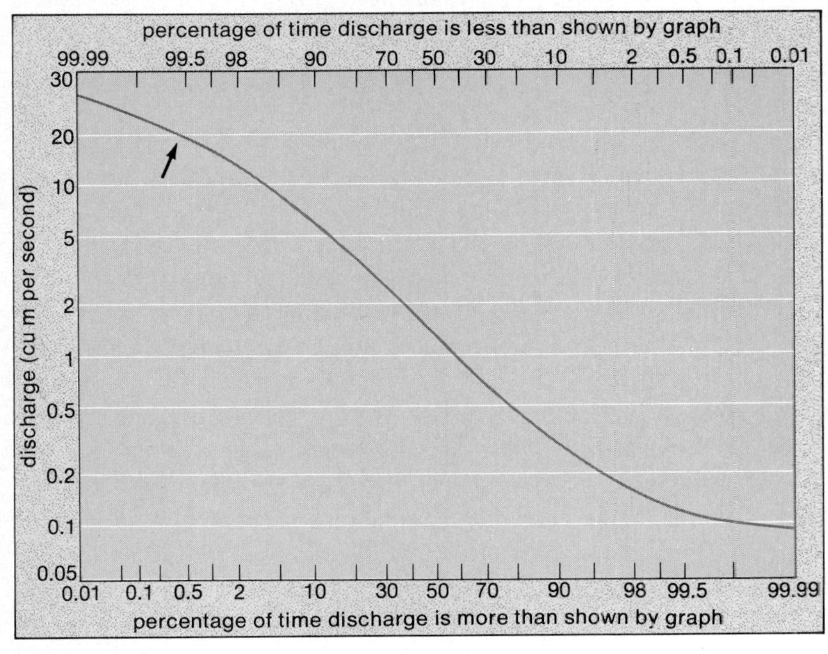

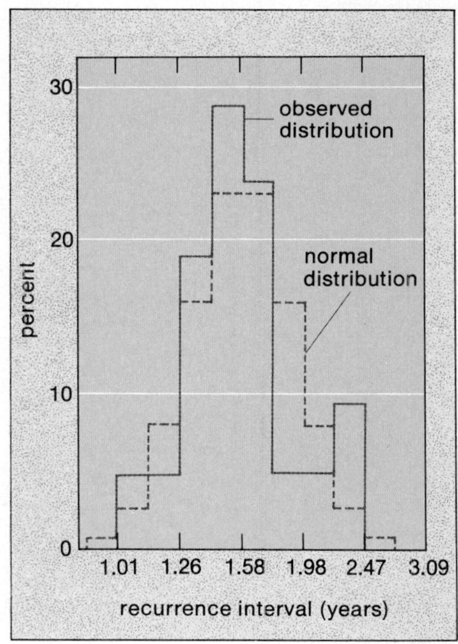

Loren McIntyre from Woodfin Camp

H. F. Garner

rating curve. Determining such a curve requires measurements of velocity of flow at intervals of depth and width as well as for various values of discharge.

Whether analyses of flooding and flood forecasting deal in terms of stage or in terms of actual discharge, they involve the concept of magnitude and frequency. In this context, all peak flows are called floods whether or not they cause inundation. The main interest, however, remains with actual or potential inundation, which can be assessed in terms both of extent and of depth. Because floods are not cyclic, harmonic analysis has no value in flood forecasting. Instead, some form of probability analysis is employed. Two chief methods are in use: the calculation of percentage probability and the calculation of the recurrence interval.

Percentage probability may be applied either to discharges or stages in general, or to peak discharges or stages. In either case, the 1% event is that which is equaled or exceeded in 1% of the observed series. For example, percentage probability has been plotted for peak stages at three stations on the Dnepr River in the Soviet Union. At the downstream station, Cherkassy, 99% of peak flows attain or exceed 0.2 meter above the gauging zero point, whereas only 1% attain or exceed 0.8 meter. The corresponding peak flow values at the intermediate station of Kiev are 0.3 and 0.85 meters, while those at the upstream station of Smolensk are 0.45 and 1.2 meters. When, as in this instance, the data produce almost linear plots, and if the peaks are

Photos, Lawrence K. Lustig

annual peaks, then the 0.1% event can be defined—that is, the peak stage expectable on the average once in a thousand years.

Percentage probability is widely used in civil engineering, but it has the disadvantage that the cumulative duration of bankfull and over-bank flows can be less than 1% or even 0.5% of the total annual flow on some streams. Hydrologists tend to favor the analysis of magnitude and frequency, which, given adequate data, can be used to forecast the probability of rare floods of great magnitude. This method can be applied not only to stream discharge but also to the forecasting of droughts, precipitation, and extremes of temperature. The precise form of analysis employed depends on the model of probability that is adopted. One of the simplest and most widely applied of these models is the Gumbel (Extreme-Value), which is summarized in the following discussion.

For purposes of magnitude-frequency analysis, the annual flood is the peak discharge in a given year (in mid-latitudes, usually the water year, October through the following September, so that winter and spring floods may be included in a single bracket). A sequential list of annual floods constitutes the annual series. Annual floods are typically distributed through time in quasi-random fashion. When a series is ranked by order of magnitude, from the largest downward, the recurrence intervals for the individual floods are computed by the simple equation $r.i. = (n + 1)/r$, where n is the number of items in the series, r the rank order, and $r.i.$ the recurrence interval—the average

(Opposite) Contrasting flow regimes of the Amazon River in Peru are shown by turbulent peak flow of the Urubamba River, a highland tributary, and the main anastomosing reach of the Amazon. Two contrasting types of desert channels are shown above. At left is the rock-strewn bed of a wash at the base of the Funeral Mts., Death Valley, California. Bankfull discharge may be attained perhaps once in 50 to 100 years or more. At right is the sand-choked channel of the Kuiseb River beneath the giant sand mountains of the Namib Desert, South West Africa. River flow must overcome extensive seepage into the bed before infrequent events can sweep the channel free.

Tom Ebenhoh from Black Star

Water year	Observed annual flood
1937	2,300
1938	3,765
1939	3,710
1940	590
1941	645
1942	2,680
1943	625
1944	2,435
1945	4,815
1946	3,455

Rank order and magnitude		Recurrence interval (years)
(1)	4,815	11.0
(2)	3,765	5.50
(3)	3,710	3.67
(4)	3,455	2.75
(5)	2,680	2.20
(6)	2,435	1.83
(7)	2,300	1.57
(8)	645	1.38
(9)	625	1.22
(10)	590	1.10

frequency over the long term of a flood of given magnitude. Thus, the 10-year flood (annual series) represents a peak value that can be expected ten times in a century; the 100-year flood can be expected ten times in a thousand years; and the average time interval between 1,000-year floods is one millennium. Nothing except improbability will inhibit two 1,000-year floods in successive years, however.

The constant 1 in the above equation ensures that the median annual flood has a recurrence interval of 2 years. Additional properties of the statistical design ensure that the mean annual flood has a recurrence interval of 2.33 years and the most probable annual flood one of 1.58 years. Where rivers are significantly deepening their channels or, conversely, filling their channels and raising their floodplains, the stage/frequency relationship and the frequency of inundation are disturbed. In other cases, the frequency of bankfull and near-bankfull flows groups itself around the recurrence interval of 1.58 years on the annual series. That is to say, river channels tend strongly to adjust to the capacity required by the most probable annual flood.

When independent peak flows inferior to annual peaks are included in the analysis, the resulting list constitutes the partial-duration series. It has the advantage of including items that would disappear from the annual series. For instance, a given river at a given station might record an annual peak of 15,000 cubic meters per second in one water year, and peaks of 25,000 and 18,000 cubic meters per second in the following year. This last item would be excluded from the annual series, even though it is greater than the preceding annual peak. Partial-duration analysis, thus, is beneficial mainly in regard to modest peaks; in it, the 1.58-year frequency of the most probable annual flood is converted to a frequency of one year, showing that bankfull stage is likely to be reached or exceeded, on the average, once a year.

When a flood series conforms strictly to the Law of Extreme Values, it is plotted as a straight line on the Gumbel probability model except for one or two items at the upper end of the series, corresponding to unusually high or unusually low discharges. The data from some stations yield graphs that flatten at their upper ends, indicating that the observed floods are approaching the practical capacity of the basin to deliver peak flows. Other stations are represented by graphs that are inflected upward, indicating a combination of one set of low peaks with another set of high peaks. The implications of distributions of this kind are, as will be seen, ambiguous.

Flood forecasting for planning and design purposes usually deals with moderately long recurrence intervals, in the range of 50–100 years. The 25-year flood represents too immediate a risk for the average design engineer, while provision against floods of recurrence interval greater than a century would, in many instances, involve impossible expense. Complete flood protection is only possible in special circumstances, such as that of the artificial Lake Kariba on the Zambezi River in southern Africa, which, with a capacity of about 158 cubic kilometers, can retain a year's runoff. Unfortunately, most discharge

Charles Steinhacker from Black Star

Wide World

records are too short for reliable long-term analysis. As a rough approximation, a record should be double the length of the recurrence interval for which a forecast is required. The desirable 200-year series needed for accurate definition of the 100-year flood is rarely available. Extrapolation beyond the actual length of record is less reliable still. Nevertheless, it can be argued that the 1947 floods on some rivers of eastern England approximated the 1,000-year peak.

Magnitude-frequency graphs that are inflected upward into dogleg form make medium-term forecasting especially difficult. They may indicate real variations in probability, or they may result from a short record that includes a set of unusually low peaks and another set of unusually high peaks. Such a contrast between the two sets may correspond to a shift in atmospheric circulation and weather patterns. For example, there is considerable evidence that, in middle latitudes, an increase in north-south (meridional) airflow at the expense of west-east (zonal) flow can change seasonal temperature and precipitation, leading to an increase in variable weather in some parts of the year. Although the weather-flood relationship is far from simple, changes in the atmospheric circulation can certainly displace the seasonal incidence of peak floods and can probably affect the flood magnitudes as well. The 15 greatest peaks of instrumental record for the Mississippi at St. Paul, Minnesota, are concentrated in the intervals 1860–1920 and 1950 onward. The interval 1920–50 may therefore represent a departure from the normal climate. It may in time be possible to refine the forecasting of floods with recurrence intervals of 50 years or less to make allowance for expected shifts in weather patterns.

Shifts in the general atmospheric circulation in low latitudes cause disruption of monsoonal and allied systems, commonly increasing the

Peak flow on the Missouri River (opposite, top) is graphically portrayed by the notice of normal bridge clearance. List of peak discharges (bottom; in cubic meters per second) for ten years of annual floods on the Ouachita River at Arkadelphia, Arkansas, illustrates ranking by order of observed magnitudes and calculated recurrence intervals. (Above left) Flash flooding on the Connecticut River produced overbank flow and hazard to trees and vegetation, whereas the results of Hurricane Agnes (right) along the Schuylkill River in June 1972 led to substantial property damage, as evidenced by these nearly submerged homes at Pottstown, Pennsylvania.

Lawrence K. Lustig

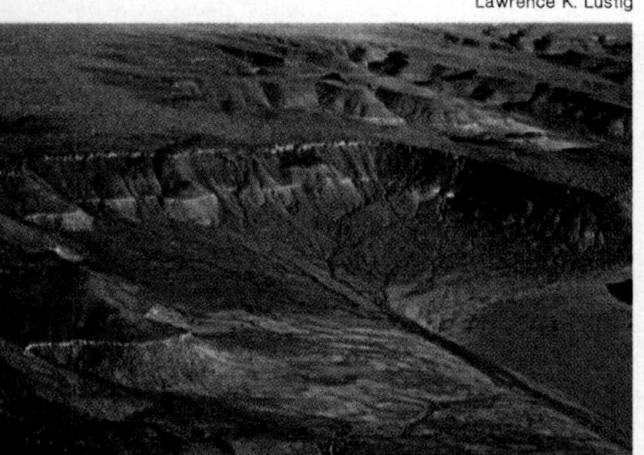

An individual drainage basin in the plateau country of the Karroo Desert, South Africa. Infrequent floods in this instance cannot be of very great magnitude because the drainage area is quite restricted.

risk of droughts in some years and of floods in others. During the last decade, the Sahel belt of sub-Saharan Africa and the semiarid-to-arid areas of inland eastern Australia experienced prolonged droughts that were terminated by extensive flooding. There is weighty archaeological evidence that the First Dark Age of ancient Egypt, which saw the destruction of the Old Kingdom, resulted primarily from successions of famine, with attendant widespread starvation and breakdown of the social order. One may reasonably infer that the famines resulted from failures, during series of years centered on 4,130 and 3,975 years ago, of the monsoon rains that feed the Blue Nile and promote the irrigation floods of the trunk channel.

Regional contrasts in liability to floods

The hazard of extensive river flooding must obviously be greatest where extensive areas adjacent to a major river channel lie below the level reached by flood crests. The Mississippi in its lower 950 kilometers, except for the delta proper, occupies an alluviated trough cut principally by glacial meltwater at times of low sea level, when glaciers were at a maximum. Cessation of meltwater discharge, the rise of sea level from a low stand perhaps 160 meters below that of the present, and—to a lesser extent—the outgrowth of the delta have caused the river to elevate its bed. In many places, the alluvial surface slopes down away from the river; in the undrained state, the lowest-lying areas constitute backswamps. Although natural levees along the channel banks have been considerably raised and strengthened by artificial means and have been supplemented by wholly artificial levees constructed on the plain, the flood risk of unusually high flows continues to be severe, with large areas endangered. Considerable success was nevertheless achieved with a floodwall built in 1955 which was estimated to have saved the city of St. Louis a potential $340 million of flood damage in 1973.

The most extreme case is that of the lower Huang Ho, which enters its alluvial plain at Meng-chin, 765 kilometers from the sea and only 120 meters above sea level. The very gentle gradient, averaging only 0.016%, is insufficient to enable the river to transport its sediment load, more than 40% of which is deposited upstream from the mouth. The deposit, equivalent to an annual depth of ten centimeters on an area of 5,500 square kilometers (about four inches on 1,250,000 acres), is in part distributed by floodwater and in part used in raising the channel bed, which after 4,000 years of artificial embanking and a far lengthier record of natural alluviation stands one to two meters above the level of the surrounding countryside. The water surface at ordinary low flows is as high as three meters above the general level of the countryside; at flood flows the difference may reach eight meters.

The outbuilding and upbuilding of the Mississippi Delta, allied with subsidence near the delta margins, have caused the Mississippi to switch its main outlet from time to time. In this case, however, the switching process usually results in the rapid reestablishment of a new main channel, and the points of diversion lie 250 kilometers or less from the river's mouth. In addition, the floodable area of the Mississippi valley upstream from the delta proper is limited by bluffs. By contrast, the lower reaches of the Huang Ho traverse a vast alluvial fan, which slopes gradually away from the main channel. Switching of the river's outlet, which always involves extensive flooding, occurs as far as 450 kilometers upstream from the coast and can displace the mouth through 335 kilometers, between the south and the north sides of the Shantung Peninsula. A deliberate breaching of the Huang Ho levees during the high water of 1938 resulted not only in a change of course but also in the displacement of an estimated six million refugees, of whom possibly half a million died.

Somewhat analogous circumstances obtain in the depositional plains of semiarid regions. The discontinuous vegetal cover of high ground in such areas permits rapid surface runoff, while the typically gentle down-valley and cross-valley slopes, combined with low channel capacities and high ratios of width to depth, ensure that heavy rainstorms cause extensive flooding. Still more extensive flooding can occur in basins that have interior drainage, if some of the streams that flow into them rise within the range of occasional tropical atmospheric depressions. Such is the case in the Lake Eyre basin of central Australia, which has been widely inundated twice in the last generation.

A case of special difficulty is that of the great north-flowing Siberian rivers (Ob, Yenisei, and Lena), on which the annual thaw occurs earlier in the southern upstream reaches than in the northern downstream areas. The frozen and ice-jammed downstream reaches are not only bordered by extensive waterlogged marshes but are also liable to disastrous fish kills if the spring meltwater is highly charged with wastes formed from humus.

The contrast between flood flow and normal flow varies widely from climate to climate. Peak flows in all regions of ephemeral (periodic) drainage can be regarded as infinitely greater than the approximate norm of zero flow. In humid regions, the steepness of the flood peak, which controls both the depth and extent of inundation and also its duration, varies with the characteristics both of drainage basins and of storms. At least 25 characteristics influence the form of the flood peak, among them basin size. Small basins, roughly coextensive with the expectable areas of individual convectional storms, are subject to rapid increase in discharge and to rapid inundation but also to the rapid draining off of floodwater when the peak has passed. Downstream reaches of large rivers, by contrast, are typically affected by slow rise and slow fall of the flood crest, where peak discharge may be maintained for some days. These contrasting characteristics have generated what is known as the upstream-downstream controversy. Small-to-moderate floods in small basins can be countered by simple conservation measures and by the construction of small dams, whereas protection in downstream reaches requires extensive and expensive engineering works.

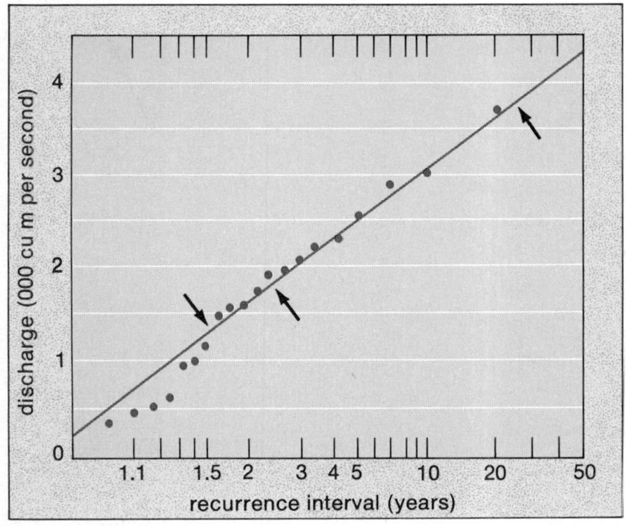

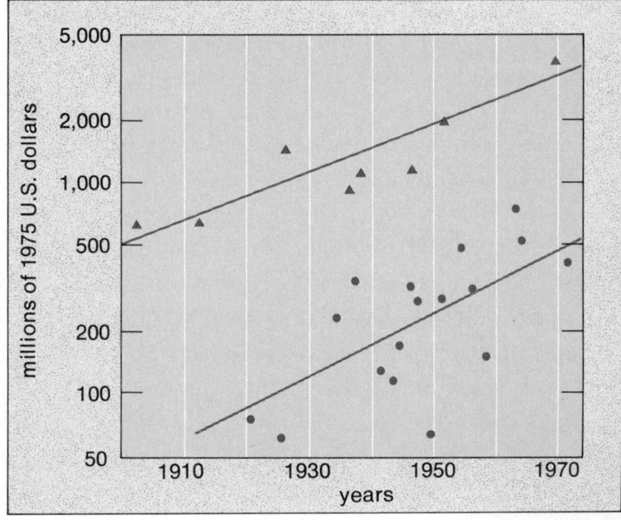

A 20-year record for the Ouachita River near Malvern, Arkansas, is plotted at left. The discharge of the most probable annual flood (left arrow) is 1,330 cu m per second, that of the mean annual flood (center arrow) is 1,810 cu m per second, and that of the 25-year flood (right arrow) is 3,825 cu m per second. Damage caused by floods in the U.S. is portrayed at right. The uppermost line is fitted to data from catastrophic floods, whereas the lower line reflects "merely disastrous" events.

Effects of floods

Most rivers with mobile beds tend to scour their channels as discharge rises to its peak value, and then to fill them in again as the flood subsides. Where the opposite sequence is followed, the rise of the peak is especially rapid. It is unusual for even very large floods to produce significant changes in channel size; when such changes do occur, they are normally rectified in a few years at the most. The exceptional 1947 floods of eastern England caused channel scour only at bridges, with deposition taking place immediately downstream. Although flooding may be accompanied by some erosion of channel banks, erosion of the floodplain surface is uncommon. Whereas the flow velocity within the channel may be 2 or 3 meters per second, the sheet of floodwater moves at 0.5 meter per second or less. Average depth of deposition on floodplains during inundation is usually measured in millimeters. Some rivers, however, are notably raising their floodplains. One example is that of the River Severn, upstream of Shrewsbury, England, which has reduced its channel capacity to three-quarters of that required to contain the most probable annual flood. In consequence, inundation has become more frequent; areas cultivated in medieval times are now frequently under water.

Flood risk and flood damage increase as flood-prone areas come under increasingly intensive use. The floodplain of India's Ganges River was, for generations, used mainly for seasonal grazing, but population pressure has caused permanent agricultural settlement to take place on it. Flood damage, mainly to crops, is currently assessed in India at an average of $1 million a year. The 1973 floods on the Indus, in Pakistan, extended over 5,000 square kilometers and ruined crops worth $250 million.

Statistics of flood damage in money terms are provided most abundantly by the U.S. The picture is not encouraging. Damage due solely to river floods, as opposed to hurricane and tidal flooding, can be separated into two categories. The damage from merely disastrous river floods, as assessed in 1975 dollars, has been increasing during the present century at a rate that predicts an event costing about $1.2 billion by about the year 2000. The corresponding prediction for truly catastrophic floods, the damage from which has been increasing more steadily than that from mere disasters, is an event causing $6,250,-000,000 damage.

Flood risk can be greatly increased by urbanization. Despite what might be expected from the swift runoff from construction sites, development may actually cause stream channels to be reduced in size because it leads to the delivery of abundant sediment. The spread of masonry and paved surfaces, combined with the extension of storm drains, may then increase the frequency of overbank flows in affected streams by a factor of five or six.

Responses to flood hazard depend on perception of the magnitude and frequency of risk. Responses to actual inundation and perceived risk are highly diverse. They include the passive, precautionary, or alleviatory measures of bearing the loss, taking out flood insurance, taking emergency action in time of actual flood, and accepting public relief. They also include structural adjustment (itself precautionary), such as the erection of levees and the flood-proofing of individual structures, regulation of land use in catchment basins and on flood-plains, and measures designed to reduce the probability that a given channel will overflow.

The probability of overflow can be reduced by the installation of flood-storage reservoirs, where these are practicable; by the cutting of bypass channels, again where practicable; by embanking; and by enlargement of a natural channel. Unfortunately, these possibilities are all to some extent self-defeating. Reservoirs may prove unable to contain the waters of unusually high floods. Bypass channels, like artificially enlarged channels, may prove similarly inadequate and in any event may serve to increase the flood risk in downstream reaches. It can be simply impractical or uneconomic to excavate and maintain a channel capable of accommodating the 100-year flood. Moreover, artificial channel enlargement in some agricultural areas of the U.S. Middle West has promoted gullying in the basins of tributary streams. Artificial embanking, as exemplified on the lower Huang Ho, acts to increase the flood risk should the banks be breached.

For the future, it seems likely that flood warning systems will be progressively improved, and that, on some rivers, defensive structures will be progressively elaborated, toward the point where the combination of annual capital cost and annual damage cost reaches a minimum. The prospect that flood-prone land will be withdrawn from high-cost or labor-intensive use appears remote. It therefore follows that acceptance of risk, allied to public relief in some countries, will remain the principal mode of adjustment to extremely high floods.

Particle Physics: A Realm of Charm, Strangeness, and Symmetry

by Henry T. Simmons

As increasingly powerful particle accelerators probe the structure of matter, the suspicion grows that the universe may be as vast in its smallest dimensions as it is in its largest.

The world of high-energy physics was thrown into ferment during 1974 and 1975 as a result of two major developments: the production of some remarkable new particles in energetic collision experiments and the possible discovery of an extremely massive particle with a single magnetic pole. Musing over such discoveries, physicists have happily commented that they are fortunate indeed to be living through the most exciting period in the history of their science. Yet, many interested persons, laymen and scientists alike, balk at the prospect of accepting still another handful of particles and concepts. Fifty years ago the proton and the electron were considered to be the fundamental particles of the universe. This comfortable idea was eliminated, however, when particle accelerators capable of higher collision energies and sophisticated detection apparatus became available. A veritable legion of oddly named particles appeared—neutrons, neutrinos, pions, muons, and antiparticles—as well as a new way of thinking about matter to explain their behavior. Recently the concepts of strangeness, the intermediate vector boson, and the colored, charmed quark, whose properties are in a state of continual revision, have added to the seeming confusion. It is fair to ask, therefore, whether the study of particle physics will ultimately yield a true understanding of what constitutes matter and energy.

VILLANI

Courtesy, Cavendish Laboratory, Cambridge, England

Compared with the sprawling, multimillion-dollar particle accelerators of the 1970s, the instrument pictured above appears primitive. Basically a voltage-multiplying device, it was used by J. D. Cockcroft and E. T. S. Walton in 1932 to bombard lithium with a beam of accelerated protons. Each lithium nucleus (three protons and four neutrons) that absorbed a proton fissioned into two helium nuclei or alpha particles (two protons and two neutrons each). A milestone of particle physics, this achievement represented the first atomic transmutation produced by particles that had been artificially accelerated.

__HENRY T. SIMMONS__ is a science writer who has treated many aspects of physics and space research.

(Overleaf) Painting by Ron Villani. Illustrations by Dennis Magdich

Fundamental probing of the atom

Recent particle discoveries and possible detection of the controversial magnetic monopole are but the latest developments in an effort that began at the turn of the century to understand the forces and the physical laws affecting the innermost structure of the atom. The scientist who pioneered exploration of this strange subnuclear world was Ernest Rutherford. Sitting in a darkened room at the University of Manchester in 1911, Rutherford was the first to perceive the core of the atom, the atomic nucleus.

He did not, of course, actually see the nucleus or even an individual atom. An atom of hydrogen is only about one billionth of a centimeter in diameter, and its nucleus is 40,000 times smaller—much smaller in fact than anything even distinguishable with modern electron microscopes. Nevertheless, Rutherford was able to develop a crude but ingenious method for investigating the atomic interior. Bombarding thin gold foil with alpha particles that were produced in the spontaneous radioactive decay of radium, he observed the positions of the faint scintillations that the particles made on a fluorescent screen after their passage through the foil. He noted that although the great majority of particles passed straight through without change, occasionally one would scatter off the foil at a sharp angle. This proved that the alpha particle had collided with something quite small, at least 10,000 times smaller than the individual atoms of gold in the foil. Moreover, it had to be extremely hard. Albert Einstein called Rutherford's technique one of "shooting sparrows in the dark," but, despite all the technological refinements of high-energy physics since 1911, it still remains the principal experimental approach for elucidating the atomic nucleus and its constituents.

Rutherford's alpha particles were actually the positively charged nuclei of helium atoms, each consisting of two protons and two neutrons. With an energy of about five million electron volts (MeV), they were far too feeble to penetrate the massive gold nucleus and produce any debris. What Rutherford observed was the result of an occasional sharp deflection of track when an alpha particle came into the immediate vicinity of the positively charged gold nucleus. Although naturally occurring alpha particles could not break up massive nuclei, by 1924 Rutherford found that they could penetrate the nuclei of most of the lighter elements up to potassium, which has 19 protons. The alpha particles drove matter from these target nuclei, thus transmuting them to elements of lower atomic number.

The use of these "bullets" of natural radioactive decay remained the mainstay of experimental nuclear physics until the 1930s, when the first cyclotrons and other man-made particle accelerators appeared. Though the work of Rutherford and other early experimentalists seems primitive today, it laid the foundations for the brilliant theoretical insights of such scientists as Niels Bohr, Louis Victor de Broglie, Erwin Schrödinger, Max Born, Werner Karl Heisenberg, Wolfgang Pauli, and P. A. M. Dirac.

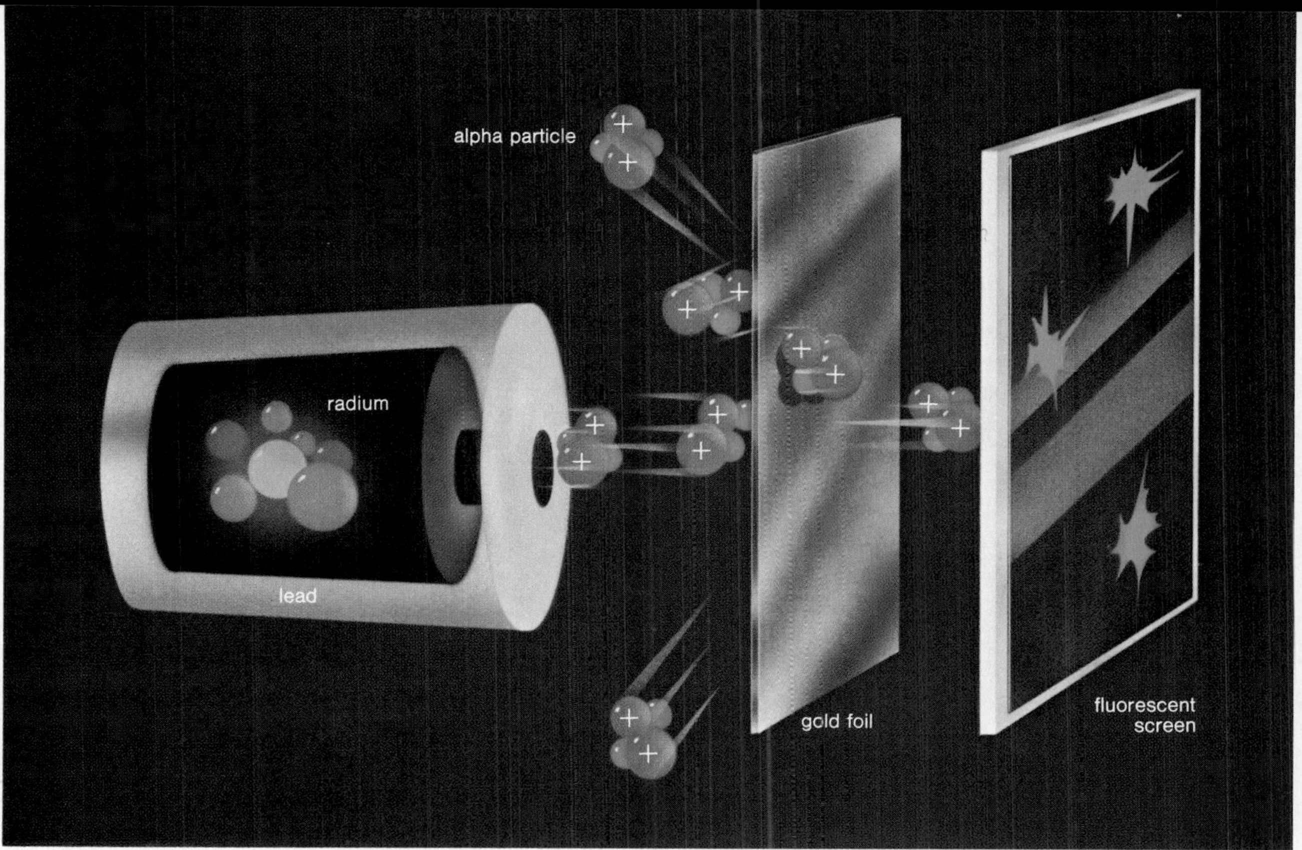

The quantum world

These theoreticians began to realize during the 1920s that the classical laws Newton originally set forth to describe the behavior of the macroscopic universe failed to account for the data gathered in subatomic experiments. What they were encountering was a quantum world, the essence of which had been identified as early as 1900 by Max Planck. He recognized that energy on an atomic scale is emitted and absorbed in distinct packets termed quanta, and this revolutionary insight overthrew the classical notion of an unbroken continuum of energy transactions from the infinitely large to the infinitely small. Reinforced by Einstein's observation five years later that light transmission also must be regarded as the migration of particle-like packets of discrete energy, Planck's conclusion led directly to Bohr's discovery in 1913 that electron orbits or vibrational states in an atom are rigidly quantized. Quantum theory was further broadened and solidified with Bohr's 1922 exegesis of the entire periodic table of chemical elements in quantum terms, with the development of quantum mechanics in the mid-1920s, and with the marriage by Dirac in 1928 of Einstein's theory of special relativity to quantum mechanics to produce a relativistic wave equation for the electron—the precursor of modern quantum electrodynamics.

Physicists soon learned that the quantum world is governed by statistical probability rather than by an exact cause-and-effect relationship. For example, of 100 radioactive nuclei with a half-life of ten years, one can predict with confidence that ten nuclei will decay in the first year, and nine in the next, but one cannot predict which ten or which nine. Similarly, an excited particle may decay to a ground state

The experimental apparatus with which Ernest Rutherford detected the atomic nucleus consisted of a radioactive substance shielded with lead except for a small hole through which a beam of positively charged alpha particles could emerge. When Rutherford placed a fluorescent screen opposite the hole in the shielding, a clearly defined spot of illumination appeared where the beam of particles struck. However, when he interposed a sheet of gold foil in the path of the beam, the spot became less evenly illuminated and occasional flashes of light could be seen at some distance from the central spot, an indication that a few particles were being strongly deflected, some in fact by more than 90°. A mathematical analysis of the pattern of illumination led Rutherford to the conclusion that, although the atom must consist primarily of empty space, at its center is a tiny, massive nucleus with a very dense positive charge distribution.

by two or more routes. Though the statistical probability of each route can be determined with great accuracy, one cannot predict which path will be taken in an individual particle decay. Hence, though quantum theory can achieve great precision in describing the behavior and interactions at the molecular, atomic, and nuclear levels, one is not allowed to see the individual components at work. The physicist's tools, necessarily made of atoms, particles, and photons of energy, are much too coarse. The act of probing a given quantum state in an atom or nucleus alters it. It is as if one tried to understand a musical note by grasping a vibrating piano string with one's fingers. Heisenberg recognized this delicate wholeness of the quantum state in his famous uncertainty principle: The more exactly the position of a particle is known, the more blurred is the knowledge of its energy.

Strong and weak nuclear forces

It was relatively simple for early scientific investigators to observe the forces that operate in the macroscopic universe—the force of gravity that draws objects to the ground, the electric sparks that make clothes crackle on a dry day, the magnetism that swings the compass needle. But it was only when physicists finally probed into the heart of the atom that they discovered the existence of two additional forces, quite alien to everyday experience but dominant in the quantum world: the strong and weak nuclear forces. The strong force, now known to be the most powerful in nature, holds protons and neutrons together in the atomic nucleus despite the immense mutual repulsion of its protons. The weak force, too feeble to hold anything together in the energetic environment of the nucleus, appears in a variety of radioactive-decay processes. The relative strengths implicit in the names of the strong and weak forces are related to the rates at which they operate. Unlike gravity and electromagnetism, which are theoretically infinite in range, the strong force is confined to an operating distance on the order of the diameter of a proton, or about 10^{-13} centimeters. When two particles that feel the strong force pass within this distance, there is a high probability that they will interact. If such an interaction does occur, it will happen characteristically in 10^{-23} second. The weak force has an even shorter range; to feel it, two particles must pass within 10^{-15} centimeters and, even then, the probability of interaction is low, less than one chance in ten billion. Additionally, the weak force operates on a time scale 13 orders of magnitude more leisurely than the strong force, requiring about 10^{-10} second.

Arranged on a scale of relative strengths, the strong force occupies the top position, followed by the electromagnetic force and then the weak force. At the bottom is placed gravity; while all particles feel the gravitational force, it is unimaginably feeble on the small scale of the atom. Yet, on the cosmic scale of the universe, it is the electromagnetic and gravitational forces that predominate. The extremely limited range of the strong and weak forces prevents them from directly manifesting themselves, despite their importance.

The four fundamental forces or particle interactions ranked according to relative strength				
	strong	electromagnetic	weak	gravitational*
particles affected	hadrons	all with electric charge	all except photon and graviton	all
particle exchanged during interaction	meson	photon	intermediate vector boson	graviton
mass of particle exchanged	about 140 MeV and greater	0	greater than 20 GeV	0
range	10^{-13} cm	infinite	10^{-15} cm	infinite
conservation principles (selected)				
energy	holds	holds	holds	—
momentum	holds	holds	holds	—
electric charge	holds	holds	holds	—
baryon and lepton numbers	holds	holds	holds	—
strangeness	holds	holds	fails	—
charm	holds	holds	fails	—
parity	holds	holds	fails	—

*The gravitational interaction reflects only those quantum principles conserved or violated in the other three interactions; it introduces no new explicit violations.

Physicists have classified particles by the way they interact with the short-range forces of the atomic nucleus. The particles that feel the weak force but not the strong force are called leptons. There are four particles in the lepton family: the electron, its more massive cousin the muon, and two massless, electrically neutral bits of energy that travel at the speed of light, the electron neutrino and the muon neutrino. Because every particle has an antiparticle, in which all quantum properties are reversed, there are four antileptons known to particle physics. For example, the positron, with one unit of positive charge, is the antiparticle of the electron, which carries one unit of negative charge.

A second group of particles, the hadrons, comprises those capable of feeling the strong force. There are two classes of hadrons: the baryons and the mesons. The baryons include the proton and the neutron, the two stable nucleons which form atomic nuclei, as well as a multitude of excited, short-lived states of the proton and neutron called hyperons. The mesons are the particles that actually carry the strong nuclear force; they are the glue that holds atomic nuclei together. As in the case of the leptons, for each hadron there is an antihadron.

There is also a separate group of particles that carry the several forces at work in the universe. As mentioned above, the meson carries the strong force; the protons and neutrons in the atomic nucleus pass mesons among themselves like balls in a game of "catch," a process of emission and absorption that proceeds at a rate in excess of 10^{17} times per second. Beyond a distance of 10^{-13} centimeters, however, the probability of a successful emission and absorption between nucleons falls off rapidly, and the influence of the strong force fails. By an apparently similar process, the weak force is maintained through

The strong nuclear force between a proton and a neutron in the atomic nucleus is thought to be mediated by the exchange of charged pi-mesons. The proton and neutron are often regarded as two manifestations of the same basic particle, the nucleon, because in the process of exchanging mesons the two particles also exchange identities.

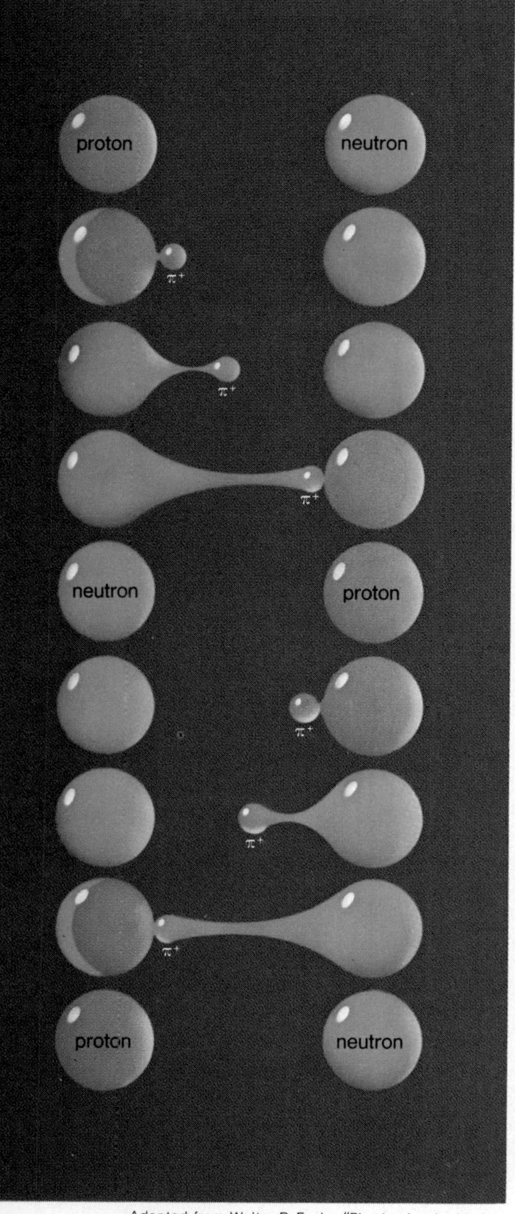

Adapted from Walter R. Fuchs, "Physics for the Modern Mind," © 1967, The Macmillan Company

the exchange of a hypothetical particle called the intermediate vector boson, whose probability of carrying the force between two weakly interacting particles is low and drops off quickly at particle separation distances beyond 10^{-15} centimeters. Another hypothetical particle, called the graviton, is considered to carry the gravitational force. The photon, the quantum of electromagnetic energy, carries the electromagnetic force.

When considered in groups or populations of its own kind, each of the four types of force-carrying particles obeys certain statistical principles that allow the individual force fields of identical particles to overlap and become concentrated. Because the graviton and photon are massless and must travel between interacting particles at the speed of light, the concentrated strengths of the forces they carry diminish only by a simple inverse proportionality with the square of the separation distance of the interacting particles; hence, a strong gravitational or electromagnetic field can exert considerable influence over a great range. On the other hand, the mesons and intermediate vector boson possess mass, a factor that causes the strengths of the forces they carry to diminish much more rapidly with distance than gravity and electromagnetism. As a result, the strong and weak forces do not appear in the macroscopic universe.

Conservation principles and quantum numbers

The extreme speeds and infinitesimally short distances that characterize subatomic quantum processes have always imposed upon the scientific investigator a severe limitation on direct experimental observation. Light travels about 30 centimeters in 10^{-9} second. Because even the weak force requires only about a tenth of this time to produce an interaction, the original participants in a weak interaction can barely be detected. In the case of swifter interactions, the original participants can only be inferred from the tracks of the secondary debris particles produced in an energetic collision or annihilation. Although this might appear to place physicists in the same plight as the six blind men trying to determine the shape of an elephant by touch, the situation is not that hopeless. Over the past 50 years, physicists have been able to deduce an important set of conservation principles by which the quantum world and its forces operate.

The most powerful conservation principle of quantum theory relates to the mass, energy, and momentum of particles. Nineteenth-century physicists observed that mass and energy are always conserved in chemical reactions, though it remained for Einstein in 1905 to propound the deceptively simple expression that describes the interchangeability of these properties in the nuclear and subnuclear world: $E = mc^2$, meaning that the energy of a system is equal to its mass times the square of the velocity of light. In a particle-decay process, a collision, or an annihilation, mass and energy can be transformed into each other or exchanged among particles; but because they must be conserved, they are never simply lost or created from nothing. For

group	particle	symbol	mass (MeV)	charge	spin	lepton number	mu-ness	baryon number	lifetime (seconds)
	photon	γ	0	0	1	0	0	0	stable
leptons	electron	e^-	.5	−1	½	+1	0	0	stable
leptons	positron	e^+	.5	+1	½	−1	0	0	stable
leptons	electron neutrino	ν_e	0	0	½	+1	0	0	stable
leptons	electron antineutrino	$\bar{\nu}_e$	0	0	½	−1	0	0	stable
leptons	muon	μ^-	106	−1	½	+1	+1	0	10^{-6}
leptons	antimuon	μ^+	106	+1	½	−1	−1	0	10^{-6}
leptons	muon neutrino	ν_μ	0	0	½	+1	+1	0	stable
leptons	muon antineutrino	$\bar{\nu}_\mu$	0	0	½	−1	−1	0	stable
hadrons: baryons	proton	p	939	+1	½	0	0	+1	stable
hadrons: baryons	antiproton	$\bar{p}$	939	−1	½	0	0	−1	stable
hadrons: baryons	neutron	n	939	0	½	0	0	+1	10^3
hadrons: baryons	antineutron	$\bar{n}$	939	0	½	0	0	−1	10^3
hadrons: baryons	omega minus	Ω^-	1,672	−1	3/2	0	0	+1	10^{-10}
hadrons: mesons	pion	π^+	137	+1	0	0	0	0	10^{-8}
hadrons: mesons		π^-	137	−1	0	0	0	0	10^{-8}
hadrons: mesons		π^0	137	0	0	0	0	0	10^{-15}
hadrons: mesons		ρ^+	750	+1	1	0	0	0	10^{-23}
hadrons: mesons	rho meson	ρ^-	750	−1	1	0	0	0	10^{-23}
hadrons: mesons		ρ^0	750	0	1	0	0	0	10^{-23}
hadrons: mesons	psi (3095)	ψ	3,095	0	1	0	0	0	10^{-20}
hadrons: mesons	psi (3684)	ψ	3,684	0	1	0	0	0	10^{-20}

Selected quantum properties and lifetimes for several subatomic particles.

example, the existence of the mysterious neutrino was inferred through an apparent violation of this rule. While observing the natural processes collectively called beta decay, in which neutrons and protons in an unstable nucleus may transform into one another, experimental physicists reported that the decay products carried less energy than the rest mass of the original nucleus. Bohr and several of his colleagues tended toward the idea that conservation of mass and energy broke down in weak decays. Pauli, however, boldly proposed in 1931 that a new particle of zero mass and zero charge—the neutrino—must be carrying off the missing energy. The neutrino was positively detected in 1956, and today it is an invaluable tool for probing hadrons because it feels only the weak force and consequently yields data uncomplicated by strong or electromagnetic interference.

Quantum properties like mass, energy, momentum, and electric charge are physically measurable and have a fairly close correspondence with everyday experience. Others are more abstract and are unique to the quantum world. All of them serve to establish family resemblances among particles, and usually each must be conserved, or remain unchanged, in particle interactions regardless of whether the original participants retain their identities or are transformed. Physicists have attempted to keep track of these quantum properties algebraically by assigning them "quantum numbers," which, like

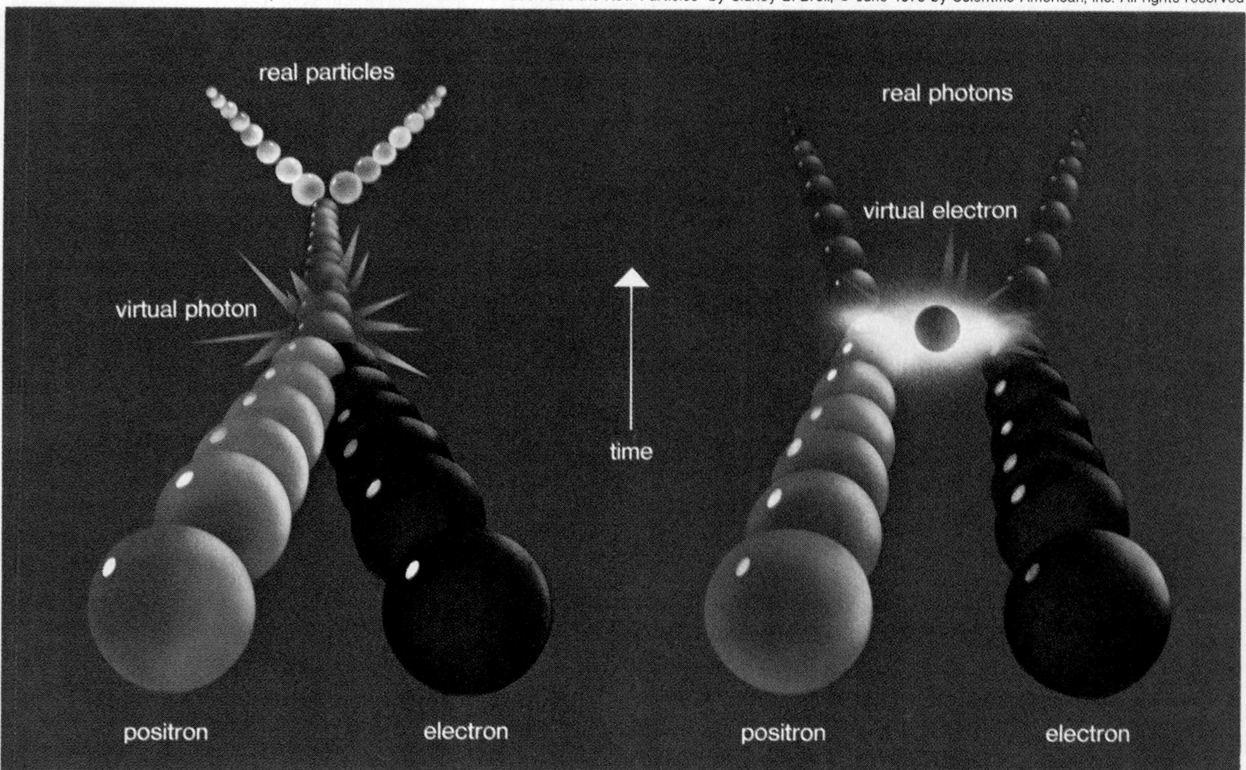

the properties they represent, must usually be conserved in particle-interaction equations. The property of "baryon number," for example, distinguishes baryons and mesons; baryons have been assigned a number of +1, antibaryons −1, and mesons 0. Because baryon number must be conserved, baryons can never transform into mesons. Thus, the proton, which is the least massive baryon, cannot decay spontaneously into a less massive meson or lepton because such a process would require the baryon number to change. "Lepton numbers" and "mu-ness" operate in the same fashion. Electrons, muons, electron neutrinos, and muon neutrinos share the same lepton number of +1. Of the four, however, only the electron and its electron neutrino share the same mu-ness of 0; the muon and its neutrino have a mu-ness of +1. An electron, therefore, can transfer its mass and charge to more massive particles and become an electron neutrino, but because mu-ness must be conserved, an electron can never transform into a muon or a muon neutrino.

Physicists have devised another important classification system for subatomic particles based on quantum numbers and the kind of quantum statistics that the particles obey. It divides all particles into two groups called bosons and fermions. Bosons are particles with distinctive quantized spins, or intrinsic angular momenta of discrete values, that can be characterized by whole-integer quantum numbers; *e.g.*, 2, 1, 0. They include all the particles that transmit the forces of the universe: the photon, the graviton, the mesons, and the vector bosons. The bosons obey Bose-Einstein statistics, an important property of which allows a population of identical particles to overlap or to occupy the same quantum state. In the case of the massless graviton and photon, overlapping builds up fields of force that can operate over extremely great distances in the macrocosmic universe. The ability to overlap also allows these two forces to become tremendously concentrated under certain circumstances. For example, the gravitational force is extremely feeble compared with the other forces in the universe, and particularly in the atomic and nuclear environment. But in a gravitationally condensed object like a stellar black hole, the gravitational force becomes so intense that it overwhelms all the other quantum forces, compressing electromagnetic radiation, matter, and space into what physicists call a singularity.

The fermions obey Fermi-Dirac statistics. This group of particles includes baryons and leptons, whose spins are characterized by half-odd integers; *e.g.*, 3/2, 1/2. In contrast to bosons, fermions are required to obey the Pauli exclusion principle, which prohibits any two identical fermions in a given population from occupying the same quantum state. For example, in the lowest energy state of the helium atom, the two orbital electrons can occupy the same innermost orbit, but their spins must be opposite to one another, since this is the only quantum property in which they can differ.

Post-World War II: symmetry violations

In 1935 the Japanese physicist Hideki Yukawa recognized the need to explain the strong force in terms of a particle that carried the force between nucleons. To confirm this theory experimentally, however, by blasting such a particle free of the strongly bound nucleus required a collision energy considerably beyond the capabilities of existing particle accelerators; hence, the verification of Yukawa's hypothesis had to wait 13 years for a meson to be created and its track photographed. Such delays made it apparent to experimental physicists of the 1930s that tools more powerful than natural radioactive decay were necessary to overcome the strong nuclear force and even briefly "materialize" an independent particle for inspection. Consequently, much of particle physics up to World War II was concerned with developing increasingly powerful atom smashers in order to penetrate into the first two layers of the quantum world—the atom and its nucleus. Postwar work, however, has been largely devoted to the third tier of matter—the individual nucleon.

Most of the conservation rules or "symmetries" at work in the atom and the nucleus apply to the individual nucleon. But postwar physics has discovered some important new principles at the higher energy levels of the newer accelerators. One is a quantum number arbitrarily called strangeness, which is possessed only by certain hadrons in excited states and which is conserved in strong and electromagnetic interactions. Even more startling was the discovery that particle interactions and decays proceeding via the weak force do not conserve all of the quantum numbers. It had long been believed that particle

Annihilation of an electron and its antiparticle, the positron, produces photons of electromagnetic energy. The electron and positron are both leptons; respectively, they have lepton numbers of +1 and −1, charges of +1 and −1, and a mu-ness of zero. When the two particles meet, their opposite quantum numbers simply cancel. Such properties as charge and lepton number are conserved by being eliminated, and momenta, being equal and opposite, yield a net momentum of zero. There remains, however, a certain total energy that cannot simply disappear. From this energy state, particles possessing nearly any combination of quantum properties theoretically can emerge, provided that the sums of their quantum numbers remain zero and the new particles retain the same total energy. If a single photon is produced from the interaction (opposite, left), it must fulfill conservation requirements by having a large energy but zero momentum, a situation impossible for a real particle. The particle thus formed is called a virtual photon because, before it can ever be detected, it swiftly decays into real particles with equal but opposite momentum. If two photons emerge from the annihilation (right), the interaction is considered to be mediated by the exchange of a virtual electron. Again, the two photons must possess a total energy equal to that of the annihilating particles and an equal but opposite momentum.

interactions were equivalent regardless of the spatial coordinate system in which they occurred; *i.e.*, that they had no preferred direction. It was thought that there would be no left-handed or right-handed preference in particle decay processes and that one process is as likely to occur as its mirror image. This property, called parity, was known to be conserved in the strong and electromagnetic interactions. But in 1957 two Chinese physicists working in the U.S., Tsung-Dao Lee and Chen Ning Yang, observed experimentally that parity was not conserved in the weak interactions. Still later it was found that the weak force failed to conserve strangeness and several other important quantum properties. In this respect, the strong force may be regarded as "High Church," because it meticulously holds to all the quantum liturgy, while the weak force is "Low Church," rather casual when it comes to observing all of the ceremony. It is thought that the neutrino, which participates only in weak interactions, is somehow responsible for this failure to conserve parity and the other quantum properties because of its distinctive left-handed and right-handed helicity.

Quark theory

One of the most bewildering results of high-energy physics since 1950 has been the creation of hundreds of new mesons and baryons. It was thought initially that these hadrons were new fundamental particles like the proton, even though their lifetimes were extremely short and they decayed into a handful of familiar, stable particles. By the early 1960s, however, physicists were exhausting the Greek alphabet in naming all the excited states and resonances produced in particle collisions and annihilations. It seemed certain that all could not be regarded as fundamental, and there was a growing suspicion that even the stable proton and neutron might not deserve this description.

When physicists become uneasy about the state of their discipline, a new theory is inevitable. In what was probably the major theoretical advance in understanding the hadrons, two U.S. theoreticians, Murray Gell-Mann and George Zweig, independently proposed in 1963 that all hadrons are composed of still smaller particles with remarkable properties. In naming these entities, Gell-Mann, who had earlier introduced the concept of the "strangeness" property of hadrons, turned to *Finnegans Wake* by the Irish novelist James Joyce, where he found the phrase "Three quarks for Muster Mark." Gell-Mann thought the name particularly appropriate as a tag for the tiny, hypothetical particles that might comprise the substructure of the hadrons.

The concept of quarks provided scientists with a new system for ordering the hadrons into families in such a way that all of their distinctive properties could be accounted for by assuming that all hadrons were composed of a very limited number of different kinds of subparticles. In its simplest form, the theory proposes that there are three basic types, or flavors, of quarks, which Gell-Mann designated *u*, *d*, and *s*, for up, down, and sideways (or strange). Their quantum properties are given in the following table.

quantum property	*u* quark	*d* quark	*s* quark
charge	+2/3	−1/3	−1/3
strangeness	0	0	−1
spin	1/2	1/2	1/2
baryon number	+1/3	+1/3	+1/3

The proton is a particle with spin 1/2, charge +1, and strangeness 0. It can be constructed with two *u* quarks and one *d* quark. To account for the proton's spin of 1/2, it is inferred that two of its quarks are "spin-up" and one is "spin-down," leaving a net spin of 1/2. The neutron, which has a spin of 1/2, charge 0, and strangeness 0, can be built in similar fashion with two *d* quarks and one *u* quark.

Quantum physics requires that all particles have antiparticles such that the two are mirror images of each other. They have the same mass and spin but opposite signs for the other quantum numbers. If all the signs that appear in the table above are reversed, the properties of three antiquarks emerge. By expanding the quarks in this fashion one can assemble all the known hadrons, the baryons as well as the mesons. Mesons can be built from a quark and an antiquark. Because only two quarks are involved, their 1/2 spins must either cancel or add; thus they produce the whole-integer spin numbers characteristic of the mesons.

More complex versions of this theory have been proposed since its introduction in 1963, including the idea that the three basic quarks are subdivided still further by quantum properties called colors. This embellishment is believed necessary because the quarks, having half-integer spins, are fermions and therefore must obey the Pauli exclusion principle. In many of the baryon and meson configurations constructed of quarks, sometimes two and even three identical quarks must be set in the same quantum alignment—an arrangement forbidden by the exclusion principle. By assuming three colors—red, yellow and blue—for each of the basic quarks, nine quarks become available, and it is then possible to construct all the known hadrons without violations. For example, the omega minus baryon, which was only found after its existence had been predicted by a precursor of the quark theory, has a charge of −1, a strangeness of −3, and a spin of 3/2. It can be constructed with three *s* quarks so arranged that they all spin in the same sense. If the three *s* quarks are made red, yellow, and blue, they then will differ in at least one quantum number and can occupy otherwise identical quantum states.

Another elaboration of the quark theory is the hypothesis of a fourth basic quark, designated *c*, which carries a new quantum property called charm. The idea of charm was proposed in the mid-1960s on simple aesthetic grounds. There are four leptons—the electron, the muon, the electron neutrino, and the muon neutrino. Like the three basic quarks, they are thought to be pointlike objects. It was felt that considerations of symmetry required that there be a fourth basic quark characterized by a new quantum number. Recently the notion of charm also came to fill a practical role in the conceptual structure of

The assignment to quarks of the quantum property of color has suggested to some physicists a theory of quark confinement that could explain why a free quark has never been observed. In the diagram on the opposite page, a baryon is represented by a bound system of three quarks, each of a different color (a), so that the baryon exhibits no net color. Several physicists have proposed that the strong force between hadrons, which diminishes rapidly with distance, is only a weak manifestation of the true strong force, which binds quarks together and remains constant with distance. In such a situation, a tremendous amount of energy, in the neighborhood of 10^{13} GeV, must be invested to force a single quark from its bound state (b). Long before this happens, however, a new process intervenes. The energy supplied to extract a quark is channeled into the formation of a quark-antiquark pair (c). The new quark returns the baryon to its original state, and the new antiquark combines with the freed quark to form a bound quark-antiquark system—i.e., another kind of particle, a meson (d). As the meson departs, its quarks shift in color with such rapidity that, like the baryon from which it was formed, the meson also exhibits no net color.

quantum physics when it was invoked to explain the absence of certain types of weak decay processes that, according to the three-quark model, should have been seen.

"Where are they?"

Time and again the quark theory has proven itself a powerful concept. Not only has it accounted for all known hadrons but it also has predicted the existence of at least one previously unsuspected hadron. Moreover, it has the additional power of forbidding precisely those hadron states that are not seen in the quantum world. "All this cannot be a coincidence," theoretical physicist Richard Feynman has observed, "yet one of the most obvious expectations is that these quarks should come apart in hard collisions with protons. Where are they? They have not been seen."

Had they been seen, they would have been readily distinguishable because of their fractional electric charge. Their wider corkscrew trajectories in the intense magnetic fields employed in bubble chambers and other detection apparatus would instantly set them apart from particles carrying a whole-integer electric charge. Neither the collisions produced by the accelerator of the Fermi National Accelerator Laboratory (NAL) near Batavia, Illinois, on stationary targets nor the still more violent proton-proton collisions produced in the intersecting storage rings of the European Organization for Nuclear Research (CERN) in Geneva, Switzerland, have been able to blast a free quark out of a hadron. Some physicists suggest that the forces between quarks rise to infinity at small distances and that a single quark can never be pried free of a hadron.

Though it may never be possible to break a quark loose from a hadron for individual scrutiny, there are several indirect techniques for probing the substructure of a nucleon. Ideal experiments for this purpose involve the electron and the neutrino. Since both are leptons, neither feels the strong force that affects the hadrons. Electrons are subject to the electromagnetic and weak force, whereas neutrinos feel only the latter. Thus the two particles can be used to develop "fingerprints" of the structure of hadronic particles without the ambiguities that arise when the strong force is manifested in particle decays.

For example, electrons have been used to probe the heart of the nucleon at the Stanford (California) Linear Accelerator Center (SLAC). Two targets were used in the experiment. One was ordinary hydrogen, whose single-proton nucleus consists of only three quarks. The other target was deuterium, a heavy form of hydrogen, with a nucleus consisting of one proton and one neutron, or a total of six quarks. Accelerated to 19 billion electron volts (GeV), electrons scattered off the two different targets in exactly the manner required if there are indeed six fractionally charged quarks in the nucleus of deuterium and only three in the proton.

Neutrinos have been used at NAL and elsewhere to probe the structure of nucleons. If the quark theory is correct, and if each nucleon

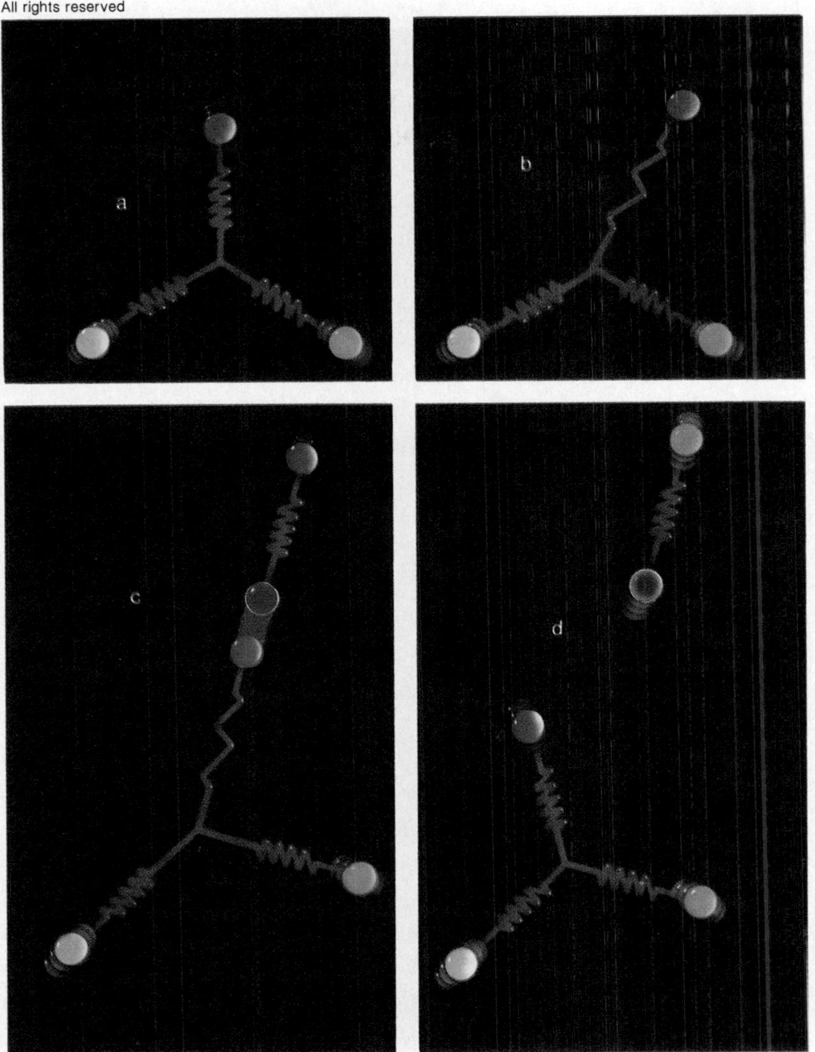

is indeed composed of three quarks with individual spins of 1/2, the neutrino and its antiparticle, with their opposite spins, would show a ratio of 3:1 in the way they scatter off the small, hard points within the nucleon. And this has been the case; in fact, the ratio has been measured and found to be 3.0 with a very small margin of error.

Although the experimental evidence by the early 1970s strongly favored the existence of quarklike objects inside the hadrons, it was becoming obvious that the picture was incomplete. Aside from the apparent necessity for color and charm elaborations of the basic theory, there was also the disquieting discovery that a fundamental prediction of the basic quark theory and its several alternative versions was not accurate. This was the prediction that in high-energy particle annihilations—for example, the collision of an electron and its antiparticle, the positron—there will be a fixed ratio of hadron events to lepton events materializing from the annihilations. Each theory predicts a different ratio, but all require that the ratios be fixed regardless of the energy level of the annihilating particles. Yet, none of these predictions has been confirmed. Even more baffling are experimental results showing that the ratio is not fixed as required by the quark theories, but that it climbs with increasing collision energies.

The psi particles

In the midst of trying to reconcile the quark theories with the lepton annihilation data came the announcement late in 1974 of two new particles produced in accelerator experiments. The first, announced simultaneously by research teams at two U.S. laboratories, had a mass of 3.1 GeV, more than three times that of the proton, the nucleus of the hydrogen atom. What startled physicists, however, was the lifetime of the particle, 10^{-20} second, or 1,000 times longer than all other particles that decay by the strong interaction force. Physicists at Brookhaven (New York) National Laboratory, under the direction of Samuel C. C. Ting, called their new particle J, and those at SLAC named it psi. Shortly after the discovery of psi, the SLAC experimenters, led by Burton Richter, William Chinowsky, and others, found a second new particle with a mass of almost 3.7 GeV and several hundred times the usual lifetime for strongly decaying particles. Because it was thought to be an excited state of the first particle, it was called psi-prime.

Thus physicists had a new mystery to consider. Why do such massive hadronic particles have such a long lifetime, and what is inhibiting their strong decay? The two readiest explanations seemed to involve color and charm. Like the individual quarks, the property of color has never been seen. The reason, according to advocates of the color theory, is that all the known baryons are composed of three quarks of different colors; consequently, they exhibit no net color. The mesons are composed of a quark and an antiquark of the same color, but the color of the two bound quarks shifts between red, blue, and yellow simultaneously and with such rapidity that the mesons, like the baryons, also have no net color. In the slow decays of the psi particles, could physicists be seeing naked color for the first time? Since the theory requires that color be conserved in strong decays, and there is no particle of lower mass to which the psi particles can transfer their color, the decays presumably go by a slower electromagnetic route that does not conserve the color property. This decay would be characterized by the emission of one or more energetic photons of gamma radiation carrying away the energy associated with the color property.

Alternatively, the charm quantum number might be present in the psi particles. Since the psi particles appear to be mesons, and the quark theory describes mesons as a bound state of a quark and an antiquark, perhaps the psi particles represent a bound state of a charmed quark and a charmed antiquark. The theory is attractive because the charmed quark is believed to be considerably more massive than the three other quarks; this consideration would tend to explain the great mass of the new particles. On the other hand, the charmed quark interpretation does not explain the long lifetime of the psi particles. Unlike the decay of a meson with net color, the quantum property of charm would cancel out in the decay of a meson composed of one charmed quark and one charmed antiquark. No special property would remain to be conserved, and hence nothing to inhibit a normal hadronic decay at 10^{-23} second by the strong route. So it is necessary to explain why charm, if present, should retard the strong decay at all. Since the notion of the charmed quark is important for other theoretical reasons, physicists are willing to entertain the idea that charm somehow requires that strong decay processes involving charm go 1,000 times slower than all other strong processes.

Fortunately, both theories of the psi particles are subject to experimental verification. Laboratories in the U.S. and Germany labored over the problem through 1975, and by autumn the evidence had clearly turned in favor of the charm hypothesis, though with many other questions still to be answered; *e.g.*, how charm inhibits the strong decay. Charm theorists had proposed several intermediate, short-lived states that, from quantum-mechanical considerations, are compelled to exist in the decay of the 3.7-GeV psi to the 3.1-GeV psi, and it appeared in these experiments that they indeed existed as part of a well-defined family of psi particles.

The magnetic monopole

The second exciting development of recent months was the reported discovery by P. Buford Price and his colleagues of the University of California at Berkeley and the University of Houston, Texas, of the long-sought magnetic monopole during a search for massive particles in the atmospheric flux of interstellar cosmic rays. They based their claim on a single dense track of a particle which passed through their cosmic ray detector on a balloon flight over Iowa in 1973. Because of the density of the ionization track of the particle through the stack of film emulsions and plastic sheets in the experimental apparatus and because there was no evidence that the particle slowed down as it passed through the stack, the investigators concluded that it was a monopole with a magnetic charge 137 times the charge on the electron. They calculated that its velocity was between one-half and two-thirds the velocity of light, and that its mass was at least 600 times that of the proton.

Some prominent physicists, notably Nobel laureate Luis Alvarez, were skeptical of the monopole claim, arguing that a massive particle like a platinum nucleus, with 78 positive charges, could have produced the track. Because existence of the monopole would have revolutionary repercussions for science and technology, most physicists preferred to withhold judgment until, at the minimum, many more such tracks have been found, or better still, until a monopole is captured for physical examination.

The controversy over the monopole runs to the heart of one of the major discrepancies in physics: the puzzling asymmetry between electricity and magnetism. Electric charge is monopolar; particles can carry a unit of positive charge, like the proton, or a unit of negative charge, like the electron. But magnetism is dipolar; there is always a north and a south pole, even in the tiny magnetic field generated by the electron. It has generally been concluded that magnetism was a secondary effect of electric charge, a force field produced by electric charges in motion. This view has been difficult to accept for some physicists, largely because of the powerful role of symmetry principles in the organization of the universe at the smallest scale. In 1931 Dirac proposed that the magnetic monopole must actually exist, and that it would be found to have a quantized magnetic charge equal to some multiple of 68.5.

If the monopole existed in nature, it would be a stable particle because quantum rules require the conservation of magnetic charge and there would be no other particles to which it could transfer all or a part of this charge. But, despite intensive scrutiny of ancient magnetized materials like terrestrial iron ores and iron-bearing ocean sediments likely to attract and capture a monopole, not one has ever been found. Studies of meteorites and lunar samples were also negative.

The ultimate particles

For many scientists who seek a return to the simple universe of Rutherford's day, the quark theory and the magnetic monopole, either separately or in combination, hold out the tantalizing carrot of fundamental and symmetrical order. If the existence of quarks and charm or some other property that endows an aesthetic structure to the hadrons can be shown categorically, physicists once again may be able to describe the universe in terms of the interaction of a handful of fundamental particles: a symmetrical arrangement of four leptons and four hadron quarks, plus the photon and a few other particles that carry the several forces found in nature.

Another approach to the problem of order was offered recently by Julian Schwinger of the University of California at Los Angeles when he proposed that hadrons are composed of more complex entities than quarks called dyons. His theory suggests that magnetic monopoles do exist but as dyons and that most of them are embedded in hadrons, where they perform the same function as quarks. Inside the hadron each dyon carries a dual charge—both the fractional electric charge required for quarks and a magnetic charge of the sort predicted by Dirac, that is rapidly exchanged among all the dyons comprising the hadron. Consequently, the hadron exhibits only a neutral magnetic charge and a whole-integer electric charge. Schwinger's particles have the important conceptual advantage that they can satisfy the Pauli exclusion principle for fermions; *i.e.*, they eliminate the need for quarks of different color. They also offer a unification of the strong and electromagnetic forces, because the enormous mass of three individual monopoles translates into the equally enormous binding energy of three dyons condensed into the structure of a baryon.

Now that physics appears to be on the verge of another major unification and simplification of the quantum world, can anyone dare to hope that this is really the last rung on the quantum ladder? Are the leptons and quarks or dyons really the ultimate building blocks of the universe? Particle physicists are divided three ways on the question. Some feel that there is no ultimate answer. They wonder whether the quantum world resembles an onion which can be peeled forever without arriving at a final layer. Sidney Drell of Stanford University suggested this view when he wrote: "It is not unreasonable to imagine that we shall someday penetrate the quark and find an internal structure there as well. Only the experiments of the future can reveal whether quarks are the indivisible building blocks of all matter, the 'atoms' of Democritus, or whether they too have a structure, as part of the endless series of seeds within seeds envisioned by Anaxagoras."

An alternative view, also grounded in Greek philosophy, is that symmetry principles or ideas are the fundamental reality, and that particles are "engineered" into existence by the particular collection of forces available at any given time and place. Thus, all particles are equal; some are permanent and others transitory, but not one is fundamental. This idea has been expressed recently by Heisenberg, who wrote: "What is really needed is a change of concepts; we will have to abandon the philosophy of Democritus and the concept of fundamental particles. Instead we will have to accept the concept of fundamental symmetries which is a concept out of Plato."

There is a third view which holds that, with the quarks, scientists have finally reached the end of the road in understanding the quantum world. According to this view, the quarks are so cunningly bound into hadrons that experimenters are forever prohibited from extracting a solitary quark. In the act of trying to do so, they simply manufacture quark pairs, a quark and an antiquark that materialize in the form of mesons, where the individual quarks are hidden from observation.

"If this interpretation of quark confinement is correct," observed Sheldon Glashow of Harvard University, "it suggests an ingenious way to terminate the apparently infinite regression of finer structures in matter. Atoms can be analyzed into electrons and nuclei, nuclei into protons and neutrons, and protons and neutrons into quarks, but the theory of quark confinement suggests that the series stops there. It is difficult to imagine how a particle could have an internal structure if the particle cannot even be created."

Perhaps there is a symmetry in the universe between its largest and smallest dimensions that inextricably intertwines the finite and the infinite. The evidence is strong that the universe exploded from an infinitely dense primal seed possibly 8 billion–18 billion years ago and that it is probably open; *i.e.*, it will continue to expand forever. In these respects, the macrocosmic universe always had and will continue to have a boundless and infinite character. One may wonder, therefore, whether the seemingly finite structure of matter is grounded on an infinitely layered quantum world whose innermost secrets, like the most distant objects in the cosmos, perpetually retreat from the inquiring human mind.

Laser-Induced Isotope Separation

by Reed J. Jensen and C. Paul Robinson

The importance of purified isotopes to nuclear power generation, medicine, agriculture, and the physical and chemical sciences has stimulated the search for more efficient, less costly separation methods.

Adam Woolfitt from Woodfin Camp

The year 1960 witnessed the first demonstration of a completely different source of light—the laser. In the nearly two decades since that occasion, various kinds of lasers have been developed, some so powerful that they yield power densities more intense than those on the surface of the Sun. So great has been the attention paid by the scientific community to the investigation and application of the lasing principle that it has been compared to the technological efforts of the U.S. during World War II to develop the atomic bomb.

As in numerous other fields of pure and applied science, the laser is finding important uses in chemistry. Typically it has been employed both to initiate chemical reactions and to study reaction dynamics. Within the last few years scientists have turned their attention to the application of laser chemistry to the classical problem of isotope separation, achieving recognition in 1975 when the separation of several isotopes was announced. From this work has emerged a new area of technology that promises to further man's knowledge of physical and chemical processes, to assist in developing new products, and to improve current fabrication techniques.

Laser principles

The laser is an instrument that produces very intense beams of highly collimated, monochromatic, and coherent light. Collimated light propagates in a straight line with little divergence; monochromatic light is made up of a pure color or single frequency; and coherent light is transmitted with all components of its electromagnetic field exactly in phase. For laser isotope separation, the most important of these properties is monochromaticity, because it enables the laser to excite one isotopic species of matter without affecting the others.

The term laser is an acronym for light amplification by stimulated emission of radiation. Though the process of light amplification by atoms or molecules is central to laser operations, the related process of light absorption is observed much more commonly in nature. The ozone in the Earth's atmosphere, for example, absorbs sunlight of certain frequencies, protecting life from destructive radiation. A beam of light can be thought to propagate as myriads of particle-like packets of energy, called photons, whose energy content is directly related to the frequency of the light. Atoms and molecules will absorb photons and become excited from a low energy level or state to a higher one, but they can accept only those photons whose energy equals the energy difference of the levels in question; *i.e.*, the photon energy must equal exactly the energy needed to raise the system from one level to the other. Thus, only light of a specific frequency will be absorbed. The absorption will persist as long as there are more atoms or molecules in the lower energy level than in the higher one. However, if a method can be devised to pump or excite a material so strongly that there are more atoms or molecules in the excited state than in the ground state, then the material will no longer absorb. Instead, some of its excited atoms or molecules will drop spontaneously to lower en-

REED J. JENSEN *is Assistant Laser Division Leader and Alternate Director and* ***C. PAUL ROBINSON*** *is Associate Laser Division Leader and Director of the Laser Isotope Separation Program. The research reported on in this article was performed at the Los Alamos (New Mexico) Scientific Laboratory under the auspices of the U.S. Energy Research and Development Administration.*

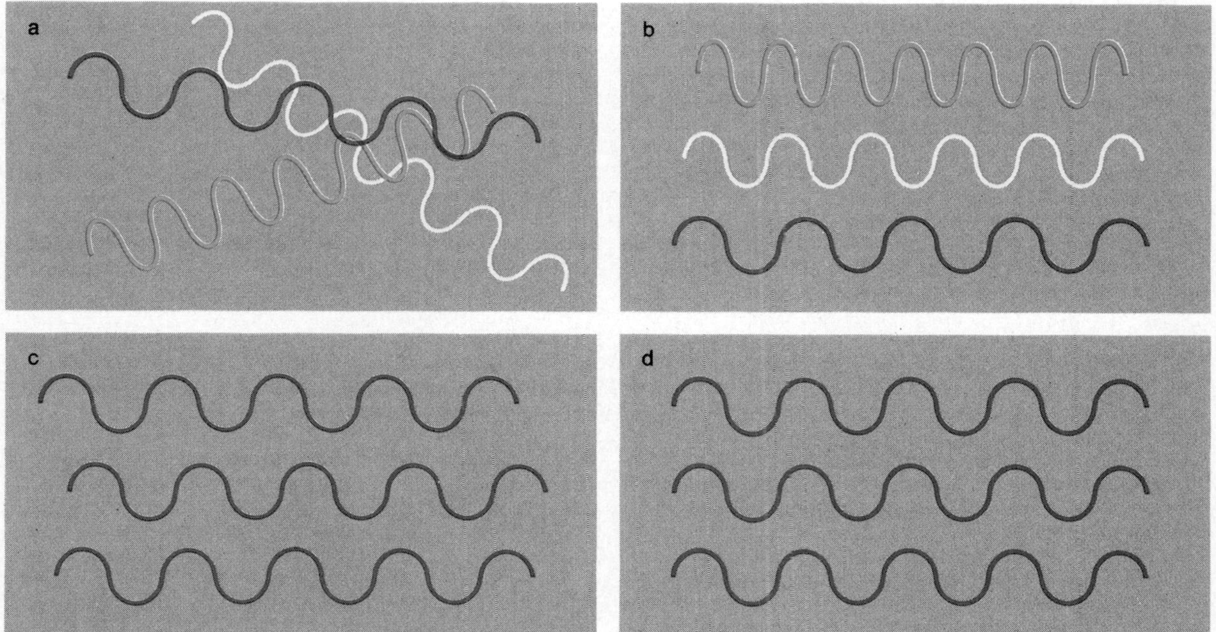

A conventional light source provides beams that are uncollimated, polychromatic, and incoherent; i.e., *their component waves tend to propagate in different directions, comprise a mixture of frequencies, and emerge from their source out of step or phase with each other (a). By contrast, the laser yields intense beams of light waves that not only travel in the same direction (b) but also are of the same frequency or color (c) and are exactly in phase (d), with the corresponding components of their electromagnetic fields aligned.*

ergy levels and emit photons of the frequency associated with their excitation energy. In turn, these photons will strike other atoms or molecules in the excited state, stimulating the emission of additional photons. The effect is aided by the use of end mirrors that repeatedly reflect the beam back and forth through the medium, allowing the beam to become highly amplified. One laser mirror is made partially transmitting to permit the beam to exit from the laser. Both mirrors are curved slightly to provide an optical cavity for the radiation.

Certain molecules with very dense groupings of energy levels are known to produce laser emission over a wide and essentially continuous range of frequencies. Some organic dyes with these characteristics have been employed in the development of tunable dye lasers. In these devices, the end mirrors are replaced by selective reflectors that can be adjusted or tuned to reflect and build up only the desired frequency. Laser tunability is important to the field of laser isotope separation, and schemes other than dye lasers have been devised to provide this tunability, including frequency mixing, optical parametric oscillation, and Raman downshifting.

Isotopes

The atom is structured around a tiny but massive central nucleus composed of a collection of protons and neutrons. Lightweight electrons move in orbits about this nucleus, and every neutral atom possesses just as many electrons as it does protons. Because atoms interact chemically through their electrons, chemical identity is determined by the number of protons. For example, an atom with six protons and six orbiting electrons exhibits chemical characteristics peculiar to the element carbon. For a given number of protons, atomic nuclei with

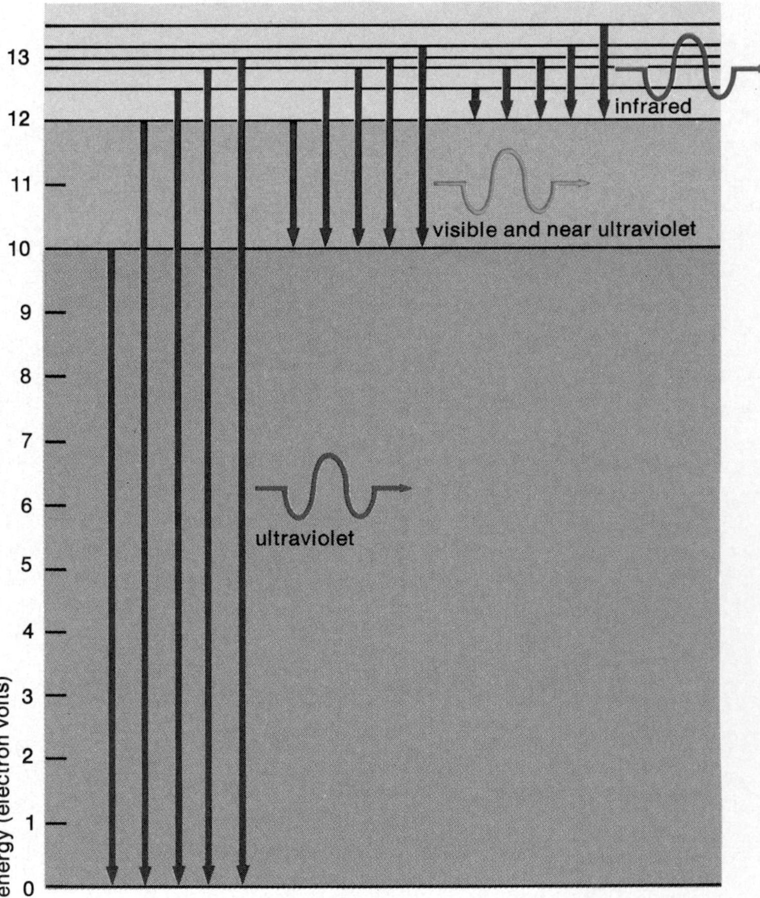

Light can emanate from atoms undergoing a process of energy absorption and emission. This process is governed by rules that allow stability to the atom only at certain quantized energy levels. Energy levels for an atom of hydrogen are depicted, with the energy scale in electron volts. In order to move either from its ground (unexcited) state or from a lower energy level to one of greater excitation, an atom must be struck by a particle whose energy is equal to or greater than that needed to make the transition between energy levels. If the incoming particle is a photon, its energy must exactly equal the transition energy. When an atom falls from a higher to a lower energy level, as indicated by the descending arrows, it emits a photon whose frequency is associated with the magnitude of the energy change involved. Hence, the greater the drop, the more energetic will be the photon emitted, and the higher will be its frequency.

differing numbers of neutrons are found to exist. Each variant form is called an isotope and is usually designated by a figure that represents the total number of protons and neutrons within a nucleus. Although chemically similar because of identical proton number, the isotopes of a given element have markedly different nuclear properties. For example, carbon-12 or ^{12}C, with six protons and six neutrons, is a very stable species; by contrast, carbon-14 or ^{14}C, with six protons and eight neutrons, is highly radioactive.

Most elements found on Earth are mixtures of several isotopes in almost unvarying proportions. Their specific abundances were determined by the primordial nuclear reactions that took place before and during the formation of the planet's crust.

Although the existence of isotopes had been known since the early 1900s, their importance in practical applications was demonstrated by the critical scientific work of World War II, which culminated in the development of the atomic bomb. The chain reaction required for the nearly instantaneous release of vast amounts of nuclear energy was found to take place for the isotope ^{235}U (92 protons, 143 neutrons), which constitutes less than 1% of the naturally occurring mixture of uranium isotopes, but not for the more abundant ^{238}U (92 protons, 146

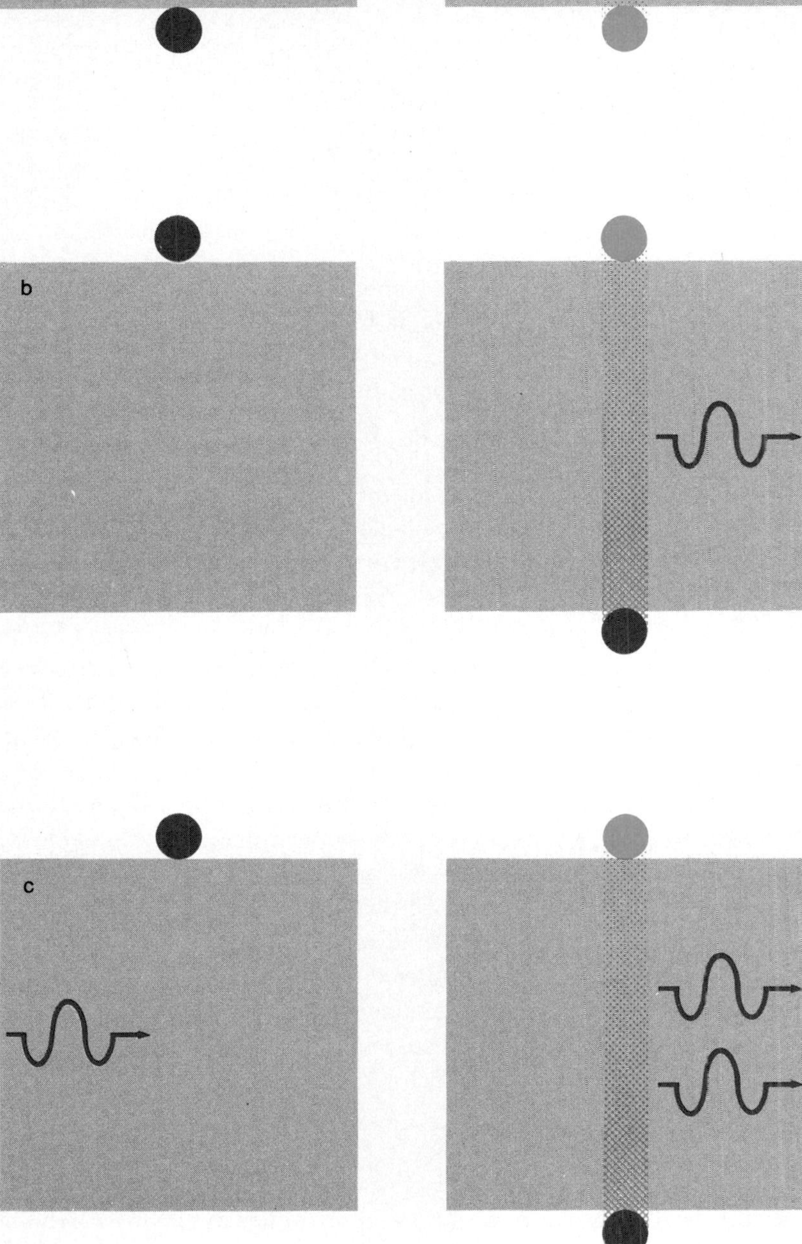

By a process known as spontaneous emission, an atom excited to a higher energy level by a photon (a) will eventually decay of its own accord (b), reemitting a photon as it does so. This emission can be stimulated to occur prematurely if the excited atom is struck by a photon of the same energy as the one about to be emitted (c). Following stimulated emission, there will be two photons of exactly the same frequency, both traveling in phase and in the same direction.

Adapted from "Lasers, Tools of Modern Technology" by Ronald Brown, © 1968, Doubleday & Co., Inc.

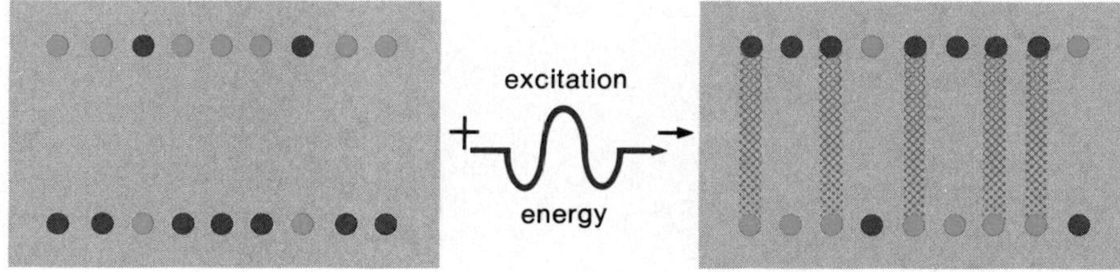

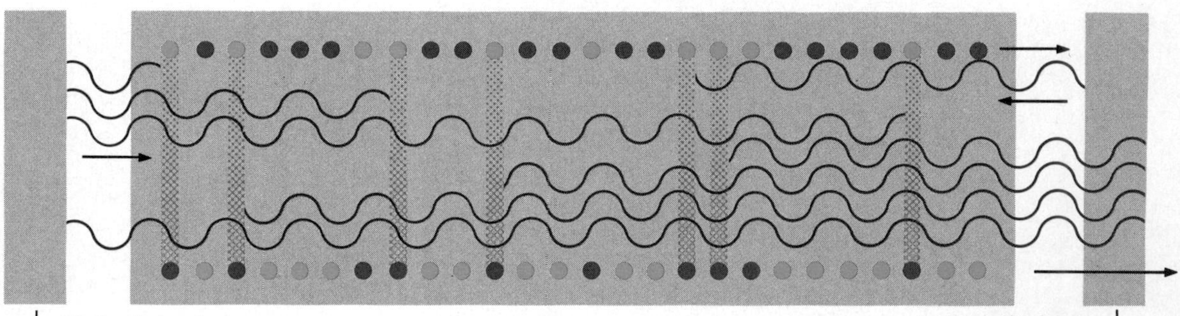

For large-scale stimulated emission or lasing action to occur, it is necessary to excite or pump the atoms of a suitable material so strongly that they will undergo a population inversion—an abnormal condition in which more atoms exist in the excited state than in the ground state (top). In such a situation, the spontaneous decay of a single atom to a lower energy level is sufficient to stimulate the decay of many others (bottom). Mirrors located at the ends of the lasing material will repeatedly reflect the beam back and forth, further enhancing the effect. If one mirror is only partially silvered, a fraction of the beam can emerge from the laser in useful form.

neutrons). Thus, before nuclear energy could be exploited, it was necessary to find a way to separate these isotopes so that a suitable mass of isotopically pure ^{235}U could be assembled.

Since the 1940s the utility of both radioactive and stable isotopes for a variety of practical applications has become widely recognized. Their uses continue to multiply as less expensive separation methods are developed. Radioactive isotopes find one of their most common applications in the fields of archaeology and geology. Assuming that the atmospheric ratio of radioactive ^{14}C isotope, which is produced continuously by cosmic rays, to the stable ^{12}C form has remained relatively constant in recent times, a plant or animal participating in the carbon-oxygen exchange cycle can be expected upon its death to fix a similar carbon-isotope ratio within its tissues. Since the amount of ^{14}C decreases predictably with time by spontaneous radioactive decay, an analysis of the carbon-isotope content of an archaeological sample, such as a bone fragment, makes it possible to assign an age to the sample and, by inference, to the rock formation in which it was found. Nonradioactive isotopic tracers also are becoming increasingly useful in medical and biological studies. Traceable fertilizers, for example, can be prepared by substituting nitrogen-15 for the normal nitrogen used in fertilizer production. After this fertilizer is applied to a test crop, samples of the plants are analyzed for the ^{15}N content in fruit, leaves, stalks, and roots. Such studies make it possible to formulate ideal chemical fertilizer mixtures for specific crops growing in diverse soils and climates.

It is as sources of nuclear energy, of course, that purified isotopes have achieved their greatest importance. Although the major world powers have already secured sufficient quantities of enriched uranium for their nuclear arsenals, the energy crisis, with its emphasis on inexpensive energy sources, has rekindled interest in enriching uranium for fuel in power plants. Such plants currently produce only 8% of U.S. electricity and 5% worldwide. As fossil fuels become scarcer, however, more extensive use of nuclear energy seems inevitable.

Scope of uranium enrichment

Any attempt to alter the naturally occurring proportions of isotopes has always been hampered by the similarity of their chemical properties. Although techniques that exploit a wide variety of approaches have been developed, most are so expensive that widespread application is prohibited. Most of the enriched uranium currently stockpiled or in use has been produced by the process of gaseous diffusion. Crushed uranium ore is treated by a process that converts the metal to its hexafluoride salt (UF_6). In gaseous form the compound is allowed to flow through barriers of a special porous membrane. Gas molecules containing the ^{235}U isotope diffuse more rapidly through the barrier than the heavier ^{238}U, resulting in a slight separation of the isotopes. This process must be repeated many thousands of times to produce useful enrichment. Diffusion plants are very large, covering almost 30 acres, and the capital cost of a typical plant is $2.9 billion. They require approximately 2,500 megawatts (Mw) of electrical power to operate pumps that drive the gas through the barriers.

A second process, using gas centrifuges, has been under development in several countries for about 15 years. As uranium hexafluoride is whirled in circles at high speeds, a heavier fraction of the gas, richer in ^{238}U, tends to collect near the outer walls of the centrifuge, where it can be drawn off. Although centrifuge plants are expected to have about the same capital plant costs as diffusion plants, they should require only about one-tenth as much electrical power. Because data are needed on the lifetimes of centrifuge equipment, it will be several years before suitable plants can be built.

Although much development will be needed before the economics of enrichment plants using lasers can be evaluated, theoretical predictions indicate that such plants could operate at less than one-tenth of diffusion plant costs and need even less electrical energy than centrifuge plants. The main reason for lower laser-enrichment costs centers on the probable use of only one or two highly selective stages, in contrast to the prolonged recycling inherent in current methods.

Laser-induced chemistry

Chemical science can be divided into two important branches: structure and dynamics. Early progress in the elucidation of chemical structures led to the dramatic development of organic chemistry in the 19th century, and since that time over one million compounds have been characterized. However, efficient exploitation of chemistry for industrial and consumer products depends not only upon a sophisticated knowledge of structure but also upon the ability to induce a high rate of reaction or chemical change.

Courtesy, Los Alamos Scientific Laboratory

Tunability of organic dye lasers makes them valuable in isotope separation procedures, where selective excitation of a desired molecular species depends in great part upon the ability of the laser to emit a precise frequency.

The rate of chemical reactions is dominated by those reactant molecules that have the most energy. In a sample of gas, not all molecules have the same energy; instead, the energy is statistically distributed over the molecules according to a law known as the Boltzmann distribution. High-energy molecules react because energy is required to bring molecules sufficiently close together and into correct arrangement for the reaction to proceed. That a barrier usually exists is known from common experience; *e.g.,* it requires the extra heat or energy of a spark to start combustion in a mixture of gasoline and air.

For molecules, vibration is an important mode of energy storage. Every molecule is in some state of vibration; its atoms are in motion about an equilibrium point, much as if they were bound together by springs. The larger the amplitude of vibration, the greater is the stored energy. Under normal conditions vibrational energy in gases is usually imparted to or removed from a molecule by collisions with other molecules. This flow of vibrational energy proceeds in a statistical fashion depending on the relative velocities of the components of a given collision. In laser-induced reactions the average vibrational energy per molecule is controlled by a laser that is continuously imparting such energy. In many cases vibrational energy can control reaction rates by providing the energy necessary to overcome the reaction barrier. In a classic experiment, when hydrogen chloride (HCl) was irradiated in the presence of atoms of potassium (K) with a laser that could excite the molecule to its first vibrational level, its reaction rate with K atoms, compared with that for cold or unexcited HCl, increased 100 fold.

The frequency of molecular vibrations between two bonded atoms depends upon the mass of the bonded atoms and the rigidity of the bond according to the equation $\omega = \sqrt{k/\mu}$, in which ω is the vibrational frequency, k is the rigidity or force constant of the molecule, and μ is

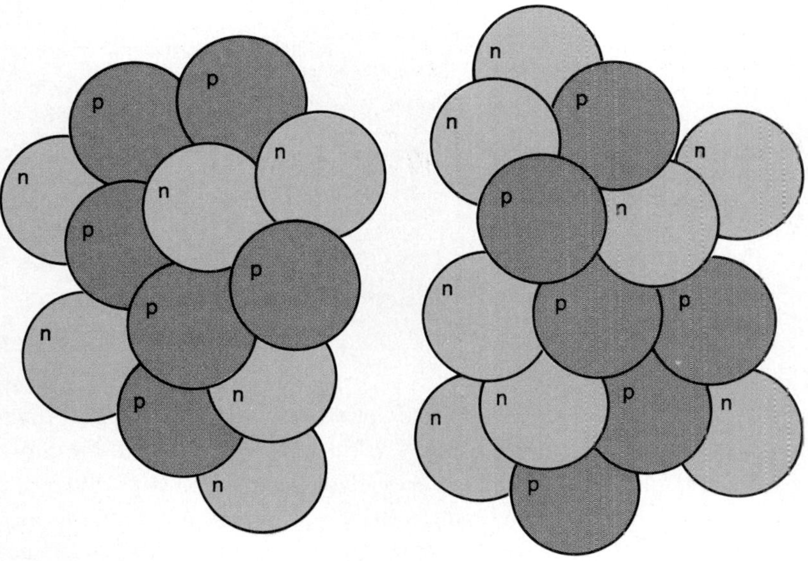

carbon-12 nucleus | carbon-14 nucleus

Although chemically similar because of identical proton number, the isotopes of carbon-12 and carbon-14 differ in the number of neutrons and thus possess markedly different nuclear properties.

a factor derived from the masses of the vibrating atoms. Thus, each chemical bond has its own specific vibrational frequency depending on the bond rigidity and the identity of the bound atoms. For example, HCl has a different vibrational frequency than either HBr (hydrogen bromide) or HF (hydrogen fluoride), and HCl can be excited selectively with a laser in the presence of these molecules. Furthermore, $H^{35}Cl$ (hydrogen chloride made from chlorine-35) has a different vibrational frequency than $H^{37}Cl$. Although this frequency difference is rather difficult to measure with classical methods, it is easily differentiated with laser techniques, making it clearly possible to excite selectively only one isotopic chemical species with a carefully tuned laser while leaving the other one unaffected.

Many important demonstrations of laser-induced chemistry have been realized. In one case, intense bursts of light from a CO_2 (carbon dioxide) laser were used to dissociate the N_2F_4 (tetrafluorohydrazine) molecule into NF_2 fragments, a reaction that proceeded more rapidly than had been achieved by any other method. In a second case, irradiation with a CO_2 laser caused the molecule B_2H_6 (diborane) to dissociate, evolve hydrogen, and reassemble in the complex molecular form $B_{20}H_{16}$. In another example, an organic molecule was rearranged by CO_2-laser irradiation via the reaction

```
CH3          H        CH3          CH3
   \        /            \        /
    C  ==  C    ---->     C  ==  C      + decomposition products
   /        \            /        \
  H          CH3        H          H
  trans-2-butene        cis-2-butene
```

$^{8}_{5}B$ 0.77 second	$^{9}_{5}B$ 8×10^{-19} second	$^{10}_{5}B$ stable	$^{11}_{5}B$ stable	$^{12}_{5}B$ 0.02 second	$^{13}_{5}B$ 0.019 second					
		$^{10}_{6}C$ 19.45 seconds	$^{11}_{6}C$ 20.3 minutes	$^{12}_{6}C$ stable	$^{13}_{6}C$ stable	$^{14}_{6}C$ 5,730 years	$^{15}_{6}C$ 2.4 seconds	$^{16}_{6}C$ 0.74 second		
				$^{12}_{7}N$ 0.011 second	$^{13}_{7}N$ 9.97 minutes	$^{14}_{7}N$ stable	$^{15}_{7}N$ stable	$^{16}_{7}N$ 7.2 seconds	$^{17}_{7}N$ 4.16 seconds	$^{18}_{7}N$ 0.63 second

mass number → 14 · proton number → 6 · half-life → 5,730 years ($^{14}_{6}C$ 5,730 years)

Stable and radioactive isotopes of boron, carbon, and nitrogen, with naturally occurring isotopes shown shaded. The chemical identity of each element is established by its proton number; hence, any attempt to alter the naturally occurring proportions of isotopes by chemical means is difficult. The mass number, i.e., the number of protons and neutrons in the nucleus, varies for each isotope of a given element, reflecting the difference in the number of neutrons that gives each isotope its distinctive nuclear properties. An important isotopic property is its nuclear stability, one measure of which is its half-life, the time required for half of the nuclei of an isotopic sample to decay into another nuclear species.

The CO_2 laser was selective in irradiating only the *trans* form. When some *cis* form was present initially, it remained unaffected. An interesting example of selective synthesis involved BCl_3 (boron trichloride) and hydrogen gases in the reaction $BCl_3 + H_2 \rightarrow HBCl_2 + HCl$. For this experiment, a CO_2 laser was used to deposit energy in the BCl_3 molecule, allowing the formation of products that would not have been evolved had the energy been supplied in the form of heat.

In each of these reactions, the rate of the reaction was accelerated by supplying, with a laser, the energy necessary to surmount the reaction barrier. Each reaction was found to be very sensitive to laser irradiation; it took much less energy to kindle the reaction with a laser than it did by thermal means.

Such results have far-reaching implications. In extending this powerful technique to even more detailed studies, it will be possible for the experimenter not only to control the rate of excitation of molecules but also to excite selected parts of complex molecules. For example, in a molecule such as

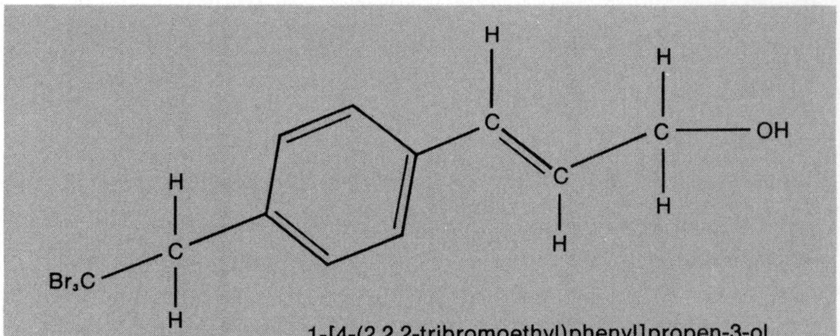

1-[4-(2,2,2-tribromoethyl)phenyl]propen-3-ol

it would be possible to deposit energy selectively in the OH group by irradiating it with a high-powered HF laser and to study the flow of vibrational energy from that point. If the molecule is irradiated very intensely, it will break at some point. The rupture point would give important clues on energy flow within the molecule.

Prior to the development of tunable lasers, studies were necessarily limited to those molecules that absorb in the frequency region of the few high-powered, single-frequency lasers then in existence. Recent laser technology, however, has made it possible to reach almost any

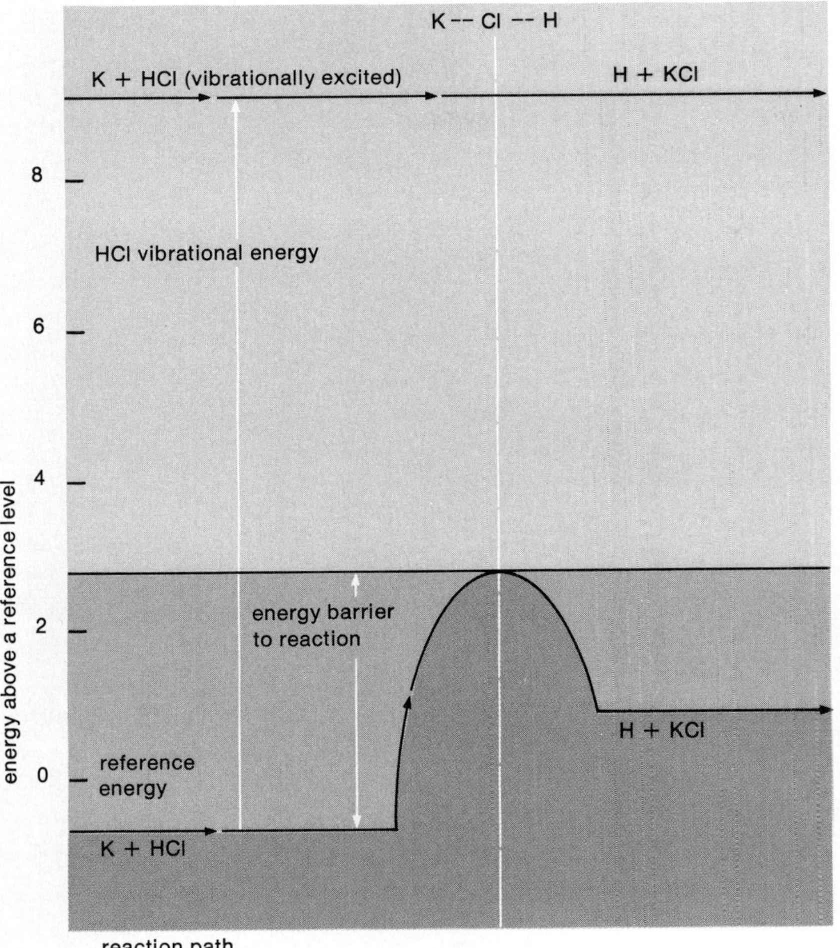

In many cases the vibrational energy of a molecule can exert control over its rate of reaction with a second species. This fact can be illustrated by an examination of two possible reaction paths of the simple reaction HCl + K→H + KCl. In the lower left corner of the figure, just below zero reference energy, is plotted the total chemical energy of the two unreacted species. For the reaction to proceed along the lower path, the energy barrier opposing the reaction must be overcome by providing extra mechanical energy to force a potassium atom sufficiently near a molecule of hydrogen chloride to form a new bond and break the old one. However, if a molecule of hydrogen chloride can be excited by laser light to its first vibrational level in the presence of potassium atoms, it will be provided with energy that far exceeds the amount necessary to overcome the reaction barrier. The reaction will then proceed along the upper path at a greatly accelerated rate.

absorption region and, thus, to select molecules without regard to the location of their absorption features. The Laser Applied Spectroscopy Laboratory, incorporating a versatile, high-power, tunable laser system, was established recently at the University of California's Los Alamos Scientific Laboratory in New Mexico to allow a variety of chemical processes to be studied at very high resolution and in detail formerly not possible.

Laser isotope separation

The high selectivity of laser-induced chemistry not only allows reactions to be driven to specific products but also allows the selection of isotopic species of reactants from a natural mixture. It is possible, therefore, to separate isotopes by selectively inducing one isotopic species to convert to a new chemical product that is easily separable from the original material. To cite a recent dramatic example of laser-induced isotope separation, isotopically selective vibrational excitation of HCl was used to cause the reaction $H^{35}Cl + Br \rightarrow HBr + {}^{35}Cl$ to proceed selectively for $H^{35}Cl$. The ^{35}Cl was then easily recovered as $^{35}ClBr$. In another experiment, both U.S. and Soviet workers

Courtesy, Lawrence Livermore Laboratory, University of California

Yellow crystals of isotopically enriched uranyl nitrate attest to the success of laser enrichment technique developed at the Lawrence Livermore Laboratory of the University of California. In one two-hour run, an experimental apparatus produced four milligrams of uranium enriched from its natural abundance of 0.7% uranium-235 to 3%, a concentration necessary for use in most U.S. nuclear reactors.

succeeded in separating isotopes of sulfur by selectively inducing the reaction $^{32}SF_6 + H_2 \rightarrow 2HF + {}^{32}SF_4$. In addition to chlorine and sulfur, isotopes of boron, hydrogen, carbon, oxygen, bromine, calcium, and nitrogen were separated successfully by laser methods.

There are three basic steps in laser isotope separation. First, a light frequency must be found that will excite only the desired isotope. Second, a physical or chemical change must be produced from the excited state. In general, this change must be as permanent as possible and must occur as rapidly as possible. Third, the changed species must be separated from the original isotopic mixture.

In separation experiments it is necessary to avoid undesirable processes, and, even with successful selective laser excitation, there are many chances to lose the selectivity. Much of the challenge of laser isotope separation, therefore, is in the development of techniques that suppress all but the desired reactions. A wide variety of physical phenomena have been exploited for laser isotope separation. These range from laser-enhanced chemical reactions to deflection of a molecular beam by the pressure of laser light as it is absorbed selectively. Although many processes are scientifically feasible, very few are useful for industrial application.

Uranium enrichment by laser

Many of the techniques being evaluated for separation of isotopes of light weight are of interest for uranium enrichment as well, although special consideration of uranium chemistry complicates the situation. An Avco-Exxon Corporation joint venture pioneered a laser-induced enrichment technique that uses atomic uranium. To produce a useful density of uranium vapor, ovens operating at temperatures of 2,500° K (2,200° C) are required. A two-step process is used: one laser selectively excites ^{235}U; a second laser ionizes the selected atoms. The electrically charged ions are deflected to a collector by electric and magnetic fields. Scientists at Avco-Exxon and at the Lawrence Livermore Laboratory of the University of California have produced measurable quantities of enriched uranium using this technique.

Laser-enrichment methods utilizing chemical properties of such uranium compounds as UF_6 are being developed at the Los Alamos Scientific Laboratory. This class of processes is intrinsically capable of high production rates from small systems. Although this technology appears to promise much lower costs for uranium enrichment, information that might simplify the production of enriched uranium for weapons is being controlled by the U.S. and other governments.

Unlike existing techniques that extract only about half of the ^{235}U, laser enrichment has the potential to remove it all. Thus, laser enrichment plants could nearly double the available fuel supply for nuclear energy. In implementing such plants, the development of powerful lasers capable of being tuned to very specific frequencies will pace the rate of progress. The research effort for these lasers is progressing in university, industrial, and government laboratories throughout the world, and goals have been established to build pilot plants using appropriate laser systems by the early 1980s.

Future applications

The high selectivity of the laser-induced methods described above should find important applications in the selective synthesis of chemical products that are difficult to prepare by existing thermal techniques. Many important pharmaceuticals and rare chemicals presently require tedious preparations. These may be simplified by substituting a laser-induced step for some of the low-yield, less specific processes.

Though waste material from nuclear reactors constitutes a potential source of such valuable elements as rhodium and plutonium, efficient recovery processes have yet to be developed. The successful application of lasers to extract these elements from waste streams would provide technical industry with an important source of radioisotopes and other rare materials.

Number Theory: The Fibonacci Sequence

by Verner E. Hoggatt, Jr.

The multiplication of rabbits, the number of petals on a daisy, and the architecture of the ancient Greeks share an intriguing relationship with a highly unusual sequence of numbers.

Man's fascination with numbers is undoubtedly millions of years old. Perhaps the interest was kindled when some prehistoric hunter, crouching in a clump of brush on an African savanna, realized that there was an astonishing connection between the two zebras he was stalking, the two hawks circling overhead, and the two fire-hardened, pointed wooden sticks he clutched tightly against his body. He began to comprehend the concept of twoness, an abstract phantom that had no place in the reality of his universe outside of the objects in which it resided. Yet twoness went beyond objects; it remained in his thoughts even when he closed his eyes and forgot about zebras and hawks. It stood alone, a solid idea. It handled easily in the mind; it could be increased, diminished, transformed, and manipulated in a variety of ways. And twoness seemed to be only a small part of a set of abstractions like itself that stretched far beyond the ability of any man to visualize, an infinity of numbers that beckoned tantalizingly for exploration. Man stepped into that infinity and number theory was born.

Since that truly momentous chain of events in the unrecorded history of human civilization, man has forever delighted in playing with numbers and has posed countless problems for himself either as intellectual challenges or as simple amusement. Often what began as a curious question or an innocent puzzle served to open new roads of mathematical and scientific inquiry that occupied the attention of great intellects for years. In modern times, interest in the properties of integers or whole numbers has developed into a branch of higher mathematics of immense complexity and abstruseness. Yet, many facets of number theory remain easily understandable and continue to hold great mystery.

Sequences of numbers

Consider, for example, some of the interesting relationships in the ordered set or sequence of numbers that must be encountered by anyone first learning the simple act of counting: 1, 2, 3, 4, 5, 6, ... Compared with a sequence of arbitrarily chosen numbers such as 10, 3, 1, 7, 2, 6, the sequence formed by counting exhibits a very definitive characteristic. One can easily discern a fixed rule of formation that determines what term is to follow a given term. In this sequence there is obviously a common difference between successive terms, a difference of +1. The sequence of integers 8, 6, 4, 2, 0, −2, . . . is likewise determined by a rule of formation. In this case each successive term is obtained by adding the common difference of −2 to the preceding term. All sequences such as these, whose successive terms vary by a common difference, are called arithmetic sequences or arithmetic progressions.

Next, consider the sequence of integers 4, 12, 36, 108, 324, ... There seems to be no common difference between these terms, but a brief examination will reveal that each term has been determined by multiplying the preceding one by 3. This common multiplier is called the common ratio because the ratio of each term to its predecessor is

VERNER E. HOGGATT, JR., *is Professor of Mathematics at San José State University and Editor of* The Fibonacci Quarterly.

Illustrations by Bill Biderbost

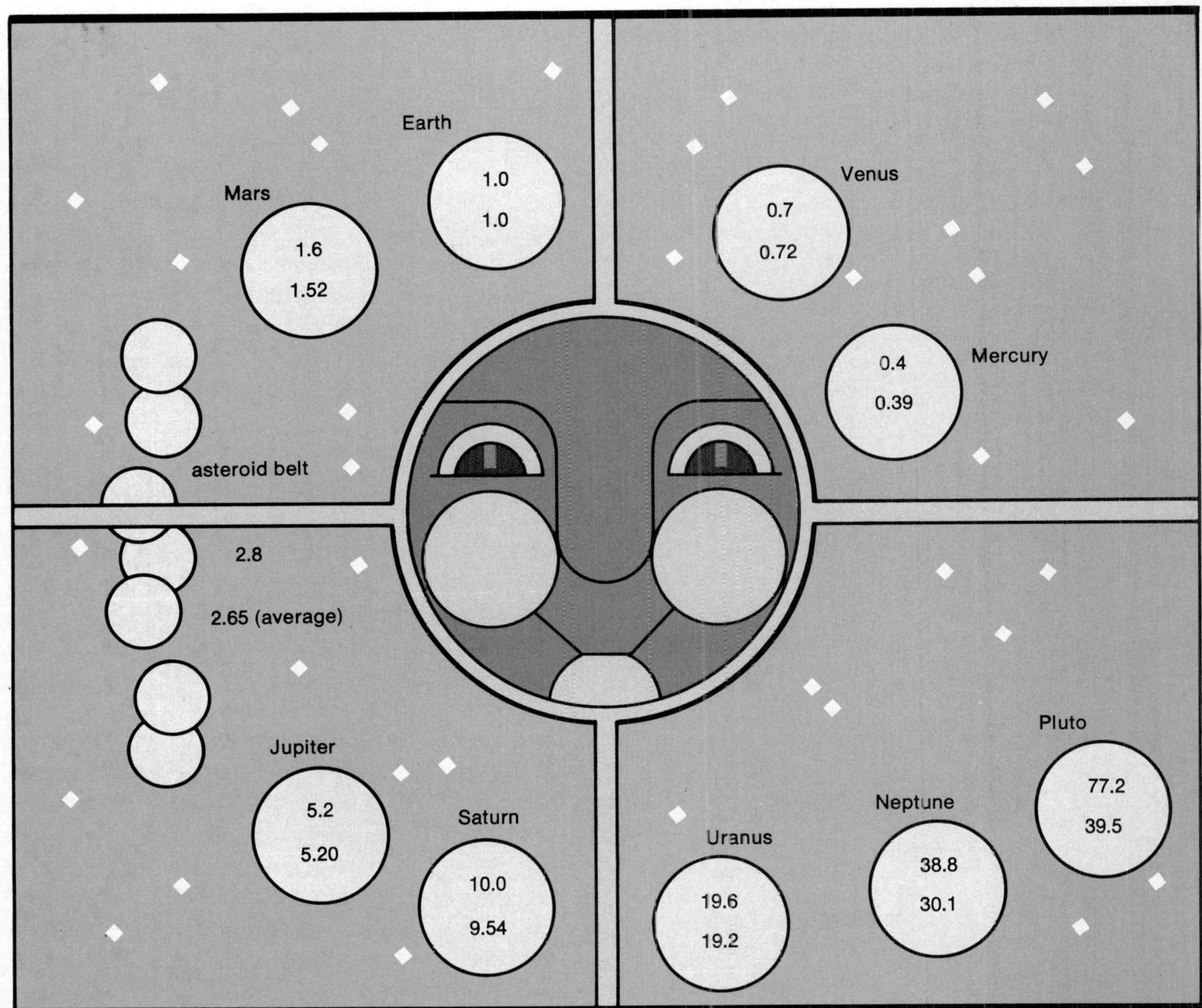

Figure 1. Mean distance from the Sun in astronomical units of the major bodies of the solar system as predicted by the Titius-Bode law (upper value) and as observed (lower value).

3 : 1. The sequence 25, −5, 1, $-\frac{1}{5}$ is also of this kind; its common ratio is easily seen to be $-\frac{1}{5}$. Ordered sets that vary from one term to the next by a common ratio are called geometric sequences or geometric progressions.

When possible, it is often useful to express sequences as general formulas. Because sequences are ordered sets, each has a first term, which can be represented by a symbol such as S_1, a second term S_2, a third term S_3, and so on. Under such a system, the natural numbers or counting numbers are used as subscripts to indicate the rank or order of the terms of the sequence. Although one could represent every term of any given sequence by symbols having numerical subscripts, for long and infinite sequences it is obviously convenient to represent any desired term, or the nth term, by a symbol such as S_n and to state a formula for finding that term. Hence, in the arithmetic sequence 8, 6 4, 2, 0, −2, . . . , S_n, . . . examined above, $S_1 = 8$, $S_2 = 6$, $S_3 = 4$, and any desired term $S_n = 8 - 2(n - 1)$. Likewise, any term of the geometric

Figure 2. A procedure known as the sieve of Eratosthenes can be employed to determine all the primes in the sequence of integers from 2 through a given number. First, every second integer of the sequence after 2 is removed, then every third integer after 3, then every fifth integer after 5, and so on. Though such a task seems laborious, it is actually quite efficient. For example, a determination of all the primes to the number 200 requires sieving only by the numbers 2, 3, 5, 7, 11, and 13.

sequence 4, 12, 36, 108, . . . , S_n, . . . can be expressed by the formula $S_n = 4(3^{n-1})$. Many sequences exist, however, that do not lend themselves to this method of formulation, and it is often necessary to resort to a second kind of formula called a recurrence formula, which expresses the term S_n with respect to one or more previous terms of the sequence. To use this method of expression, it is apparent that one must begin at some point in the sequence with one or more defined values; for arithmetic and geometric sequences, statement of any one term is sufficient. For example, to express the arithmetic sequence 8, 6, 4, 2, 0, −2, . . . , S_n, . . . as a recurrence formula, the first term S_1 is defined as 8; the second term $S_2 = S_1 - 2$, $S_3 = S_2 - 2$, and $S_n = S_{n-1} - 2$. For the geometric sequence 4, 12, 36, 108, . . . , S_n, . . . , its first term $S_1 = 4$ and $S_n = 3S_{n-1}$.

Sequences even in their simplest forms often have played key roles in the acquisition of scientific knowledge. Of much importance to the progress of astronomy in the 18th and 19th centuries were the numbers known collectively as the Titius-Bode law. Discovered in 1766 and popularized in 1772 by two German astronomers, the law was based upon the set of numbers beginning with 0 and continuing with the first eight terms of the geometric sequence 3, 6, 12, 24, 48, 96, 192, 384, . . . , S_n, ... When 4 was added to each term and the results divided by 10, the new sequence so obtained (*see* figure 1) corresponded empirically quite well with the observed mean distances from the Sun of the known principal planets of the solar system when considered in astronomical units; *i.e.*, in terms of the mean Sun-Earth distance, or about 93 million miles. The discovery of Uranus in 1781 with a mean distance from the Sun in close agreement with the prediction of the law stimulated the search for an undiscovered planet at a predicted 2.8 astronomical units, which culminated in the discovery of Ceres and the other members of the asteroid belt. Though the Titius-Bode law broke down after the discoveries of Neptune and Pluto and, as yet, has no solid theoretical foundation, many scientists cite evidence of other numerical relationships among members of the solar system and expect that eventually some sound physical reason for the Titius-Bode law will emerge.

Prime numbers

Any consideration of sequences of numbers would be incomplete without some discussion of one that has baffled and delighted mathematicians concerned with number theory at least since the time of Euclid in the 4th century B.C. Like the sequence generated by counting, this sequence can have as many terms as one wishes to add and begins 2, 3, 5, 7, 11, 13, 17, 19, ... Examination shows it to be neither an arithmetic nor a geometric sequence, yet there is something unique about its terms: Each is divisible only by itself and by ±1. For this reason, the terms of this sequence are called prime numbers.

Greek mathematicians contributed a great deal to man's knowledge of prime numbers. Euclid offered an elegant proof that there existed

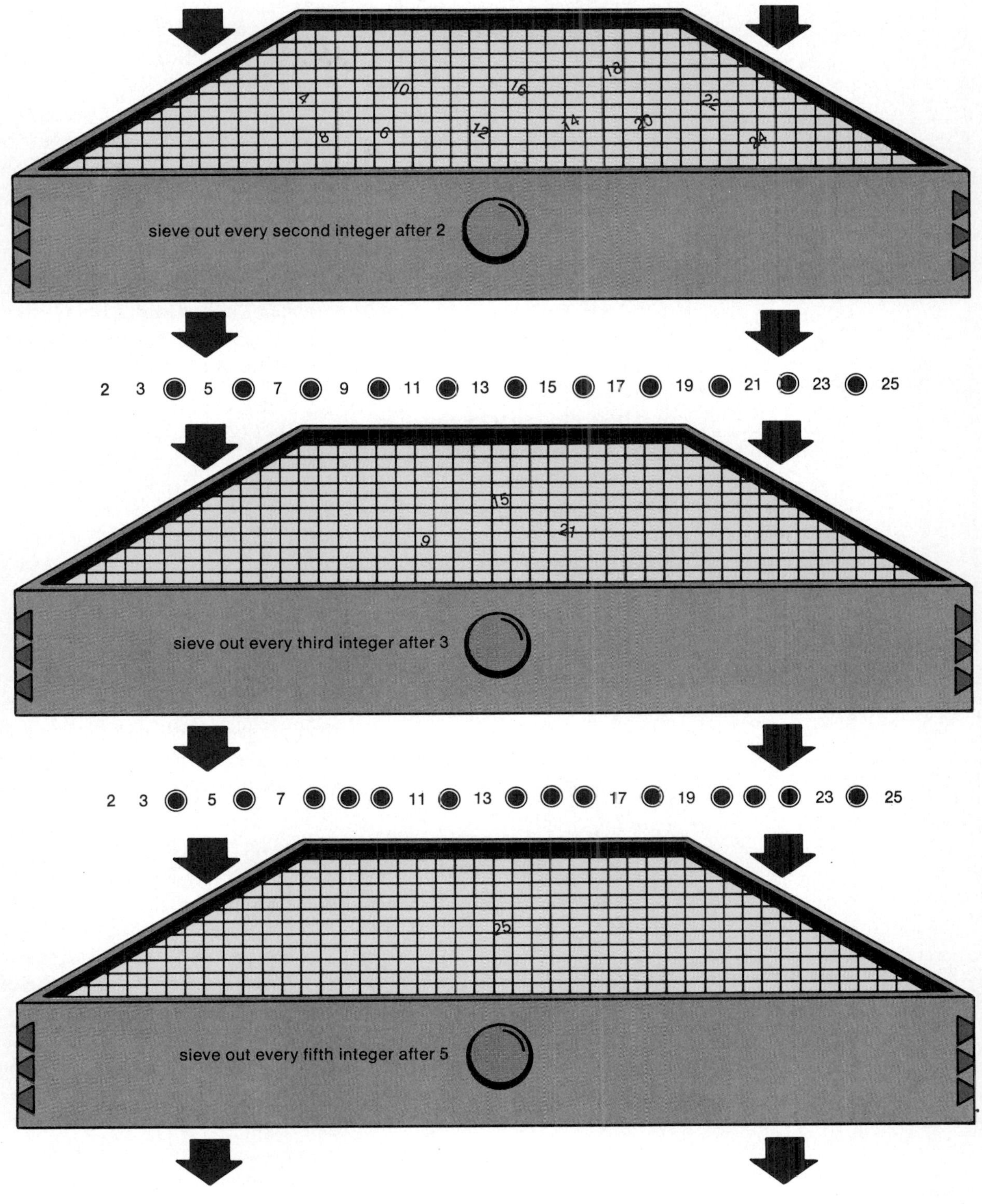
2 3 4 5 6 7 8 9 10 11 12 13 14 15 16 17 18 19 20 21 22 23 24 25
4
10
16
18
22
8
6
12
14
20
24
sieve out every second integer after 2
2 3 5 7 9 11 13 15 17 19 21 23 25
15
9
21
sieve out every third integer after 3
2 3 5 7 11 13 17 19 23 25
25
sieve out every fifth integer after 5

infinitely many primes. About 250 B.C. Eratosthenes of Cyrene, a contemporary of Archimedes, succeeded in developing a technique of sieving out all but the primes in the sequence of integers from 2 up to any given number without having to test them with divisors (*see* figure 2). Though this technique accomplished its goal, it certainly did not provide any insight into a formula for expressing a prime number in terms of preceding prime numbers. Since then, mathematicians have sought in vain for a proof that all or even part of the sequence of prime numbers can be expressed by a general formula. A number of formulas that seemed at first to yield only primes were later shown to be failures when some of the large unwieldy numbers they generated were proved to be composite—*i.e.*, factorable into two or more primes—by the sophisticated methods of later generations. Such discouraging findings eventually induced most number theorists to give up that particular line of attack on the primes.

The Fibonacci sequence

One of the most intriguing sequences of numbers to tickle the curiosity of man is called the Fibonacci sequence. This infinite ordered set of integers, 1, 1, 2, 3, 5, 8, 13, 21, 34, . . . , F_n, . . . , first became known in the early 13th century when the Italian mathematician Leonardo of Pisa, also called Fibonacci, published a recreational puzzle whose solution generated the sequence. That it follows a rule of formation is easily determined because each term, excluding the first two, can be found by adding the two terms immediately preceding it. Hence, any desired term of the sequence can be expressed by the recurrence formula $F_n = F_{n-1} + F_{n-2}$ when two starting values such as $F_1 = 1$ and $F_2 = 1$ are stated. This relationship between F_n and its preceding terms, however, does not need to be understood for the solution of Leonardo's problem and apparently went unrecognized by Leonardo and by subsequent mathematicians until about 1600.

Over the years, and especially since the mid-20th century, the properties of the Fibonacci numbers and a related sequence, the Lucas numbers ($L_n = L_{n-1} + L_{n-2}$ given $L_1 = 1$ and $L_2 = 3$), have been extensively studied, resulting in a large body of literature and the formation of an association to promote, correlate, and publish further research. Like the Titius-Bode law, Fibonacci numbers seem to have strong correspondences with many natural phenomena. They appear in unusual and unexpected places, ranging from the structure and life processes of organisms to the classical theories of aesthetics, and continue to be a dynamic and exciting field of inquiry, raising many questions with each newly discovered relationship.

In 1202 Leonardo, who was probably the most distinguished mathematician of the Middle Ages, published an important mathematical work, *Liber abaci*. Though it performed an invaluable service by spreading throughout Europe the use of Hindu-Arabic numerals in arithmetical operations, the book is best remembered for its presentation of the following puzzle:

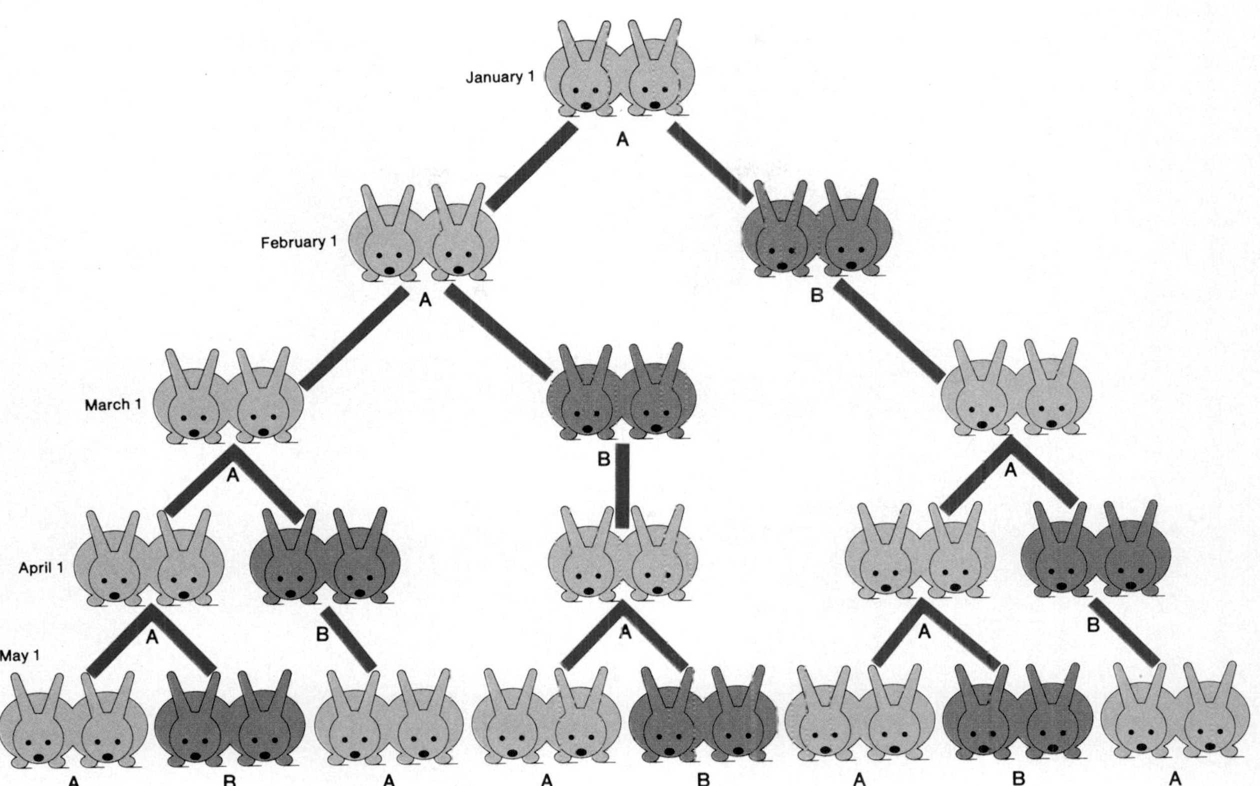

A certain man put a pair of rabbits in a place enclosed on all sides by a wall. How many pairs of rabbits will be born there in the course of one year if it is assumed that every month a pair of rabbits produces another pair, and that all new rabbits begin to bear young two months after their own birth?

Figure 3 (top and bottom). The solution to Leonardo of Pisa's mathematical problem concerning the multiplication of rabbits (see text).

To solve the puzzle one may look at the situation in this systematic fashion: (1) There is one pair of rabbits in an enclosure on January 1. (2) This pair will produce another pair on February 1 and on the first day of every month thereafter. (3) Each new pair will mature for one month and then produce a new pair on the first day of the third month of its life and on the first day of every month thereafter. A chart, as in figure 3, can be made to keep an account. Let A represent an adult pair of rabbits and B a baby pair. On January 1 there is only an A; on February 1 there is the original A and a baby pair B; on March 1 there is the original A, a new B produced from the original A, and the B that was born on February 1 and has matured into an adult pair A.

If this process is continued until the next January 1, the chart will be completed as shown. From the final number in the last column of the completed chart, which represents both the original pair and all pairs born during the course of a year, it can be shown that the answer to Leonardo's problem is 376 pairs. The solution itself, however, is incidental. The columns of numbers that were generated in arriving at the solution hold the real interest because, aside from the 0 in the second column (actually the F_0 term) and the absence of 1 in the third column, each of the three columns comprises the first dozen terms or so of the Fibonacci sequence: 1, 1, 2, 3, 5, 8, 13, . . . , F_n, ...

date	number of A's	number of B's	total number of pairs
January 1	1	0	1
February 1	1	1	2
March 1	2	1	3
April 1	3	2	5
May 1	5	3	8
June 1	8	5	13
July 1	13	8	21
August 1	21	13	34
September 1	34	21	55
October 1	55	34	89
November 1	89	55	144
December 1	144	89	233
January 1	233	144	377

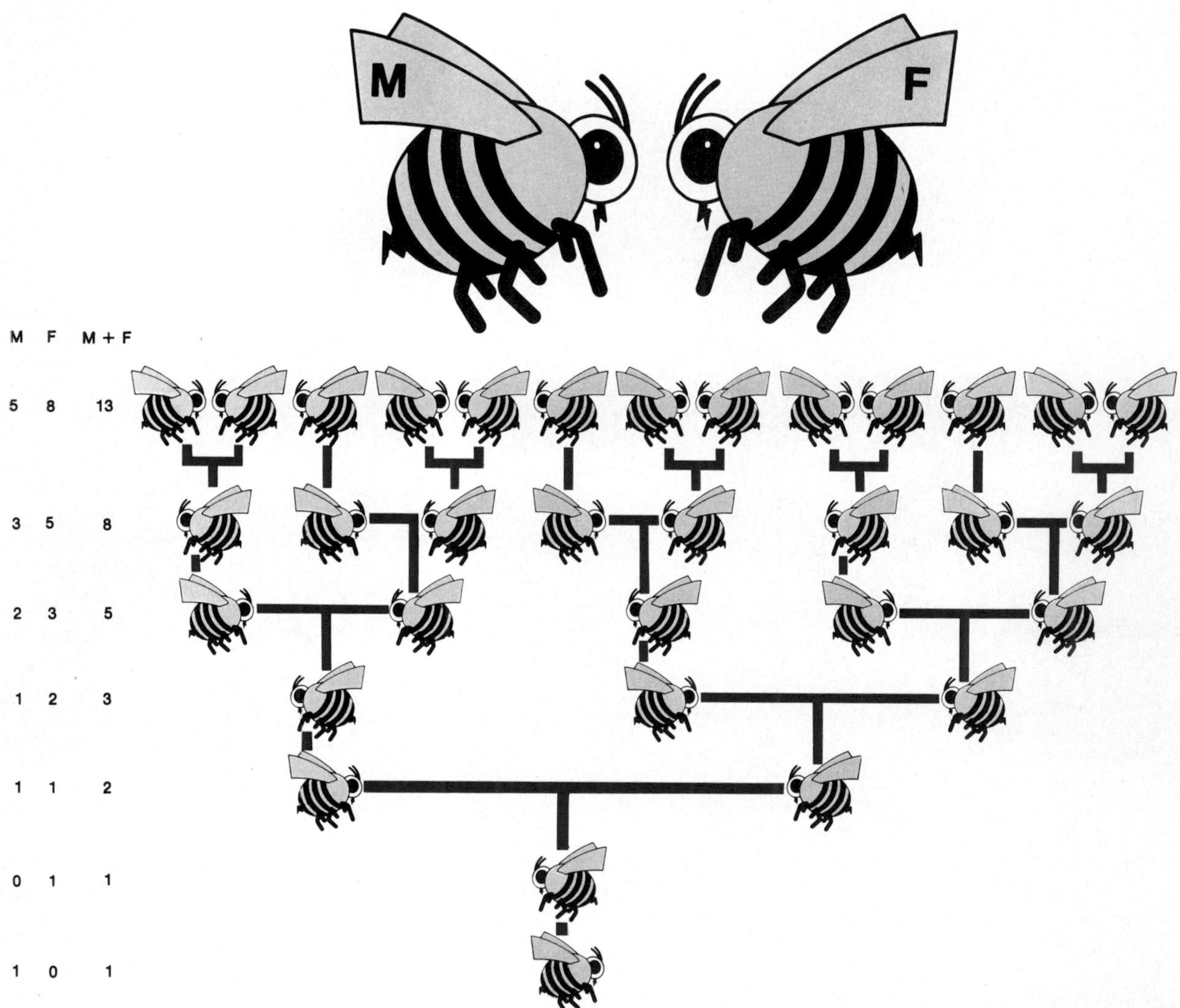

Figure 4 (above). The Fibonacci numbers as manifested in the genealogy of male bees (see text). Figure 5 (opposite, top). Petal counts of several composite flowers and of flowers with true petals. Figure 6 (opposite, bottom). The seed-bearing bracts of a pinecone grow outward in a spiral pattern from the point of the cone's attachment to the branch. In the example depicted, the spirals curving left or counterclockwise total 13, a Fibonacci number, and the spirals curving right total 8, the adjacent smaller Fibonacci number.

The genealogy of male bees

While it is true that Leonardo's rabbit problem is a manufactured situation, one of the most striking features of the Fibonacci numbers is their repeated manifestation in nature. One such example can be developed by tracing the genealogy or natural descent of the male honeybee (*Apis mellifera*). In the honeybee society, the female of the species is dominant; all fertilized bee eggs develop into females, who either take their places among the worker caste or, with special dietary treatment, become queens. The male bee or drone develops from an unfertilized egg, a method of reproduction common among lower forms of life; as such, the male bee has no father.

Given this information, one can construct a genealogical chart or family tree, tracing the ancestry of a male bee to any desired past generation, as shown in figure 4. Each M represents a male bee, which has only a female parent. Each F represents a queen, a female who necessarily arose from both a male and a female parent. As the chart is built up and numbers of ancestors are tallied, it becomes apparent that the male bee has a Fibonacci number of male, female, and total ancestors at each generation in time.

Seed patterns and petal counts

The Fibonacci numbers make another and perhaps more mysterious appearance in the seed arrangement of many species of plants and in the petal counts of various flowers. Asteraceae, the composite family of flowering plants, comprises about 930 genera and 20,000 species, which are distributed throughout the world. The flowering heads of members of the family have central disks composed of many small flowers, called florets, which are ordered on the disks in regular patterns. One member, the sunflower (genus *Helianthus*), grows to 12 feet and may develop heads over a foot in diameter. When the disk florets of the sunflower mature to seeds, their spiral arrangement in the head becomes clearly visible. A typical count of these spirals may give 89 steeply curving to the right, 55 more shallowly to the left, and 34 again shallowly to the right. The largest known specimen to be examined had spiral counts of 144 right, 89 left, and 55 right. The significance of these two sets of numbers is that they are consecutive terms of the Fibonacci sequence. In the disk florets of some species of daisy, there are often 34 right spirals and 21 left spirals, again consecutive Fibonacci numbers. Such patterns are not an inviolable law of nature but, nevertheless, are curiously persistent.

On the heads of composite flowers, the structures commonly called petals that surround the central disk are actually strap-shaped florets. For each head, a count of these florets will consistently yield a Fibonacci number or at least one very close to a Fibonacci number. Hence, a desperate lover, trying to divine the sympathies of his or her beloved by playing "loves me—loves me not" on the head of a daisy, very likely will pluck 21, 34, 55, or 89 petals before an answer is forthcoming. In addition to members of Asteraceae, a number of noncomposite flowering plants, with true petals, also exhibit Fibonacci numbers in their petal counts. A table giving petal counts of a selection of both types is shown in figure 5.

petal count	composite family	other plants
2		enchanter's nightshade
3		lily and iris
5	wall lettuce	buttercup
		wild rose
		larkspur
		columbine
8	squalid senecio	delphiniums
	field senecio	bloodroot
	cosmos	
13	ragwort	double delphiniums
	corn marigold	
	mayweed	
	cineraria	
21	heleniums	
	aster	
	black-eyed susan	
	chicory	
	hawkbits	
34	field daisies	
	plantain	
	pyrethrums	
55	african daisies	
	heleniums	
	Michaelmas daisies	
89	Michaelmas daisies	
233		double peony
377		ice plant

Pinecones and pineapples

Bracts, the tablike or scalelike structures that make up the main body of a pinecone, are considered to be modified leaves that have been crowded together on a very short stem. Like the seeds in sunflower heads, bracts form sets of spirals that emanate from the point of the cone's attachment to the branch. Two sets of oppositely directed spirals can be observed, one steep and the other more gradual. A count of the steep spirals will reveal a Fibonacci number and a count of the gradual ones often will be the adjacent smaller Fibonacci number or, if not, then the next smaller Fibonacci number (*see* figure 6). Some idea of the pervasive occurrence of these numbers is apparent in the results of one investigation of 4,290 cones from ten species of pine found in California, from which it was determined that only 74 cones or 1.7% deviated from the Fibonacci pattern. Similar Fibonacci patterns have been found in the cones of other conifers such as coast redwood, mountain redwood, Douglas fir, and mountain hemlock.

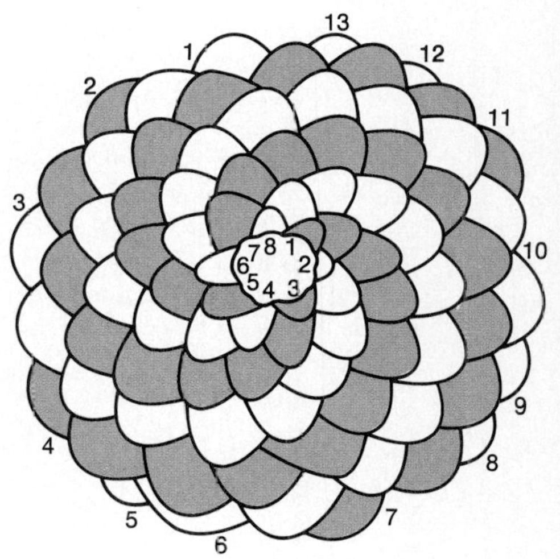

From "Fibonacci and Lucas Numbers," Verner E. Hoggatt, Jr., © 1968, Houghton-Mifflin Co., Boston

In the representation of the pineapple above (top), scales are numbered in an order based on the nearness of each scale along the major axis of the pineapple to its leafy crown. Three sets of spirals are evident; for instance, beginning with scale 0, one shallow spiral ascends along scales 0, 5, 10, . . . ; the second ascends more steeply along scales 0, 13, 26, . . . ; and the third ascends in the opposite direction along scales 0, 8, 16, ... If the surface of the pineapple is laid out as a plane (bottom), three spirals of numbered scales can be seen emerging from each scale at the base of the pineapple. These form arithmetic sequences with common differences of 5, 8, and 13, three consecutive Fibonacci numbers.

The pineapple offers another example for consideration. Like pinecone bracts, pineapple scales are patterned into spirals, and because they are roughly hexagonal in shape, three distinct sets of spirals can be counted. One examination of 2,000 pineapples yielded no deviations from the Fibonacci pattern; every set of spirals on every pineapple totaled to a Fibonacci number.

Phyllotaxy

Phyllotaxy, the arrangement of leaves on the stem or branch of a plant, is a botanical phenomenon whose mathematical precision first caught the attention of the famous 17th-century astronomer Johannes Kepler. Recent reexamination has proved it to be yet another natural residence for numbers of the Fibonacci sequence. Consider the section of stem from an elm tree, illustrated in figure 7. With leaf 0 as a starting point, all leaves are counted up the stem until a leaf is reached that is directly above the starting point. For elm, that number virtually without exception will be found to be 2, a Fibonacci number. In addition, to reach the point directly above the starting point takes one turn around the stem; 1 is also a Fibonacci number. These phyllotactic numbers are often expressed as a ratio of the number of turns to the number of leaves. Thus, for elm this ratio is 1/2. By contrast, to find a leaf on the stem of a cherry tree that is directly above the starting point, one must count 5 leaves and make 2 turns around the stem, for a ratio of 2/5. Performing the same operation on the stem of a pear tree gives a ratio of 3/8.

Figure 7 also provides a table illustrating the phyllotactic ratios of several plants. In every case, the ratio is made up of Fibonacci numbers. Obviously, the information in this table is far from exhaustive; the examination of an estimated 250,000 species of flowering seed-producing plants and 700 nonflowering ones poses a formidable task. To date, however, cumulative results indicate a Fibonacci phyllotactic ratio in an overwhelming number of species. Non-Fibonacci ratios are rarely, if ever, seen except where damage or abnormal growth has modified the leaf arrangement.

Fibonacci numbers and divine proportions

Thus far, this discussion of Fibonacci numbers has touched upon a few diverse curiosities among the lower forms of life, entities apparently governed by some common, well-hidden mathematical principle, but by one still far removed from any direct connection with the thought, culture, scientific progress, motivations, and high aspirations of a rational, thinking creature such as man. What if, however, one were to suspect that the Fibonacci numbers in some fashion also had succeeded in cutting to the very core of a definitively human characteristic—aesthetic judgment? What if, at least several millennia in the past, some highly creative civilization felt a tug on its artistic sensibilities that was subtly connected with that unique sequence of integers beginning 1, 1, 2, 3, 5, 8, 13, . . .?

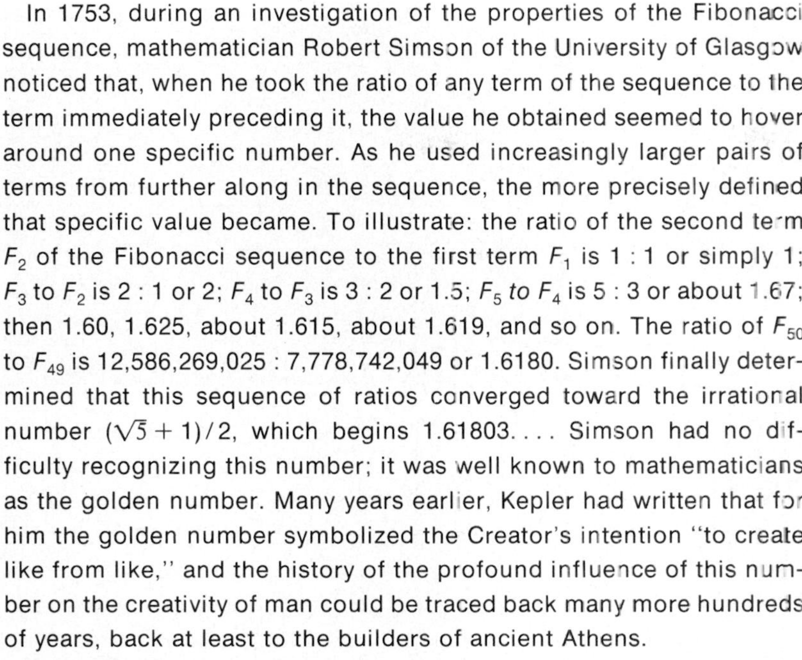

In 1753, during an investigation of the properties of the Fibonacci sequence, mathematician Robert Simson of the University of Glasgow noticed that, when he took the ratio of any term of the sequence to the term immediately preceding it, the value he obtained seemed to hover around one specific number. As he used increasingly larger pairs of terms from further along in the sequence, the more precisely defined that specific value became. To illustrate: the ratio of the second term F_2 of the Fibonacci sequence to the first term F_1 is 1 : 1 or simply 1; F_3 to F_2 is 2 : 1 or 2; F_4 to F_3 is 3 : 2 or 1.5; F_5 *to* F_4 is 5 : 3 or about 1.67; then 1.60, 1.625, about 1.615, about 1.619, and so on. The ratio of F_{50} to F_{49} is 12,586,269,025 : 7,778,742,049 or 1.6180. Simson finally determined that this sequence of ratios converged toward the irrational number $(\sqrt{5}+1)/2$, which begins 1.61803. . . . Simson had no difficulty recognizing this number; it was well known to mathematicians as the golden number. Many years earlier, Kepler had written that for him the golden number symbolized the Creator's intention "to create like from like," and the history of the profound influence of this number on the creativity of man could be traced back many more hundreds of years, back at least to the builders of ancient Athens.

In their science, art, architecture, philosophy, and all other phases of human activity, the classical Greeks, perhaps more than any other culture, sought to incorporate unifying principles of beauty and perfection. Even such a basic task as dividing substances into smaller parts was infused with an underlying philosophical concept of perfect proportions. Probably what began as an unexplainable aesthetic "feeling of rightness" eventually became solidified into a practical statement: The smaller part is to the larger as the larger part is to the whole. Called the divine proportion by later writers, this concept passed easily into the mathematics of geometry.

The most fundamental application of the divine proportion was used by the Greeks from about the 5th century B.C. and is known today as the golden section. Consider the case of a line AB divided somewhere by a point C, as in figure 8. To discover its golden section, one must locate point C on AB in conformance with the divine proportion such

phyllotactic ratio	plants	common trees
1/2	grasses	elm
1/3	grasses, wild blackberry, fiddle-neck	beech, hazel
2/5	mustard, common groundsel	toyon, coast live oak, madrona, California bay, pepper tree, poplar, holly, manzanita, apple, plum, cherry, apricot
3/8	petty spurge	weeping willow, pear, locust (thorn phyllotaxy)
5/13	bottlebrush, ocotillo (thorn phyllotaxy)	pussy willow, almond
5/8	cholla cactus (spine phyllotaxy)	

From "Fibonacci and Lucas Numbers," Verner E. Hoggatt, Jr., © 1968, Houghton-Mifflin Co., Boston

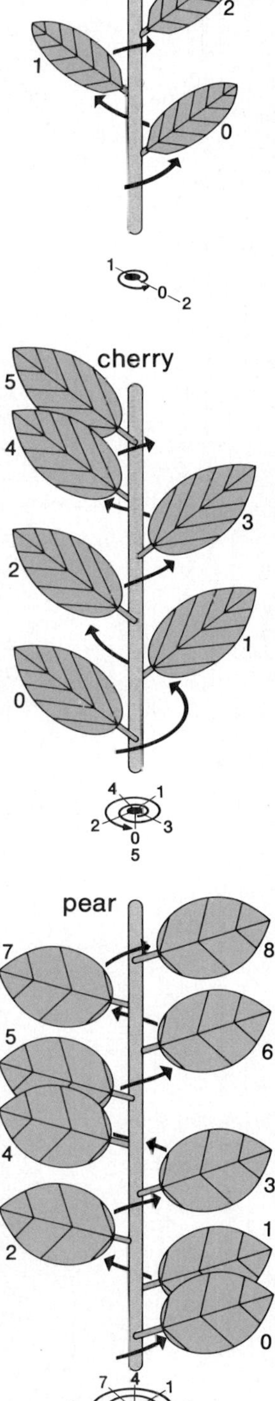

Figure 7. Phyllotaxy for elm, cherry, and pear trees and phyllotactic ratios for several plants (see text).

Courtesy, the Metropolitan Museum of Art, New York, bequest of Stephen C. Clark, 1961

Extensive use of the golden section can be discerned in the oil painting "La Parade" by the French Neo-Impressionist artist Georges Seurat (1859–1891).

that the ratio of the larger part AC to the smaller part CB is equal to the ratio of the whole line AB to the larger part AC; *i.e.*, AC / CB = AB / AC. By trial and error, one can easily estimate the numerical value of this ratio as about 5 / 3 or 8 / 5, because 5 is to 3 approximately as 8 is to 5. In the 15th century the Italian mathematician Lucas Pacioli derived from the golden section the quadratic equation $x^2-x-1=0$ whose positive root defined the value of this ratio precisely. It was this same value that Robert Simson three centuries later suddenly rediscovered in the ratio of successive terms of the Fibonacci sequence: $(\sqrt{5}+1)/2$ —the golden number!

Since its conception, the divine proportion has given rise to other so-called perfect proportions. A golden rectangle can be constructed with short and long sides in the ratio of the golden section (*see* figure 8). If a square is cut off one end of the rectangle, the remaining rectangle possesses the same golden proportions as its parent rectangle and is thus itself a golden rectangle. Such a figure appears as an influence of design throughout classical Greek art and architecture. The main measurements of many buildings of antiquity, including the Parthenon in Athens, are governed by golden sections and rectangles. There is even some evidence in the Great Pyramid at Giza, Egypt, constructed more than a thousand years before the rise of Greek civilization, that the ratio of the height, slope, and base of the structure may have been calculated from the golden section. Harking back to classical themes and techniques for their inspiration, many artists of the Renaissance incorporated the divine proportions in their works. The 20th-century architect Le Corbusier developed a scale of proportions which he called the Modulor, based on a human body whose height is divided in golden section commencing at the navel, as in figure 9.

Through the divine proportion, the Fibonacci numbers once again can be seen to manifest themselves in nature. If a figure such as 10 is constructed for a succession of diminishing golden rectangles and if circular arcs are drawn through the squares so formed with the sides of the squares as radii, the resulting involuted curve approximates the logarithmic spiral. An aesthetic form admired by many diverse cul-

Figure 8. Line divided according to the golden section and used to construct a golden rectangle.

$$\frac{AC}{CB}=\frac{AB}{AC}=(\sqrt{5}+1)/2$$

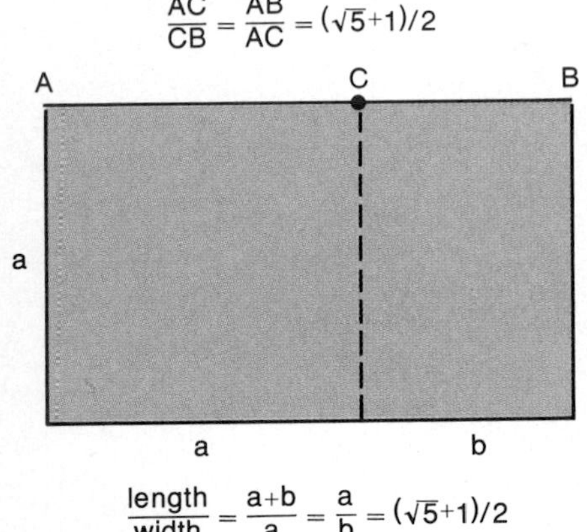

$$\frac{\text{length}}{\text{width}}=\frac{a+b}{a}=\frac{a}{b}=(\sqrt{5}+1)/2$$

tures, it appears repeatedly in such objects as snail shells and animal horns and, as was discussed above, in flower heads and pinecones.

The visual arts, including painting, sculpture, architecture, and other forms that deal with what is pleasing to the eye, have their auditory counterpart in musical composition. Whereas art must be concerned with the division of planes and volumes, music achieves some of its goals by a similar division of time, using notes of various duration and spacing. It is not surprising, therefore, that patterns which can be expressed mathematically in terms of Fibonacci relationships have been discovered in the works of several composers, including Palestrina, Bach, Beethoven, and Bartok. Moreover, a number of 20th-century works, notably Ernst Krenek's *Fibonacci Mobile*, have been deliberately structured upon Fibonacci proportions.

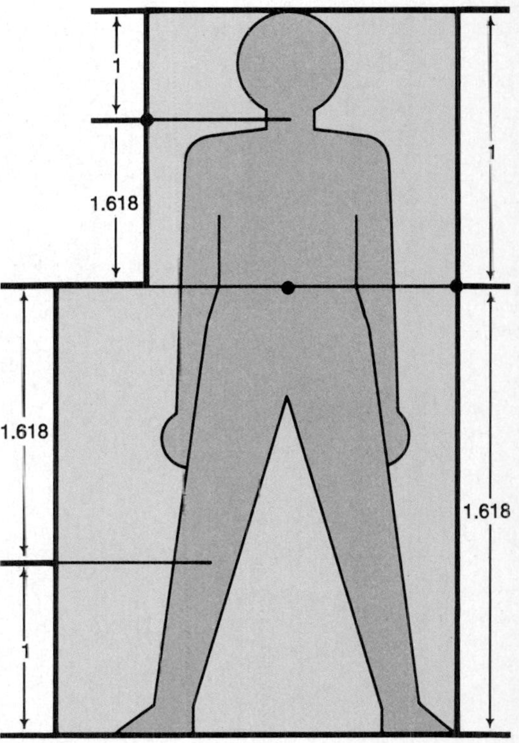

Figure 9. The Modulor, a scale of proportions developed by the architect Le Corbusier.

The connecting thread

The investigations aimed at deepening and expanding man's knowledge of Fibonacci relationships have also given rise to theories and studies that have attempted to explain why one particular mathematical abstraction should manifest itself in so many very real ways. Mathematical models have been offered that describe plant phyllotaxy in terms of a reaction to "contact pressure," an evolutionary trend toward maximizing sunlight on leaves and access to the atmosphere. That the tabs of pinecones are highly modified leaves may make them submissive as well to the same principle.

In the 19th century the German physicist and psychologist Gustav Fechner delved into experimental aesthetics, seeking to determine by actual measurements which shapes and dimensions were aesthetically pleasing. In a series of experiments involving human subjects, Fechner and experimental psychologist Wilhelm Wundt accumulated evidence indicating that most people unconsciously do favor golden dimensions when selecting pictures, greeting cards, mirrors, parcels, and other rectangular objects. At face value, this discovery has been heavily exploited by commercial manufacturers in their package and label designs, by retailers in their store displays, and in other areas of business and advertising. More profoundly, however, it suggests that, as man's first object of aesthetic contemplation and his first model for ideal beauty, nature itself in its many mathematical subtleties may have been unconsciously perceived and assimilated in a way that has affected man's aesthetic judgment even to the present day.

Figure 10. Involuted curve derived from a succession of diminishing golden rectangles.

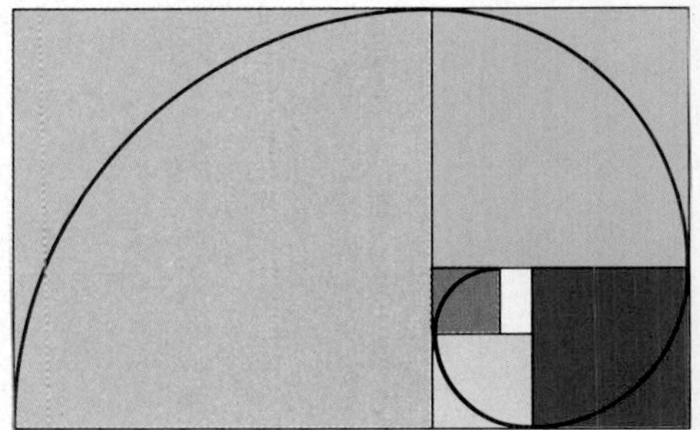

Bees, sunflower heads, pinecones, phyllotaxy, the Parthenon, and Fibonacci numbers—what thread can conceivably run through them all? Are they like the terms in the sequence of prime numbers, each displaying a characteristic common to the sequence but stubbornly refusing to be unified into a formula encompassing them all? Or will they prove to be like an arithmetic sequence, each related to a basic concept and separated only by some ultimately explainable difference? Many lines of investigation appear promising but solutions are not yet in sight.

The Jojoba: Cinderella Crop for the '70s?

by Noel D. Vietmeyer

The seeds of a desert shrub produce oil that may bring prosperity to impoverished Indians and save the sperm whale from extinction.

A scruffy shrub called jojoba, which grows in obscurity in deserts that straddle the westernmost section of the U.S.-Mexican border, is causing great excitement in scientific circles. The oil in its seeds is so unusual, so radically different from the seed oils of all other plants, that it is being eagerly seized by researchers and by industry for testing in scores of products and processes. The results appear to be so promising that the oil has been dubbed "liquid gold" by Thomas Miwa, a prominent researcher at the U.S. Department of Agriculture laboratories in Peoria, Illinois.

Jojoba oil may well become a new materials resource for the world. It is a renewable resource, produced by solar energy in desert areas where sunlight is abundant. It can probably be produced profitably in marginal, unproductive areas where uncertain rainfall makes growing conventional crops too risky. And, because of a remarkable coincidence, it is likely to save the sperm whale from extinction.

With all this promise, it is surprising that jojoba oil was not developed decades ago. But, as with many scientific discoveries, researchers never adequately brought it to the attention of those who possessed the authority and funds to move it out of the laboratory and into the marketplace. The essential details of jojoba's significance and value were announced to the world in 1935, but by 1936 they were largely forgotten. Then, in 1943, under wartime pressures for raw materials, jojoba development was taken up with vigor, but the end of the war removed the incentive and the main jojoba plantation was destroyed. Since then jojoba has periodically been rediscovered, but it was not until 1971 that scientists made connections with administrators who could establish a well-funded project.

In that year, researchers contacted the Office of Economic Opportunity (OEO), which was anxiously searching for ways to improve the economic situation of American Indians living on parched, barren reservations in Arizona and California. During a nine-month period, scientists met with OEO staff. They found that jojoba was already growing luxuriantly on much of the Indian land in the Southwest.

OEO's interest was tweaked by this finding, and in June 1972 the agency funded a jojoba conference so that a broad sampling of jojoba knowledge—much of it unpublished or buried in 20-year-old reports—could be exposed to modern scrutiny. More than 150 researchers and industry representatives gathered at the University of Arizona in Tucson. They examined all the known information on jojoba and searched for any seemingly insurmountable difficulties that could defeat attempts to develop the plant into a crop for Indians. None appeared.

The conference convinced OEO that its funds would be wisely spent on a jojoba development project. In addition, the conference scientists were profoundly affected by what they too learned from their colleagues. Deciding to maintain the spirit of scientific cooperation that the conference had initiated, they instituted an informal international committee of jojoba researchers. It has been a major factor in subsequent jojoba developments.

NOEL D. VIETMEYER *is a Professional Associate at the U.S. National Academy of Sciences.*

(Overleaf) Painting by Eraldo Carugati. Photos, courtesy, Noel D. Vietmeyer

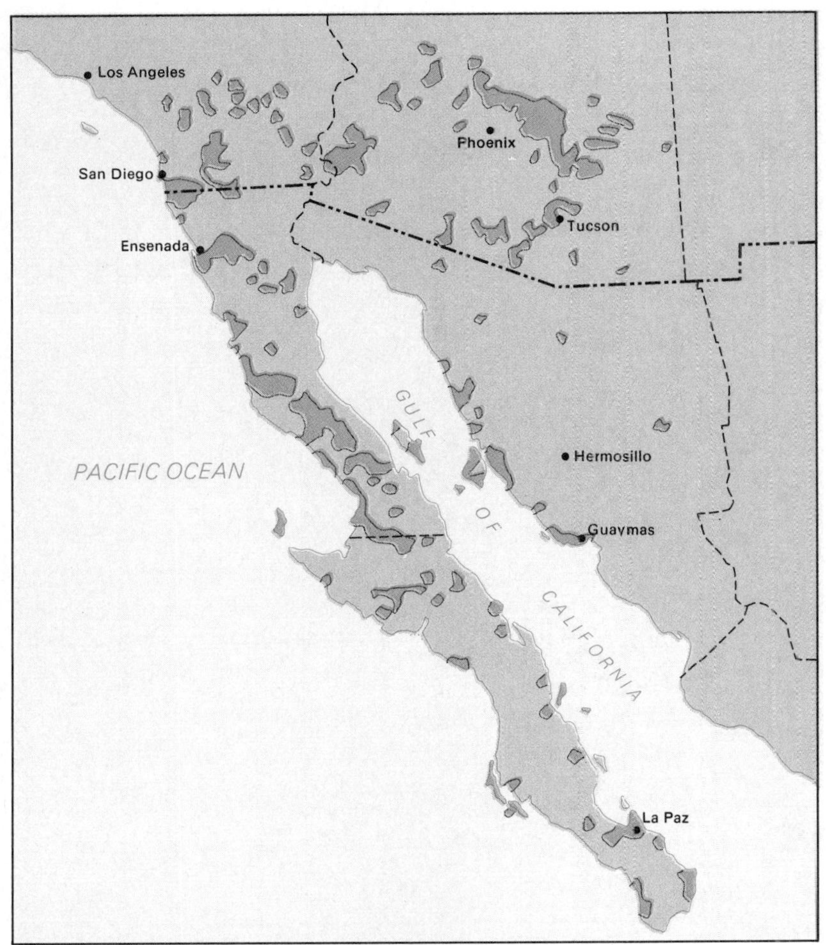

Green shaded areas on map indicate the distribution of jojoba in desert areas of the southwestern U.S. and northwestern Mexico. Jojoba bushes grow alongside saguaro and cholla cacti on a hillside in the Sonoran Desert near Superior, Arizona.

Flower on a male jojoba plant, which produces pollen rather than seeds.

The jojoba plant

In 1769 a Spanish expedition led by the famous explorer-missionary Junípero Serra, traveling northward along the Baja California coast, found Indians cooking with oil from the seeds of a bush they called "ho-ho-wi." The Spaniards wrote it as "jo-jo-ba" (pronounced ho-*ho*-ba). In 1822 when British naturalist H. F. Link landed in the area, he collected specimens of the same desert shrub. Link and his specimen boxes proceeded to China and then returned to England. On the way the box containing the jojoba specimens became mixed with his Chinese collection, and the shrub was eventually designated as species *chinensis* of the genus *Simmondsia*. Jojoba still suffers this geographical misnomer.

Patches of jojoba are scattered throughout Baja California, the southern half of Arizona, the California desert east of Los Angeles and San Diego, and along almost 1,000 miles of the western half of Mexico's mainland. Millions of jojobas dot the slopes and valleys in these areas, occupying elevations between 2,000 and 4,000 feet.

Few plants can survive the inhospitable conditions of the region, with its sparse rainfall (in places less than five inches per year), poor soils, and temperatures reaching as high as 115° F (46° C). But jojoba has built-in survival mechanisms: it needs no water during the summer, when rain almost never falls; it is covered by a film of wax that virtually seals its pores (stomata), thereby reducing the evaporation that in the desert's low humidity would rapidly shrivel it up; it possesses great tolerance to temperature change and can easily withstand the desert's high daytime and low nighttime temperatures; and it shows remarkable resistance to salt, which is commonly found in desert regions and is lethal to most plants. The jojoba plant not only survives in such a hostile environment but also has a natural life span in excess of 100 years.

Jojoba is a woody, upright bush, ranging from two to fifteen feet tall. The leaves are leathery, gray-green ovals, an inch or so long. Individual plants are either male and have pollen-bearing flowers, or are female and have tiny, green, inconspicuous, leaflike flowers, which, when pollinated, develop into fruit. The fruit is greenish and about the size of an olive. On maturing, the outer husk dries out, splits, and peels away, exposing the brown-skinned seeds. The seeds are usually single, although some occur in clusters of up to five. Desert aridity stunts jojoba and yields can be less than one pound of seed per bush, but the average is five pounds. When a shrub is cared for, it can yield more than twelve pounds of clean, dry seed. Some desert animals eat the seed, and in 1976 investigations were under way to determine whether the seed meal (the residue left after extracting the oil) could be used for livestock feed.

Adapted from "Products from Jojoba: A Promising New Crop for Arid Lands," National Academy of Sciences, 1975

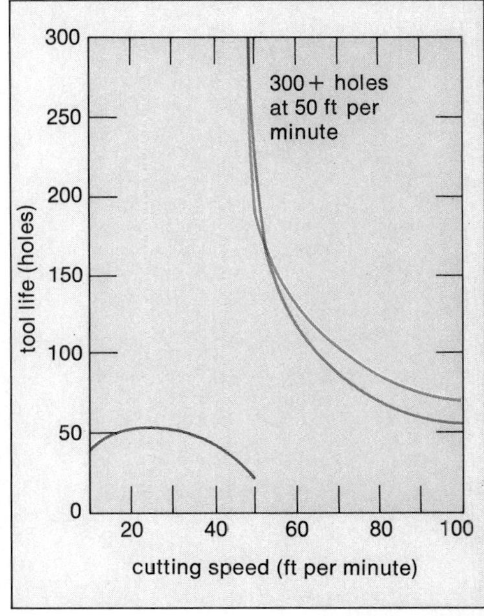

Lubricant properties of 25% sulfurized oil from sperm whales (upper curve) and from jojoba seeds (middle curve) are seen to be quite similar in a graph of tapping tool life and cutting speed. The curve at lower left shows the graphed relationship for a base mineral oil.

Jojoba oil

Jojoba seeds are about half oil. The chemical structure and molecular shape of this oil are quite different from those of other vegetable oils. The oils found in peanuts, corn, soybeans, cottonseeds, and olives are ester combinations of glycerol and fatty acids, but in jojoba oil the glycerol (a molecule with three alcohol groups) is replaced by a much larger linear molecule with just one alcohol group. Perhaps jojoba produces its unique seed oil to help its seedlings survive and grow in the desert's harshness.

Only one other organism produces a similar oil in any quantity, the sperm whale. Why one animal and one plant left the mainstreams of evolution to produce their unique substances, which are liquid waxes instead of fats, is not known. That the two developed virtually identical oils is one of nature's great coincidences.

Both jojoba oil and sperm oil are polyunsaturated molecules (that is, they contain double bonds between some of the carbon atoms). But there are small structural differences between the two; jojoba oil molecules are slightly larger (the sperm oil alcohols and acids are 14, 16, and 18 carbon atoms long, while jojoba oil alcohols and acids are 20 and 22 carbon atoms long).

Conventional vegetable oil equipment and simple refining produce the oil as a pure, uncontaminated material. It is clear like water and has no smell. Seeds growing as far apart as central Arizona and the southern tip of Baja California have been proved to have oil with the same chemical structure.

Industrial potential

Jojoba oil can become a basic feedstock for the chemical industry. Its double bonds and ester group (a chemical structure formed when an acid combines with an alcohol) make it reactive to several chemical transformations, including hydrogenation, isomerization, sulfurchlorination, epoxidation, sodium-reduction, and hydrolysis. All of these reactions produce new chemicals with properties that are important to industry.

In 1975 a distinguished panel of the National Academy of Sciences concluded a two-year analysis of jojoba oil and of its value to industry. The unusual properties of the oil, they concluded, open new vistas for many industries. For the automotive and textile industries jojoba oil could become a new lubricant; for the cosmetic industry it offers a colorless, odorless oil with remarkable resistance to oxidation and rancidity. Jojoba's antioxidation characteristics also suggest for it uses requiring a stable vegetable oil, such as a carrier for pesticides and plant hormones, a waterproofing agent, and a softener for leather and adhesives. Jojoba oil can be hydrogenated (the process by which vegetable oils are "hardened" into margarine) to produce a sparkling, crystalline, white solid that has a wide array of potential uses, ranging from waxes for polishing cars to bright-burning candles that do not sag in hot weather. For these applications it can replace the increasingly scarce and expensive carnauba wax, candelilla wax, and beeswax. When it reacts with sulfur chloride, jojoba oil forms compounds with properties ranging from oils to rubbery solids. These could find applications in the manufacture of such products as lubricants, varnishes, synthetic rubber, linoleum, and printing ink.

The double bonds that are in each half of the jojoba molecule all have a *cis* orientation (the carbon atom chains are connected to the same side of the double bond), but appropriate catalysts can flip them into the *trans* orientation (with the carbon atoms connected to opposite sides of the double bond), producing a mixture in which four out of five are *trans* and the remainder *cis*. The physical result is that the clear, liquid, natural oil becomes thick, like face cream. This "jojoba cream" is a recent discovery and is likely to result in the development of a series of new products.

This reaction demonstrates how a change in molecular shape can change the properties of chemicals. Enzymes are particularly sensitive to changes in molecular shape. For example, the lipase enzymes used by the human stomach to hydrolyze vegetable oils and animal fats so that they can be digested will not hydrolyze jojoba oil. It passes through the digestive system largely unutilized, and because of this

A handful of jojoba seeds, which yield approximately half their weight in oil.

jojoba oil, in principle, offers the food industry a no-calorie vegetable oil for such purposes as diet salad dressings and diet mayonnaise.

If, as seems likely, jojoba products become widely used, there will be demand for large quantities of the oil. In the U.S. this would benefit Indians in the Southwest; in other arid lands throughout the world it would benefit rural farmers who also now eke out a risky living on marginal land; and in the oceans it would benefit the sperm whale.

Benefit to American Indians

In the summer of 1975 more than 1,600 Indians harvested the seeds from jojoba growing wild on their reservations. It was the third year in a row that they had braved the desert sun to do this. Indians have been in the forefront of the effort to develop and exploit jojoba. During the three years after the scientists' conference, the jojoba program expanded so as to become the largest Indian business enterprise in history. As one Indian explained, "We feel that we can run this ourselves. We know the plant; we know the environment. This is one project we can handle our own way. It is not like a factory or an industry—it's part of our culture."

Indians formed jojoba cooperatives in Arizona and California. Through these cooperatives they organized and managed the annual jojoba harvests, established plantations (almost 700 acres of jojoba are expected to be in the ground by 1977), and sold seeds, oil, and candles decorated with Indian designs.

Much of the tribal land in Arizona and California was left to the Indians because it was unproductive and of little commercial value. Reservations such as the San Carlos Apache are among the most impoverished sections of the U.S., with approximately 90% of the work force unable to find a job. Perhaps the jojoba plant can bring to these reservations an economic viability that they never before enjoyed.

A jojoba plantation developed by the University of California. A cultivated bush may yield about twelve pounds of seeds, whereas a wild plant will yield about five pounds on the average.

Benefit to arid lands

In the 1950s jojoba caught the attention of Israeli researchers who were seeking productive plants that could survive the harshness of the Negev Desert and the salt of the Dead Sea region. They collected jojoba seed from Arizona and, since then, have been patiently cultivating the crops and improving their yield and characteristics. By 1975 Israel had the best-producing jojobas, and, with a grant of $1.4 million from the government of Israel and from the World Bank, the nation's researchers began establishing large jojoba plantations. Scientists in Australia were also undertaking extensive jojoba research. There, the government planted experimental plots in three states to test the plant's potential as a crop for Australian Aboriginals, who suffer from much the same economic difficulties as the Indians in the U.S.

Jojoba-based shampoos and cosmetics are popular throughout Mexico. They contain oil harvested from wild plants, but Mexican authorities have been watching the jojoba developments in the U.S., and they are now establishing extensive plantations in the region contiguous with California and Arizona.

Benefit to sperm whales

Sperm oil is so important to industry that to supply demand in 1976 one sperm whale was killed every 29 minutes. This bluff-nosed leviathan with its underslung jaw full of teeth had become the most hunted whale species. Sperm-whale meat is almost inedible; oil is the commercial incentive behind sperm whaling.

This one-year-old jojoba bush will require several more years before it produces seeds containing marketable oil.

It is a testament to the unusual chemical properties of sperm oil that in the face of adverse public opinion, nations continue to send fleets of ships armed with helicopters, radar, and harpoons with explosive grenades to far-flung corners of the globe to obtain it. It is ironic that the relatively accessible jojoba bush produces the same thing. Its home, scorched by the desert sun, is far removed from the cool depths frequented by the sperm whale. Nonetheless, the product that each produces is so similar that jojoba could probably be marketed as "desert-grown sperm oil." Experiments on sulfurized jojoba oil determined that it has lubricating properties equivalent to those of sulfurized sperm oil. For example, both substances improve to an almost equal degree the load-carrying capacity of lubricants under conditions of extreme pressure.

Producing oil from the jojoba seed is likely in the future to be less expensive than searching for whales across the globe. But as of 1976 the jojoba could not compete economically with the whale. That will only happen when the jojoba is brought under cultivation. A strong start was being made with the plantations in California, Arizona, Mexico, and Israel. Jojoba is a slow grower, however, and plantation-grown oil will not enter the marketplace until about 1980. When it does, most industries probably will choose jojoba instead of sperm oil because it will be cheaper. Then the incentive for slaughtering sperm whales will have been removed. The species will be able to live unmolested and to reproduce until its numbers are once more adequate to keep it from extinction.

The Coral Reef Community

by Peter W. Glynn

The coral reef is home to an astonishing array of plant and animal species. An understanding of their biological interrelationships is necessary if man is to protect them from his own activities.

Douglas Faulkner

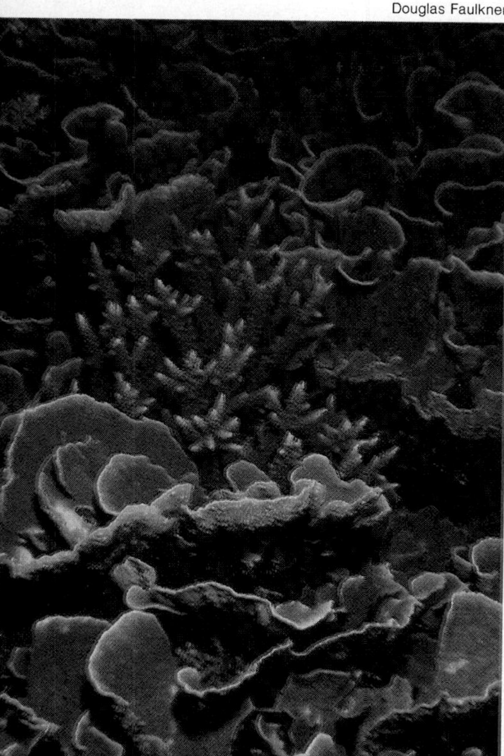

Mixed species of coral carpet the sea bottom of many shallow marine waters.

PETER W. GLYNN is a Biologist at the Smithsonian Tropical Research Institute, Balboa, Canal Zone.

(Overleaf) Photograph by Nicholas de Vore III from Bruce Coleman Inc. Illustrations by John Youssi

Coral reefs, like tropical rain forests, rank among the Earth's most complex biological systems. Any given coral reef contains a multitude of plant and animal species, representing a wide range of life forms. The association and interaction of these species to obtain food, shelter, and other requisites comprise the coral reef community.

When biological communities are compared globally, it is found that those with high numbers of species usually occur in tropical latitudes, where climatic conditions are relatively stable. Evidently, species-rich communities tend to develop in environments characterized by steady and predictable physical conditions over periods of time sufficiently long to allow the evolution of complex specializations and interdependencies. On the other hand, where climatic conditions are severe or unpredictable, species are subject to the limitations of their physical environment; consequently, complex interdependencies are less commonly developed.

With evolution strongly directed toward biological relationships in coral reef communities, it is not surprising to find a large number of interactions among species. It will be of interest to explore the variety of such interactions and, as they are presently understood, to examine their effects on coral survival and reef growth.

General features of coral reefs

Simply defined, coral reefs are rigid structures formed in shallow, warm seas from the calcareous skeletal remains of corals, coralline algae, and other organisms. Skeletal deposition takes place primarily in the living community of organisms present on the sunlit outer surfaces of reefs. The internal structure of reefs may consist of coral skeletons undisturbed since growth, but more commonly it is made up of dislocated skeletons and skeletal sand and debris that have been compacted and cemented together. In other words, the usual building materials undergo significant movement and alteration in the building process. Some coral reefs grow rapidly, building vertically at the rate of 10–15 meters every 1,000 years. However, a more common growth rate for that span of time is 1–5 meters. Organisms with the potential to build reefs also may occur on a surface of other than their own making and may just manage to replace themselves from generation to generation. In this instance, reef building obviously is not taking place, though such distinctions often require careful field study.

The kinds of corals that contribute most significantly to reef building are known technically as scleractinian, or stony, corals. Of the class Anthozoa (phylum Cnidaria or Coelenterata), they are closely related to sea anemones, but, unlike anemones, the outer tissue layer of stony corals deposits a calcareous skeleton. Other coelenterates that form calcareous deposits are the hydrocorals and the soft and horny corals. Hydrocorals produce massive skeletons and are often important in reef building; the majority of soft and horny corals produce only minute calcareous spicules. The basic functional unit of stony corals is the polyp, an animal having a hollow cylindrical body

Courtesy, The Geological Society of America

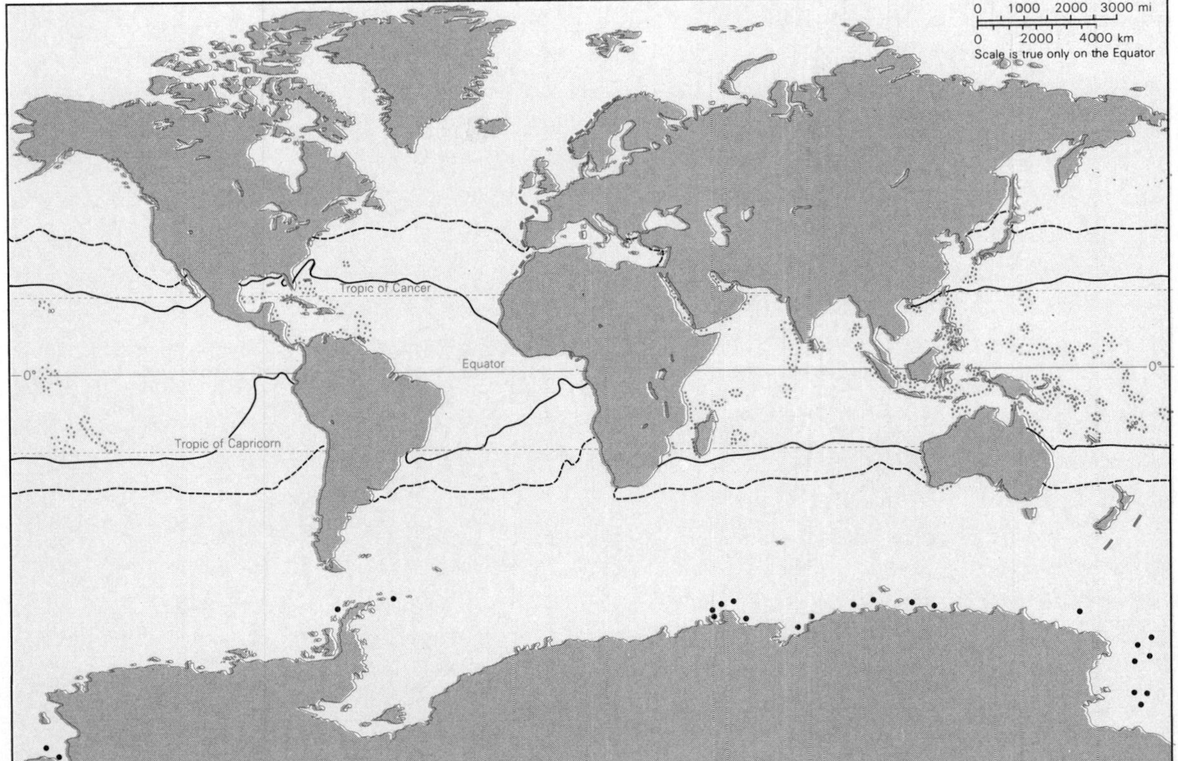

Deepwater coral banks or thickets
Localities for solitary corals, Antarctica only
Coral reefs
Extended 70°F isotherm (summer)
Restricted 70°F isotherm (winter)

Worldwide distribution of corals and their relationship to the temperature of the sea surface. Note that reefs are basically restricted to warm waters, whereas solitary corals and deepwater species occur in high latitudes.

and a ring of tentacles surrounding a mouth that functions as the only major opening to a saclike gut. Like other coelenterates, stony corals are armed with potent stinging cells, or nematocysts, that are used for food capture as well as for defense.

Corals can reproduce both sexually and asexually. Through sexual reproduction, egg and sperm unite to produce a free-swimming larva that eventually settles and develops into a polyp. Asexual reproduction occurs through budding, the outgrowth of an extension of the adult body cavity. The daughter polyps remain attached to the original polyp and, by constant addition of new buds, colonies are formed that in some species can reach several meters in diameter and attain approximately 1,000 years in age.

It should be noted that there are two kinds of stony corals: those that build true reefs and those that do not. The reef-building corals—called hermatypes from the Greek *hermatos*, "mound"—contain within the inner tissue layer a symbiotic alga that speeds the calcification process. Corals that do not form reefs, the ahermatypes, lack symbiotic algae, calcify relatively slowly, and do not play an important role in reef building. Ahermatypic corals are not confined to warm or sunlit waters and may form extensive coral thickets or banks in calm deep water in the higher latitudes. For example, coral banks occur commonly along the Norwegian coast between 180 and 270 meters in depth. Some of these deepwater banks cover as much as 1.6 square kilometers of sea bottom and extend vertically from 30 to 60 meters above the surrounding seafloor. These banks and thickets are not

(Top left, right, bottom left) Jane Burton from Bruce Coleman Inc.; (bottom center, right) Anne L. Doubilet

(Top left) Individual zoanthid soft coral polyps attached to dead coral skeleton; (top right) colony of the soft coral Goniopora. *(Bottom, left to right) Photosynthetic green algae evolving oxygen on reef surface; colony of soft coral polyps; branches of stinging fire coral.*

true reefs because they are not rigid and compact in structure and probably could not withstand the incessant wave action in shallow seas. Coral reefs are more enduring; they also grow more rapidly than ahermatypic coral banks and cover far greater areas of the sea bottom.

The coral-algal partnership

Symbiosis, the living together of different species in close or intimate association, is encountered in a variety of forms in rich tropical communities. Of all the symbiotic relationships present in coral reefs, the most fundamental involves the association of the coral animal host with algal cells, the vegetative stages of a dinoflagellate. This coral-algal association probably originated in middle Triassic time, about 200 million years ago, when scleractinian corals first began to build coral reefs similar to those now present in the tropical seas. The symbiotic algae, commonly called zooxanthellae, depend on light for photosynthesis; for this reason hermatypic corals flourish only in shallow waters, seldom deeper than about 100 meters. Coral and zooxanthellae assist each other in various ways. Some beneficial effects arising from this partnership are greatly accelerated skeletal formation; the availability of algal photosynthetic products to the coral host; and the efficient recycling of nutrients between partners. Since both symbiont and host benefit from this association, it is described as a mutualistic form of symbiosis.

Although it is not clear how zooxanthellae enhance calcification, their role in this process is undoubted. The fastest rates of skeletal formation occur when conditions are most favorable for photosynthesis: that is, during midday hours under cloudless skies, in clear water, and when the Sun's rays are most nearly vertical. It can be demonstrated experimentally that, when zooxanthellae are removed from corals or when corals are fed and maintained in darkness, calcification rates are greatly reduced.

Currently one of the most active areas of research in coral physiology centers on the nutritional relationship of the coral host and its symbiotic algae. The investigation of this problem is complicated because corals have evolved a variety of feeding mechanisms that are developed to varying degrees in different species. In some coral species, particulate food such as zooplankton is captured actively by the tentacles; in others, it is passively trapped in a sheet of mucus and then conveyed to the mouth along ciliary tracts. Many coral polyps even extrude parts of their digestive system to feed on proteinaceous material adjacent to the colony. The structures involved in this kind of feeding may exit through either the mouth or the body wall and have both digestive and absorptive functions. Organic matter dissolved in seawater also can be absorbed directly through the body wall. It has been established that zooxanthellae make available to the coral host large amounts of the materials they synthesize. In some cases the coral host can actually stimulate the release of algal photosynthetic products. Because there is some outward movement of these latter

(Top and center) D. R. Robertson; (bottom) Herbert Lanks from Black Star

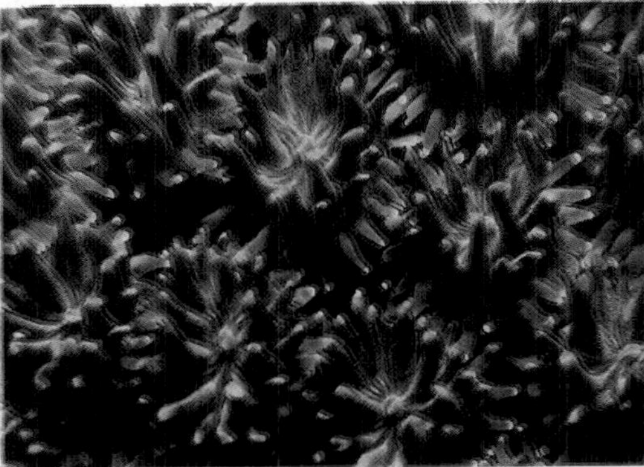

(Top to bottom) Polyps of Galaxea, *a hermatypic stony coral; tentacle mass and mouth of the ahermatypic stony coral* Tubastraea; *branching colony of Gorgonian corals.*

Douglas Faulkner

(Above) Hollow masses of Dictyosphaeria cavernosa, *a bubble alga, shroud the skeletal remains of once-flourishing reef corals near Coconut Island, Kaneohe Bay, Hawaii. (Opposite page) Generalized diagram of cnidarian nematocyst or stinging cell. Nematocysts are located among the cells of the epidermis or outer surface layer of the polyp tentacle (a). Produced in a specialized cell called the cnidoblast, each nematocyst consists of a capsule (b) with one end inwardly folded (invaginated) and extended into a long, spirally wound hollow thread. A cover or operculum normally covers the capsule. A minute spine, the cnidocil, projects from the free surface of the cnidoblast into the water. When mechanically disturbed, the cnidocil triggers the thread to turn inside out (evert) and discharge outwardly with great force (c). Nematocyst threads of coral polyps inject a poison that paralyzes their prey.*

two sources of nutrients, however, their overall importance as food has proved difficult to determine.

Several lines of evidence indicate that zooxanthellae play an essential role in the conservation and recycling of nutrients. For example, it has been found that the exchange of such nutrients as phosphates between corals and the surrounding medium is much less in hermatypic corals than in ahermatypes, and less in hermatypes with zooxanthellae than in those from which the algae were removed. It has also been shown that hermatypes have less nutrient excretion than most other marine animals. The efficient cycling system of corals is of utmost importance in allowing for the vigorous building of reefs in nutrient-poor waters.

Competitive interactions

Competition for space can be keen on coral reefs. Because of their symbiotic zooxanthellae, reef-building corals must find and hold a position on the reef that receives adequate illumination. This location must receive a supply of waterborne food as well. Although many kinds of organisms favor such locations, attention will be concentrated on competitive interactions involving free-living algae and corals. Two effective strategies employed by many corals to insure sufficient exposure are rapid growth and a spreading growth form. Rapid growth is especially advantageous on newly opened surfaces where many different kinds of pioneering species are competing for establishment. In mature coral communities species with spreading colonies, which more completely intercept light and discourage understory growth, are commonly present.

To document in detail the competitive interactions between two species is a difficult task, and often complicating this are the effects of stressful or suboptimal conditions that can alter the competitive capacity of the organisms concerned. Free-living algae, for example, generally have a greater tolerance than do corals for extreme variations in temperature, sedimentation, freshwater dilution, and nutrient enrichment. Where such conditions begin to interfere with the life processes of corals, the balance tips in favor of algal growth. In some instances a marked change in water quality will lead to the sudden death of corals. It is important to bear in mind, however, that the character of a community also can be transformed by more indirect effects of competition. Lamentably, large tracts of the majestic coral reefs of Kaneohe Bay, Hawaii, were destroyed in this manner. The combined effects of dredging, poor land management, and improper sewage disposal changed the once clear bay water into a sediment-laden and nutrient-enriched medium. These new conditions favored the proliferation of a bubble alga that first gained a competitive advantage on the reef in the early 1960s and that has since replaced reef corals extensively in the middle and south regions of the bay. The Kaneohe Bay coral community has been thoroughly transformed; different sectors are now dominated by algae, sponges, tunicates, sea cucumbers,

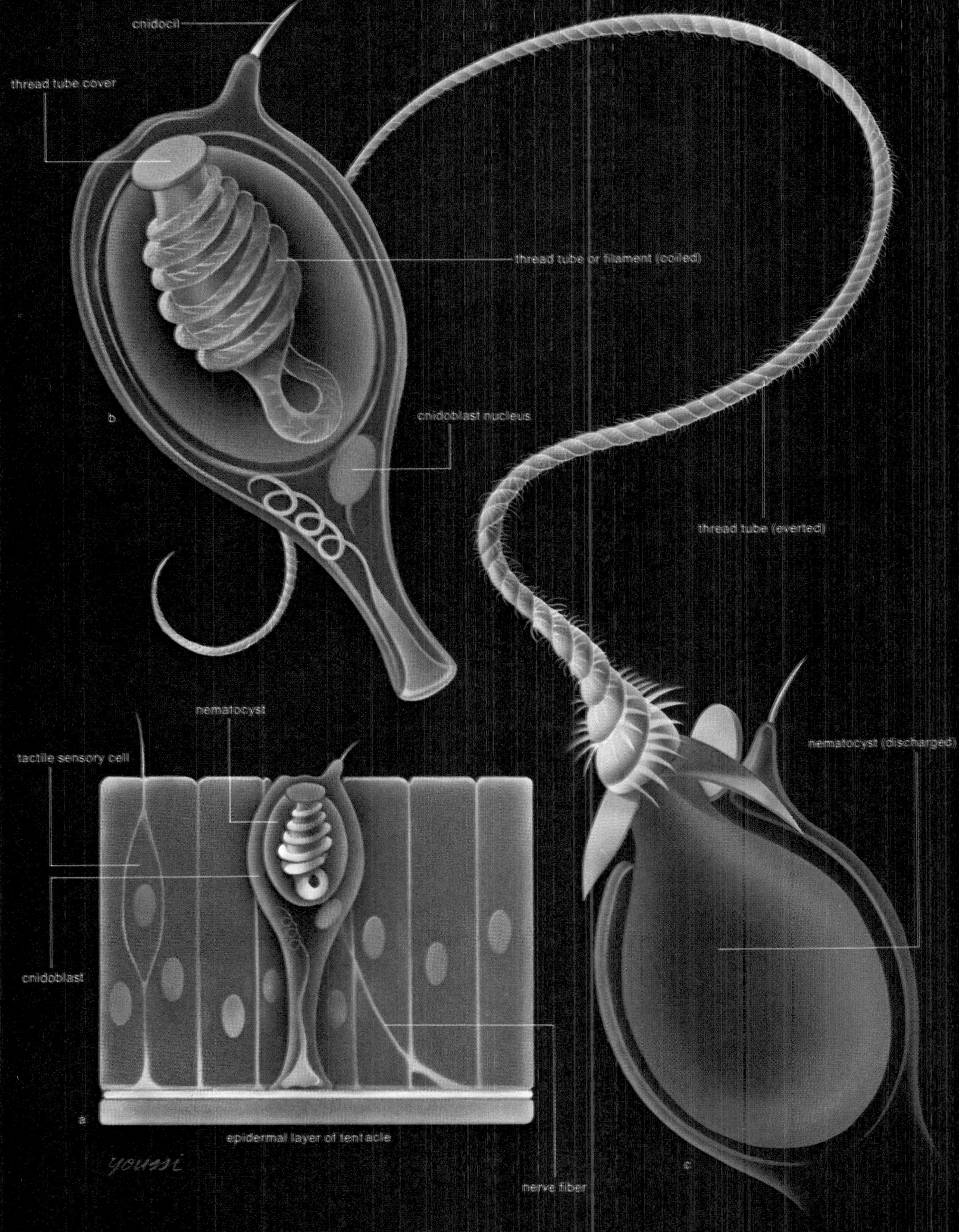
cnidocil
thread tube cover
thread tube or filament (coiled)
b
cnidoblast nucleus
thread tube (everted)
nematocyst
tactile sensory cell
nematocyst (discharged)
cnidoblast
a
epidermal layer of tentacle
youssi
c
nerve fiber

(Top) Nicholas Smythe; (center) Douglas Faulkner; (bottom) David Doubilet

and other animals that require large amounts of particulate food. Recovery of these reefs will probably require several decades following the restoration of the environment to its initial condition.

Competition for space is manifested in other ways as well. Coral-reef biologist Thomas F. Goreau observed that rapidly growing coral species do not always fully dominate reef surfaces that appear suitable for them. These observations posed an intriguing and ecologically important question: What factors are preventing these corals from occupying more space than they do? Part of the answer was found by Judith Lang while studying at Goreau's field station on the northern coast of Jamaica. She observed that when two different coral species are placed side by side so that their polyps touch, one of the corals will extrude mesenterial filaments onto the other that digest all of the tissues within reach. Lang termed this interaction interspecific aggression. Further study of all of the major hermatypic corals in the Caribbean disclosed a pecking order or aggressive hierarchy in which the dominant aggressors destroy their subordinates. Similar interactions have been observed among corals of the Indo-Pacific region, indicating the general nature of this phenomenon. The most aggressive species tend to be small in size and have massive or encrusting colonies. Thus, small and slowly growing corals can compete successfully with rapidly growing and spreading species by destroying and possibly even assimilating their tissues in the process. Some large massive corals that grow moderately fast and are relatively high on the aggressive scale form prominent buttress systems on Caribbean reefs and probably owe their success to these two attributes. Weak aggressors, like the leaf and hat corals, are abundant more commonly in deep-reef zones where competition with other corals is less intense.

Modification by erosion and sedimentation

At the same time coral reefs are actively building, they are also subject to the incessant destructive forces of a variety of organisms in the reef community. Bioerosion, the erosive activity of living organisms, is more significant than mechanical abrasion by waves and currents. Organisms involved in the erosion of reefs are represented by species of bacteria, fungi, and blue-green algae and by such animals as sponges, worms, barnacles, mollusks, echinoderms, and fish. Bioerosion greatly accelerates the production of calcareous sediments, which extend from the reef core to form rubble islands, beaches, sand aprons, muddy lagoons, and other features that contribute to the expansion and influence of the carbonate reef environment. In addition, the holes, tunnels, and chambers formed via bioerosive processes allow for greater water movement within the reef and provide special habitats for a rich cryptic fauna occupying the interior of reefs.

Reef fish alone can produce extraordinary amounts of sediment. Measurements of the fine sand produced by fish on a Pacific reef near the island of Saipan indicate a production of four to six metric tons per hectare (10,000 square meters) per year. On a Bermudian reef, sand production by fish was estimated at about two metric tons. One species of puffer fish feeding on live coral in Panama can generate 0.2 metric ton of sand and gravel per hectare per year. Further analysis in Panama disclosed that the combined sediment production of a puffer, an egg shell (a gastropod mollusk of the genus *Jenneria*), and two species of hermit crab can exceed six metric tons. This estimate, which does not include erosion by other living organisms, represents about one-third of the annual coral growth. The puffer and hermit crabs bite and scrape the tips of coral branches; the egg shell, however, grazes on living coral tissues, leaving intact coral skeletons that are often of cobble and pebble size. The dead coral colony is further eroded by grazing and burrowing animals and eventually becomes overgrown with algae and other organisms. When complete erosive budgets become known and are compared with skeletal production rates, it may well be shown that the drain of bioeroded material in some coral communities is the principal factor preventing active reef construction.

While feeding on corals or extracting animals living in coral colonies, fishes and invertebrates can significantly alter the appearance and even the spatial arrangement of coral populations. Fish, crabs, and sea urchins that graze on the tips of ramose, or branching, corals can greatly modify their strength and extent of branching. Like plant pruning, the cropping of coral branches discourages apical growth, and consequently the colony becomes more robust in form. Triggerfish bite into live corals in order to remove burrowing mollusks. Often the coral fragments broken loose continue to live and grow, and in some cases they form a fairy ring of biscuit- or potato-shaped masses that surrounds the parent colony. The branches of ramose corals broken loose by fish may continue to grow and form new colonies, sometimes far removed from the parent colony. Corals are also moved incidentally by fishes and conchs foraging through bottom sediments. This frequent overturning and tumbling permits some corals to grow fairly uniformly over all surfaces and also to develop oddly shaped spherical colonies.

Flip Schulke from Black Star

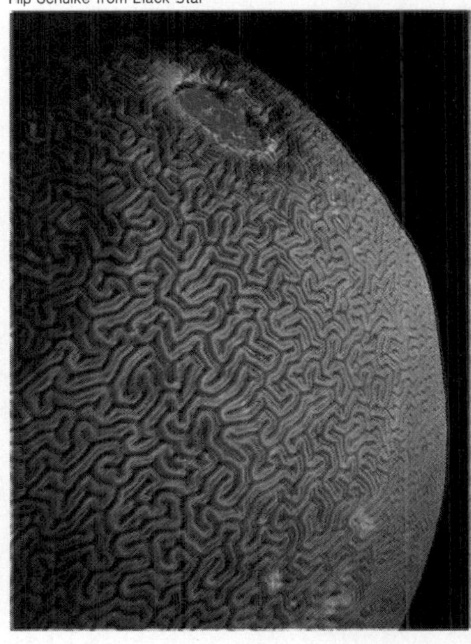

Interspecific aggression occurs between the colonies of two species of coral in contact (opposite, top). The dominant species on the right has extruded mesenterial filaments near the point of contact and has begun extracoelenteric digestion of the subordinate species. Populations of slate-pencil urchins (center) and puffer fish (bottom), both typical inhabitants of the reef community, can strongly influence the nature of the community through their feeding activities. (Above) Circular growth of red sponge lends a cyclopean appearance to a spherical colony of brain coral.

Coral-reef herbivores

Probably of far greater consequence to community structure than the forms of modification discussed above is the constant pressure of feeding on shallow-reef surfaces exerted by herbivores, especially fish and sea urchins. Two important fish families that are found on coral reefs throughout the world are the parrot fish, which rasp the surfaces of reef rock or ingest sediment, and the surgeon fish, which remove algae by browsing. Active during the daylight hours, large schools of these fish can be seen foraging over the reef shallows, especially on plant-encrusted surfaces. Some schools of parrot fish are mixed; that is, they comprise several different species and often include carnivorous fish. When these schools of mixed species are feeding, both plant and animal food resources are exploited simultaneously. Secretive

Photos, Douglas Faulkner

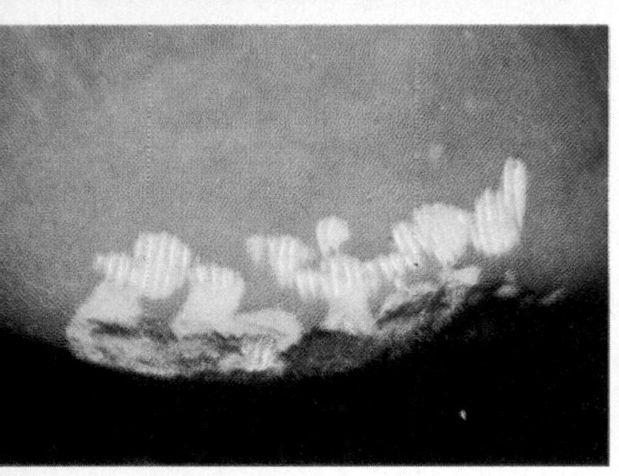

Gouges in the calcareous skeleton of a coral colony (bottom) testify to the efficiency of the hard, powerful beak of the parrot fish. An individual coral polyp (opposite page) depicting a variety of feeding mechanisms: capture of particulate food by tentacles; extracoelenteric feeding via extrusion of mesenterial filament through the body wall; direct absorption of dissolved organic material; and absorption of photosynthetic products of symbiotic zooxanthellae. Also shown is the coenenchyme, which in colonial corals extends to unite neighboring polyps.

and normally inaccessible invertebrates and small fish stirred up by fish herbivores are preyed upon in turn by fish carnivores.

Experimental studies have shown that the intense feeding activities of fish herbivores can influence markedly the nature of shallow-reef communities. In Australia reef surfaces that were covered with cages to exclude fish herbivores developed thick carpets of algae; the algal growth nearby, where fish were free to feed, remained sparse. The bands of bare sand that so commonly separate reefs and sea-grass beds in the Caribbean Sea are, in part, a result of grazing fish. Parrot fish and surgeon fish feed more often on sea grass nearest reefs so that they can seek shelter in the reef when predaceous fish approach. Consequently a band of sand, sparse in sea grass, is formed along the edge of reefs. The placement of an artificial reef in a lush sea-grass bed, constructed by John E. Randall for purposes of experimental verification, resulted in the formation of a sand band in little more than eight months. In addition, John C. Ogden and co-workers found that sea urchins feeding on sea grass around reefs can form circular sand bands or halos. Because the urchins concentrate near reefs and not on open grass bottoms, it is possible that, like fish herbivores, they too use the reef as a refuge from predators. There is some evidence that overfishing by man may be responsible for the dominant role of sea urchins in clearing the bottom of sea grasses. Where the halos were observed—*e.g.*, St. Croix and the U.S. Virgin Islands—fishing pressure has greatly reduced competition from fish herbivores and the population of natural predators of sea urchins, thus allowing the urchin to grow in numbers and influence.

Reef herbivores can also have an indirect effect on the settlement and survival of coral larvae. For example, young corals often must compete with rapidly growing algae. Hence, on surfaces largely occupied by plants, the survival rate of coral can be greatly diminished. However, if a sea urchin or fish herbivore removes plants from a suitable surface, corals will find it easier to settle and grow. It was generally assumed that herbivores feeding on plant-encrusted surfaces are not very discriminating and tend to remove all of the available food—plants and coral alike—with no advantage to the coral. Work during 1975, however, suggested that at least some reef herbivores do have a special discriminatory capacity and can select the kind of food they want from the smorgasbord array of available items. A clear example was observed by Charles Birkeland on settling plates set out on coral reefs along the Caribbean coast of Panama. After algae and corals had settled on the plates, parrot fish began to rasp the surfaces of the developing community in their search for food. The characteristic beak marks of the parrot fish were scattered over all surfaces except those where young coral colonies were present. These observations indicate that preferential feeding or active avoidance by a herbivore can favor the survival of young corals.

Some territorial reef herbivores exercise yet another kind of control over reef-bottom populations. The damselfish, a pugnacious reef fish

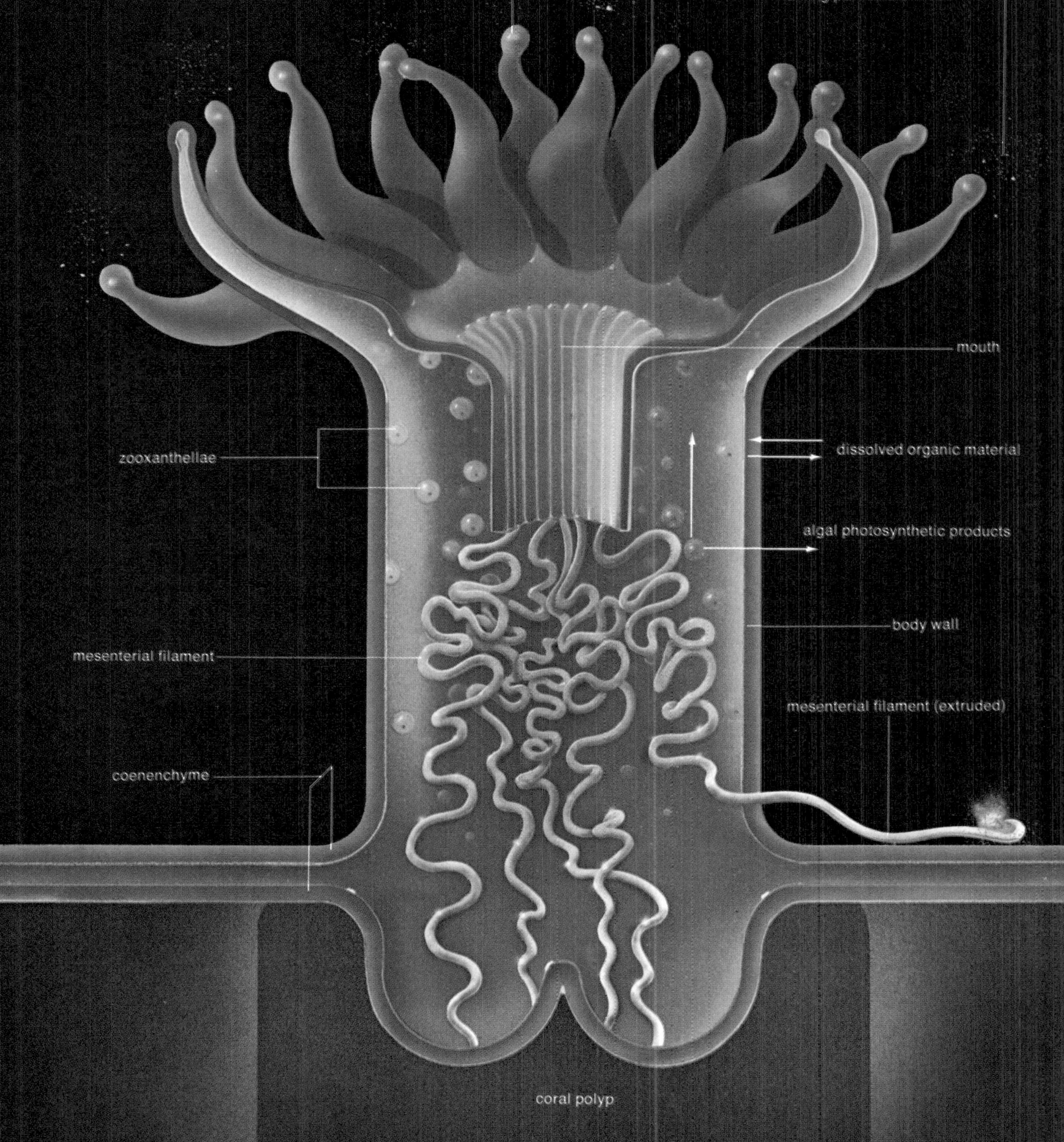
mouth
zooxanthellae
dissolved organic material
algal photosynthetic products
body wall
mesenterial filament
mesenterial filament (extruded)
coenenchyme

coral polyp

Douglas Faulkner

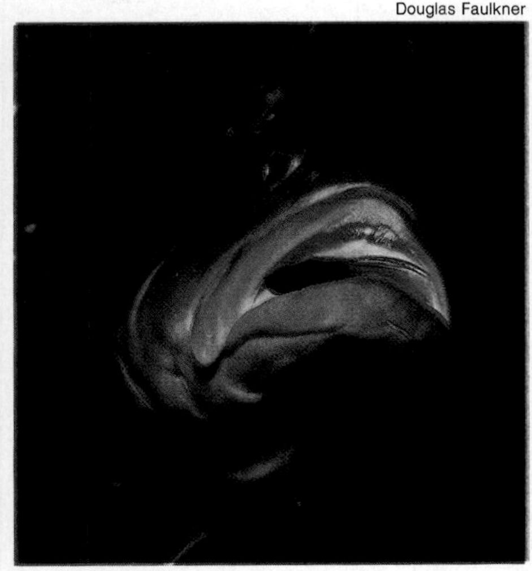

In an example of cleaning symbiosis, Nassau grouper allows fish cleaner, a neon goby, to remove parasites and infected tissue from its mouth cavity.

usually no larger than a human hand, defends its territory to protect an abundant supply of algae. Carnivorous fish swimming into these territories are largely ignored, but potential algal grazers are immediately expelled. Damselfish that display this kind of behavior are, in a sense, farmers. They care for their algal gardens, even removing debris and herbivorous snails, and harvest the crop as needed. Such activities can have very decided influences on community structure. For instance, the plants growing in the farmed territories are more abundant and luxuriant than those outside. In addition, few corals and other animals that compete with algae are found in gardens defended by damselfish, and it is suspected that the farming activities of the damselfish allow the algae to compete more effectively with animal species. Thus, by controlling the kinds of fish that are allowed into its territory, even a relatively small fish can promote plant growth and thereby limit the success of corals.

Cleaning symbiosis

Also present on reefs is a guild of animals, the cleaning symbionts, that actively remove ectoparasites, diseased tissues, and other unwanted materials from fish hosts. The best-known cleaners are species of shrimp and fish, and the focal points occupied by such forms are termed cleaning stations. An indication of the extent of this relationship is evident in the complex behavior of both cleaner and host. Cleaners are usually brightly colored, bear striking pigment patterns, and select obvious perches on the bottom—*e.g.*, near large sea anemones or atop coral heads or promontories—where they perform characteristic movements to accentuate their visibility. Movements used to attract fish depend, of course, on the species of cleaner. In shrimp, body swaying and a to-and-fro motion of the antennae are often used. Elaborate displays are employed by some fish cleaners, involving oscillatory swimming movements and fin erection. Although the fish hosts being serviced are many times the size of the cleaners, the cleaners are seldom if ever eaten. While being cleaned, the host commonly exposes to the cleaner the parts of its body that are badly parasitized or infected. The mouth and gill covers are held open so that the cleaners can move freely in and out while performing their important services. If alarmed, some fish hosts will even signal to the cleaners to retreat.

To gain some idea of the importance of fish cleaners in affecting the distribution of reef fish, L. B. Slobodkin and L. Fishelson located all cleaning stations along a reef front in the Gulf of Eilat (Aqaba), Israel, and observed the numbers and kinds of fish frequenting them. Thirteen stations were found along a 600-meter stretch of the reef. About twice as many kinds of fish and over ten times the number of individuals were observed at the cleaning stations as in surrounding areas. This particular symbiotic relationship again emphasizes the importance of biological processes that can influence the distribution patterns of species in reef communities.

Freckled grouper accommodates a member of a species of cleaner shrimp on its snout.

Douglas Faulkner

Studies of a Panamanian coral reef community revealed that the combined feeding activities of a puffer, two species of hermit crab, and an egg shell can produce sediment in excess of six metric tons per hectare per year. Such bioerosive modification significantly enhances the quality of life in and around the reef environment and encourages development of a rich diversity of species within the community.

puffer
0.2 metric ton
per hectare
per year

hermit crab
1 metric ton
per hectare
per year

egg shell
5.1 metric tons
per hectare
per year

sand and gravel

sand and gravel

pebbles and cobbles

John H. Harding

The crown-of-thorns starfish

Until the mid-1960s the prevailing opinion among biologists who study coral reefs was that living coral colonies, protected by their stinging cells and mineral skeletons, were seldom eaten in significant amounts by reef animals. Then, to the astonishment of both the lay and scientific communities, it was announced that the crown-of-thorns starfish, *Acanthaster planci*, was destroying massive sections of coral reefs on the Great Barrier Reef of Australia. Further reports indicated that *Acanthaster* also had infested coral reefs in many different parts of the Indo-Pacific region, including Sudan, Sri Lanka, the Ryukyu Islands, Guam, the Palau Islands, the Marshall Islands, and Hawaii.

To provide some indication of the magnitude of an infestation and the amount of coral mortality involved, it is instructive to cite some of the quantitative results obtained by Robert Endean and co-workers in their investigations of the starfish invasions in Australia. Before infestation, the reefs most seriously affected contained an average surface coverage of about 35% live coral. It was determined that a starfish could eat about ten square meters of coral per year. An average sized reef of ten million square meters whose coral cover had been killed in 2.5 years thus was infested by some 140,000 starfish, or 140 individuals per hectare, compared with a normal density of 0.06 starfish per hectare. The main reefs affected were those present along the middle third of the Great Barrier Reef system, encompassing a stretch of some 640 kilometers. These findings clearly indicated an ecological imbalance of major proportions. For some reason, a factor or combination of factors that previously had held the population size of *Acanthaster* in check had been negated.

(Left) Nicholas Smythe; (right) Douglas Faulkner

(Opposite page) Crown-of-thorns starfish, Acanthaster planci, *feeding among colonies of branching coral. Coral crab symbiont (above left), partly hidden by coral, defends against approaching* Acanthaster. *(Above)* Acanthaster *devouring coral colony.*

The study of different reef systems subsequently has shown that *Acanthaster* varies greatly in abundance and effect throughout the Indo-Pacific region. It is absent or rare on many, and perhaps the majority of, reefs; elsewhere it is moderately common or at times abundant and preys heavily on corals; and finally, in parts of the Great Barrier Reef and other heavily infested regions, it has killed the majority of reef-coral populations. This range of effects, coupled with the absence of reliable information on the abundance of *Acanthaster* before 1960, greatly complicates the analysis of this problem.

Several hypotheses based on population control through predation have been advanced to explain the sudden increase in the crown-of-thorns starfish. In Australia triton shells are known to eat *Acanthaster* and some indications suggest that their removal by collectors has effectively eliminated an important source of starfish mortality. It has also been proposed that, because corals feed on the larvae of *Acanthaster*, a mass reduction in the amount of live coral cover on reefs would enhance the recruitment of juvenile starfish. A less direct explanation has implicated the effects of such chemical pollutants as insecticides. If these selectively kill the predators of larval *Acanthaster*, the recruitment and subsequent population growth would be greatly favored. Common to these three explanations is the difficulty that infestations cannot be correlated in all cases with the different causative agents proposed.

Thomas F. Dana and William A. Newman recently have hypothesized that *Acanthaster* infestations can be formed simply by the redistribution of starfish in existing populations. They found that infestations were correlated with the frequency of typhoons in the regions under study and reasoned that coral reef destruction caused by typhoons will induce populations of *Acanthaster* to migrate and aggregate in search of new food supplies. The starfish migrate laterally toward undisturbed reefs where corals are plentiful, gaining in number and destructive capacity as they go.

There is some historical evidence that high population densities of *Acanthaster* are not a uniquely recent phenomenon, a fact which would de-emphasize man's influence. Nor is *Acanthaster* the only marine animal that becomes immensely abundant and outstrips its food supply. Other species of starfish have been known for years to proliferate on oyster and mussel beds in Europe and along the east coast of North America. More recently, sea urchins have undergone a population explosion on the coast of southern California, where they have eliminated and are now preventing the reestablishment of kelp beds. Additional examples of marked population fluctuations could be given for other echinoderms, mollusks, and polychaete worms. All of these animals share a similar mode of development; myriads of eggs are produced and fertilized externally, and larval development proceeds freely in the water column. Before settling, the larvae characteristically suffer heavy mortality from predation and other causes. Under favorable conditions, however, enormous numbers of larvae do survive and settle. This kind of life history is the reason why many species of marine animals fluctuate so greatly in numbers from year to year. Peter J. Vine proposed a natural-causes theory, which takes into account the natural properties of resurgent populations, to explain the sporadic occurrence of *Acanthaster* infestations. He reasoned that a successful settlement could produce an effective seed population. *Acanthaster* normally tends to form aggregations, and this behavior would favor further breeding success that could accelerate into a local population explosion. It is possible that conditions favoring the survival of starfish larvae are met more frequently near centers of human habitation, with man providing an additional positive influence toward population growth. This explanation is consistent with the frequent occurrence of infestations near centers of human activity and does not conflict with occasional outbursts of the starfish in more remote areas.

While large numbers of *Acanthaster* typically have a devastating effect on reef corals, it is also important to recognize that even small numbers of the starfish can alter the structure of coral communities. Field observations show that *Acanthaster* prefers to feed on certain coral species and avoids others. A combination of several factors—for example, colony shape, stinging-cell potency, and location on reef—discourages the starfish from mounting certain corals. Other corals are protected by symbiotic crabs and shrimps that bite, jerk, and snap at the starfish and thus prevent it from mounting and feeding. Over a period of time such selective feeding reduces the numbers of some corals and favors those of others, thereby modifying the structure of the reef community.

The animal and plant populations on coral reefs overrun by the crown-of-thorns starfish undergo a rapid and profound change. As the bulk of the coral cover disappears, so do the other organisms directly or indirectly dependent on live corals. Fleshy and calcareous algae and soft corals typically overgrow the dead coral skeletons. It should be noted that while *Acanthaster* kills the living tissues of corals, it does not destroy coral skeletons or the foundations of reefs. On the other hand, many of the erosive activities discussed above continue and the reef frame begins to diminish in size. The proliferation of calcareous algae, which bind and strengthen the dead coral framework, partially counteracts this trend. Reef recovery begins when larvae from surviving resident corals and from more distant populations settle and colonize reef surfaces. The recruitment and growth of corals is at best a slow process, requiring perhaps several decades to reach a condition comparable to that before infestation. Increased competition with algae, soft corals, and other organisms that have preempted the space formerly occupied by corals further prolongs recovery. Additionally, if a moderately large number of starfish remain in the area and continue to feed on young settling corals, the effects could indeed be long lasting.

A serious concern

It is apparent that the success of reef-building corals depends on many biological factors: symbiotic algae, competitive interactions, bioerosion, movements of reef animals, substrate conditioning by herbivores, predation, and symbiotic crustaceans, to name but a few. An alteration of conditions affecting any one of these factors could have a marked effect on coral survival. As has been shown, activities as remote as the use of pesticides, the collecting of triton shells, or an occasional typhoon could conceivably contribute to a starfish infestation and a radical change in the species composition of reef communities. The complexity of such interactions, where disturbances at one level could produce widespread and magnified effects elsewhere in the community, are not to be dealt with lightly. Only by continuing research efforts into the various population processes that shape and regulate coral-reef community structure will man be better prepared to understand the response of coral reefs to a world increasingly affected by his activities.

FOR ADDITIONAL READING

R. E. Johannes, "Pollution and Degradation of Coral Reef Communities," in *Tropical Marine Pollution*, E. J. Ferguson Wood and R. E. Johannes, eds. (Elsevier Scientific Publishing Company, 1975).

O. A. Jones and R. Endean (eds.), *Biology and Geology of Coral Reefs*, vol. i: Geology 1 and vol. ii: Biology 1 (Academic Press, 1973).

D. R. Stoddart, "Ecology and Morphology of Recent Coral Reefs," *Biological Reviews* (vol. xliv, 1969, pp. 433–498).

C. M. Yonge, "The Biology of Coral Reefs," in *Advances in Marine Biology*, vol. i, F. S. Russell, ed. (Academic Press, 1963).

Biotoxins: The Poisons of Life

by Bruce W. Halstead

Once employed in primitive weapons of the soldier and hunter, the poisonous substances produced by living organisms are being intensively investigated for their pharmacological benefits.

Then raising the cup to his lips, quite readily and cheerfully he drank off the poison. . . . He walked about until, as he said, his legs began to fail, and then he lay on his back, according to the directions, and the man who gave him the poison. . . pressed his foot hard and asked him if he could feel; and he said, No; and then his leg, and so upwards and upwards, and showed us that he was cold and stiff. And he felt them himself, and said: When the poison reaches the heart, that will be the end. He was beginning to grow cold about the groin. . . . In a minute or two a movement was heard. . . ; his eyes were set, and Crito closed his eyes.

More than 2,300 years ago, Plato's careful eyewitness account recorded this execution by poisoning of his friend and teacher Socrates. His nearly clinical observations of the progression of symptoms make it clear that poison hemlock (*Conium maculatum*), which induces a general paralysis, was used in the poison cup, rather than water hemlock (*Cicuta virosa*), a painful convulsant. By the standards of many cultures Socrates' death was humanely administered, reflecting in his executioners both a sense of compassion and dignity and a discriminating awareness of toxic substances and their effects.

Predating even the ancient Greeks, however, the story of man's interest in poisonous plants and animals, and in the active biological principle or biotoxin produced by these organisms, can be traced at least to the dawn of written language. The records of vanished cultures as diverse in place and time as those in Mesopotamia, Egypt, Palestine, ancient China, classical Greece and Rome, and the Islamic state of the Middle Ages all testify that man had always found it valuable to know not only the means for surviving a poisonous bite or sting but also the techniques for turning such profound effects against an enemy or quarry. In fact, the English word toxic is rooted in the Greek *toxon*, "bow," and *toxikon pharmikon,* "poison for smearing arrows." Additionally, because of their dynamic and mysterious effects, toxic substances often became deeply integrated into the religion of a culture and its superstitious and magical lore.

It was, however, only after Renaissance thinking had penetrated the intellectual gloom of the Middle Ages that systematic studies were made in a manner consistent with modern toxicology. Many studies were published in the vernacular and became influential because they called attention to the injudicious use of medicinal plants and some of the effects of poisons on animals. The early 19th century saw the founding of experimental toxicology in the appearance of *Traité de toxicologie* (1814) by the French physician M. J. B. Orfila, who investigated inorganic agents and plant poisons, as well as the venoms of snakes, spiders, wasps, bees, and poisonous fish. In 1888 it was first demonstrated that the disease-producing agent of a microorganism in man was a toxic substance released by bacteria; in 1890 tetanus toxin was discovered. These kinds of investigations helped to establish the character and direction of much modern research—the detection and chemical identification of toxic substances and the determination of their biological effects, not only on organs or whole systems of cells but also on individual cells and their parts.

***BRUCE W. HALSTEAD** is Director of the International Biotoxicological Center at the World Life Research Institute.*

(Overleaf) Painting by Kerig Pope. Illustrations by Richard Keane

Survival of the deadliest

Knowledge of the evolutionary significance and development of most biotoxins is still poorly understood. Generally, in the struggle to survive, members of every kingdom developed chemical agents to procure food (*e.g.*, snakes, jellyfish, octopuses), to defend against predators (bees and some fish), or to discourage competition (the production of antibiotic substances by microorganisms). Some toxins are synthesized by the organism whereas others are obtained from its diet. For example, the cardiac toxin found in the monarch butterfly comes from the milkweed food of its larval caterpillar. Its presence apparently explains why birds refrain from eating the monarch.

It is logical to suppose that the same struggle for survival would equip organisms with natural defenses against these poisons. Instinct helps certain animal species avoid contact. Regurgitation and purgation reactions reject poisonous substances from the body and are common symptoms exhibited by poisoned animals. At the cellular level special biochemical processes produce antibodies that attack poisons and detoxify them by altering their chemical structure or blocking their mode of action.

In the case of man, painful experience rather than instinct has taught him to respect biotoxic substances. Together with these lessons his intelligence has enabled him to reach into the arsenals of other species, borrowing their practices and, in some instances, the chemical substances themselves. With increased awareness of the destructive effects of man-made poisons (petrochemicals, pesticides, and heavy metals) on the environment, scientists have grown more appreciative and more interested in the unique, biodynamic properties of the naturally occurring poisons. They are not studied merely to learn which should be avoided or which can take life. Many of these biotoxins can be of great pharmacological benefit. Their final effects, good or bad, are usually a function of dosage.

The meaning of toxicity

Without a consideration of dosage, it is scientifically impossible to assign a useful meaning to the term toxic. The classical definition referred only to poisons of microbial origin with high molecular weights that stimulated the production of antibodies. Recently the definition has included poisons from higher animals and plants. Even more broadly, however, any substance can be considered toxic if used in sufficient quantity. Obviously, then, one property of a toxin must be that it is harmful to sensitive hosts in relatively small amounts. The relationship between amount and effect for many toxins has been determined with test animals and can be depicted on a graph that plots dose against response. The steepness of the slope of the linear portion of the curve for any given toxin thus becomes an indicator of its potency. It should be noted, however, that this indicator need not coincide with two other properties of toxins; namely, the rapidity with which the effect is produced and the possibility of death of the host.

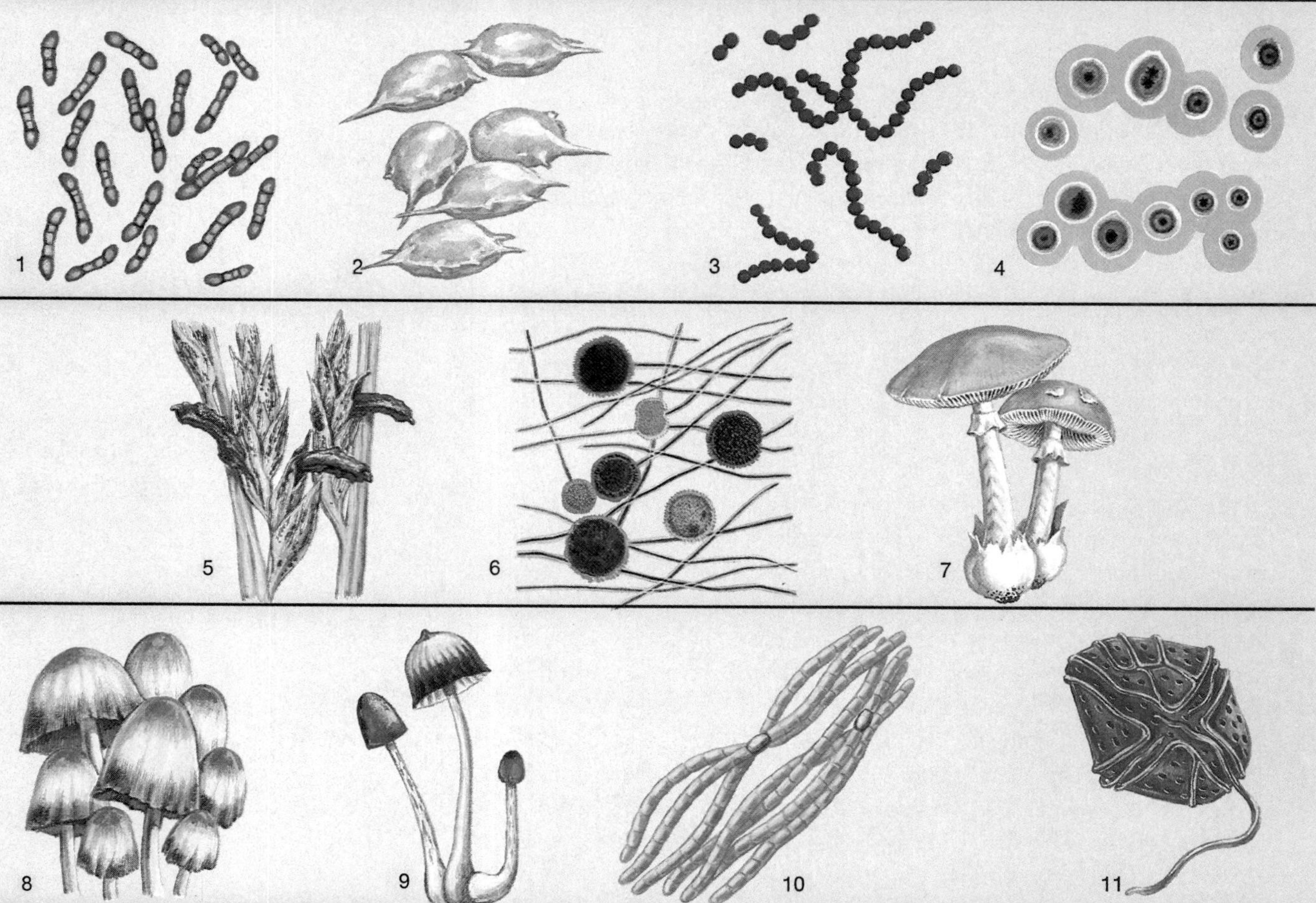

Selected toxic bacteria, fungi, and algae

organism	toxin and comments
1. *Corynebacterium diphtheriae* bacterium	diphtheria toxin (protein); responsible for most of the symptoms of the disease in man
2. *Clostridium botulinum* bacterium	botulinus toxin (protein); one of the most toxic substances known; average lethal dose 0.00015 milligrams per kilogram of body weight
3. *Streptococcus pyogenes* bacterium	streptolysin O, erythrogenic toxin (protein); causes scarlet fever, strep throat, other "strep" infections
4. *Escherichia coli* strains, *Salmonella* species bacteria	conventional endotoxins, a complex of protein, lipid, and polysaccharide; sources of common food poisoning
5. *Claviceps purpurea* (ergot) microfungus	complex of alkaloids and other molecules; source of LSD; causes ergotism, vomiting, nervous disorders, abortions
6. *Aspergillus flavus* microfungus	aflatoxin complex (at least 16 known toxins); damages liver, kidneys; among most potent liver-cancer causing agents known
7. death cup or death cap *(Amanita phalloides)* mushroom	amanitine, phalloidine; painful symptoms; fatality rate up to 90%; no known antidote
8. inky cap *(Coprinus atramentarius)* mushroom	toxin unknown; edible when immature, although some people experience giddiness and gastrointestinal upset; alcoholic beverages seem to intensify symptoms
9. Mexican hallucinogenic mushroom *(Psilocybe mexicana)*	psilocybin, psilocin; causes euphoria, loss of sense of distance and size, and hallucinations; used since Aztec times for religious ceremonies
10. *Aphanizomenon flos-aquae,* a freshwater alga	toxin is perhaps a polypeptide neurotoxin; causes mass mortality in fish, convulsions in many animals, and inflammation of intestinal tissues in man
11. *Gonyaulax tamarensis,* a dinoflagellate (placed by some taxonomies in the animal kingdom)	saxitoxin (paralytic shellfish poison); causes tingling, burning, weakness, thirst, muscular paralysis; passed to man through shellfish that have fed on species

In most instances the dose-response curve is plotted from values that measure the death of the host. The 50% lethal dose (LD_{50}), or the dosage that will result in the death of 50% of the test animals, has been established as one of the most useful ways to estimate toxicity. The LD_{50} of a toxic substance is given in grams, milligrams, or micrograms of that substance per kilogram body weight of test animal. Similar classification systems provide values that attempt to estimate such quantities as the minimum lethal dose or the probable lethal dose, and some have been integrated into a descriptive system of six classes of toxicity, ranging from the supertoxic (class 6; probable lethal dose less than 5 milligrams per kilogram; *e.g.*, digitoxin, botulinus toxin) to the practically nontoxic (class 1; probable lethal dose more than 15 grams per kilogram).

Biotoxins can be conveniently grouped into three major categories: microbial toxins, produced by bacteria, algae, fungi, etc.; phytotoxins, produced by higher plants; and zootoxins, produced by animals. The geographical distribution of poisonous organisms varies greatly. Poisonous microorganisms seem to be found everywhere; poisonous plants and animals are found most abundantly in warm temperate and tropical regions.

Microbial poisons

Many species of microbial life produce biotoxins. Various attempts to classify these poisons have been hampered by a lack of precise knowledge concerning the biological nature and phylogenetic relationships of the species and the chemical complexity of the poisons.

The poisonous agents produced by bacteria are often segregated into two groups—exotoxins and endotoxins. The exotoxins are composed of protein and usually do not contain nonprotein substances; most of them also stimulate the formation of antibodies. In bacterial cultures the exotoxins may appear during the declining phases of growth. In some cases they are released at the time of normal destruction of the cells after death. They are less stable in the presence of heat than endotoxins; in many instances, food that had become contaminated and extremely toxic because of improper home canning procedures was unsuspectingly made safe because it had been boiled before eating. The toxicity of many exotoxins is high. The protein causing botulism, for example, has been known to produce severe symptoms from a mouthful of food merely chewed but not swallowed. Exotoxins are neutralized by homologous antibodies, which are active agents in blood serum produced by a process involving the bacteria against which the serum is to be used.

Endotoxins are complexes of proteins, polysaccharides (large molecules built up of numerous sugars), and lipids (fats). The protein part determines the quality of being reacted against as a foreign substance. The polysaccharide part places limits on the types of antibodies that can react with the endotoxin molecule. Some of the lipids possibly determine the toxicity. Endotoxins are derived from the bacterial cell wall and are released only upon the death of the cell. They are not neutralized by homologous antibodies and are relatively stable in the presence of heat.

Fungi are plantlike organisms that do not contain any chlorophyll. About 85,000 species exist, of which 55,000 are microscopically small. The ability of certain fungi to produce toxic effects has long been recognized. A classic example is the series of "epidemics" known in the Middle Ages as St. Anthony's fire, caused by the infestation of rye with the fungus *Claviceps purpurea*, or ergot, which transformed the grain into a complex of alkaloids and other toxins, including LSD. New toxic fungi are still surprising man. In 1960 the eruption of a mysterious malady of turkeys in England, called turkey X disease, was traced to moldy peanuts used in turkey feed; this resulted in discovery of complex substances called aflatoxins. And this in turn led to a renewed interest in fungal toxins or mycotoxins. Now recognized as potential cancer-producing agents that become involved in man's food supply, fungal poisons are proving important in the study of environmentally caused cancer.

Of the more than 30,000 species of large mushrooms, only a few are poisonous when eaten. About 90% of the deaths attributed to toxic mushrooms, or toadstools, results from eating members of the genus *Amanita*. No single rule or test method exists by which the toxicity of a mushroom can be determined. The most poisonous species closely resemble some of the most prized edible species; in addition, toxicity within a given wild species may vary from one set of ecological conditions or geographic locality to the next. Moreover, although some mushrooms that are poisonous when fresh are edible when cooked, dried, or preserved in some other way, others remain poisonous in spite of all procedures. It has also been observed that some people may become poisoned by eating mushrooms that apparently do not affect others.

Algae comprise some of the smallest forms of chlorophyll-bearing organisms. Two major groups produce toxins—blue-green algae and dinoflagellates. The approximately 1,500 species of blue-green algae are very primitive, resembling bacteria in a number of ways. Widely distributed, they live in all parts of the aquatic environment that receive light, including waters perpetually near freezing or boiling. Some strains of a species are toxic whereas others of the same species are not. Water blooms (particular times of concentration) of blue-green algae have been responsible for the deaths of fish, waterfowl, cattle, horses, and other animals. They have also been implicated in human poisonings.

Dinoflagellates, numbering about 1,100 species, are one-celled organisms that rely upon various inorganic nutrients in the water and upon photosynthesis. They constitute an important part of plankton, the broth of minute animals and plants that is the primary food supply in the oceans. Planktonic blooms, called red tides because they so color the water, are often associated with weather disturbances that may bring about changes in water masses or upwellings. During these periods large numbers of toxic dinoflagellates may be ingested by shellfish, thus causing poisons to accumulate in their digestive glands. Man and animals may in turn be poisoned by eating the shellfish. Certain species of dinoflagellates are capable of producing some of the most toxic substances known. The two species most commonly involved in human poisoning are *Gonyaulax catenella* along the Pacific coast of North America and *G. tamarensis* along its Atlantic coast. The poison, variously called paralytic shellfish poison, mussel poison, and saxitoxin, is a complex nonprotein, nitrogen-containing compound. There is no specific treatment or antidote for it, and the best protection is rapid detection followed by effective public-health quarantine measures.

One of the more recent blooms of red tide occurred along the New England coast in September 1972, resulting in a temporary ban on shellfish sales along the New England seaboard and the recalling of interstate shipments of fresh and frozen mussels and clams. Despite all efforts, hospitals in the affected region reported dozens of cases of paralysis from shellfish poisoning, some so severe that iron lungs were needed to supplement failing chest muscles. Less severe respiratory irritation also resulted from the inhalation of toxic products in the windblown spray from bloom areas.

Poisonous plants

Most of the higher plants that produce toxins are angiosperms, or flowering plants; but, of their more than 200,000 species, only a small percentage are recognized as poisonous. This percentage may be classified according to the chemical nature of their toxic agents, their phylogenetic relationship, or their botanical characteristics. The following classification, based on toxic effects, is also useful: (1) plants poisonous to eat, (2) plants poisonous on contact or injection, (3) plants producing photosensitization, an abnormal reaction to sunlight, and (4) plants producing airborne allergies.

Phytotoxins comprise a large range of biologically active substances, including alkaloids, glycosides, resins, toxalbumins, and a group of structurally undetermined molecules. Alkaloids, found in an estimated 10% of all plant species, are bitter-tasting, nitrogen-containing compounds with the ability to react with acids to form salts. Most of the 5,000 alkaloids characterized thus far cause a strong physiological reaction when administered to an animal; common examples are caffeine, nicotine, cocaine, and atropine.

Among the most interesting alkaloidal toxins of historical and modern significance is curare, obtained from several tropical American plants of the genera *Chondodendron* and *Strychnos*. Curare is considered poisonous only upon injection because, when merely ingested, it is absorbed slowly from the alimentary canal and has little effect. Crude preparations, which contain several bioactive compounds, have been used for centuries by South American Indians to poison arrow

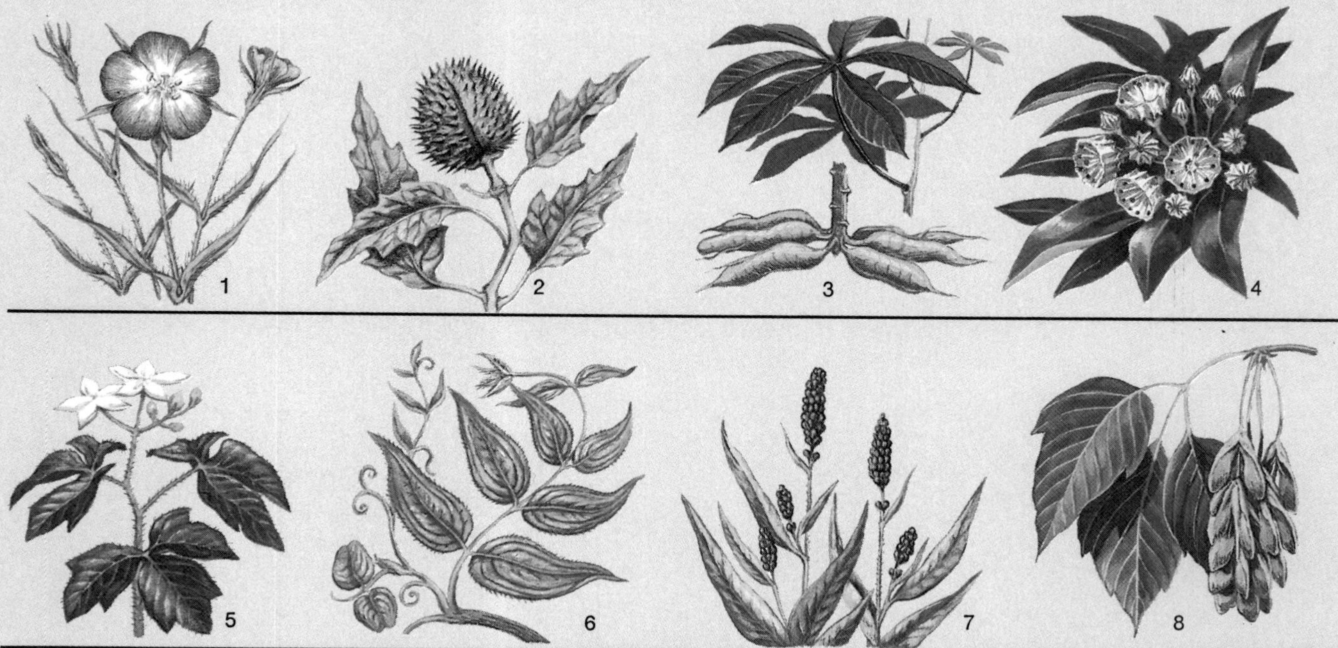

Selected toxic plants

organism and distribution	toxin and comments
1. corn cockle *(Agrostemma githago);* North America, Europe	githagin, agrostemmic acid; poisonous to eat; milled seeds sometimes contaminate wheat flour, resulting in pain, burning, and prickling sensations in lower extremities, and increasing paralysis
2. jimson weed *(Datura stramonium);* temperate, tropical regions	hyoscyamine, atropine, and other alkaloids; poisonous to eat; symptoms include dryness of skin, mouth, throat, cyanosis, mania, vomiting; alkaloids used as sedatives, stimulants, antisecretory agents; children often poisoned by seeds or flowers; smoked in certain American Indian ceremonies to produce hallucinations
3. cassava *(Manihot esculenta);* tropics	glycosides capable of producing cyanide compounds; poisonous to eat; must be detoxified before plant can be used as food
4. mountain laurel *(Kalmia latifolia);* North America	andromedotoxin; poisonous to eat; causes watering of the eyes, nose, and mouth, dizziness, vomiting; children poisoned by eating leaves, sucking flower nectar; poisoned honey reported from beehives near large stands of *Kalmia*
5. spurge nettle *(Jatropha stimulosa);* North America, Europe, Asia	toxin unknown; poisonous by contact; stinging, itching, temporary skin eruption produced by stinging hairs that inject toxin into skin
6. *Strychnos toxifera,* a woody vine; tropical America	curare (tubocurarine and other alkaloids); poisonous upon penetration of skin; produces muscle relaxation and ultimately respiratory failure; used as arrow poison and for blowgun darts
7. buckwheat *(Fagopyrum sagittatum);* North America, Europe	fagopyrin; ingestion by farm animals causes deposit of photosensitizing pigment in skin; sunlight then produces skin irritation, eyelid swelling, prostration
8. box elder *(Acer negundo);* Northern Hemisphere	oleoresin and a water soluble antigen; causes hay fever and a form of dermatitis; some species of poplar, ash, and elm produce same allergic reactions

and spear tips. In modern medicine the constituent alkaloids of curare find use as powerful muscle relaxants in surgical anesthesia and during prolonged examinations of the throat, rectum, and other body cavities that need to be held open. Their properties also assist in the relief of various muscle contractions and convulsions, such as in back spasms of the sacroiliac region and in severe cases of tetanus.

Some glycosides, which are compounds that yield sugar and nonsugar (aglycones) components when hydrolyzed (chemically degraded by water molecules), are extremely toxic to animals. The toxicity of the cassava, an important food plant of the American tropics that has to be rendered nonpoisonous by a complex refining system, is due to a glycoside. Resins are heterogeneous assemblages of complex compounds that differ widely in chemical properties. Some are physiologically very active, causing irritation to nerve and muscle tissue. Toxalbumins are highly toxic protein molecules found only in a small number of plants. Ricin, a toxalbumin from the castor bean (*Ricinus communis*), is one of the most toxic substances known.

Poisonous animals

Animals that produce toxic substances are widely distributed throughout the approximately one million species of the kingdom; the only major group that seems to be exempt is birds. The percentage of animals that are poisonous is not known, but there are probably many more inhabiting the marine environment than are found on land.

Zootoxins can be divided into: (1) oral poisons, those poisonous when eaten; (2) parenteral poisons or venoms, those produced by a specialized poison gland and administered by means of a venom apparatus that is able to penetrate the skin of the victim; and (3) crinotoxins, those produced by a specialized gland but merely released into the environment, usually by means of a pore. Oral zootoxins are generally thought to be small molecules. Most venoms are apparently large molecules, usually a protein or a substance in close association with one. Little is known about the biological or chemical properties of most crinotoxins, although much recent research has concentrated on the structure and synthesis of certain frog and toad crinotoxins.

It is difficult to generalize when dealing with poisonous animals because they involve such a complex assemblage of unrelated organisms. Although there are poisonous animals in the polar regions, the greatest number and variety inhabit warm temperate or tropical areas. One of the most dangerous of all toxic organisms probably is the sea wasp (*Chironex fleckeri*), which inhabits the northeast coastal waters of Australia. By means of its stinging apparatus this jellyfish is able to inflict death in an adult human in less than one minute. Among the least suspected of poisonous animals is the polar bear. Although its flesh is edible, its liver contains vitamin A in concentrations that can cause severe intoxication in humans. Polar explorers living off the land have poisoned themselves inadvertently and died because of their unfamiliarity with the toxicity of polar bear liver.

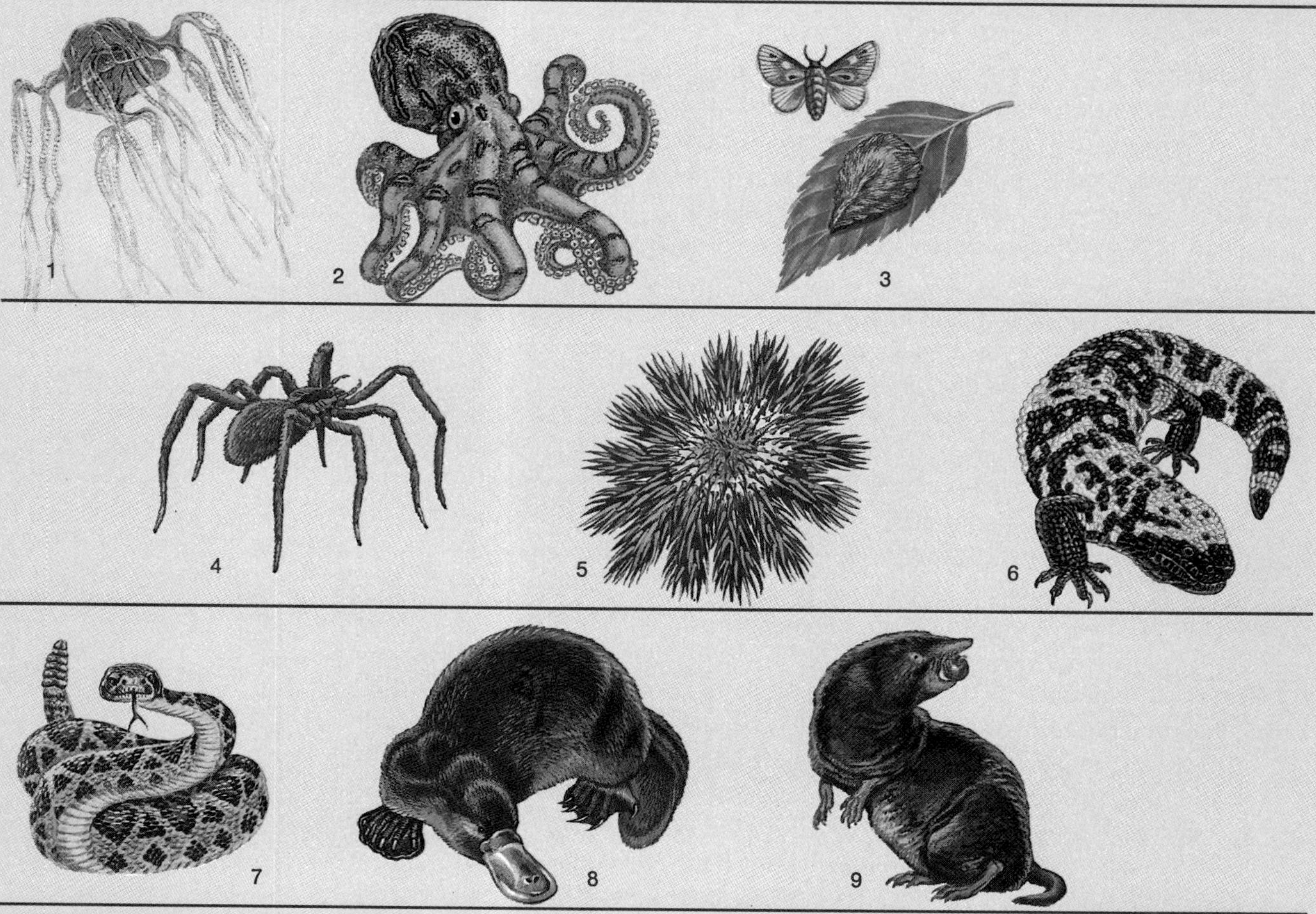

Selected animals that inflict a venom

organism and distribution	toxin and comments
1. sea wasp *(Chironex fleckeri),* a jellyfish; northern and northeast Australia	a cardiotoxin; poison ejected from stinging cells on hanging tentacles; in man causes large wheal-like skin lesions, circulatory failure, rapid death in high percentage of cases; probably the most venomous of all creatures; about 9,000 species of the phylum (includes jellyfish, anemones, corals) secure prey by means of such specialized stinging cells
2. spotted octopus *(Octopus maculosus);* Indo-Pacific, Indian Ocean	cephalotoxin, a neuromuscular poison; causes increasing numbness and paralysis, frequent deaths
3. puss caterpillar *(Megalopyge* species); U.S., Latin America	toxin unknown; stinging hairs are associated with poison glands; contact produces intense itching, pustules, fever, swelling, paralysis; six-day recovery period
4. brown spider *(Loxosceles* species, including brown recluse); U.S., South America, Europe, Asia	hyaluronidase and toxins that destroy tissue cells and dissolve red blood cells; causes blistering, bleeding into the skin and internal organs, fever, cardiovascular collapse, sometimes death
5. crown-of-thorns starfish *(Acanthaster planci);* Indo-Pacific	nature of toxin unknown; spines produce painful wound, swelling, numbness, paralysis
6. Gila monster *(Heloderma suspectum),* a reptile; southwestern U.S., northern Mexico	heloderma venom, primarily neurotoxic; all teeth are venomous; causes respiratory distress, cardiac failure; no antiserum available
7. eastern diamondhead rattlesnake *(Crotalus adamanteus);* southeastern U.S.	cephalinase and other enzymes; bite produces edema, bleeding into the tissues, anemia, yellow vision, death; antiserum available; *Crotalus* species handled in Hopi Indian snake-antelope ritual
8. duck-billed platypus *(Ornithorhynchus anatinus),* a mammal; Australia	protease and other enzymes; poison inflicted by means of venom-charged spur on hind leg of male, producing immediate intense pain, swelling; no known deaths
9. short-tailed shrew *(Blarina brevicauda);* North America	toxins that affect nervous system and break down protein; lower incisor teeth associated with poison glands; bite produces localized pain; no known deaths

Some very complex biotoxicological relationships are found in the marine environment. Certain marine biotoxins, such as those which cause ciguatera, or tropical fish poisoning, apparently originate in marine plants—perhaps a toxic form of blue-green algae—that are ingested by plant-eating animals. The toxins are then passed on to flesh eaters and eventually to man. The convoluted route by which this is accomplished has eluded scientists for many years. With the buildup of toxic industrial chemical pollutants in the marine environment, the problems of toxicity in marine organisms are becoming more serious. There is evidence that under certain conditions chemical pollutants may trigger biotoxicity cycles in marine organisms. The outbreaks in Japan in the 1950s of Minamata disease, a poisoning caused by mercury compounds in industrial waste discharged into Minamata Bay in Kyushu, were the result of such a cycle: microorganisms, algae, shellfish, and fish ingested or absorbed the highly toxic compounds and were in turn consumed by humans, causing a number of deaths.

"In poison there is physic"

This fragment of Shakespearean dialogue summarizes the thinking of those involved in directing the present and future efforts of modern biotoxicology. Drug companies are showing progressive interest in poisonous organisms as sources of pharmaceuticals. F. Hoffmann-La Roche & Co. has recently established a multimillion-dollar institute of marine pharmacology just north of Sydney, Australia, to investigate marine biodynamic substances, including biotoxins, and their applications to medicine. Some company laboratories have already reported finding antibiotics and antiviral and anti-inflammatory agents. Recent studies backed by the U.S. National Cancer Institute and the Sea Grant program resulted in the testing of marine biodynamic substances derived from sponges, corals, sea whips, and sea anemones. Extracts of 1,665 specimens tested by scientists at the University of Oklahoma produced 104 agents active against leukemia in mice and 30 against human cancer cells in tissue culture.

Poison from the skin glands of certain toads (genus *Bufo*) has been used in Asia for several thousand years for treating such afflictions as sores, inflammations, and toothache. By the 1970s its bioactive agents had been identified as specific steroid hormones. Some of these were found to have an effect on heart action equal to digitoxin and local anesthetic effects 90 times those of cocaine. One steroid has already been used in man as a respiratory stimulant; another seems to inhibit certain types of cancer. With such promising virtues a need to obtain these substances in large quantities has arisen, and several routes of chemical synthesis are under development.

Because of the environmental hazards caused by many of the synthetic pesticides, attention is now being directed toward degradable biotoxins as a source of new pesticides. One interesting line of related research in weed control is making use of the phenomenon of antibiosis or allelopathy, the natural process by which plants produce and

Selected animals poisonous to eat

organism and distribution	toxin and comments
1. sea cucumber *(Holothuria* species); tropical Pacific	holothurin; most sea cucumbers are edible, though toxic species difficult to identify; can be fatal
2. whelk *(Neptunea* species); Europe, Pacific region	tetramine; causes nausea, diarrhea, intolerance to light, dryness of mouth; toxin believed localized in salivary glands
3. Greenland shark *(Somniosus microcephalus);* Arctic regions	toxin unknown; flesh poisonous, may be detoxified by drying; ingestion can cause death
4. moray eel *(Gymnothorax javanicus);* Indo-Pacific	ciguatoxin; toxicity common to more than 400 species of tropical reef fish; toxin apparently originates in marine plants, contaminates fish otherwise edible
5. bluefin tuna *(Thunnus thynnus);* temperate and subtropical seas	saurine (a histamine-like toxin) produced in flesh when fish allowed to stand at room temperature for several hours; symptoms resemble a severe allergy: headache, nausea, throbbing of blood vessels, massive red welts, itching; responds to antihistamines
6. leatherback turtle *(Dermochelys coriacea);* temperate and tropical seas	chelonitoxin (chemistry unknown); produces nausea, diarrhea, foul breath, skin rash, sloughing of skin, enlargement of liver; fatality rate high
7. sei whale *(Balaenoptera borealis);* worldwide	toxin unknown; livers of many marine mammals toxic; antihistamines used in treatment
8. polar bear *(Ursus maritimus);* Arctic	liver contains high concentrations of vitamin A, possibly other toxic substances; ingestion causes intense headaches, nausea, muscle cramps, visual disturbances

1

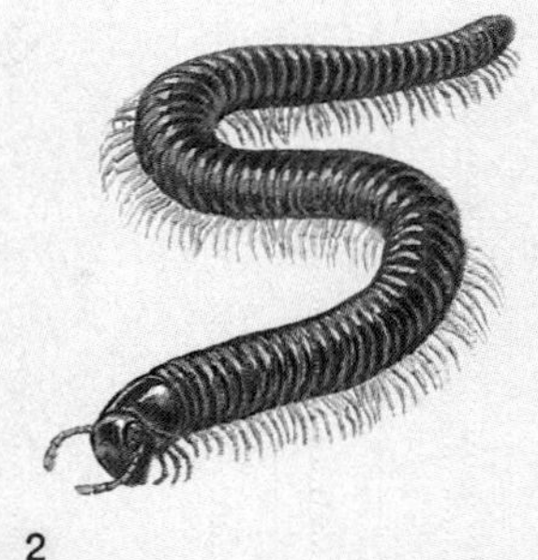

2

3

Selected crinotoxic animals

organism and distribution	toxin and comments
1. red moss sponge *(Microciona prolifera);* eastern U.S. coastal waters	toxin unknown; contact produces a chemical irritation of the skin, swelling, blisters, pustules
2. millipedes (species of *Orthoporus, Rhinocricus, Julus,* and *Spirobolus);* temperate and tropical regions	toxin unknown; fluid distasteful to enemies is exuded or squirted from body pores up to distance of about 30 in.; contact with skin causes burning, pigmentation; contact with eyes can cause temporary blindness
3. tree frogs (some species of *Hyla* and *Phyllobates);* northern South America and Central America	batrachotoxin, steroidal alkaloids, histamine, and other toxins; skin secretions produce a burning sensation when handled; ingested poison causes vomiting and abdominal pain; used by Indians as arrow poison

secrete toxic substances that inhibit the growth of other plants. Until lately the very existence of allelopathy had been only reputed; it is now considered to be a rather common process. In a recent experiment under laboratory conditions, the growth of two economically important weeds potted with cucumber plants was retarded by as much as 34 and 22%. Collected nutrient solution that had leached through the cucumber potting medium also was found to inhibit weed growth and verified the reasoning that the cucumbers did indeed produce a toxin inhibitory to certain weeds. It is hoped that such natural competitive abilities can be bred back into domesticated crop plants that may have lost them during the years of artificial selection.

Biotoxicologists are continuing their research on the basic detection and identification of new and little known poisons. Such substances periodically make their debut at unsuspected times and places, as in outbreaks of animal or human intoxications. This type of research might be classified as epidemiological in nature and is of great importance because of its bearing on food resources and the development of valuable molecular structures that might otherwise have been overlooked. It often requires a great deal of sleuthing in order to properly document the causative agent. The ultimate goal, however, in most biotoxicological research is the ability to chemically characterize the molecular structure of these poisons. This knowledge is vital to their proper control and utilization—for whatever purpose.

Superconductivity in Electronics

by Wilhelm Anacker

The extraordinary behavior of certain materials when cooled to temperatures approaching absolute zero is contributing to the design of electronic devices of unprecedented speed, sensitivity, and flexibility.

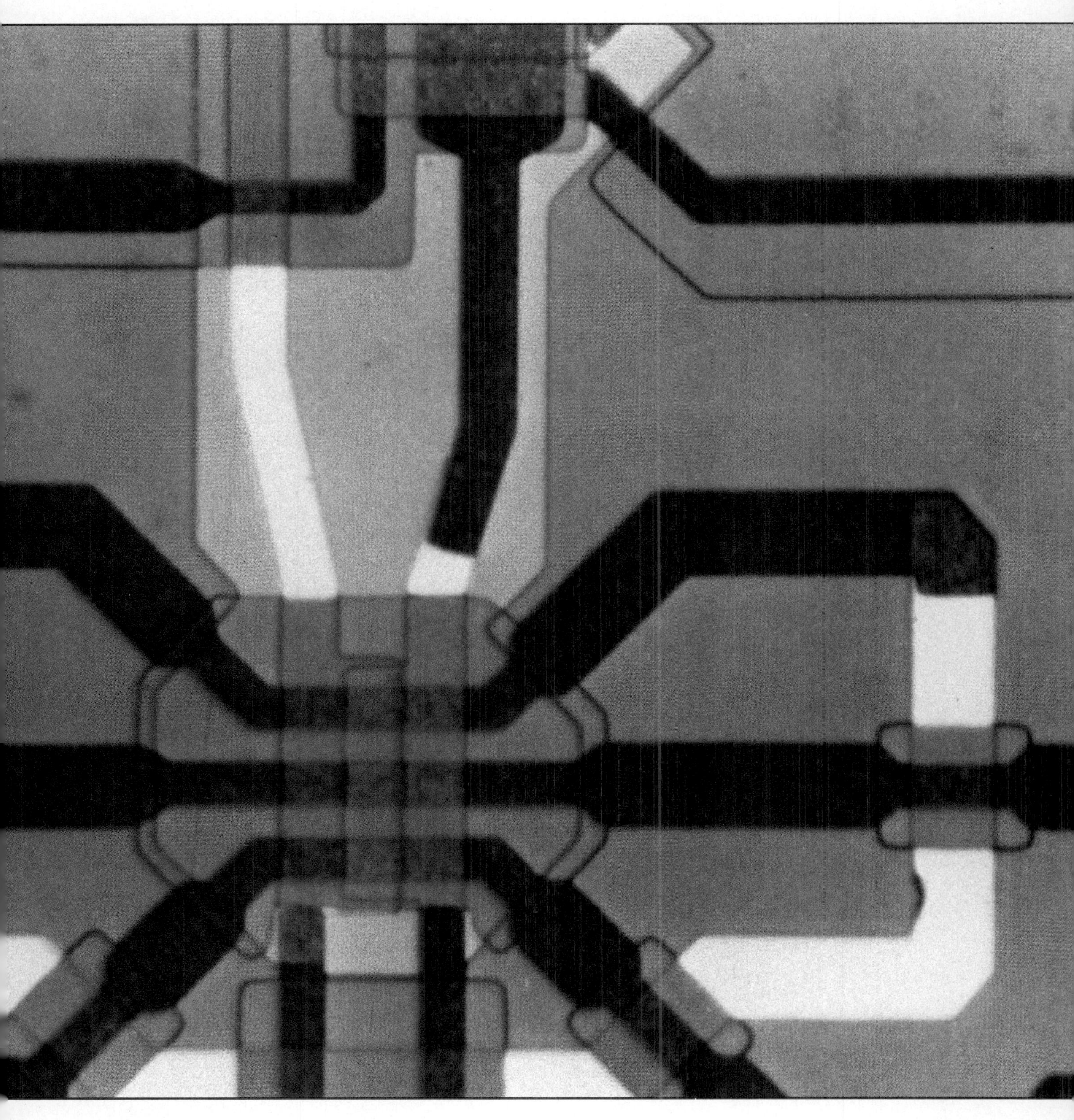

Superconducting electronics is a young branch of electronics that began in the late 1950s and early 1960s after a quick succession of successful theoretical predictions and significant discoveries. Prior to this period, the field had been considered an esoteric scientific phenomenon and laboratory curiosity at best. The reawakening of interest coincided with efforts to use superconductors for electromagnets with unprecedented field strength and with the development of powerful, reliable refrigeration systems for cryogenic applications needed in the U.S. space program.

Superconductivity is an extraordinary state of solid matter attained in certain materials and characterized by a complete absence of electrical resistance and exclusion of magnetic fields. The superconducting state exists only when the material is cooled below a critical temperature, typically a few degrees above absolute zero (0° K or −273° C), and when the magnetic field surrounding the material is below a critical strength. Samples or devices are most conveniently brought to ultracold temperatures by immersion in liquid helium kept in a thermally well-insulated cryostat. Helium has the lowest boiling point of all liquids—a frigid 4.2° K under atmospheric pressure, and even lower temperatures under reduced pressures—and will maintain its boiling temperature until all the liquid evaporates.

Superconductivity had eluded satisfactory explanation since its discovery in 1911 by the Dutch physicist Heike Kamerlingh Onnes when he observed that frozen mercury lost its electric resistance abruptly and completely once its temperature had dropped below about 4.2° K. Its mysteries persisted almost 50 years until John Bardeen, Leon Cooper, and John Schrieffer succeeded in 1957 in developing a comprehensive theory of superconductivity. Now known as the BCS theory, it explains, in principle, all known properties of the superconducting state and the mechanism causing the normal-to-superconducting transition. Still another five years passed before Brian Josephson discovered a new and central aspect of superconductivity. Called the Josephson effect, it not only provided a better basic understanding of the phenomenon but also laid the foundation for modern superconducting electronics.

The nature of superconductivity

Thousands of metallic elements and alloys and a few semiconductors are now known to be superconductors, among them such common metals as aluminum, tin, lead, and niobium. It is very difficult, however, if not impossible, to prove the complete absence of electrical resistance conclusively by experiment. At best, an upper limit of 10^{-22} ohm for the residual resistance of a superconductor has been established, with no indication that the resistance is not indeed zero. A superconductive material behaves exactly like a normal metal as long as its temperature is above its critical temperature. Below this temperature it may be superconducting, depending on the magnetic fields and electric currents to which it is subjected. More precisely, the

WILHELM ANACKER *is Manager of the Exploratory Cryogenic Technology Department at the Thomas J. Watson Research Center of IBM.*

(Overleaf) Courtesy, Institute for Electrical Engineering, photograph, J. H. Greiner

superconducting state exists in a superconductor only as long as its temperature, the magnetic field around it, and the electric current through it remain below critical magnitudes. The range in which superconductivity exists may be illustrated by a three-dimensional coordinate system of temperature, magnetic field, and electric current. A curved surface spanning the critical magnitudes of temperature, field, and current marks the normal-to-superconducting transition; the volume enclosed by the surface represents the superconducting state and the space above it the normal conducting state. Each superconductor has its own specific critical temperature, critical field, and critical current, the latter being related to the critical field. The highest critical temperature definitely established is about 23° K, but critical temperatures for most superconductors are below 5° K.

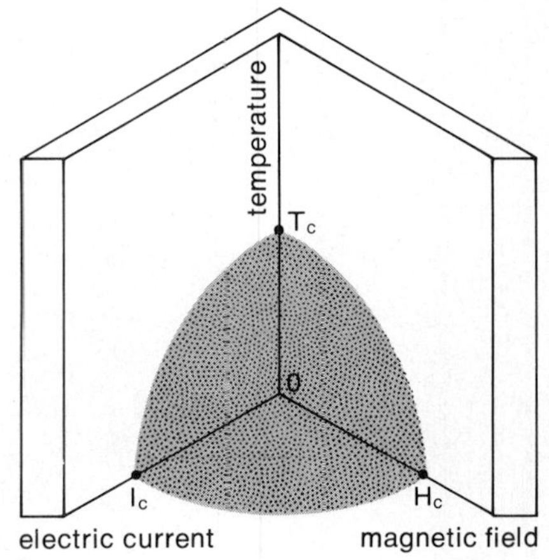

Superconducting and normally conducting states of a superconductor plotted on a three-dimensional coordinate system. The points I_c, T_c, and H_c denote critical magnitudes of current, temperature, and magnetic field that cannot be exceeded if the superconducting state is to exist (see text).

Initially, superconducting properties were believed to be restricted solely to the absence of resistance or to perfect conductivity. Though it had been observed that magnetic fields did not penetrate into a superconductor, the effect was attributed to circulating currents in the material that generated opposing magnetic fields and persisted due to the lack of resistance. In 1933 the validity of this explanation was tested for the first time by Walther Meissner and R. Ochsenfeld with surprising results. They found that a magnetic field present in a superconductor just above its critical temperature was not maintained, as would be expected from the perfect-conductivity model, but that it was in fact expelled from the interior of the superconductor during the normal-to-superconducting transition. This field exclusion, or Meissner effect, has since been regarded as another fundamental property of the superconducting state. Field exclusion is equivalent to perfect diamagnetic behavior; *i.e.*, the formation of a magnetic dipole equal in strength and opposite to the direction of the applied magnetic field. Diamagnetic behavior of superconductors has been demonstrated lucidly in an experiment in which a permanent bar magnet floats in helium gas above a superconducting lead bowl, the magnet held floating against the force of gravity by the repulsive forces between the diamagnetic bowl and the ferromagnetic bar.

A consequence of the Meissner effect is the trapping of magnetic flux, or lines of magnetic induction, around superconducting rings. When a ring is cooled below its critical temperature while a magnetic field is present, the flux will be expelled out of the material of the ring but not out of the hole in its center. When the external field subsides while the ring remains superconducting, the flux threading the hole cannot escape; instead its flux lines close around the ring. Because magnetic flux and electric currents are inevitably linked, a circulating current is established in the ring as well; it will circulate persistently and maintain the flux indefinitely. In one experiment a persistent current actually circulated for more than 2½ years without power consumption; this occurred in a ring of lead that was immersed in liquid helium during a futile search for residual electric resistance in a superconductor.

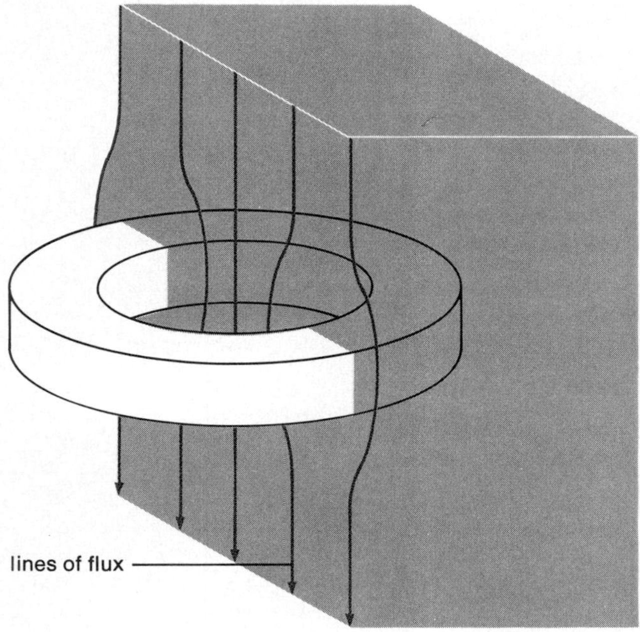

external magnetic field removed

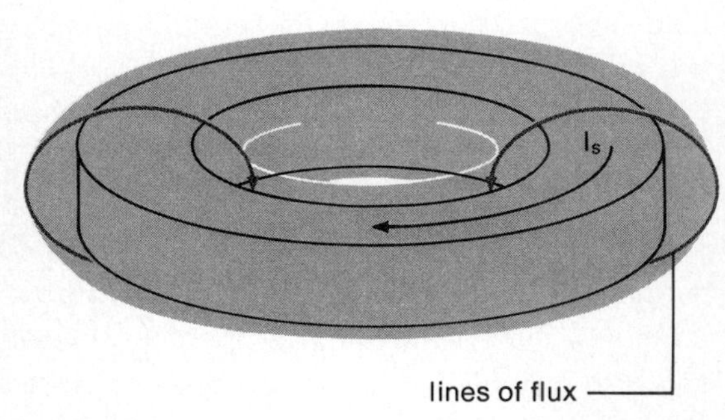

When an external magnetic field is present during the transition of a superconducting ring from its normal to its superconducting state (left), magnetic flux is expelled from the ring but not from the hole in its center. When the external field is removed (right), the flux lines threading the hole close in on themselves. Established in the ring is a circulating current I_s that will maintain the flux indefinitely.

It had long been suspected that perfect conductivity, diamagnetic behavior, and other superconducting properties were caused by "superconducting electrons" of some kind. Their real nature, however, as well as the mechanism of their formation, remained obscure until explained by the BCS theory. The theory identified these superconducting electrons as paired electrons, called Cooper pairs. These are produced by an intricate process involving phonons; *i.e.*, vibrations or deformations in the crystal lattice (atomic arrangement) of the superconducting material. Specifically, the motion and charge of one electron deforms the lattice and thereby affects the motion of another electron. When the temperature is low enough and both electrons possess the proper energy as well as equal and opposite momentum and spin, the electron-phonon interaction leads to a weak binding force between the two electrons, with a slightly lower total energy than the sum of the energies of both electrons when unpaired. The number of Cooper pairs or their density (the number per unit volume) is zero at and above the critical temperature and increases gradually as the temperature is lowered toward absolute zero, until at 0° K (were that state ever to be reached) all electrons that conceivably could have become Cooper pairs would have done so. The number of single electrons that can participate in the Cooper pair formation diminishes, of course, at the same rate as the number of Cooper pairs increases. Below the critical temperatures, the disappearance of single electrons from a region of quantum or energy states that they occupy in the electronic structure of a superconductive metal in the normal conducting state creates what is termed the superconducting energy gap. This region of energy states in the electronic structure becomes depleted of electrons and is unable to accept electrons.

The electron-phonon interaction enables the electrons of Cooper pairs to "feel" each other over thousands of interatomic distances. Millions of Cooper pairs in fact can overlap in space. Furthermore, their motion is strongly correlated, so that the total Cooper pair system in a superconductor may be viewed as a rather rigid, or incompressible, electron fluid that reacts as a whole to perturbations imposed on it anywhere. Thus, when a low-energy disturbance, such as an electric current below the critical value, is applied to a superconducting material, it is transmitted without effect on the quantized state; *i.e.*, without experiencing resistance.

Since Cooper pairs act as a quantum-mechanical system, they can be described by a wave function which satisfies a Schrödinger equation. A Schrödinger equation is the basic differential equation in wave mechanics. Its solutions, called probability waves or wave functions, assign to particles a wave nature with quantum or energy states. For example, an electron may be viewed either as a sine wave with quantized wavelength in a crystal or as a particle encircling an atomic nucleus in a quantized orbital. Transition from one quantum state to another can only occur when discrete quanta of energy of the right magnitude, *e.g.*, photons or phonons with the right wavelength, are absorbed or emitted, and when the quantum state into which the transition is attempted is not occupied by another electron. Energies of lesser amount are rejected and do not perturb the prevailing quantum state at all. All Cooper pairs in a superconductor possess the same energy and therefore obey a single "macroscopic" Schrödinger equation. In a sense, a superconductor may be looked upon as a giant atom that is orbited by a single particle system composed entirely of Cooper pairs.

A further consequence of the quantum-mechanical nature of the Cooper pair system is the requirement that magnetic flux always be trapped in discrete magnitudes or superconducting flux quanta. The fundamental unit of the superconducting flux quantum is represented by the ratio of two fundamental physical constants, $h/2e$, in which $2e$ denotes the electronic charge of a Cooper pair and h Planck's constant of energy quantization. This relationship is of central importance to the Josephson effect.

Tunnel devices and "weak" superconductors

In developing his theories on superconductivity, Josephson studied the interaction of Cooper pairs in two superconductors in close proximity but not in intimate contact. The structure he investigated consisted of two superconductors separated from each other by a tunnel barrier; *i.e.*, an extremely thin insulator, not thicker than about 30 Å (angstrom; 1 Å = 10^{-8} cm). These structures are known as tunnel devices; they have peculiar characteristics even at room temperature. An insulator between two metals can be viewed as a potential barrier for electrons, similar to a dike between two water reservoirs. In the view of classical physics, electrons would have to acquire sufficient

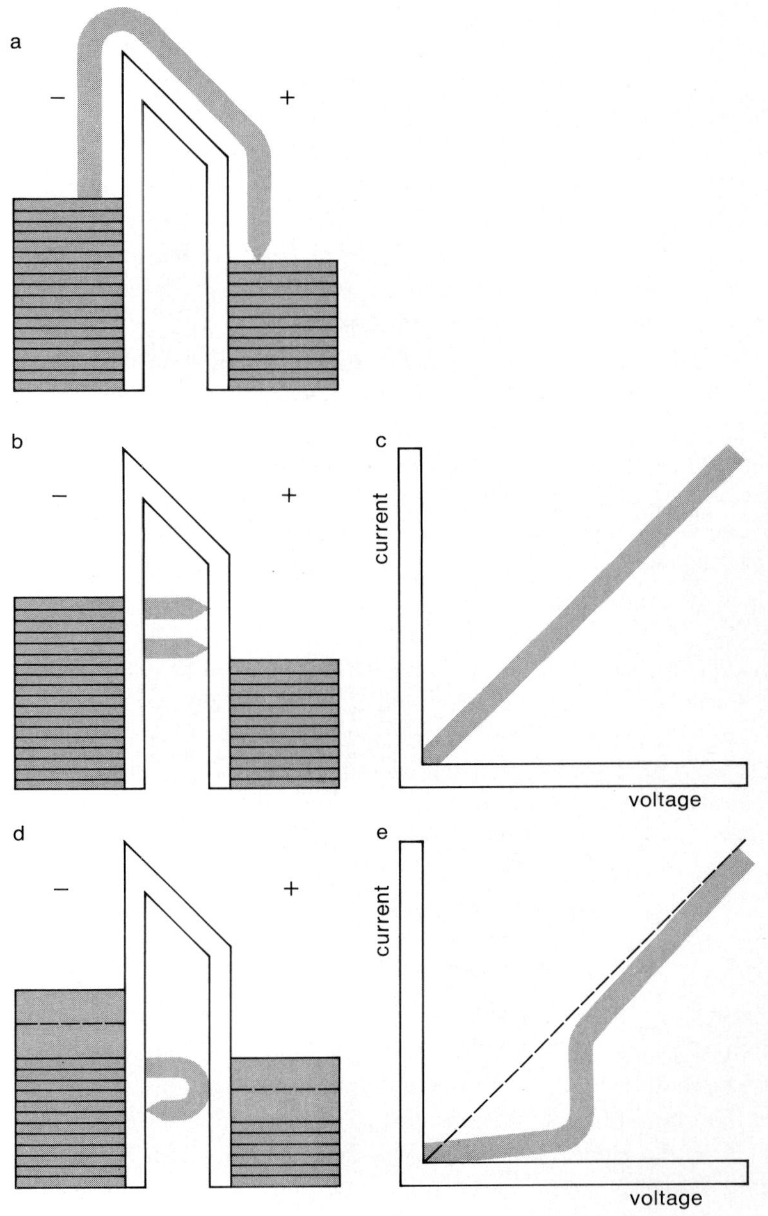

energy to surmount the potential barrier of the insulator—regardless of how thin it is—so that they can flow from one metal to the other, just as water must be raised in height with an input of energy to surmount the dike. From the quantum-mechanical view of physics it is, however, quite possible, even probable, for electrons to penetrate or tunnel through a thin insulator or tunnel barrier without having to acquire the energy necessary to surmount it. The water reservoir analogy must, therefore, be abandoned.

Electron tunneling is due to the quantum-mechanical wave nature of electrons and can be deduced from their Schrödinger wave function. The wave function of a single electron is sinusoidal inside a metallic crystal, but instead of ending abruptly at its surface, it extends beyond it, although rapidly diminishing in magnitude with distance from the surface. The local magnitude of the wave function at any point inside and outside of the metal is related to the probability for the electron to exist at that point at any time. If a second metal is located close enough, the electron in question may actually exist with a finite probability at the surface or inside the second metal; it then can become part of the electron assembly of the second metal.

With a voltage applied between the metals in the normal conducting state, single electrons will tunnel from the negative to the positive potential, causing a flow of tunnel current. These electrons have not surmounted the potential barrier but have tunneled through, without raising their energies. However, an electron from a given energy state can tunnel only into an equivalent energy state that is allowed and not occupied by another electron. If the tunnel structure is of the superconductor-insulator-superconductor type and is kept below its critical temperature, the energy states in the superconducting energy gap are in fact not allowed. The tunnel current, therefore, is severely suppressed in superconducting tunnel devices for small applied voltages. When the voltage is raised to a specific magnitude related to the magnitude of the superconducting energy gap, the tunnel current rises rapidly, since allowed and unoccupied energy states above the gap are becoming available. For still larger voltages, the superconducting energy gap gradually loses its influence on the tunneling process; the tunnel current approaches that of a tunnel device which conducts normally. Superconducting tunnel devices have directly confirmed the existence of the superconducting energy gap postulated by the BCS theory; they are also useful for electronic applications.

Even more intriguing than the tunneling of single electrons in a superconducting tunnel device are the reactions of the Cooper pair system predicted by Josephson. In analogy to the single electron tunneling behavior, it was to be expected that the Cooper pair densities, which are also described by wave functions, would not end abruptly at the surface of the superconductors, but would extend into the tunnel barrier, albeit strongly attenuated. The tunnel barrier, therefore, might be considered a "weak" superconductor with low Cooper pair density that might possibly not be able to exclude magnetic and electric fields as completely as ordinary superconductors were known to do. Furthermore, since both superconductors possess their own Cooper pair system that are completely independent when the superconductors are far apart but merge into one when they are in intimate contact, a specific mutual or phase relation of their wave functions might also exist across the tunnel barrier. The phase relation of two sine waves with equal wavelengths is a well known concept in electronics; it denotes the lag of one wave behind the other in time or space. In summary, it became relevant to ask how electric and magnetic fields might interact with the Cooper pair system and their wave functions.

Josephson answered these questions with three fundamental equations that exploit the importance of the phase difference between the

Representation of a tunnel barrier between two superconducting metals. All electronic energy levels (stacks of lines) at the left of the barrier are raised with respect to those on the right by the application of a voltage to the metals. The classical view of physics requires that, for a current to flow, the energy of the electrons must be raised sufficiently to surmount the barrier (a). In the quantum-mechanical view, single electrons can tunnel through the barrier without a change of energy (b); electric current transported by tunneling electrons rises in proportion to the applied voltage (c). When the metals are in the superconducting state (d), electrons cannot tunnel until energy levels filled with electrons below the energy gap in the superconducting metal on the left align with energy levels void of electrons above the energy gap in the superconducting metal on the right. In this situation, the tunnel current (e) is suppressed for small applied voltages until the two energy gaps are sufficiently displaced relative to each other by the application of a specific voltage, whereupon the current rises rapidly. For even larger voltages, the current rises gradually until it approaches that of a normally conducting tunnel device.

Cooper pair wave functions of the superconductor on both sides of the tunnel barrier. The first equation relates the current flowing through a weak superconductor to the sine function of this phase difference. The second relates the phase difference to the time integral of the voltage applied across the weak superconductor. In conjunction, these two equations state that a current will oscillate (and radiate electromagnetic waves if the frequency is high enough) in the weak superconductor with a frequency proportional to the magnitude of the applied constant voltage as long as the voltage is applied. Conversely, when electromagnetic radiation is applied to a weak superconductor, a constant voltage proportional in magnitude to the frequency of the radiation is developed across the barrier. The relationship between frequency and voltage is expressed by the equation $f = 2eV/h$, in which f is the frequency, V the constant voltage, and $2e/h$ the inverse of the superconducting flux quantum ($2e/h = 4.84 \times 10^{14}$ hertz/volt). Thus, for one millivolt of constant voltage across the barrier, the current oscillates at 4.84×10^{11} hertz.

Josephson's equations include the case of zero voltage, where formalistically an oscillating current with zero frequency—*i.e.*, a direct current—can flow. A direct current flowing through a tunnel barrier without a voltage across the barrier indicates that the barrier behaves like a superconductor. At all applied voltages the amplitude of the oscillating current, whose value is also the maximum direct current or threshold current at zero voltage, depends on the tunnel barrier,

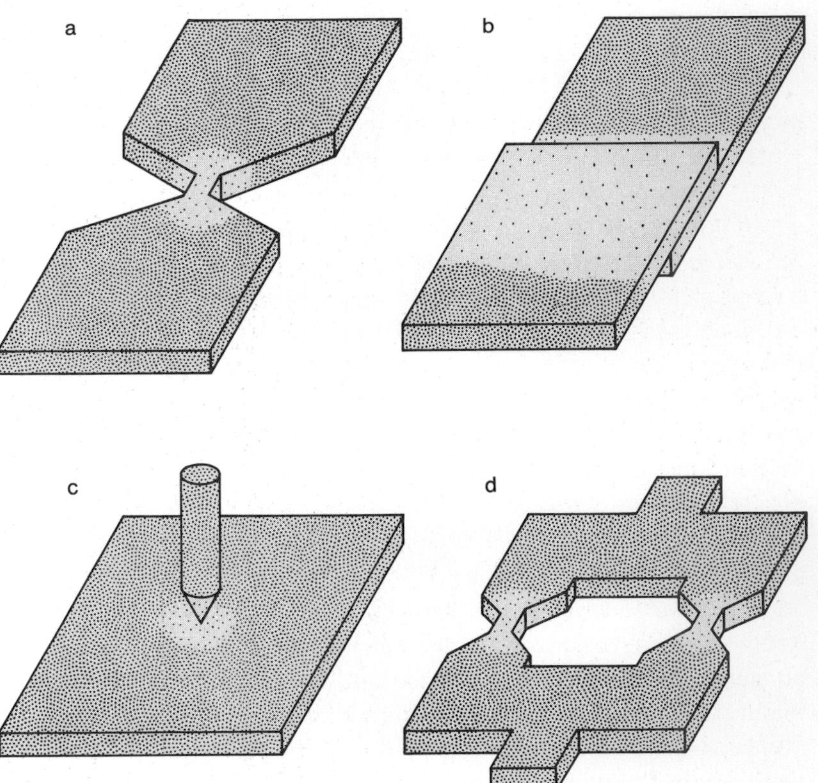

predominantly on its thickness, and on the superconductors. The threshold current is much smaller than the critical current that would destroy the superconductivity in the adjoining ordinary superconductors. When an external current in excess of this threshold current is applied, it is transported by single electron tunneling and requires a finite voltage across the tunnel barrier, which acts in this situation as a normal conductor.

Josephson's third equation links the phase difference along a weak superconductor with the integral of magnetic flux penetrating it. In conjunction with Josephson's first equation, it predicts that the magnitude of the current through the tunnel barrier is sinusoidally modulated along it by an applied magnetic field of constant magnitude and that the superconducting flux quantum ($h/2e$) determines the period of the modulation. This indicates that circulating currents are set up in weak superconductors whenever magnetic fields are applied. These currents impair the ability of a tunnel barrier to carry external direct currents without a voltage across the barrier; its apparent threshold current can, therefore, be modulated by magnetic flux.

Josephson's predictions are not restricted to tunnel devices but are valid for a variety of other structures comprising weak superconducting regions, notably thin superconducting films with geometrically constricted sections and superconducting point contacts made with thin superconducting wires pressed slightly against a superconducting plate or cylinder. When a weak superconductor is incorporated in a superconducting ring, its threshold current is affected not only by magnetic fields applied to the weak superconductor itself but also by those applied to the ring.

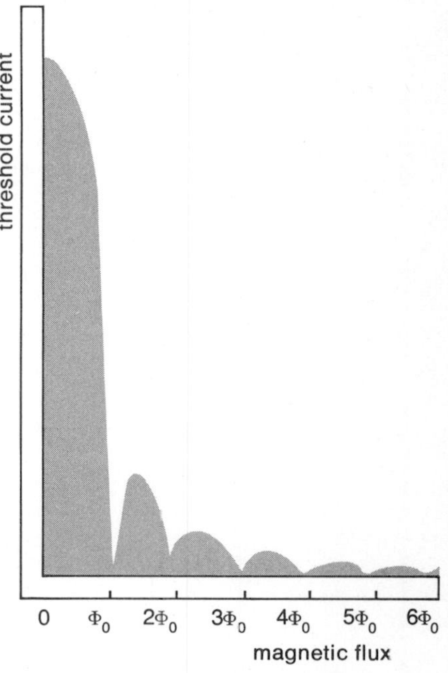

As described by Josephson's first and third equations, circulating currents are set up in weak superconductors whenever external magnetic fields are applied; these can impair the ability of a tunnel barrier to carry external currents without a voltage across the barrier. The value of the threshold or maximum zero-voltage current is impaired totally whenever a single flux quantum Φ_0 or its multiples are enclosed in the weak superconductor (above). The threshold current has its major maximum value when the enclosed flux is zero. On the opposite page are four typical structures of weak superconductors: (a) geometric constriction in a superconducting thin film; (b) two superconductors separated by an extremely thin insulator; (c) point contact made by lightly pressing a superconducting wire or screw against a superconducting plate; and (d) quantum interference device consisting of two thin-film, constriction-type superconductors.

Superconducting electronics

Early attempts to apply superconductivity in electronics had focused on switching devices, such as the cryotron for computer circuits and amplifiers, in which the resistance in a superconducting thin film was locally restored by application of magnetic fields above the critical value. Although cryotrons did operate as predicted, they were too slow to compete successfully with transistor circuits. Other devices such as the bolometer, for the detection of low-level electromagnetic radiation, have been developed and used successfully. In this apparatus the heat generated by the radiation to be detected is used to raise the temperature of a thermally isolated thin film slightly above its critical temperature and, in consequence, to raise the resistance of the film substantially.

Modern superconducting electronic devices, however, rely mainly on the Josephson effect and have superseded most superconducting electronic devices that used perfect conductivity and the Meissner effect alone. Based on the mode of operation, application, and specific structures, superconducting electronic devices can be grouped into three categories; these are magnetometers, computer circuits, and microwave devices.

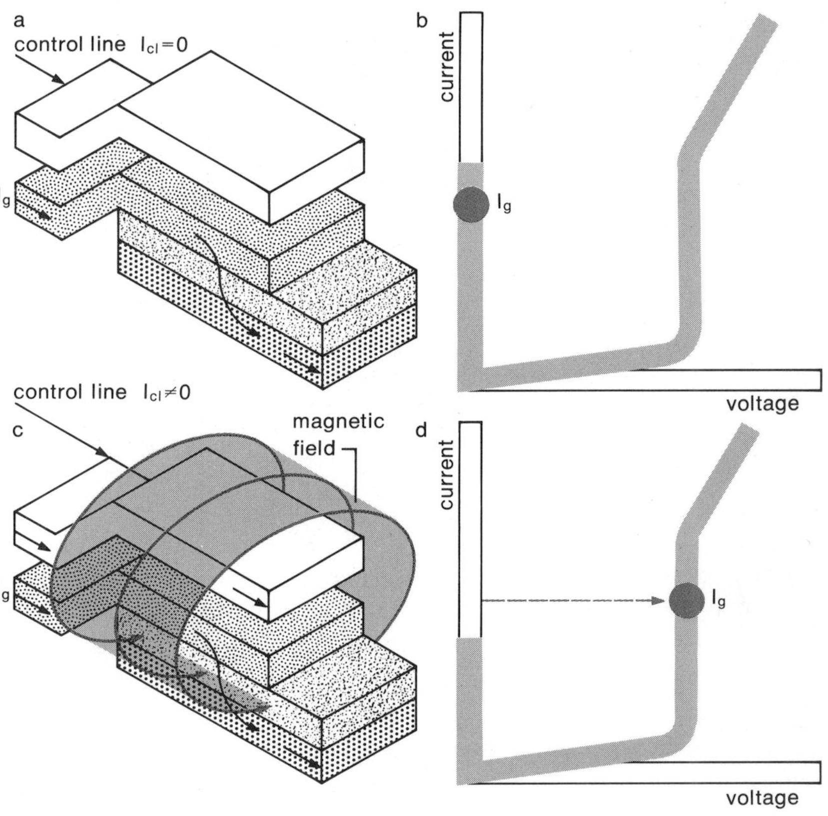

Tunnel device with magnetically coupled control line for use in computer switching circuits. When the control current I_{cl} is zero (a), the threshold current is greater than the external current I_g flowing through the tunnel device; hence, I_g flows at zero voltage (b). When I_{cl} is not zero (c), the threshold current is partially suppressed to a value below the external current I_g. In this case, the current I_g is transported by single-electron tunneling at a finite voltage (d).

Magnetometers

Magnetometers are instruments that measure magnetic fields. Perhaps the best developed of all superconducting electronic instruments to date, they employ superconducting quantum interference devices, Squids for short. Although they are exceedingly simple in structure, in conjunction with ordinary radio frequency circuitry Squids provide instruments with unprecedented sensitivity. A typical Squid consists of a niobium thin-film superconducting ring about seven millimeters wide with a precisely dimensioned constriction one-half micrometer wide and a few micrometers long. The ring is deposited around a small quartz rod less than two millimeters in diameter. When a Squid is exposed to a magnetic field, a circulating current, which tends to counteract the applied field, is set up in the ring. The current collapses periodically whenever it exceeds the current threshold of the constriction-type weak superconductor, admitting or releasing single superconductive flux quanta in the process. When coupled with instruments that record the instances of current collapse and measure the circulating currents during the periods between current collapse, the Squid becomes an extremely sensitive indicator of minute magnetic-field changes even in the presence of much stronger superimposed fields. The sensitivity can be improved still further by coupling flux transformers with large pickup loops (analogous to antennas) to the Squid. Squid magnetometers respond quickly; they are eventually limited by the electronic noise in their own electronic

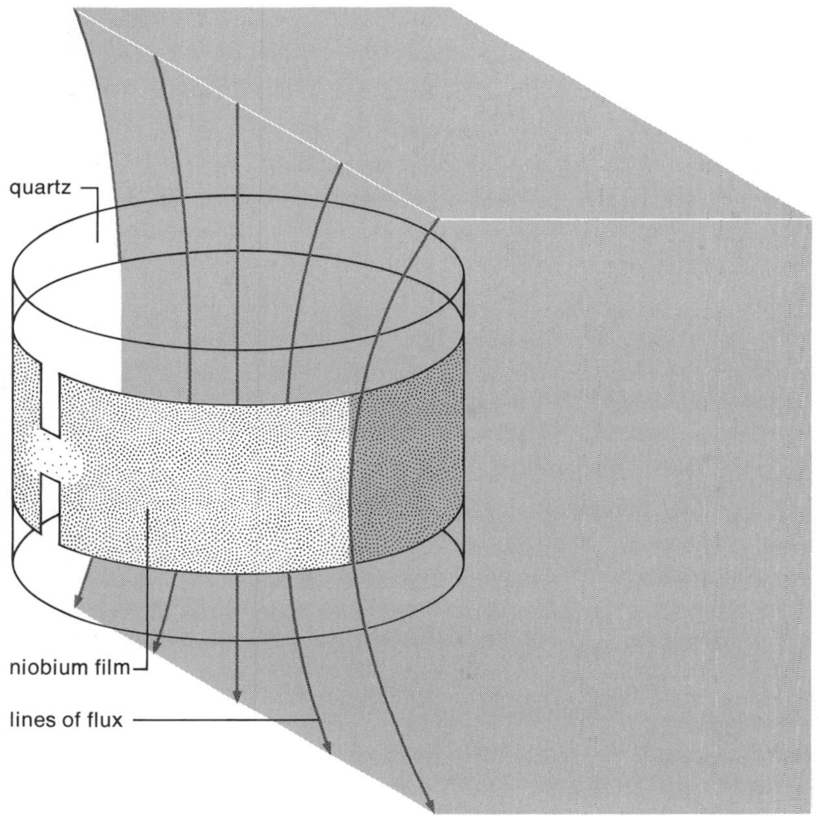

Superconducting quantum interference device, or Squid, as employed in a magnetometer consists of a geometrically constricted thin niobium film deposited around a quartz cylinder. Magnetic flux applied to the cylinder generates circulating currents in the ring that collapse periodically when the threshold current of the constriction is reached.

circuitry, which is related to the system's response time and to the absolute temperature of its components. Typical Squid magnetometers with response time of one second can detect magnetic fields as small as 10^{-10} gauss, or approximately five-billionths of the magnetic field of the Earth.

The applications of Squid magnetometers are many and diversified. Used in medicine they can monitor the magnetic activity of the human heart and brain without the need to attach electrodes to a patient. The magnetocardiograms and magnetoencephalograms so obtained are already known to reveal useful medical information not contained in conventional electrocardiograms and electroencephalograms. In biochemistry Squid magnetometers can detect weak magnetic signals generated by nuclear magnetic resonance (NMR) in minute biological samples weighing typically less than 100 micrograms to identify constituents and structures of chain molecules. In geoscience they have been used to measure the faint residual magnetism in rocks, including rocks from the Moon, for the reconstruction of ancient magnetic and geological events. Incorporated into gravimeters and accelerometers they assist magnetically in determining minute elongation or motion of heavy spherical or cylindrical masses caused by changes in gravity or by minute tremors, as in the search for gravitational waves or in the prediction and early warning of imminent earthquakes. Ordinary superconductors play a role in gravimeter design as well; in an experiment similar to the floating magnetic-bar demonstration, scientists have proposed to suspend a 5000-kilogram metal cylinder coated with a superconducting film over superconducting coils by persistent circulating currents. The long-term accuracy of this suspension arrangement is thought to be superior to that of a mechanical spring suspension system. Arrays of Squid gradiometers (which measure changes in magnetic fields with distance) have already been shown to allow the determination of distance, direction, orientation, and strength of magnetic dipole sources, *e.g.*, of trucks several hundred feet away, by measuring their magnetic field gradients in three dimensions. Squid magnetometers may also serve as magnetic receivers in special communication links through media such as the sea or Earth that are impenetrable to short-wave radio signals. Finally, Squids have been used as ultrasensitive voltmeters for the measurement of voltages as small as 10^{-16} volts in a procedure that first transforms the voltage to be measured by a small precision resistor into a current, which is then used to generate a magnetic flux in a coil coupled to a Squid.

Computer circuits

Ultrafast computer circuits are another potentially wide ranging application for weak superconductors in the form of tunnel devices. The transition between the superconducting state, where current is transported by Cooper pairs, and the "resistive state," where current transport is by single electron tunneling, is extremely fast, taking place in less than 10^{-11} second in miniaturized devices a few micrometers in length and width. The energy consumed and dissipated into heat in the switching process in these small devices is also very small. The combination of both characteristics holds promise for these devices in ultrafast computer systems to supersede transistor circuits that switch more slowly and generate much more heat. The difference is significant since switching and storage circuits must be packed very densely to transform fast device switching characteristics into ultrafast computer system performance. Dense packaging, however, is hampered by the need to remove the generated heat from the volume of the package. Already becoming increasingly difficult to achieve in fast transistor circuits, this appears to be no problem for superconducting computer circuits.

Superconducting computer circuits are typically composed of thin-film tunnel devices, resistors, and striplines and are deposited like integrated transistor circuits on such substrates as wafers and chips, though by a different fabrication process. Switching circuits are composed of tunnel devices with magnetically coupled control lines nearby. Control current signals generate local magnetic fields that pass through the tunnel barrier and thereby modulate its threshold current. A fixed-value bias current through the tunnel barrier can then be either below or above the actual current threshold. In the former case, the bias current is carried by Cooper pairs without a voltage across the barrier; in the latter case, it is carried by single electron tunneling with a finite voltage. Output striplines are connected across the tunnel barrier and used as control lines for other switching circuits; they comprise thin film resistors and are made to form a network for performing desired logic functions. The current in these lines depends on the voltage across the tunnel device and is either zero or has a finite value, commensurate with binary "yes-no" logic.

The time to perform logic functions in a computer is determined by the transition time needed by a switching device to change its state and by the time a signal needs to propagate to another switching circuit. The propagation time is determined by the distance between the circuits and by the effective propagation delay in the line connecting them. The distance between the circuits is minimized in superconducting computers by dense circuit package; the effective propagation delay is minimized by using properly terminated superconducting transmission striplines, which minimize the reflection, attenuation, and distortion of signals.

A storage circuit is made of a superconducting ring incorporating one or more tunnel devices. Binary information is represented by trapped flux and persistent current circulating clockwise or counterclockwise without power consumption in the ring. The direction of current circulation and, consequently, the stored information can be changed by temporarily reducing the apparent threshold current of a proper tunnel device in the ring. An external current is supplied in coincidence. The combined action reverses the direction of the trapped flux and the circulating current. The stored information is read, *i.e.*, the actual current direction is sensed, by another tunnel device coupled to the ring as a kind of magnetometer. The reading process does not disturb the trapped flux and the stored information is therefore retained. Experimental integrated superconducting computer circuits have already been built and operated faster than other computer circuits.

Potential performance improvements of future superconducting computer systems by perhaps 100 times over transistor circuits have been projected. If they become a reality, they may alter dramatically the outlook on the scope of certain computer applications; *e.g.*, in modeling and simulation, image processing and pattern recognition, and data management and information retrieval. Data processing of weather satellite pictures showing cloud formation and temperature profiles of the whole globe may serve as an illustrative example. Vital meteorological data sets can be obtained to date from a single weather satellite for the whole globe and with high resolution every 12 to 24 hours. Rapid evaluation of these data—recognition of atmospheric changes in these pictures, extraction of pertinent information from them, and simulation of the dynamics of the atmosphere by computers in real time—could lead to reliable global weather forecasting. At present, however, the largest computers would need many hours to analyze even part of the available data. A hundredfold increase of computer speed might allow forecasting to become feasible and practical on this global scale.

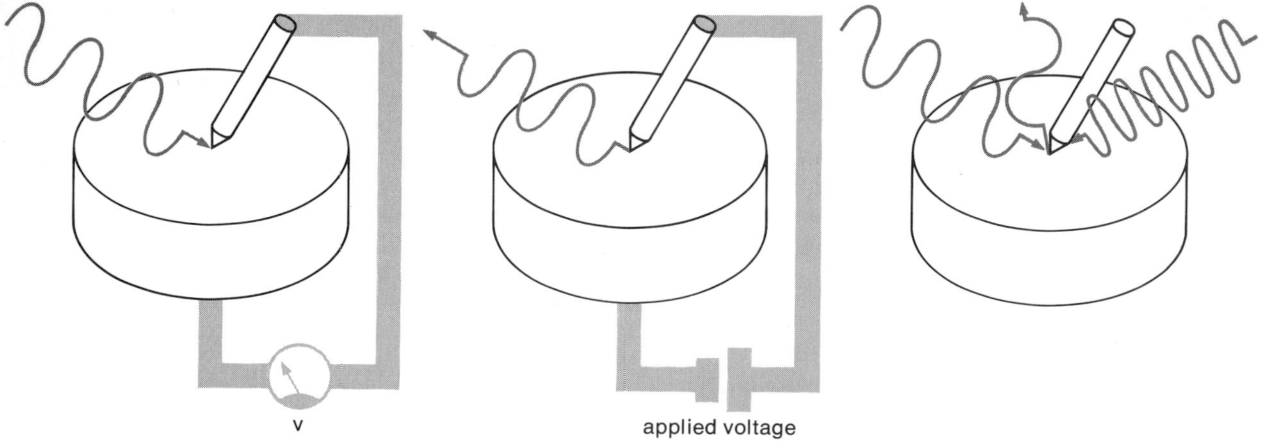

Point-contact weak superconductors find principal applications in microwave technology as (left to right) detectors, generators, and mixers (see text).

Microwave applications

A classical domain of electronics has been the manipulation of electromagnetic radiation and oscillations. Almost the entire frequency spectrum, from ultralong radio waves to ultraviolet light, has been explored and is being employed in a variety of applications in such fields as radio astronomy, laser technology, infrared detection, radar, and metrology, as well as in communications. Electronic means for the generation, radiation, detection, mixing, and stabilization of radio signals are the fundamental ingredients; both ordinary and weak superconductors have unique potential in all of these capacities. The most versatile devices are again weak superconductors, usually in the form of Josephson point contacts. A typical device consists of a thin niobium wire or sharply pointed niobium screw pressed slightly against a tin plate or cylinder. When a voltage is applied between the niobium wire and the tin cylinder, an oscillatory current is set up in the point region with a frequency precisely related to the magnitude of the voltage. When the frequency of the oscillating current is high enough, radiation of electromagnetic waves takes place. In this operation mode the point contact acts as a tunable oscillator and microwave generator in which the frequency can be adjusted by the applied voltage over a wide range, from essentially zero to 400 gigahertz (1 gigahertz $= 10^9$ hertz) and quite likely beyond. Point contacts find much use at millimeter and submillimeter wavelengths, a range too short for conventional microwave electronics and too long for lasers.

Conversely, point contacts can be operated as detectors since incoming electromagnetic radiation develops a voltage across the contact precisely related to the frequency. In this operation mode, a point contact is able to detect microwave signals with power levels as small as 10^{-14} watts. The frequency response extends into the terahertz (10^{12} hertz) frequency or submillimeter wavelength range where sensitive detectors with fast response are scarce. Point-contact microwave detectors have already been used in radio astronomy for the detection of wide-band radiation from the Sun and Moon and from space.

When two microwave signals with different frequencies arrive at the same point contact, the contact acts as a mixer; that is, it generates oscillating currents at frequencies that correspond precisely to the sums and differences of the incident frequencies and their harmonics (integral multiples of the incident frequencies). Mixing is used extensively in communication electronics to shift signals up or down on the frequency scale. Josephson point contacts permit mixing in an extremely wide wavelength range and can internally generate high order harmonics. For example, laser-generated infrared radiation with a frequency of 3.8×10^{12} hertz has been successfully mixed in a point contact with the 401st harmonics of a microwave signal with a frequency of about 9.6×10^{9} hertz.

Several intriguing applications of point contact microwave devices do already exist. One has been the determination of the ratio $h/2e$, *i.e.*, Planck's constant h divided by twice the electronic charge e, with the unprecedented accuracy of better than one part in 100 million. The ratio was determined by measuring simultaneously and precisely both the frequency of incident radiation and the voltage across a point contact. The ratio of both was then equated with Josephson's frequency-voltage relation $V/f = h/2e$. A procedure based on this measurement has been adopted recently as the new voltage standard by the National Bureau of Standards. Standard voltage is now defined as the constant voltage across a point contact when a highly stabilized and precisely measured microwave signal is coupled into the weak superconductor and is at least ten times more accurate than the electrochemical standard cell in use before.

Future outlook

Although theories have been verified, fundamental properties determined, and possibilities and potentials assessed, superconducting electronics is, to a large extent, still in its infancy. The physicists seem to have successfully delivered this strange and exciting child, and the engineers have helped it to its first toddling steps. The initial results are encouraging, but much more work needs to be done. What might happen when magnetometers, microwave devices, and computer circuits are combined has not yet been thought through. Reflecting on the ultrasensitive, ultraaccurate and ultrafast qualities of superconducting electronics, one is tempted to attribute to it the capability of greatly advancing the state of the art in electronics—perhaps even by a quantum step.

Science Year in Review Contents

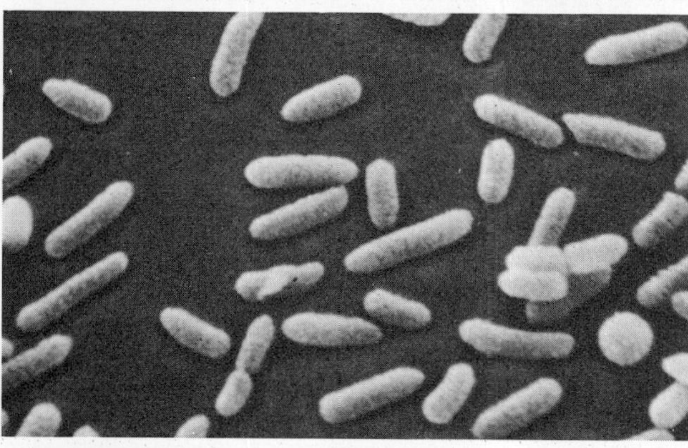

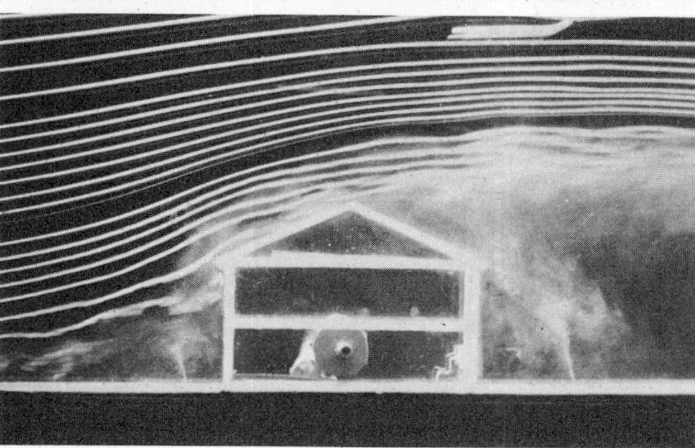

Contributors to the Science Year in Review

Joseph Ashbrook *Astronomy.* Editor, *Sky and Telescope,* Cambridge, Mass.

Fred Basolo *Chemistry: Inorganic chemistry.* Professor of Chemistry, Northwestern University, Evanston, Ill.

Louis J. Battan *Earth sciences: Atmospheric sciences.* Director, Institute of Atmospheric Physics, University of Arizona, Tucson.

Harold Borko *Information sciences: Information systems.* Professor, Graduate School of Library and Information Science, University of California, Los Angeles.

Robert E. Boyer *Earth sciences: Geology and geochemistry.* Professor of Geological Sciences and Education and Chairman of the Department of Geological Sciences, University of Texas at Austin.

D. Allan Bromley *Physics: Nuclear physics.* Henry Ford II Professor and Chairman, Department of Physics, Yale University, New Haven, Conn.

Charles H. DePuy *Chemistry: Organic chemistry.* Professor of Chemistry, University of Colorado, Boulder.

F. C. Durant III *Information sciences: Satellite systems.* Assistant Director, National Air and Space Museum, Smithsonian Institution, Washington, D.C.

Robert G. Eagon *Life sciences: Microbiology.* Professor of Microbiology, University of Georgia, Athens.

Samuel H. Fraerman *Medical sciences: Orthopedic surgery.* Professor of Medicine, Northwestern University, Evanston, Ill., and Diplomate, American Board of Orthopaedic Surgery.

John Paul Froehlich *Information sciences: Communications systems.* Assistant Professor of Electrical Engineering, University of Hartford, West Hartford, Conn.

David R. Gaskell *Materials sciences: Metallurgy.* Associate Professor of Metallurgy, University of Pennsylvania, Philadelphia.

Robert Geddes *Architecture and civil engineering.* Dean of the School of Architecture and Urban Planning, Princeton University, Princeton, N.J.

Thomas C. Hanks *Earth sciences: Geophysics.* Geophysicist, U.S. Geological Survey, Reston, Va.

Robert Haselkorn *Life sciences: Molecular biology.* F. L. Pritzker Professor and Chairman of the Department of Biophysics and Theoretical Biology, University of Chicago.

L. A. Heindl *Earth sciences: Hydrology.* Executive Secretary, U.S. National Committee for Scientific Hydrology, U.S. Geological Survey, Reston, Va.

John F. Henahan *Chemistry: Applied chemistry.* Science writer, former Senior Editor, *Chemical and Engineering News.*

Richard S. Johnston *Space exploration: Manned flight.* Director of Life Sciences, NASA Johnson Space Center, Houston, Texas.

Lawrence W. Jones *Physics: High-energy physics.* Professor of Physics, University of Michigan, Ann Arbor.

John Patrick Jordan *Food and agriculture: Agriculture.* Director, Colorado State University Experiment Station, Fort Collins.

Lou Joseph *Medical sciences: Dentistry.* Manager of Media Relations, Bureau of Public Information, American Dental Association, Chicago, Ill.

Mina W. Lamb *Food and agriculture: Nutrition.* Professor emeritus, Department of Food and Nutrition, Texas Tech University, Lubbock.

John G. Lepp *Life sciences: Zoology.* President, Hire Education—JGL Associates (Consultants), South Charleston, W.Va.

Howard J. Lewis *U.S. science policy.* Director, Office of Information, National Academy of Sciences.

Melvin H. Marx *Psychology.* Professor of Psychology, University of Missouri, Columbia.

Raymond Lee Owens *Anthropology.* Assistant Professor of Anthropology, University of Texas, Austin.

Richard K. Pefley *Mechanical engineering.* Chairman, Department of Mechanical Engineering, University of Santa Clara, Calif.

Willard J. Pierson *Earth sciences: Oceanography.* Director, Institute of Marine and Atmospheric Sciences, City University of New York.

Froelich Rainey *Archaeology.* Director of the University Museum and Professor of Anthropology, University of Pennsylvania, Philadelphia.

Anthony Ralston *Information sciences: Computers.* Chairman and Professor, Department of Computer Science, State University of New York at Buffalo.

Byron T. Scott *Medical sciences: General medicine.* Assistant Professor of Journalism, Ohio University, Athens, and Editor, *Medical Communications.*

Mitchell R. Sharpe *Space exploration: Space probes.* Historian, Alabama Space and Rocket Center, Huntsville.

James S. Shirk *Chemistry: Physical chemistry.* Associate Professor of Chemistry, Illinois Institute of Technology, Chicago.

Albert J. Smith *Life sciences: Botany.* Associate Professor of Biology, Wheaton College, Wheaton, Ill.

Frank A. Smith *Transportation.* Senior Vice-President, Transportation Association of America, Washington, D.C.

J. F. Smithcors *Medical sciences: Veterinary medicine.* Editor, American Veterinary Publications, Santa Barbara, Calif.

William E. Spicer *Physics: Solid-state physics.* Professor of Electrical Engineering and Materials Science and Engineering, Stanford University, Stanford, Calif.

Lynn A. Steen *Mathematics.* Professor of Mathematics, St. Olaf College, Northfield, Minn.

Norman M. Tallan *Materials sciences: Ceramics.* Acting Chief, Processing and High Temperature Materials Branch, Air Force Materials Laboratory, Dayton, Ohio.

Kenneth E. F. Watt *Environment.* Professor of Zoology and Environmental Studies, and Research Systems Analyst in the Institute of Ecology, University of California, Davis.

James A. West *Energy.* Associate Assistant Administrator, International Energy Affairs, Federal Energy Administration, Washington, D.C.

Frederick Wooten *Optical engineering.* Professor of Applied Science, University of California, Davis.

Anthropology

During 1975 anthropologists added to the many honors received by Raymond Firth and Sol Tax, whose contrasting careers (the former is best regarded as a theoretician and the latter as a man of action) provide a convenient focus for reviewing the present state and future prospects of anthropology.

Raymond Firth. To Firth, whose honors during the last 50 years included knighthood in 1973, being selected to write the 1975 *Annual Review of Anthropology*'s opening essay, "An Appraisal of Modern Social Anthropology," may have been a mundane assignment. He carried it out, however, with the same perceptiveness and lucidity that had long been his hallmarks. With Ashley Montagu, E. E. Evans-Pritchard, and a few others, Firth helped to form Bronislaw Malinowski's first seminars at the London School of Economics in October 1924. At that time relevant anthropological literature and anthropologists were in limited supply, and a determined effort was made to address books (such as Firth's classic, *We, The Tikopia*, 1936) to a nonspecialized audience. This helped to earn anthropology a reputation as the most literary and least jargonistic of the social sciences, although that reputation has since suffered considerably.

Throughout his long career Firth has also been in the forefront of development in the discipline. In some 20 books and innumerable articles, he addressed himself to a wide range of topics: art, economics, physical anthropology, religion, social organization, history of anthropology, family, and kinship. His contribution to the recent focus on symbolism, *Symbols: Public and Private* (1973; paperback edition, 1975), is a thoroughly researched, meticulously documented, lucidly written work which seeks to establish a balanced anthropological perspective on the subject. Firth thinks that anthropologists are in a good position to make a valuable contribution to the study of symbolism because they are well equipped to make empirical statements about symbols, to see the relationship between a symbol and that which it represents. However, he feels that the current dramatic spurt of interest has left some scientific untidiness in definition and methodology.

In his book he carefully explores the definition and use of symbols by Greek and Roman writers, the proto-anthropologists of the Romantic movement in the late 18th century, early anthropologists such as Lewis Henry Morgan, Sir Edward Tylor, Franz Boas, and others, as well as modern anthropologists of the "symbolic idiom" which became widespread after World War II; *e.g.*, Claude Lévi-Strauss, David Schneider, Ronald Berndt, Mary Douglas, Terence Turner, Nur Yalman, Victor Turner, Hortense Powdermaker, Edmund Leach, George Homans, Meyer Fortes, and S. F. Nadel. He examines, among other things, the relation between public and private symbols, food symbolism, hair symbolism, bodily symbols of greeting and parting, and Christian symbolism. Throughout, he emphasizes the necessity of relating symbolic studies to empirical data that can be examined by others.

Firth continues his discussion of symbolism in his *Review* article, where he comments favorably on Victor Turner's use of the "root paradigm"

A small air hammer chips away at a stone matrix within which a child's skull is embedded. The skull, found by Donald Johanson in Ethiopia, may be as much as 3.5 million years old; its teeth and jaws resemble those of the genus Homo, *true man.*

developed in Turner's *Dramas, Fields and Metaphors: Symbolic Action in Human Society* (1974). Yet, despite his admiration of the possibilities inherent in symbolic anthropology, he maintains a skepticism about some of its grander claims. Symbolic anthropology, he argues, "would be meaningless without an accompaniment of studies of functions, structures and processes"; or, more succinctly, "A focus on the world of ideas should complement, not replace, a focus on the world of action."

Sol Tax. A citizen of that world of action, Tax was born in 1907 and earned his Ph.D. at the University of Chicago in the depths of the Great Depression—a fact that may bear some relation to the activism that has marked his career. This activism became pronounced in the 1940s after Tax came to the department of anthropology at the University of Chicago. He seemed to act as a catalyst for projects that were needed and useful, though perhaps distracting to scholars whose principal payoff was in the printed word. Colleagues were constantly surprised by the wide range of his involvements. Because he undertook so many projects, some occasionally were roughly executed or could not be completed—Tax once told a friend that he saw himself as a cup which accumulated things to do and that sometimes the less important projects overflowed—but to many his contribution to the profession is much greater than the sum of his projects.

Certainly one of Tax's most enduring contributions was the founding of *Current Anthropology*, which more than any other journal serves as a truly worldwide forum. One of the fundamental features of that journal is the "*CA* treatment": prior to publication, articles submitted to *CA* are sent to fellows of *Current Anthropology* for their comments, and these are printed, together with a reply by the author to his critics, along with the original article. Fittingly enough, the December 1975 issue gave the "*CA* treatment" to two articles by Tax and to a report on a symposium held in 1974 at Panajachel, Guatemala, the site of Tax's famous study *Penny Capitalism* (1953). All three articles focus on Tax's ideas about "action anthropology."

Tax's early work was fairly standard ethnographic research, but after coming to the University of Chicago he became concerned with taking a more activist role in assisting the people he had studied. Perhaps he was reacting to the intensely academic focus at Chicago. It is conceivable that in a different academic environment he either would have felt less need for activism or would have moved more aggressively in the direction of the dominant mode of activism, *i.e.*, "applied anthropology." In any case, convinced that the supposed inability of scholars to assist directly the people among whom they worked was a "self-fulfilling prophecy," Tax set out to demonstrate that it need not be so.

Action anthropology. Tax directed his Fox Project, among the Mesquakie Indians of Iowa, for a decade beginning in 1948. The approach that he and three generations of graduate students took was described as action anthropology: "an activity in which neither the 'learning' nor the 'helping' could become dominant." He asserted that the "best if not the only way to test a hypothesis concerning a group's perception of a situation is to change the situation in terms of the hypothesis." In so doing he and his students came to see themselves as guided by three values: (1) The value of truth; "We were concerned to remain anthropologists and not become merely propagandists." (2) The value of freedom; "Freedom in our context means usually freedom for individuals to choose the group with which to identify, and freedom for a community to choose its way of life." (3) "A third value—or is it a principle of operation?—is a kind of Law of Parsimony which tells us not to settle questions of value unless they concern us." Applying the last of these three principles meant that the Fox Project took no position as to whether or not the Mesquakie Indians should be assimilated into white society.

The Fox Project was a limited success. Fred Gearing concluded in his *The Face of The Fox* (1970) that "although we learned, we were unable to help." Tax observes in his recent *CA* article, "The Mesquakie continue to have the same problems as in 1948, presumably from the same causes. But they are still Mesquakie people, for better or for worse, and this is the alternative that they clearly chose."

However the Fox Project may be assessed, two other projects which Tax considers to have been carried out in ways compatible with the principles of action anthropology have been impressive successes. The Cornell University Vicos Project in Peru, which helped extremely depressed plantation serfs to organize their own development, became the model for national development in the region. Similarly, the Comilla Village Aid Society in Bangladesh, which began under the leadership of Aktar Hameed Khan, did so well in training villagers to help themselves that it also became a model for national rural development.

Perhaps the strongest argument of the reviewers in *Current Anthropology* is that it is time to alter the pattern whereby anthropologists might be presumed to guide change to include more than the interest of the single small community, as conceived by action anthropology.

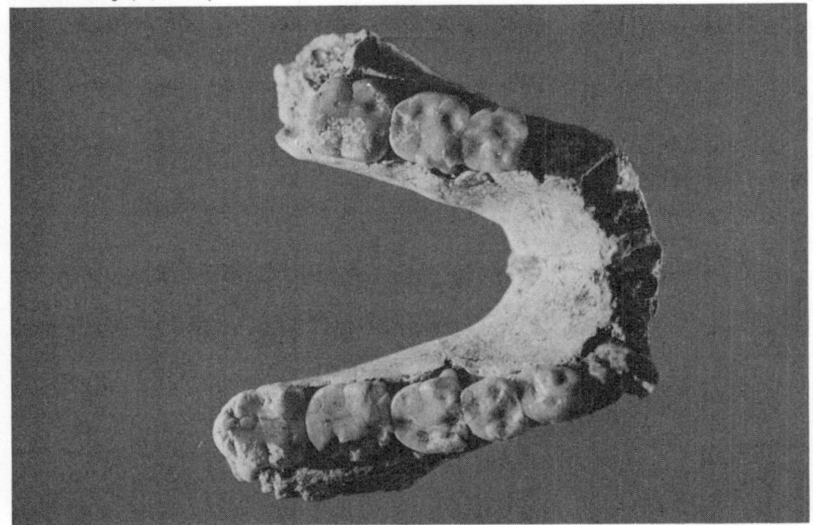

Fossil of the lower jaw of an adult of the genus Homo *was dated by the potassium-argon method at between 3,350,000 and 3,750,000 years old. Mary Leakey and co-workers found the jaw and other remains of early man at Laetolil in Tanzania.*

The reviewers could also have pointed out that the success of the Vicos and Comilla projects was assured not alone because they realized the values of the downtrodden but because men in political power supported those same values and actively promoted their cause. Clearly the Vicos Project was rescued from certain destruction by local landlords because its organizers had formed linkages with persons of great leverage in national politics. The Comilla situation is more complex. Initially the project was ignored by the most powerful landlords; thus in its earliest years it was spared a power confrontation. In addition, the poorest peasants in what was then East Pakistan were Bengalis, but the economic and political power structure was almost entirely controlled by West Pakistanis. When Bangladesh became independent in 1971, significant portions of the old power elite were replaced by Bengalis sympathetic to their downtrodden brethren, for whom Comilla had been one of the few bright spots.

Richard N. Adams pioneered among anthropologists in arguing that the key to development is the control of power on a national, not simply a community, basis. He made this clear in his case study of Guatemala, *Crucifixion by Power: Essays on Guatemalan National Social Structure, 1944–1966* (1970), and an important purpose of his *Energy and Structure: A Theory of Social Power* (1975) is "the construction of concepts and a series of related theoretical propositions" for the study of social power. Adams also contributed an essay to *Rethinking Modernization* (1974), edited by John J. Poggie, Jr., and Robert N. Lynch, which reconsiders the nature of development and the anthropological contribution to it. The book reports a symposium held at the University of Rhode Island in 1971 which may suggest a shift on the part of anthropologists to a much broader canvas. As the editors conclude

> Almost all the cases of modernization that made up the focus of our discussion in the symposium involved some kind of major economic and social transactions. In fact, practically all problems of modernization, because of the extreme delocalization and worldwide articulation that has now developed, involve complex interdependencies in different sectors of economic market systems, as well as interactions among different political institutions among nation states.

State of the discipline. The production of increasing numbers of anthropologists, together with a downward shift in birthrates, was resulting in a severe limitation of academic job opportunities. A thoughtful article on the topic by R. G. D'Andrade, E. A. Hammel, D. L. Adkins, and C. K. McDaniel in the December 1975 issue of the *American Anthropologist* concludes that "from 1977 onward the number of academic positions for anthropology Ph.D.s will decrease yearly until it reaches a 'worse than zero' situation in 1987 and 1988." Whereas only about 5% of all anthropologists with Ph.D.s were working outside academia in 1976, projections indicated that after 1982 over two-thirds would be doing so.

The activities of the American Anthropological Association were permeated with concern about the projected job crisis. The executive committee opened a placement service for nonacademic positions for 1975–76 and urged departments of anthropology to offer work experience in a variety of public service areas. It was decided to include in future meetings a panel on public issues and to set up a task force to identify research possibilities "of both practical relevance and theoretical interest in the contemporary world."

—Raymond Lee Owens

Archaeology

Two issues concerning archaeological research became increasingly prominent during the year. One had to do with the relevance of such research in a period of accelerating economic, social, and political crises, and the other with the scale and elaboration of archaeological excavations in a period when economic development, deep plowing, construction, and trade in antiquities were destroying many more important archaeological sites than were being excavated systematically.

The significance of ancient history to an understanding of human affairs always has been a concern of archaeological writers, but the current upheaval in political affairs, the struggle for food and other resources, and the shift in centers of wealth and influence gave added urgency to the search for an explanation of the present in a study of the past. Thus, Loren Eiseley, a writer and archaeologist, took a searching look at our times in mass media publications; Barry Cunliffe worked with the British Broadcasting Corporation in producing a documentary on the rise and decline of the Roman Empire; and Ivor Noël Hume raised the whole issue in an address to the annual meeting of the Societies for Historical and Underwater Archaeology in Philadelphia. It seemed inevitable that this trend would accelerate.

Increasingly meticulous methods of excavation and specialized systems of interpretation, requiring more time not only for excavations in the field but for publication of results, appeared to be coming into conflict with the need to save significant archaeological sites before they are destroyed. Many archaeologists believed there was no point in excavating at all if it was not done properly, while others pointed to the irrevocable loss of knowledge about the past taking place everywhere while money, time, and manpower were concentrated on relatively few systematic excavations. As yet there appeared to be no obvious compromise in such a conflict of values.

Asia. Theodore Wertime of the Smithsonian Institution, Washington, D.C., writing in the *Washington Post,* described the site of Ban Chiang in northeast Thailand as "now possibly the most exciting and controversial arena of archaeology anywhere in the world." The second season of excavation there by the Thai Department of Fine Arts and the University Museum, University of Pennsylvania, under the direction of Pisit Charoenwongsa and Chester Gorman, was completed in the fall of

Courtesy, The University Museum, University of Pennsylvania

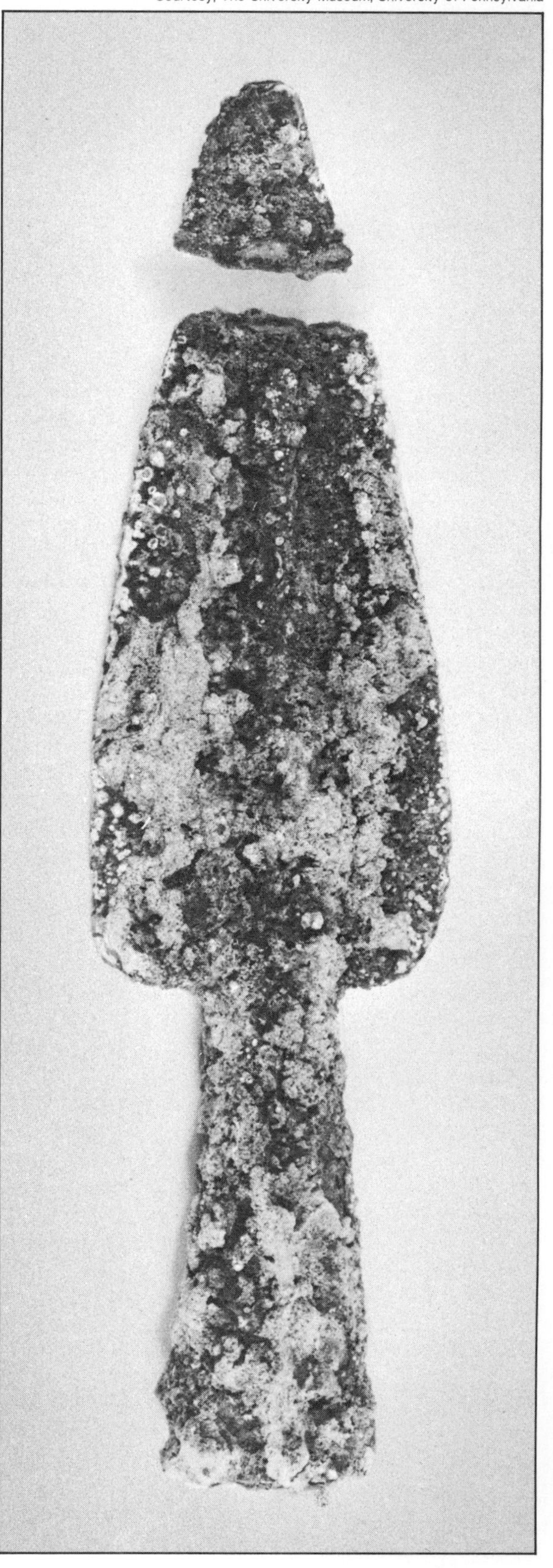

Bronze spearhead was found in a grave at the lowest levels at Ban Chiang in northeast Thailand. Bronze objects at these levels date from about 3600 B.C.

1975, and the collections (21 tons) were being studied at the University Museum.

The crucial issue remained the age of the bronzes found at the lowest levels at the site and not yet dated by radiocarbon and thermoluminescence (TL). There were, however, adequate charcoal and skeletons from the lowest levels, and these were to be dated by the University Museum's Applied Science Center for Archaeology (MASCA), an operation financed by the National Science Foundation. The oldest carbon-14 dates so far established at Ban Chiang for bronze were on the order of 3600 B.C., but TL dates on pottery of unknown level extended to about 4600 B.C. William Schauffler, a member of the team, and Pisit Charoenwongsa also made soundings in other sites in the area, and it appeared certain that they were all part of a major cultural complex.

The significance of Ban Chiang was its relevance to a growing controversy about the origin of the techniques basic to civilization. Current evidence indicated that there were at least three centers of origin, in Southeast Asia, the Near East, and Central and South America, where the basic requirement for civilization, agriculture, developed at a very early period. Metallurgy, presumably another basic element in civilized living, was known in all three regions, but scholars had long assumed that it originated in the Near East. One of the major puzzles about the newly discovered center for agriculture and metallurgy in Southeast Asia was the lack of evidence of the flowering of fully developed civilization in the region until quite recent times. Many years of research would be required to understand what appeared to be a process of change that was quite different from that in western Asia.

The worldwide sale of painted pottery from northeast Thailand (described by William Honan in the *New York Times Magazine)* reached such proportions that members of the Ban Chiang expedition reported more than a score of sites totally destroyed by pot hunters, even during the two seasons of work there, and it was questionable whether the prehistory of the region could be recovered before the sites were gone. Once again, this pointed up one of the major problems of contemporary archaeology: with limited time, funds, and trained personnel, how does the archaeologist decide between meticulous excavation standards and the loss of knowledge? The age and the nature of all these sites is unknown, and there was no assurance that systematic sampling of a few sites could result in more than a distorted historic record.

Further dramatic discoveries were reported from China. At Sian, capital of China during the Chou, Ch'in, Han, Sui, and T'ang dynasties (11th century B.C. to 10th century A.D.), an immense structure (700 by 200 ft) buried 15 to 20 ft underground was discovered by well diggers 3,500 ft east of the outer walls of the old city. Excavation during 1975 exposed a fantastic army of some 6,000 terra-cotta warriors with their horses and the remains of wooden chariots. Horses and men are life-size, startlingly lifelike, and arranged in a phalanx in 11 symmetrically balanced corridors. The men are armed with real swords, spears, and crossbows and were in a standing position as if in the order of attack. No two warriors look alike. Chinese archaeologists said the figures were related to the tomb of the emperor who ruled from 221 to 210 B.C., although the tomb itself is in the inner part of the old city and not yet excavated.

Life-size terra-cotta figures of warriors and horses were excavated from a burial pit at Sian in northwest China, near the tomb of an emperor who ruled from 221 to 210 B.C.

UPI Compix

Eastfoto from Sovfoto

Jug of 23-carat gold, part of the Thracian Treasures from Bulgaria Exhibit shown at the British Museum in London. It is believed to date from 300 B.C.

Another outstanding discovery was a very well-preserved body, like the woman in the now famous Han period tomb discovered a few years earlier. This is a man buried in 167 B.C. (a date given on a bamboo tablet found in the tomb) in a red preservative fluid in the inner of a series of three coffins. The find was made in the town of Chinancheng in Hupeh Province, central China. The tomb contained 500 burial objects, including lacquer and bronze utensils, copper pins, clothes, headgear, and shoes. Excavations at the site of ancient Panlung near Wuhan, also in Hupeh Province, uncovered the foundations of another Shang dynasty palace, dating from 3,400 years ago. The find demonstrated that during its early stages the Shang dynasty had spread its control southward from the Yellow River to the Yangtze.

In Burma the earthquake of July 8, 1975, badly damaged many of the pagodas and monasteries at Pagan, famed royal capital in the period from the 11th to the 13th century. More than half of the 500 major temples were partially destroyed, with severe damage to sculpture and frescoes. These must be shored up in order to prevent further disintegration.

Near East. The archaeological mission of the University of Rome reported the discovery of 15,000 clay tablets at Tell-Mardikh in Syria which, when deciphered, will have a profound effect upon knowledge of the 3rd millennium B.C. in Syria and Mesopotamia and will aid in the study of the Sumerian language. The texts are written in syllabic cuneiform, but in two different languages; one is Sumerian and the other an unknown Semitic tongue. Some are diplomatic documents and literary texts, which are relatively rare in collections of clay tablets but extremely important in terms of historic details. There are also grammatical texts and students' workbooks annotated by their teachers. One of the religious texts contains the oldest known version of the famous Gilgamesh poem. The bulk of the tablets were inscribed about 2300 B.C. and were found in a room off the great royal audience hall of what was at that time the city of Ebla.

Europe. In September 1975 Greek archaeologists announced the discovery of the oldest known shipwreck in the world. It was found by Peter Throckmorton off the island of Hydra and was dated by the pottery found in the remains as Early Bronze Age II (about 2500 B.C.). Presumably it was a Cycladic ship since all the pottery brought up by divers was characteristic Cycladic ware. It was not known whether enough of the ship remained to allow a reasonable reconstruction of its design. All that is known of Early Bronze Age Cycladic ships comes from sketches found on pottery, and the design of any ship of this age would be of great concern to nautical history in general.

In Yugoslavia construction of a dam exposed the remains of a city extending about two miles along the Tisza River near Senta, close to the borders with Romania and Hungary. Excavations, under the direction of Laszia Szekeres, exposed walls, pit dwellings, granaries, and other structures. Pieces of burial urns and other objects convinced Szekeres that the city was occupied during the 1st century A.D. by Sarmatians, one of the nomadic hordes who, like the Huns, plagued the Romans on their frontiers in Europe. The great settlement clearly shows that the Sarmatians were not purely nomadic. Remains of burials suggested to the archaeologists a remote possibility of discovering the famed burial of Attila, king of the Huns.

In October 1975 Soviet archaeologists reported the accidental discovery, by construction workers, of rich burial chambers near Anapa on the Black Sea coast. Two stone-roofed chambers containing sarcophagi were mentioned; some had been plundered but at least two yielded treasures including golden crowns, silver vessels, and a huge iron sword with a gold-encrusted handle. Frescoes, with the colors still fresh, represent everyday scenes of the period and also Greek legends and mythical heroes. The tombs are dated to the 2nd century A.D. and may be associated with the so-called Bosporus State.

The "Thracian Treasures from Bulgaria" exhibit, which opened at the British Museum in January 1976, focused worldwide attention on archaeological research in Bulgaria. The most complete prehistoric chronological scale in Europe was worked out at Varna, and a recent excavation in one of the mounds near Varna uncovered a Chalcolithic necropolis, dating to the 4th millennium B.C., which should drastically alter our conception of this period in Europe.

America. The age and the place of origin for the domestication of corn (maize) was again in question as the result of a discovery in Ecuador. Donald Lathrap, University of Illinois, excavating a site near Real Alto, found evidence that corn was grown there about 3000 B.C. It was generally believed that corn was first domesticated in Mexico about 3000–4000 B.C., but if corn was grown in Ecuador, several thousand miles to the south, at roughly the same time, real questions arose as to the age and the place of origin. The site contains grinding stones, corn beer mugs, pottery decorated with corn-kernel impressions and teeth badly worn down by the grit from the grinding stones. Significantly, 80–90% of the animal bones in the refuse are of Virginia white-tailed deer, a species known to thrive where there are Indian cornfields.

Mexican, U.S., and Canadian archaeologists, using computer analysis, concluded that Teotihuacán in the Valley of Mexico was a great urban center, not only a ceremonial center as originally thought, and the focus of a far-flung empire stretching from the plains of Mexico to the mountains of Guatemala during the period from A.D. 400 to 700. Windowless one-story apartment compounds housed about 100 persons each, with no particular separation of classes. Similar conclusions were developing in the study of the great Maya city of Tikal, although in this case there was a significant class system.

In both cases researchers were asking why the cities died. René Millon of the University of Rochester speculated that Teotihuacán's inhabitants may have abandoned the city because it had become "a clumsy giant—too unwieldy to change with the times." Others believed that the people desecrated the temples and abandoned the city in a rage against the gods for permitting a famine.

A somewhat different look at the decline of Mayan civilization was emerging in a study of the

Remains of a 5,000-year-old oval house, about 25 by 35 feet, were found in Ecuador. Holes are from poles used for room divisions and hammocks, and rectangular area is from an earlier excavation.

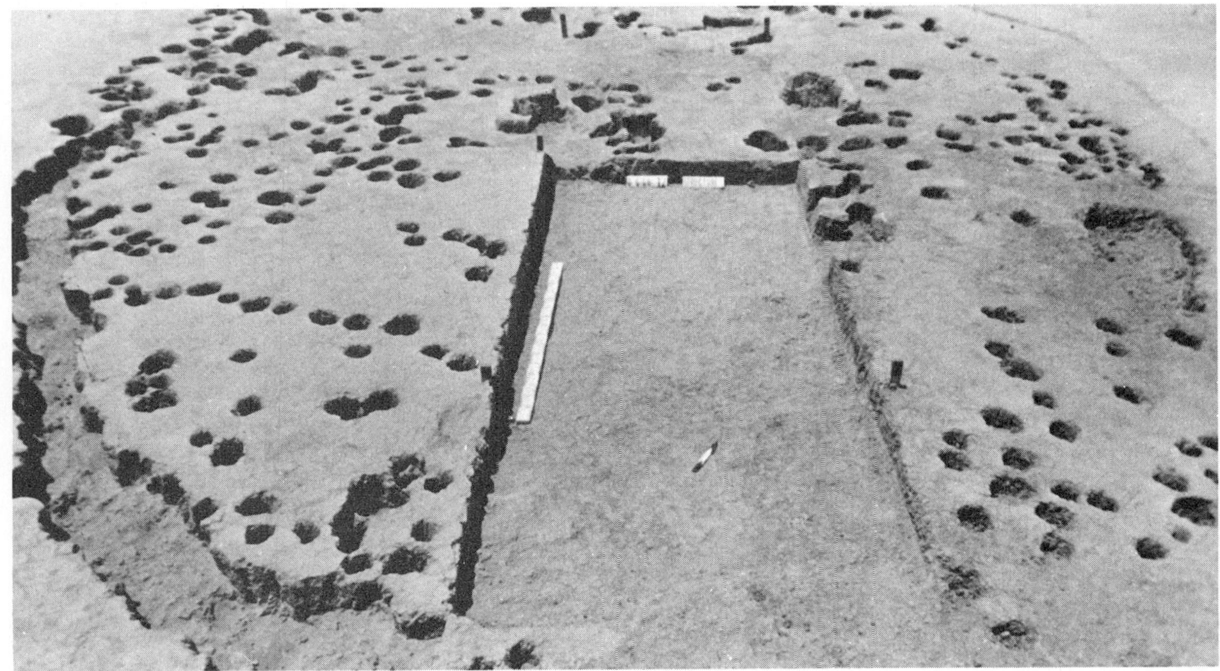

Courtesy, Donald W. Lathrap, University of Illinois

island of Cozumel off the coast of Yucatán, where Jeremy Sabloff and William Rathje were documenting the development of a new kind of trading empire based on the island by a group of Chontal-Maya-speaking people known as the Putun. By A.D. 1224 Chichén Itzá had been abandoned, and Toltec power was gone from Yucatán. The Putun on Cozumel were in a unique position to develop a coastal sea trade in cocoa beans, salt, cotton cloth, and many other raw materials. Columbus, on his fourth voyage in 1502, met what probably was a Putun trader in a dugout, eight feet wide and as long as a galley, propelled by 20 paddlers. The accepted theory is that by the 12th century classic Mayan civilization had long been in a state of decay, and this was looked upon as a general social decline. It was now proposed that, after the decline of centralization and the elite class, the commoners' standard of living rose.

Africa. The most recent in the series of discoveries of fossil bones of men and manlike creatures made in Tanzania, Kenya, and Ethiopia during the past few years was announced by Mary Leakey in October 1975, when radioactive dating gave an age of 3,750,000 to 3,350,000 years for fossil bone deposits excavated in an arid region 25 mi S of Olduvai Gorge. The deposits included the jaws and teeth of at least 11 individuals who appear to be of the genus *Homo*. Like earlier finds made in Kenya by Richard Leakey and in Ethiopia by Donald Johanson and Maurice Taieb, these differ markedly from *Australopithecus*. Mary Leakey believed they confirm the opinion that this "ape-man" was an offshoot of the hominid line, a contemporary of early man who died out. No stone tools were found with the bones, but the search for them continued.

Techniques. MASCA was planning to announce a breakthrough in the search for ways to preserve ancient mud brick and stone monuments. Darrel Butterbaugh of MASCA had developed two chemical compounds, one to preserve brick and adobe structures, and the other stone monuments and buildings. Both are inexpensive and simple to apply. Field tests were made at Chaco Canyon in the southwestern U.S., at Hasanlu in Iran, and in Philadelphia. At Hasanlu experiments were also being conducted in the production of mud bricks and mud plaster in which the chemical compound had been mixed. It appeared that such material can be used commercially in areas where mud brick or adobe is still a major building material. Unlike other chemicals used in previous experiments, which formed a thin crust that peeled off in time, the new compounds deeply penetrate adobe and stone. An additional advantage is that they do not change the color of the structure.

—Froelich Rainey

Architecture and civil engineering

To promote "appreciation of the aesthetics of steel bridges and to honor the architectural excellence of modern bridge design," a national competition is held each year by the American Institute of Steel Construction to select beautiful new bridges. Because there are many types of steel bridges, the 1975 awards were given in seven categories: long span (bridges with a span over 400 ft in length), Daniel Webster Hoan Memorial Bridge, Milwaukee, Wis., designed by Howard Needles Tammem & Bergendoff; medium span, high clearance (bridges with a vertical clearance of 35 ft or more and a span between 125 and 400 ft in length), Kentucky Highway 312 Bridge over Laurel River Lake, designed by the U.S. Army Corps of Engineers, Nashville District; medium span, low clearance (bridges with a vertical clearance of less than 35 ft and a span between 125 and 400 ft in length), South Fork Flathead River Bridge, Montana, designed by Morrison-Maierle; short span (bridges having no span more than 125 ft in length), Laceyville Bridge, Laceyville, Pa., designed by Gannett Fleming Corddry and Carpenter; highway grade separation bridges, South Weber Interchange, Davis County, Utah, designed by the Utah Department of Highways; movable span bridges, Burlington Northern Bridge No. 117.35, Beardstown, Ill., designed by Howard Needles Tammem & Bergendoff; and special purpose bridges, Pedestrian Bridge over U.S. 27 at Pine Knot, McCreary County, Kentucky, designed by the Kentucky Bureau of Highways, Division of Bridges.

The Daniel Webster Hoan Memorial Bridge is a vivid expression of structural form. The type of structure, a tied arch, was selected primarily because of the necessity to eliminate an external horizontal thrust that could not be sustained by the poor lateral resistance of the site's underlying foundation materials. The center section of the stiffened tie girder is suspended from the primary arch by bridge strand hangers, and the end sections are supported above the arch rib by box posts. The primary arch, the rectangular vierendeel bracing (an open-web truss with vertical members but no diagonals), and the tie girder are all made of welded steel box girders, brightly painted to differentiate the arch (gold) and the tie (blue). The excellence of the design comes from the elegant geometry of the structural form and from the aesthetic of welded steel sculpture.

Clearly, it is not size that makes the Hoan Bridge notable; its 600-ft span is not comparable to the world's longest arch bridges, the 1,652-ft Bayonne

Courtesy, American Institute of Steel Construction

The Daniel Webster Hoan Memorial Bridge in Milwaukee, Wis., designed by Howard Needles Tammem & Bergendoff, won a 1975 architectural award given by the American Institute of Steel Construction.

Bridge or the 1,650-ft Sydney Harbour Bridge. Rather, the Hoan has the beauty of a line drawing that gives a sense of easy movement in space. The movement is strongly communicated by the curved lines of the arch and by the straight-as-an-arrow line of the tie. The contrast between the straight line and curved lines is striking.

The year 1975 marked the half-century anniversary of two milestones in the modern movement of architecture. In 1925 Le Corbusier was building the Pavillon de L'Esprit Nouveau at an international exposition in Paris, while in Dessau, Germany, Walter Gropius was starting construction of the new school building of the Bauhaus. Although different in form and use, these two buildings seem in retrospect to share some common aspects. First, the vision of each architect, as expressed in his building, was one of optimism and hope about the possibilities for 20th-century culture. Second, by means of the buildings the architects sought linkages between the arts and sciences, and between design and industry. And third, the structures conveyed specific ideas about the buildings, furniture and equipment, and cities of an industrial society.

In sharp contrast to the modern movement, the fountainhead of official architecture had been the outgrowth of the École des Beaux-Arts in Paris. In the U.S., for example, French academic classicism served as the formula for most of the state capital buildings, banks, railroad stations, and the "City Beautiful Movement" in urban development. But, as Ada Louise Huxtable wrote in the *New York Times*, "ultimately, the Beaux Arts was the casualty of a revolution in structural technology and massive changes in society and the economy. It is hard, even for would-be revivalists, to rationalize the logic and costs of the grafting of classical forms and orders intrinsic to masonry construction onto the totally different requirements of modern steel and concrete. It is equally hard to fit the straitjacket of academic classicism on the many new building forms of the 20th century." Nevertheless, the Museum of Modern Art in New York City, which had been a pioneer in the modern movement, sponsored in 1975 a major exhibition of 19th-century drawings by students at the École des Beaux-Arts. The intention of the exhibition was to provoke critical thought in the continuing evolution of modern architecture.

In 1975 the design awards of the American Institute of Architects and the Royal Institute of British Architects provided insights into the architectural principles and the social context of modern architecture. Because the awards are not categorized according to the size or type of project, one notes

Courtesy, Kimbell Art Museum, photograph by Ezra Stoller

Kimbell Art Museum in Fort Worth, Texas, the last design completed by Louis I. Kahn before his death in 1974, won a 1975 Honor Award from the American Institute of Architects.

with interest the distribution of the successful building types: in the U.S., three awards were given to office and commercial buildings, two to art museums, and one each to a private house, a high school, a housing complex, and an urban renewal development center. In Great Britain two awards were given to arts centers, and one award each to an office building, a private house, a vicarage hall, a library, and a college residence hall.

In neither country were 1975 awards given to building types that would have been the proudest achievements in other periods, such as a church or urban palace of the Renaissance, or even a railroad station or industrial warehouse of 19th-century England. In both countries, the design awards for houses in 1975 were given to small residences that were similar in their aesthetic intention to create a contrast between the building and its romantic, wooded landscape. In both countries the office building designs attempted to develop floor plans and building shapes that were more complex, geometrically, than the simple rectangular cube that is conventional today. And, in the total number of 1975 awards, buildings for the arts seemed to offer the best opportunities for modern architecture to demonstrate excellence in design.

In the U.S. in 1975 the art museum was a widely discussed building type. A number of significant new museums and expansions were completed, giving rise to debates among the public and critics. The Kimbell Art Museum in Fort Worth, Texas, designed by the late Louis I. Kahn, was almost universally admired for the qualities of its interior lighting, from both natural and artificial sources. The vaults over each gallery were carefully designed to capture and admit light, a principle of architecture that has been honored since classical times. On the other hand, two other major museum designs (the Hirshhorn Museum and Sculpture Garden in Washington, D.C., designed by Gordon Bunshaft of Skidmore, Owings & Merrill, Architects; and the Lehman Wing of the Metropolitan Museum of Art in New York City, designed by Roche and Dinkeloo) were the subjects of widespread debates. The architectural issues involved their scale (especially the large-scale interiors of

Courtesy, Herman Hertzberger, photo by Aerophoto Schiphol

Office building of the Central Beheer Insurance Cooperative Society in Apeldoorn, Neth., won praise for the close relationship of its structural form with the social organization of those working inside it.

the Lehman Wing) and their contexts (especially the relationship of the Hirshhorn to its landscape setting on Capitol Mall).

The design of office buildings received critical appraisal in 1975. In the British *Architect's Journal*, Francis Duffy wrote a carefully studied evaluation of one of the most innovative designs of office buildings in recent years, the Central Beheer Insurance Cooperative Society offices in Apeldoorn, Neth. The building serves as the workplace for approximately 1,000 people, and the design sought to establish a personal identity and a sense of community for everyone. The building is especially interesting because its structural form is closely related to the social organization of those working within it.

The structure consists of a cluster of square modules of open space, defined by a grid of split columns that follow the main circulation routes. There is close integration between the structural system, the environmental control services, and the spatial organization. The building is, therefore, a kind of matrix of structure, equipment, and space. In terms of its historical precedents, the Central Beheer is indebted to the ideas of Louis Kahn (especially his Richards Medical Research Building in Philadelphia) and Aldo van Eyck (especially his Children's Orphanage in Amsterdam).

—Robert Geddes

Astronomy

A new minor planet, provisionally known as 1976 AA, was accidentally discovered on Jan. 7, 1976, by Eleanor Helin of Hale Observatories with the 18-in. Schmidt telescope on Palomar Mountain in California. Her photographic plate of a region in the constellation Gemini showed a long trailed image, evidently of a 13th- or 14th-magnitude minor planet near the Earth and moving across the sky at about 2° per day. Mrs. Helin rephotographed this speeding object with the same telescope on the next three nights, and additional observations were obtained at other observatories in California, Massachusetts, and Japan.

Calculations showed that 1976 AA has the smallest orbit of any asteroid known, being the first to have a mean distance from the Sun that is less than that of the Earth. At perihelion it is 73 million mi, slightly beyond the orbit of Venus. Its aphelion is 106.1 million mi, somewhat greater than the Earth's, and its orbital period is only 347 days. Thus, 1976 AA is an extreme example of the Apollo-type asteroids, which are defined as those that come to perihelion inside the Earth's orbit; it was the 20th of these to be discovered.

Because the orbit of 1976 AA is inclined 18.8° to the plane of the ecliptic (the plane of the Earth's orbit extended to meet the celestial sphere), there is no current possibility of a collision with the Earth. In fact, the least separation during the January 1976 close approach was about 11 million mi. In the course of tens of thousands of years, however, the gravitational attraction of the planets may change the orbit of 1976 AA enough for it to intersect the Earth's. The probable effect of colliding with the Earth would be to make an explosion crater about 20 mi across.

Streak of light is the asteroid 1976 AA as photographed on its discovery date, Jan. 7, 1976. The asteroid has the smallest orbit of any of the minor planets.

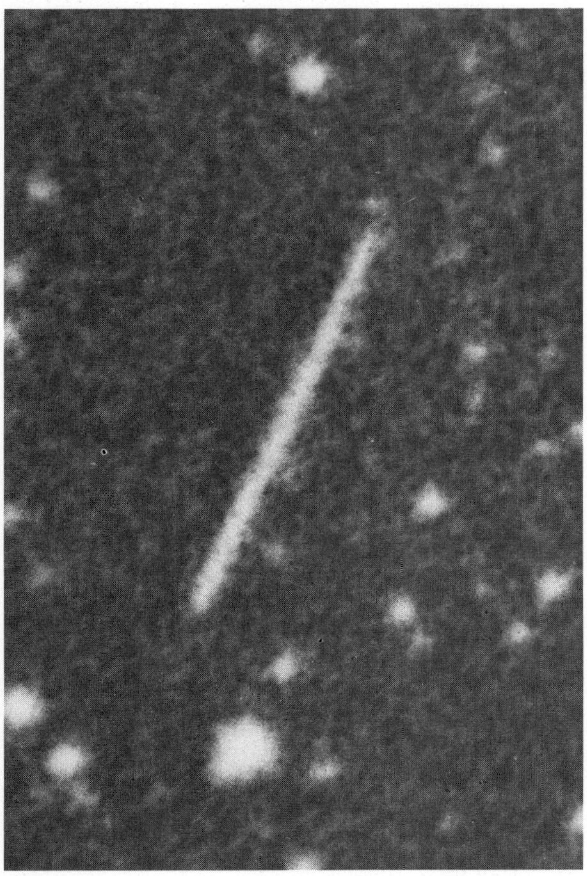

Courtesy, Eleanor F. Helin and Hale Observatories, California Institute of Technology

Physically, 1976 AA is a rock less than one mile in diameter. Like the other Apollo asteroids, it is possibly the nucleus of an old comet, as suggested by Eugene M. Shoemaker, planetologist at the California Institute of Technology. Spectrophotometric observations at the University of Arizona and the University of Hawaii showed that the surface of 1976 AA resembles the asteroid Icarus.

Shoemaker urged that 1976 AA be considered an attractive destination for an unmanned or manned space mission, being easier to reach than any other celestial body except the Moon. Because this asteroid makes 20 revolutions of the Sun while the Earth is completing 19, its close approaches to the Earth recur at 19-year intervals. The next favorable opportunity for either Earth-based observations or a space mission to 1976 AA will come, therefore, about 1995.

By an odd coincidence, another fast-moving asteroid (1975 YA) had been discovered with the same telescope at Palomar on Dec. 27, 1975, and was also (but erroneously) announced to be the first asteroid with a smaller orbit than the Earth's. Later calculations showed that its mean distance from the Sun is 121 million mi and its period 1.48 years. Even so, 1975 YA is of considerable interest, being an Apollo-type asteroid that passed within approximately five million miles of Earth near the date of discovery.

Tests of general relativity: radio deflections. During 1975 astronomers reported several observational tests of Einstein's general theory of relativity that strongly supported it rather than any of several competing theories of gravitation. The most conclusive experiment involved the bending of radio waves passing near the edge of the Sun.

In principle, the experiment is analogous to the well-known deflection of starlight near the Sun. According to general relativity, this deflection is 1.75 sec of arc radially outward for a star at the Sun's limb and less for stars farther from the Sun. The deflection of starlight has been measured at total solar eclipses, but only to low accuracy; a much better result can be obtained by measuring the deflection of radio waves from a quasar (quasi-stellar source) lying in nearly the same line of sight as the Sun.

The most decisive result of this kind was obtained by Edward Fomalont and Richard Sramek of the U.S. National Radio Astronomy Observatory in West Virginia. They used four large radio telescopes spaced up to 35 km (22 mi) apart as an interferometer array, with which they could determine small angular changes in the relative positions of three celestial radio sources. Fomalont and Sramek found a deflection equal to 1.015

Adapted from "Radio-Wave Deflection Experiments Confirm Einstein" by Marian S. Rothenberg, "Physics Today," p. 17, April 1975, © American Institute of Physics

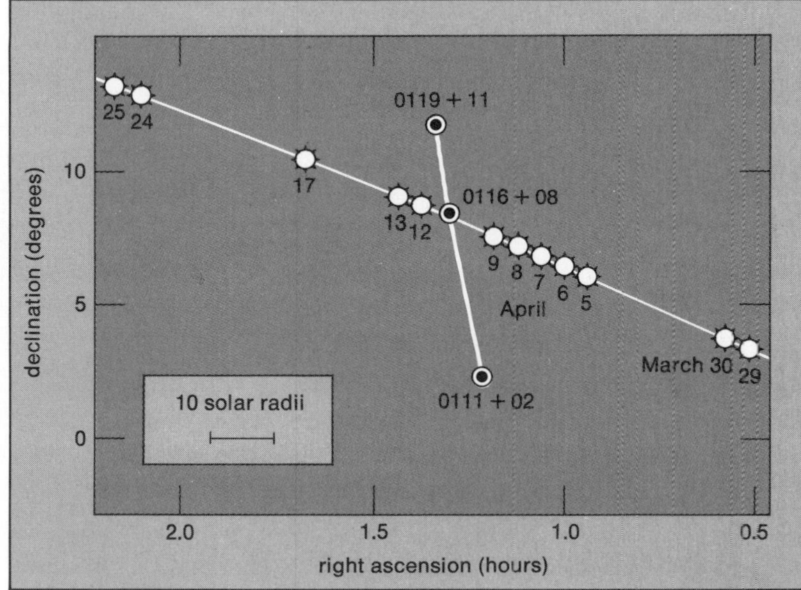

Positions of three radio sources and the Sun at noon were determined by Edward Fomalont and Richard Sramek by using four radio telescopes as an interferometer array. By measuring the angular changes in the relative positions of the radio sources they found a value for the deflection of the radio waves that supported Einstein's general theory of relativity.

(±0.011) of that predicted by Einstein's general relativity.

A similar but less accurate result, 0.99 (±0.03), was obtained by scientists at the Massachusetts Institute of Technology, the Haystack Observatory in Westford, Mass., and the U.S. National Aeronautics and Space Administration (NASA) using an array of four radio telescopes in Massachusetts and West Virginia. In a third such experiment, Dutch investigators at Westerbork Radio Observatory in The Netherlands found a deflection of 1.038 (±0.034) from quasar measurements with radio telescopes 1.5 km (0.9 mi) apart. The results of all three experiments did not differ significantly from the value unity (1.000) corresponding to general relativity.

The most widely publicized competitor to Einstein's theory is the scalar-tensor theory of gravitation proposed by Carl Brans and Robert H. Dicke in 1961. It predicts a deflection of radio waves (or light) near the Sun that is about 0.95 that predicted by Einstein. This is significantly smaller than the three experimental values. This evidence, though in itself not entirely conclusive, supports the validity of Einstein's theory rather than that proposed by Brans-Dicke.

Tests of general relativity: the Sun's shape. Einstein's theory of general relativity weathered another crucial test in 1975 with the finding that the true shape of the Sun is spherical rather than slightly oblate, as required by the Brans-Dicke theory of gravitation. Until the advent of the Brans-Dicke theory, the strongest observational evidence for general relativity was that it exactly predicted the observed advance of the perihelion of Mercury's orbit (the continual turning of the major axis of Mercury's orbit through the orbital plane so that the perihelion moves progressively around the Sun, about 9 min 34 sec of arc per century). The Brans-Dicke theory predicted a perihelion advance that was significantly smaller than the observed value. Dicke noted that this discrepancy would be removed if the Sun were somewhat flattened at the poles. He and other Princeton University scientists built an ingenious and precise scanning device to determine the oblateness of the solar disk and in 1967 reported that the Sun's polar diameter was about 45 parts per million shorter than its equatorial diameter—exactly the amount required by the Brans-Dicke theory.

The Sun's shape was remeasured at the University of Arizona by Henry A. Hill and his collaborators, who used a refined scanning method. A solar image is formed at the base of a special solar tower telescope, whose focal length is held constant to one part in 10^7. The imaged Sun's edge is scanned photoelectrically through two slits at the end of a diameter, and then the pair of slits is turned to measure the perpendicular diameter. The difference between the two solar diameters can be found to about 0.001 sec of arc.

In this way Hill's group measured a vanishingly small difference between the polar and equatorial diameters (10 ±7 parts per million). Moreover, the Arizona work showed that at times the equatorial rim of the Sun is slightly hotter than the polar rim, which can produce a small, spurious oblateness. The earlier Princeton scanning device was not specifically designed to distinguish between such temperature differences and true oblateness.

(Top) © 1975, courtesy, Ben Mayer; (bottom) courtesy, Noël Argue and Mark Rayne, University of Cambridge

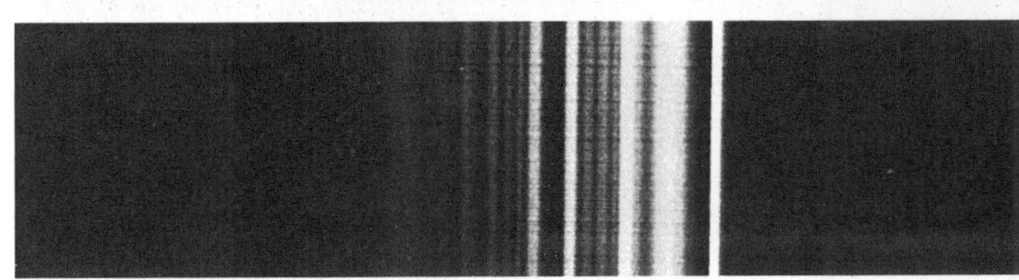

The exploding star Nova Cygni, photographed near its maximum brightness on Aug. 30, 1975, achieved an increase in light greater than any previous nova. Below is the hydrogen emission spectrum of the nova, as obtained by the 17/24-inch Schmidt telescope at the University of Cambridge Institute of Astronomy.

If the Arizona measurement is correct, the Sun is spherical and the advance of Mercury's perihelion confirms Einstein's general theory of relativity rather than the Brans-Dicke theory. Also, if the Sun is spherical, it is no longer necessary to postulate (as Dicke did) that its interior is rotating more rapidly than its surface layers. This differential rotation had been invoked to explain why the Sun could be as flattened as Dicke thought while its observed rotation period is nearly a month long.

Nova Cygni 1975. An exploding star in the constellation Cygnus was a major preoccupation of astronomical observers in the fall of 1975. Nova Cygni 1975 (also called V1500 Cygni) reached maximum light on August 31 at visual magnitude +1.8. This was the brightest new star since Nova Puppis in 1942, which attained +0.5.

The unexpected appearance of the brilliant star, which changed the pattern of a familiar constellation, was independently noted by hundreds if not thousands of persons throughout the world. The earliest known sighting was in Japan on the evening of August 29 by an amateur astronomer, Kentaro Osada. At that time the star was of the third magnitude. During the next few hours, as nightfall spread westward across Asia, Europe, and North America, the brightening nova was seen by increasing numbers of observers.

This new star was fainter than magnitude 21 on photographs taken at Palomar Observatory about 1950; its rise in light by a factor of at least 40 million is the greatest known for any nova. (Supernovae can have ranges this large, but the spectrum of the Cygnus object was unmistakably that of a nova and not a supernova.) The large range suggests that V1500 Cygni is a "virgin" nova, a star exploding for the first time. Another unprecedented feature was the stepwise character of the early increase in light; photographs taken at observatories in the Soviet Union and Argentina showed that the star lingered near magnitude 16 for at least 18 days in August. By August 29, however, the nova was brightening by half a magnitude per hour. This steep rise immediately before discovery was unusually well documented by sky patrol photographs (mostly taken by meteor observers) in California, Maryland, and Canada.

Several astronomers announced that Nova Cygni was the intrinsically most luminous nova on record, with an absolute magnitude of about −10 at maximum light. This conclusion was based on the extreme rapidity of the star's fading during the week after maximum. The empirical relationship between luminosity and rate of decline is based, however, on a study of normal novae in the Andromeda galaxy (Messier 31) and may not be valid for such a peculiar object as Nova Cygni. If the absolute magnitude at maximum was in fact −10, the approximate distance of the nova from Earth is probably somewhere between 4,000 and 6,000 light-years.

Of considerable theoretical significance was the discovery by P. Tempesti in Italy and by R. H. Koch and C. W. Ambruster in Pennsylvania of a small rhythmic brightness oscillation superimposed on the light curve of the fading nova. The amplitude of this 6.6-hour modulation decreased from 0.11 magnitude in early September to 0.03 by October 2. This light variation indicates that the nova is a close binary star, whose light was shining through the luminous expanding shell of gas ejected during the explosion. It had previously been known that dwarf novae and also old novae, after they have faded to their original faintness, generally are close binary systems. Nova Cygni is the first case of a nova where such behavior has been observed near maximum light. By mid-January 1976 Nova Cygni had faded at a continually slackening rate to magnitude 10.

X-ray nova in Monoceros. The sky as observed at X-ray wavelengths is noteworthy for strong, temporary, localized sources, which were frequently detected during the extensive surveys made with X-ray telescopes aboard the British Ariel 5 satellite, the U.S. SAS-3, and the Dutch ANS. It was estimated that up to 100 of these short-lived X-ray transient events may occur each year in our galaxy.

A particularly instructive example was discovered on Aug. 3, 1975, by scientists at the University of Leicester, England, who were monitoring the Milky Way with Ariel 5. Named by them A0620-00, this new source in the constellation Monoceros was initially quite weak. However, during the following days it strengthened rapidly until by August 14 it had become the brightest X-ray source in the sky, as measured at wavelengths from 0.7 to 6 Å (angstroms). It was about four times as strong as Scorpius X-1 and more than 20 times stronger than the Crab Nebula. According to the British investigators, the rise to maximum X-ray intensity was entirely due to "soft" photons of wavelengths greater than 1.2 Å; the observed flux of photons below 1.2 Å actually diminished during this period.

A0620-00 was also observed with the SAS-3 satellite, beginning on August 8, by scientists at the Massachusetts Institute of Technology. They also noted the dramatic increase in the soft X-ray flux and the subsequent gradual decline. Soviet astronomers reported observations on September 10 with the X-ray telescope aboard the Salyut 4 satellite. On that date A0620-00 was still more intense than Scorpius X-1 in soft X-radiation of 1.2–20 Å.

Radio astronomers at Arecibo (Puerto Rico), Nançay (France), Jodrell Bank (England), and Effelsberg (West Germany) were able to detect A0620-00 as a weak radio source at wavelengths between 6 and 31 cm. Through the combined efforts of X-ray and radio astronomers, the position of A0620-00 in the sky was determined to an accuracy of better than half a minute of arc. This made possible the optical identification of the X-ray nova.

In this position, in a rich field of faint stars, McGraw-Hill Observatory astronomers discovered on August 15 a 12th-magnitude star that had been as faint as 20.5 in 1955, when this region of the sky was photographed at Palomar Observatory. Thus the X-ray nova is also an optical nova, and so has received the variable star designation V616 Monocerotis. Fading was very slow, from visual magnitude 11.4 at the beginning of September to 13.2 in late January 1976. The optical spectrum is that of a very hot star. Examination of early Harvard College Observatory sky photographs shows that V616 Monocerotis had a novalike outburst in November 1917, when it brightened to magnitude 12, but otherwise remained faint.

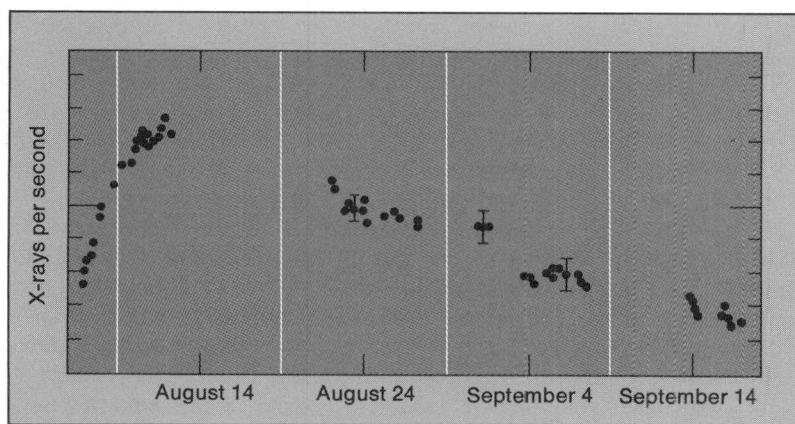

Adapted from information obtained from the SAS-3 Group

A0620-00, an X-ray nova in the constellation Monoceros, was found also to be an optical nova by astronomers at the McGraw-Hill Observatory. As an X-ray source, A0620-00 briefly became the brightest in the sky at wavelengths from 0.7 to 6 Å.

Courtesy, S. Christian Simonson III, University of Maryland

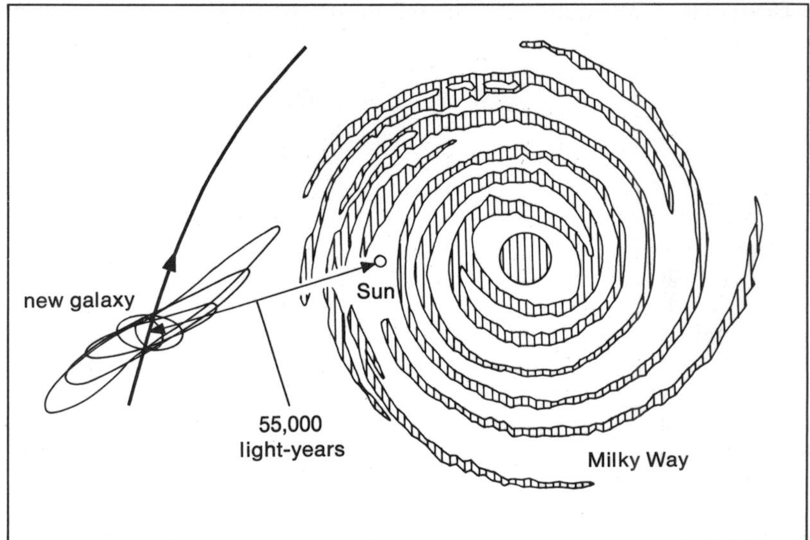

Nearest neighbor galaxy to our own Milky Way was discovered in 1975 by S. Christian Simonson III, using the 21-cm radiation of neutral hydrogen. The gravitational effect of the Milky Way deforms the much smaller neighbor, tearing off its outer parts. The arrow indicates the path of the galaxy's orbit.

Most astronomers regard optical novae as very close binary star systems, with orbital periods of less than a day, in which material flows in streams and spurts from the more massive component star to a white dwarf companion. To explain the X-ray emission of the Monoceros nova, it was suggested that a similar model applies but with a neutron star or a black hole companion instead of a white dwarf. The inflowing matter would be greatly accelerated by the strong surface gravity of the neutron star or the accretion disk of the black hole and would collide with it at high velocity, resulting in a copious release of X-rays.

Compact infrared source. Astrophysicists at the California Institute of Technology identified a very small, extremely hot star that they believe is exploding and giving birth to a planetary nebula. This object first came to astronomers' attention as a strong localized source of infrared radiation, detected by U.S. Air Force Cambridge Research Laboratory scientists in an infrared sky survey with rocket-borne instruments. Known as CRL 618, this source is in the constellation Perseus.

On long-exposure photographs, two small round nebulae, about 7 sec of arc apart, lie on each side of the position of CRL 618, which is itself invisible. The infrared brightness of CRL 618, as measured at various wavelengths from 1.65 to 3.4 microns, indicates a temperature of about 275° K (2° C). Yet the optical spectrum of the two flanking nebulae, as observed with the 200-in. Hale telescope, shows numerous bright lines of hydrogen and ionized metals, implying a high temperature.

The Caltech investigators interpreted CRL 618 as consisting of a hot central star (temperature about 32,000° K) surrounded by a shell of ionized gas 0.25 sec of arc in diameter, which is in turn embedded in a dense dust cloud. This dust is heated by the star within it, and the thermal radiation from the dust produces the infrared source, which is 0.4 sec in diameter. The star and the ionized gas shell are blocked from direct view by the dust, but at the sides the dust is thinner, allowing the light from the star and gas shell to illuminate the flanking nebulae.

The ionized gas shell was regarded by the Caltech scientists as probably a tiny planetary nebula, smaller than any of the approximately 600 already known in the Milky Way. Each planetary nebula consists of a very hot, old star, surrounded by an expanding shell of gas that is expelled when the star collapses, after burning the hydrogen fuel in its core. After some tens of thousands of years, the gas shell has expanded and faded to invisibility. Thus, the tiny planetary nebula that is associated with CRL 618 is evidently the youngest one known; its central star has been caught in the act of ejecting it.

An alternative but less likely interpretation of CRL 618 is that the dust cloud and ionized gas are collapsing and that the star is in the process of formation. This alternative is not favored by the investigators, as CRL 618 lies a considerable distance beyond the central plane of the Milky Way and thus out of the regions where star formation generally occurs. Moreover, in the immediate neighborhood of CRL 618 there appear to be no other nascent stars.

Nearest neighbor galaxy. Because interstellar dust and the dense star fields of the Milky Way cover about 10% of the sky, a nearby dwarf galaxy lying in that area could easily be undetectable even with powerful photographic telescopes. Radio astronomer S. Christian Simonson III of the

University of Maryland announced during the year that he had discovered just such a nearby galaxy.

His finding resulted from detailed mapping of the sky conducted by several radio astronomers using the 21-cm radiation of neutral hydrogen. The main features on such maps are the hydrogen clouds of our galaxy. Simonson called attention to an anomalous region in the constellation Gemini, close to the central line of the Milky Way, where the hydrogen gas has large velocities of approach to the Earth (some 90 km per sec relative to surrounding areas). Also, for about 40° westward along the Milky Way across Auriga there is a chain of rapidly approaching hydrogen clouds.

Simonson interpreted these features as due to the hydrogen gas in a previously unsuspected close neighbor galaxy in Gemini. In that part of the Milky Way, interstellar dust dims background objects by three or four magnitudes, which is sufficient to blot out the main body of that galaxy optically but not enough to hide its nucleus, if any. Because long-exposure photographs seem to show no indication of such a nucleus, the new galaxy must be of low mass.

According to Simonson's estimates, the new galaxy is a dwarf irregular system roughly 17,000 parsecs (55,000 light-years) distant. This is so close that it must be a satellite of the Milky Way, nearly in contact with it, and greatly deformed by tidal effects. The total mass of the satellite galaxy may be only about 10^8 suns, or roughly 0.0005 the mass of the Milky Way.

Simonson's galaxy appears to be the nearest of all galaxies, at about one-tenth the distance of the Magellanic Clouds. Simonson pointed out that it is also the first galaxy ever to be found solely on the basis of 21-cm-line radio observations.

—Joseph Ashbrook

Chemistry

Advances in chemistry during the year reflected a growing desire among some scientists and many funding agencies to conduct exploratory research in fields directly related to world needs. Polymer science contributed heavily to this end with new organic and inorganic materials that possessed unusual and highly desirable properties. Research in bioinorganic chemistry probed the complex relationships between metal atoms and the molecular structures directly involved in the respiration of living organisms and in the disease-causing abilities of certain bacteria. Newly invented laser-based spectroscopic techniques allowed the chemist to extract information on the molecular composition of very small regions of space, facilitating such projects as air-pollution analyses. One of the most interesting events of the year was the announcement of the synthesis of a complete mammalian gene, an achievement certain to be of great importance to genetic studies and the health sciences.

Inorganic chemistry

During the past year research in inorganic chemistry continued to contribute toward a better understanding of some of the intimate details of reaction mechanisms and of the photochemistry of metal complexes. In addition, good progress was made in long-range exploratory research on problems related to world concerns.

Polymer research. The commercial production of synthetic polymers is a large and important segment of the world's chemical industry. These materials provide thousands of such products as man-made fibers, polyethylene bags, and polystyrene cups. Most of the materials are organic polymers with carbon-carbon backbones in the form of long separate chains or cross-linked chains. For example, polyethylene consists of long chains of carbon atoms with two hydrogens on each carbon; polystyrene has the same structure, but a phenyl group replaces a hydrogen atom on every other carbon.

In spite of their great success, however, organic polymers have some serious deficiencies. Few can withstand temperatures above 150° C (300° F), and few remain flexible or rubbery over a wide range of temperature. For these reasons, automobile tires with nylon cords tend to bump on cold morning drives, and plastic garden hoses cannot be unwound in the winter. Because of such drawbacks, years of research have been devoted to an attempt to discover inorganic polymers that can withstand high temperatures and that can maintain their desirable properties during conditions of both heat and cold.

Several years ago Dow Corning Corp. and General Electric Co. manufactured silicone polymers with flexible and elastomeric properties over a temperature range of −30° to 250° C. This achievement represented a definite improvement over organic polymers, and the silicones have been widely employed where such special properties are required. Silicone polymers have a backbone of alternating silicon-oxygen bonds. At temperatures above 250° C, however, the silicone chains break to form rings, resulting in the loss of their flexible and elastomeric properties.

In recent years the search to find better inorganic polymers has gained momentum. Much of the work has involved investigations of polymeric

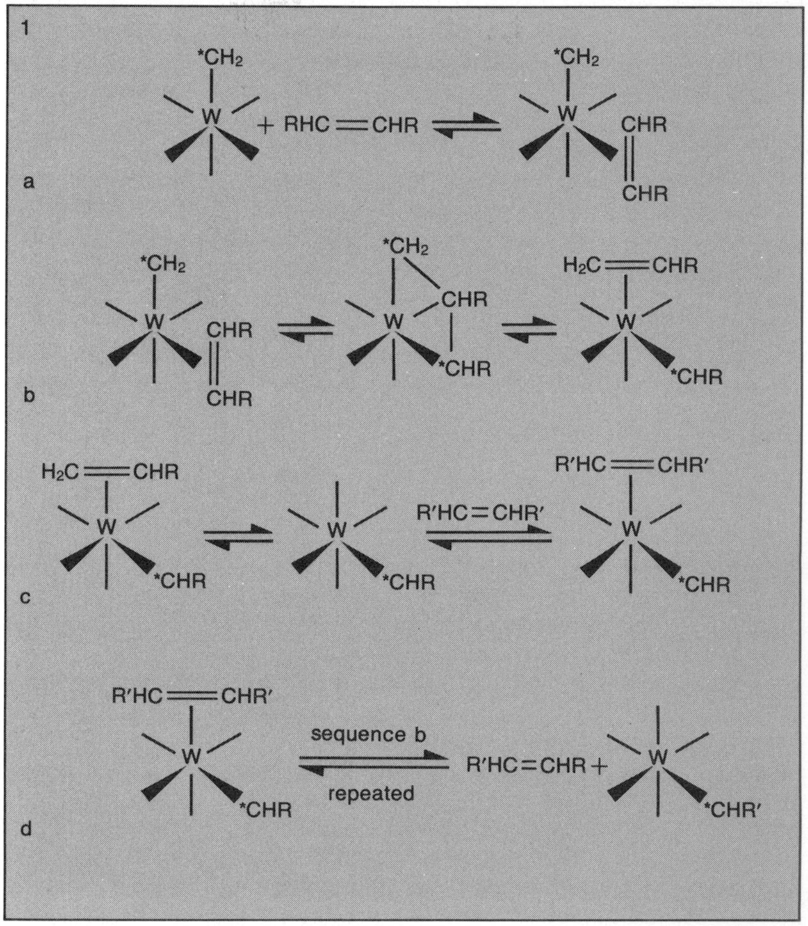

1

a
*CH2 W + RHC=CHR ⇌ *CH2 W CHR=CHR

b
*CH2 W CHR=CHR ⇌ *CH2 W CHR *CHR ⇌ H2C=CHR W *CHR

c
H2C=CHR W *CHR ⇌ W *CHR ⇌ (R'HC=CHR') R'HC=CHR' W *CHR

d
R'HC=CHR' W *CHR ⇌ (sequence b repeated) R'HC=CHR + W *CHR'

systems of such elements as boron, silicon, nitrogen, phosphorus, sulfur, oxygen, aluminum, tin, and germanium. To cite one notable example, Harry R. Allcock and co-workers at Pennsylvania State University succeeded in modifying the laboratory curiosity long known as "inorganic rubber" to obtain polymers with highly useful properties. For further information, see *Applied chemistry*: *Phosphazene polymers*, below.

For some time chemists have been intrigued with the possibility of endowing inorganic and normally nonpolymeric materials with the flexibility and moldability of organic polymers. Metal oxides, for example, possess magnetic, electric, and catalytic properties that make them extremely valuable in the fields of electronics technology and chemical synthesis. Unless dispersed as powders, however, they are generally hard and brittle. Recently a group of inorganic chemists from Georgetown University, Washington, D.C., led by Michael T. Pope and Carl O. Quicksall, reported the synthesis of a new class of organometal oxide complexes that, with further research, may lend themselves to polymerization.

The new molecules cannot be considered organometallic compounds because they do not contain metal-carbon bonds. More properly they are organic derivatives of heteropoly anions, which in turn are clusters containing 6–18 metal atoms and 40–60 oxygen atoms and carrying charges of −3 to −12. Often these anionic clusters are viewed as fragments of metal oxide lattices because they exhibit magnetic and other properties characteristic of such lattices. Pope's team prepared smaller, anionic derivatives of heteropoly anions with attached methyl, ethyl, phenyl, and other organic groups, and then obtained and characterized a number of their crystalline salts. The scientists believe that it should be possible to link numbers of these ionic molecules together to form organic polymers with metal oxide domains, which in turn could lead to thermoplastic materials with unusual metallic properties.

Catalysis. Manufacturing processes for approximately 90% of all industrially synthesized chemicals depend upon the use of catalysts that permit reactions to take place at moderate conditions of temperature and pressure. Most of the commercial

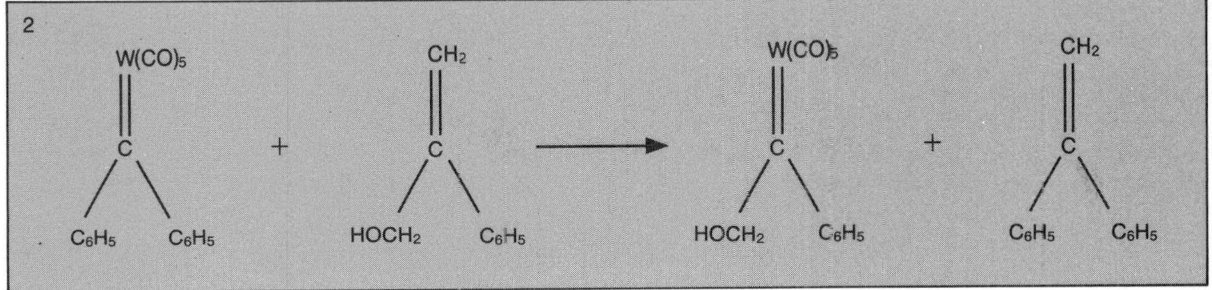

processes are of the heterogeneous catalysis type, in which the reactants are adsorbed and activated on the surface of the solid catalyst. Over the last decade considerable research has been directed toward the application of homogeneous catalysts. These materials are generally transition-metal organometallic compounds with the ability to bind and activate various types of substrates. For example, a plant recently built by the Monsanto Chemical Co. for the production of acetic acid employs a homogeneous catalysis process with the overall reaction $CH_3OH + CO \rightarrow CH_3COOH$; the homogeneous catalyst used is a rhodium complex in the presence of iodine. This process, which amounts to the conversion of coal into vinegar, is well understood in terms of some fundamental reactions of such organometallic compounds.

Extensive investigations of homogeneous catalysis were under way in many laboratories throughout the world. Much of this research was being directed toward increased comprehension of the detailed mechanisms of the catalytic processes. During the past year major progress was made toward a better understanding of the mechanism for the olefin metathesis reaction

$2RHC = CHR' \rightleftarrows RHC = CHR + R'HC = CHR'$,

in which R and R′ represent different generalized substituents attached to the carbon atoms.

The net transformation comprises scission of a carbon-carbon double bond (C = C), which is generally extremely difficult to accomplish. Yet with an appropriate catalyst, this reaction can be made to proceed very rapidly even at ambient temperatures.

Although several mechanisms had been proposed for this reaction, with little experimental verification, the research groups of Charles P. Casey at the University of Wisconsin and Earl L. Muetterties at Cornell University apparently found support for a metal carbene mechanism of the type shown in (1). A carbene is a highly reactive species of carbon atom that has been deprived of two of its usual four bonds and, thus, possesses two unbonded electrons. Attached to the tungsten (W) catalyst depicted in (1) is the carbene methylene (CH_2). The asterisk in the figure denotes the location of an unbonded electron. Evidence in accord with this mechanism is the proven existence of carbene compounds, and, of greater importance, the experimentally observed role of a carbene compound of tungsten in an actual metathesis reaction (2).

Stable metal carbenes were discovered a decade ago by Nobel laureate Ernst O. Fischer and coworkers at the Technical University in Munich, West Germany. Until recently, however, no one had been able to produce even the simplest metal carbene ($M = CH_2$) required in the proposed metal

3
R
H—N
CONH
CONH
$(CH_2)_5$
$(CH_2)_2$ $(CH_2)_5$
$(CH_2)_2$ $(CH_2)_n$
R′
N—C
N—C
N—C
O O O O O O
Fe^{III}

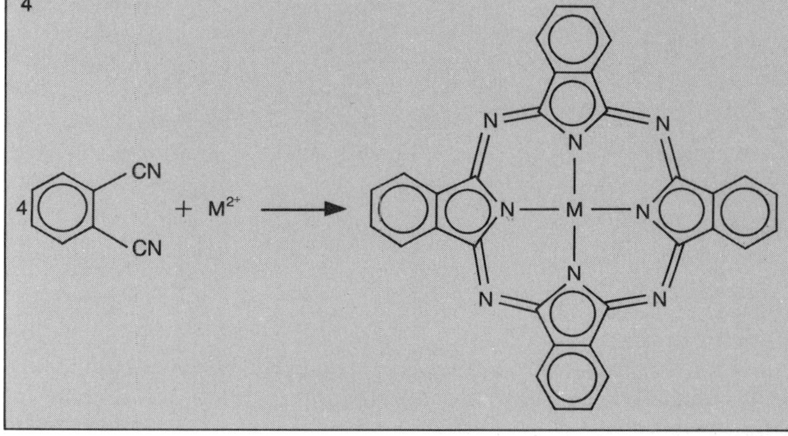

carbene mechanism (1) for olefin metathesis. Therefore, it was highly significant that Richard R. Schrock of the du Pont Chemical Co. was able to prepare and characterize the first simple metal carbene. This was accomplished by the use of a strong base to remove a proton from a methyl group attached to tantalum according to the reaction $[Ta-CH_3]^+ + NaOCH_3 \rightarrow Ta=CH_2 + CH_3OH + Na^+$. (Other groups attached to the tantalum atom are not shown.) The discovery of this simple metal carbene lent further support for the mechanism proposed for olefin metathesis. These findings should aid in the design of better and more stereospecific catalysts for such reactions.

Bioinorganic chemistry. During the past year inorganic chemists continued to work enthusiastically and effectively on biologically related problems in which metals play an important role, including oxygen-carrying complexes of vital importance to animal respiration (see *1976 Yearbook of Science and the Future* Year in Review: CHEMISTRY: *Inorganic chemistry).* For example, research groups at the University of Southern California, Los Angeles, and at Northwestern University, Evanston, Ill., independently observed the first oxygen-carrying manganese-porphyrin complex. This is an important discovery because the manganese dioxygen adduct (porphyrin-Mn-O_2) behaves in a manner markedly different from oxyhemoglobin and oxymyoglobin. Extensive studies were made of the manganese oxygen carrier.

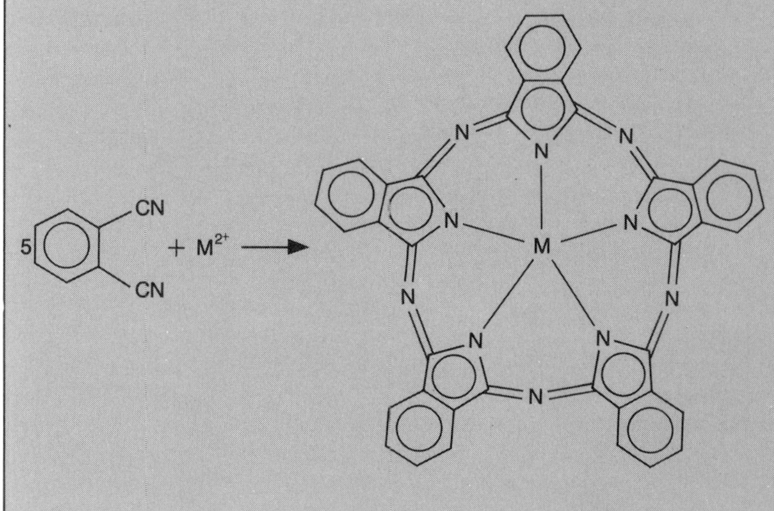

Some of the iron sequestering and transport agents (the siderochromes) found in microbial life are polyhydroxamic acid. The linear ferrioxamines (3) are produced by several species of *Streptomyces* and *Nocardia* bacteria. Three of the hydroxamate groups coordinate to iron, thereby forming a stable octahedral ferric complex. The pathogenicity of certain infections apparently is associated with microbial iron transport, and some ferrioxamines are potent antibiotics, whereas others are growth factors.

Questions regarding the structure-function relationships of the siderochromes were under investigation by John B. Neilands and Kenneth N. Raymond at the University of California, Berkeley. One approach used in this study was the replacement of iron(III) by chromium(III) in the hydroxamate complex. The iron(III) complexes are kinetically labile, whereas the chromium(III) complexes are kinetically inert. Because of this, it is possible to isolate and characterize the stereoisomers of the chromium(III) complexes. The properties and biological activities of these isomers were studied in an attempt to determine what structure factors are most important in the behavior of the siderochromes within the parent organism.

Another field of bioinorganic chemistry that has been of interest for several years centers on the role of metal ions as templates in the cyclization and condensation reactions that produce macrocyclic metal complexes. Biological examples of such compounds include the iron-porphyrin complex heme of hemoglobin and the cobalt complex of Vitamin B_{12}. Considerable research of this type has been done by Daryle H. Busch and his students at Ohio State University, Columbus. In related work, Tobin J. Marks and his students at Northwestern University reported the fascinating observation that increasing the size of the metal template yields an expanded macrocyclic ligand by allowing a greater number of subunits to coordinatively cyclize. Normal metal phthalocyanine complexes had been known for a long time and are readily prepared (4) using such metals as nickel and copper from the first row of transition elements in the periodic table. Marks prepared superphthalocyanine metal complexes (5) by the same reaction, but with a much larger metal ion, specifically one of uranium, UO_2^{2+}. This pronounced tendency of the uranyl ion to form an expanded phthalocyanine suggests that it is ideally suited to construct five-membered macrocyclic porphyrin analogs using template reactions. Further research was in progress, and attempts were to be made to prepare other supermacrocyclics.

—Fred Basolo

Organic chemistry

The study of organic chemistry has been greatly aided by the fact that most organic molecules are chiral; they differ from their mirror images in the same way that a left and a right hand differ. The presence of chirality (from the Greek word meaning hand) in an organic molecule usually is due to the presence of one or more chiral carbon atoms; *i.e.*, carbon atoms surrounded by four different substituent groups in a tetrahedral arrangement. Chemists had hardly finished celebrating the centennial of the recognition of the tetrahedral nature of carbon-atom bonding (by two young students in 1874) when its importance again was emphasized by the awarding of the 1975 Nobel Prize for Chemistry to two scientists whose work has greatly extended and enlarged man's knowledge of the utility of this concept (*see* Year in Review: SCIENTISTS OF THE YEAR). Vladimir Prelog, a lifelong experimenter in the field of stereochemistry, was a major contributor to the development of the internationally recognized system of nomenclature by which the chirality or handedness of a molecule is designated. John W. Cornforth made extremely subtle use of chirality to study the way in which living systems construct complex organic molecules from simple precursors. For example, he prepared molecules containing the chiral methyl group, whose chirality depends upon the substitution for normal hydrogen of its heavier isotopes, deuterium (D) and tritium (T). When these isotopes are substituted, one of the two isomers (called enantiomers) of acetic acid can be prepared (1). When one enantiomer is fed to a living system, the fate of the methyl group can be followed as it is incorporated into organic molecules of ever increasing complexity. The study of biosynthesis, as this process is called, continues to be investigated by organic chemists, and the methods pioneered by Cornforth and Prelog play important roles in it.

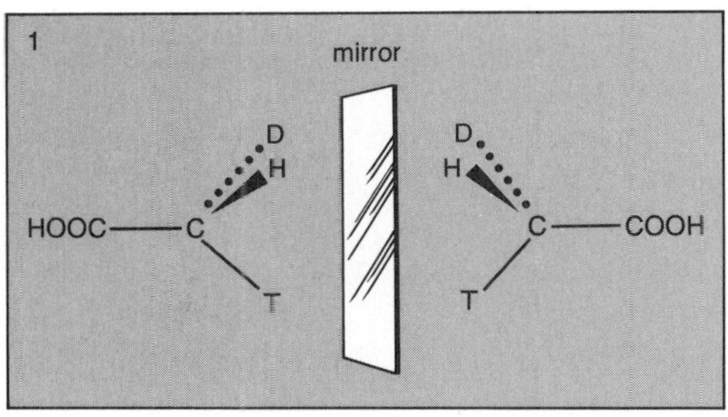

Organic synthesis. Electroorganic chemistry, the study of the reactions that organic compounds undergo when subjected to the action of an electric current, has undergone something of a renaissance in recent years. There is important industrial potential in these studies; in many ways electricity is the ideal chemical "reagent," adding nothing but electrons to the solution of reactants and so eliminating the need for the removal and disposal of acids or other chemicals that are usually used to bring about reactions. In one electrochemical

2

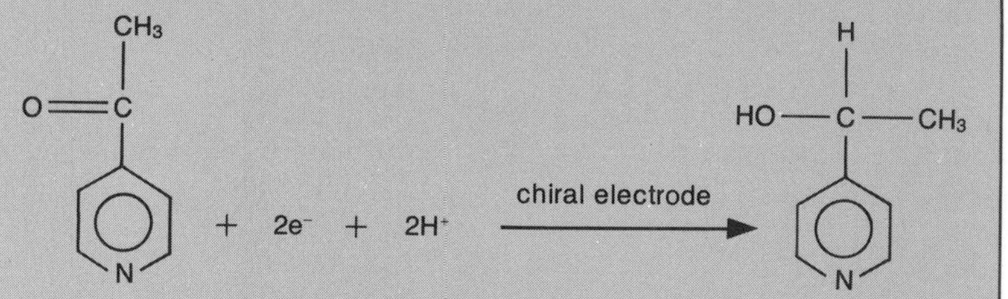

$$\text{4-acetylpyridine} + 2e^- + 2H^+ \xrightarrow{\text{chiral electrode}} \text{HO—CH(CH}_3\text{)—pyridine}$$

experiment, Larry L. Miller and co-workers of Colorado State University showed that a chiral electrode can be produced by chemically binding naturally occurring chiral amino acids to the surface of a graphite electrode. This chiral electrode may be used to carry out an electrochemical reduction to generate a chiral carbon atom from an achiral starting compound (2). The product is formed with a predominance of one of the two possible enantiomers of the product. The experimental results established the principle of chemical modification of the surface of electrodes and gave promise of tailoring electrodes for specific chemical purposes.

Chemists are also learning to use enzymes in some of their more difficult syntheses. The preparation of a single enantiomer of a chiral molecule can be a tedious task by standard organic reactions, but in favorable cases an enzyme can do the job quickly and easily. For example, enzymatic catalysis of the hydrolysis of the achiral diester (3) results in its conversion to the chiral acid. Being itself chiral, the enzyme is able to distinguish between the two ester groups, which are indistinguishable to the usual, achiral reagents. Acid or base catalysis of this same reaction gives an equal mixture of the two enantiomers, resulting from equal attack on each ester group.

Because the reactions of most easily available inorganic chemicals with organic substrates have been extensively studied over the years, it is surprising that a promising new reagent has been found. The chemical, potassium superoxide (KO_2), was shown by several groups in recent months to be an extremely useful reagent for the synthesis of alcohols and peroxides. Superoxides have not been useful in the past because of their insolubility in most organic solvents. This insolubility can be overcome by the addition of crown ethers, large cyclic molecules that combine with the potassium ion and literally drag the superoxide into nonpolar organic solvents. In such solvents the superoxide ion, O_2^-, is extremely reactive and promises to be a useful weapon in the arsenal of the synthetic chemist. Among the exotic new reagents developed during the year, the hydrides of the metal zirconium seem highly promising. Jeffrey Schwartz at Princeton University demonstrated how these reagents can be used to convert unsaturated molecules to alcohols.

Organic synthesis of molecules of practical interest progressed rapidly. Compounds of the general structure shown in (4), known as peroxyacyl nitrates, are constituents of photochemical smog and are known to have harmful effects on living organisms. A detailed study of these compounds has been hampered by the lack of a general method of synthesis. This problem was remedied by R. Louw, working at the University of Leiden, Neth. In contrast to the lower members of the series (in which R is a methyl or ethyl group), analogs of higher molecular weight are reasonably stable and their chemical and biological properties can be studied without special techniques.

In a remarkable biochemical synthesis, Argiris Efstratiadis and co-workers at Harvard University constructed a complete mammalian gene through a multistep procedure that reproduced the exact

3

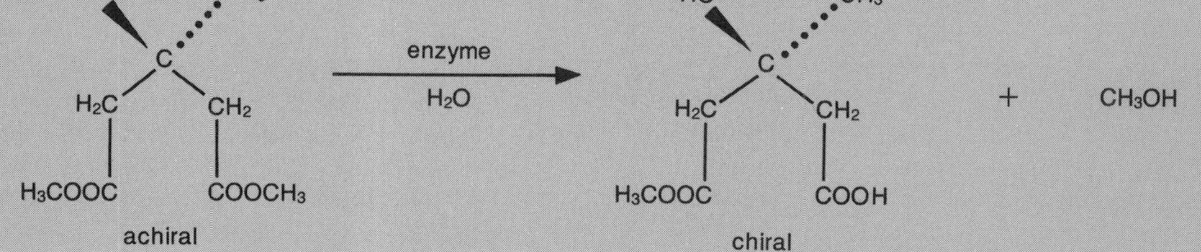

$$\underset{\text{achiral}}{\text{HO(CH}_3\text{)C(CH}_2\text{COOCH}_3\text{)(CH}_2\text{COOCH}_3\text{)}} \xrightarrow[H_2O]{\text{enzyme}} \underset{\text{chiral}}{\text{HO(CH}_3\text{)C(CH}_2\text{COOCH}_3\text{)(CH}_2\text{COOH)}} + CH_3OH$$

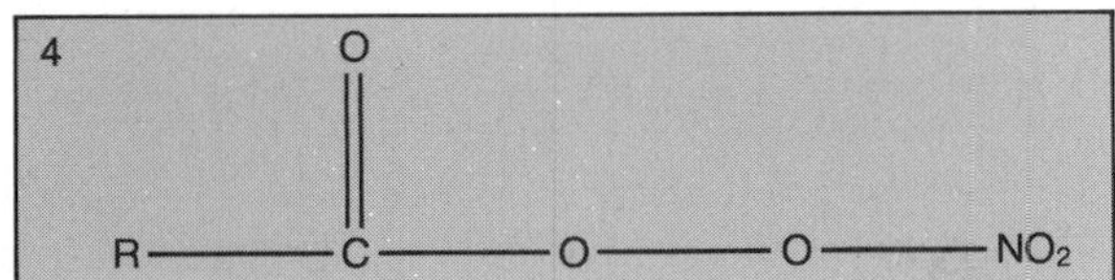

chemical sequence of subparts or nucleotide units found in a specific molecule of rabbit DNA. The biochemical-synthesis system of a living cell uses information stored in coded form on the strands of DNA as the directions for assembling all the proteins for which the cell has responsibility. The essence of the information code is the specific sequence of nucleotide units comprising the complementary double strands of each DNA molecule. Instructions for protein synthesis are copied from a DNA strand onto a "messenger" molecule, called messenger RNA. The messenger RNA carries this information to the site of protein synthesis, where its code is "read" by the cell's protein-building units, the ribosomes.

Efstratiadis and his colleagues began their synthesis with a purified strand of messenger RNA that carried information for the synthesis of rabbit hemoglobin. Employing a nucleotide broth as a source of genetic material, they added an enzyme commonly called reverse transcriptase to reverse the normal function of messenger RNA. Hence, rather than serving as a set of instructions for protein synthesis, the messenger RNA became a pattern for the construction of a copy of the strand of DNA from which it originally had taken its existence. From a knowledge of the strand's spatial arrangement and through the use of other enzymes, the group was able to extend one end to form a complementary strand of DNA and thus complete the double-stranded molecule that constituted the rabbit hemoglobin gene.

The synthesized gene contained 650 nucleotide units per strand and, as such, was comparable in size to a human gene. Primarily chemical methods of gene synthesis, as exemplified by the pioneering work of Nobel laureate H. Gobind Khorana and his team from the Massachusetts Institute of Technology, have not yet reproduced genes of this length, but in 1973 a bacterial gene containing 126 nucleotides per strand was prepared. Khorana's approach took the form of a chemical synthesis of short segments of DNA that subsequently were joined with appropriate enzymes in an overlapping fashion until the entire double strand was produced. The two synthetic methods—one basically biochemical, the other basically chemical—are complementary.

The gene synthesis techniques of the Harvard and MIT teams have laid the foundations of an important new method for genetic study. Through the application of recently developed gene transfer techniques, artificially synthesized genes could be inserted into rapidly multiplying bacteria to make large amounts of both the gene and its specified product. It also would be possible to make slight modifications in the gene and thus in its product, allowing investigations highly relevant to the fields of health and disease, organic chemistry, and molecular biology.

Theoretical studies. In addition to improving methods for the synthesis of complex organic molecules, chemists have continued their studies on how reactions occur and upon the physical properties of molecules. These studies have as their ultimate goal the prediction of chemical and biological properties in advance of synthesis.

It has been recognized for some time that the solvent in which a reaction occurs can play an important, perhaps decisive, role in the result. In order to study the reaction itself, unaffected by the solvent, chemists have examined reactions between ions and neutral molecules in the gas phase. In one approach, fundamental studies with very simple molecules and ions were carried out in order to obtain detailed knowledge about a reaction. Richard J. Cross, Jr., at Yale University prepared a beam of methyl cations and allowed it to interact in the gas phase with a beam of ethylene molecules. The ionic products were scattered in a way that gave information about the detailed mechanism of the reaction. Equally noteworthy were an increasing number of investigations on relatively large and complex molecules and ions. A characteristic observation in solution is that electron-attracting groups like fluorine greatly speed up attack by bases on a benzene ring. However, J. M. Riveros of the University of São Paulo, Brazil, recently demonstrated that in the gas phase no such acceleration is observed. The solvent must be treated as an integral part of the process in any reaction in solution.

Computers continue to play an important role in organic chemistry, both in performing fundamental calculations on the structure and stability of organic molecules and in assisting in the design of syntheses and even in the elucidation of reaction pathways. For example, Khorana's gene synthesis mentioned above was aided greatly by computer searches for repeating sequences of nucleotides, permitting the same nucleotide segments to be used in several parts of the synthesis. Paul R. Schleyer of Princeton University employed a computer to determine the most probable mechanistic pathway of a complex rearrangement of a polycyclic hydrocarbon. The rearrangement (5) proceeds under remarkably mild conditions and in high yield because the product is a segment of the very

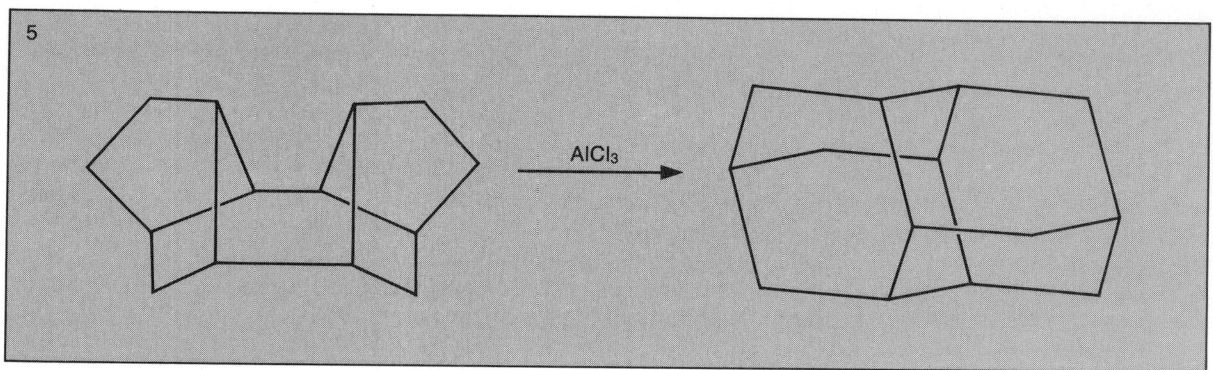

stable diamond lattice. While there were an estimated 40,000 possible isomeric intermediates, a computer analysis uncovered a nine-step, low-energy pathway that provided a framework for experimental investigations, such as the isolation of intermediates or isotopic-labeling studies.

—Charles H. DePuy

Physical chemistry

Physical chemistry is the study of the fundamental properties of chemical compounds and chemical changes. It includes studies of the rates of chemical reactions and the conditions for chemical equilibrium. Most of the properties of bulk matter can be related to the structure of the individual molecules and the interactions between these molecules; thus, much of physical chemistry currently involves the determination of these structures and interactions.

Within the past few years there has been rapid progress in several branches of physical chemistry. For example, the ability to understand and control the rates of chemical reactions has increased swiftly, and several new laser-based techniques have assisted in the exploration of fast and very fast chemical processes. In addition, the laser has provided a sensitive analytical tool for studying the products of chemical reactions at very low concentrations and has aided spectroscopists in determining many details of molecular structure. There has also been an impressive advance in understanding the nature of the surface of solids and of the chemical reactions that occur on such surfaces. Studies of gas-solid interactions seem certain to lead to new and more efficient catalysts.

Laser spectroscopy. Among the most promising new spectroscopic techniques is stimulated or coherent anti-Stokes Raman scattering. CARS, as the technique is sometimes called, is based on an optical effect that occurs with intense laser sources. Two lasers are required, at least one of which is tunable. When the two lasers, with light of frequencies ν_1 and ν_2, are focused on a sample, certain nonlinear effects occur, one of which is the generation of light at a frequency $2\nu_1 - \nu_2$. When the difference in frequency of the two lasers is equal to a Raman active vibration of the molecule (ν_R), then the nonlinear mixing is greatly enhanced and the light at $2\nu_1 - \nu_2 = \nu_1 + \nu_R$ becomes quite intense. By scanning one laser over a reasonable frequency range, it is possible to observe the vibrational spectrum of a sample.

While the theory is complex, the experiment is rather simple. Two lasers are directed on a sample, and one of the lasers is tuned. When the frequency difference of the incoming beams reaches a vibrational frequency of the sample, a third intense, collimated beam is produced. Because this stimulated beam travels in a direction different from the incident beams, it can easily be observed by blocking off the incident beams with a black beam stop.

This technique is far more efficient than ordinary Raman spectroscopy. About one-thousandth of the incident power is converted to light at the frequency $2\nu_1 - \nu_2$. Thus, stimulated anti-Stokes Raman scattering can be used for studying vibrational spectra of quite dilute solutions. It can also be used effectively with fluorescing samples, a frequent stumbling block in ordinary Raman studies. The volume of sample needed in this technique is quite small, making it possible to obtain spectra from a small region of space. Hence, stimulated anti-Stokes Raman scattering can be used for pinpoint air pollution analysis, for mapping the chemical composition of flames, or for deducing the components in the regions near active electrodes in solution. The lasers can be pulsed with pulse widths less than 20 nanoseconds (billionths of a second) so that Raman spectra of molecules with lifetimes of only a few nanoseconds can be observed with great sensitivity.

Laser chemistry. Laser-induced chemistry has been the subject of intense, well-funded study. Generally, a laser is used to populate excited states of a molecule that react more rapidly than the ground state. Such techniques have great practical importance. They can be used to prepare pure isotopic samples. They also can be used to purify materials chemically. When laser isotope separation can be done commercially, it will lower substantially the cost of nuclear-fuel reprocessing as well as the cost of isotopes for medical and research purposes.

Several successful isotope-separation schemes have been patented. Boron, nitrogen, hydrogen, and sulfur isotopes have been separated using laser techniques. Uranium-235 has been separated from a natural isotopic mixture by scientists at Lawrence Livermore Laboratory and Avco-Exxon Laboratories using a two-step photoionization technique. Although most current isotope-separation techniques require the sample to be in the gas phase where the isotopic splittings are more easily resolved, Robin Hochstrasser of the University of Pennsylvania was able to dissociate *sym*-tetrazine selectively in a solid solution at 1.6° K and to produce enriched ^{13}C and ^{15}N in high purity.

Another technique for isotope enrichment was used by Vladimir S. Letokhov and his associates at the Russian Institute of Spectroscopy in Moscow, who found that they could selectively dissociate isotopes of SF_6 with an intense infrared laser. The actual process is not yet clearly understood. When illuminated by sufficiently strong laser beams, a molecule apparently can absorb enough infrared photons, up to 200 photons for each SF_6 molecule, to raise it well above the dissociation limit. The molecule then dissociates, or more likely explodes, into a variety of small fragments, each of which can undergo secondary reactions and can be separated from the reaction mixture. (For a comprehensive treatment of laser isotope-enrichment research, *see* Feature Article: LASER-INDUCED ISOTOPE SEPARATION.)

Water structure. The connection between the properties of individual molecules and the properties of macroscopic collections of molecules is obviously crucial to physical chemistry as a practical science. The connection is not always a simple one, however, especially for liquids. Liquid water plays an especially important role in chemical and biological processes. The structure of individual water molecules and their intermolecular interactions have been understood for some time, but they are sufficiently complex that only recently did it become possible to calculate the properties of liquid water from a knowledge of the properties of the molecules.

Liquid water can be treated in several ways, of which molecular-dynamics calculations are the most pictorial. In these calculations, a model is taken for the water molecule, and a force law between the molecules is assumed. The paths of each of the molecules are then calculated from classical mechanics in much the same way as one would calculate the paths of billiard balls on a pool table. A few years ago Aneesur Rahman of Argonne National Laboratory and Frank Stillinger of Bell Telephone Laboratories calculated the paths of 216 water molecules at conditions simulating room-temperature water and using a simplified model of water interactions. Recently, using better models of water interactions, they performed additional calculations that describe the

Courtesy, Lawrence Livermore Laboratory, University of California

Experimental laser-ionization apparatus developed at the Lawrence Livermore Laboratory of the University of California successfully produced milligram quantities of uranium that was isotopically enriched to 3% uranium-235.

liquid as a random network of water molecules arranged in an approximately tetrahedral configuration. There is much disorder in the liquid; the networks are highly strained and bear little similarity to anything resembling a small ice crystal. The structure evolves continuously.

The current models are sufficiently refined to permit determination of the molecular vibrations and intramolecular vibrations in the water networks. Diffusion occurs in water as a continuous process and not in discrete hops as had been proposed. These calculations provide a detailed understanding of liquid water and supply information on aqueous liquids that is unavailable from experiments. Similar calculations were being done on aqueous solutions and on fused salts.

Touchy-Feely I. Chemists have long held the notion that chemical and even biochemical behavior can be derived from theory. The computational problems, however, that arise for large molecules and for reactions between all but the smallest molecules are formidable. In a novel approach to these problems, Kent R. Wilson of the University of California, San Diego, built the first version of an instrument called Touchy-Feely. This instrument provides the chemist with visual and tactile interactions with a molecular model. The model is a component of a computer system that is capable of solving the classical equations of motion for several hundred atoms. The chemist can choose a molecule in a solvent, for example, and watch the evolution of the interaction between molecule and solvent. He can then reach in and manipulate any atom he wishes, feel the forces on that atom, and guide the molecular system into a chemically interesting configuration or structure.

This approach helps alleviate a major obstacle to calculations on large molecular systems; it enables the chemist to concentrate on interesting combinations of atomic positions and momenta and to eliminate the overwhelming fraction of configurations where no reaction occurs. It allows him to use his intuition to steer the molecules into likely configurations for reaction. Because he can also feel the forces acting on the model atoms he is manipulating, the chemist's intuition is improved. This technique can reduce the search through an impossibly large number of a priori configurations to a few reasonable ones that are likely to lead to reactions. Those few configurations can then be used for detailed calculations.

Fast chemical kinetics. Spectroscopists studying events on the picosecond time scale (10^{-12} second) continued to uncover new phenomena using a variety of ingenious experiments. Studies in liquid solutions progressed the farthest, especially studies of the way in which molecules in solution move and exchange energy with the liquid that surrounds them.

The equipment and techniques for these experiments were sufficiently improved to allow exciting pulses as short as 0.3 picoseconds to be used to excite and study processes on the subpicosecond time scale. For example, W. Kaiser in Munich, West Germany, measured the time for a vibrationally excited ethyl group in a coumarin dye to give up its energy to its carbon tetrachloride solvent. He found the relaxation time was 1.7 picoseconds with a resolution of 0.6 picoseconds. In another study, Charles V. Shank and E. P. Ippen measured with similar accuracy the time required for a carbon disulfide molecule (CS_2) to rotate in liquid CS_2. These studies of molecular reorientation and relaxation in liquids as well as earlier studies on the rate at which molecular fragments can escape each other in solution are important to understanding the mechanisms of solution reactions.

—James S. Shirk

Applied chemistry

A number of recent advances in theoretical and research chemistry found applications in the field of health and disease. A new class of polymer with remarkable biocompatibility and high permeability to oxygen was employed successfully in experimental lung-support systems, and a therapy involving high-pressure hydrogen held out the possibility of a cure for certain human cancerous tumors. An interesting highlight of the year was the announcement of the discovery of several compounds that act as odor counteractants.

Phosphazene polymers. Over the past 50 years, exploitation of the ability of carbon to combine prolifically with itself and with other atoms to form large molecules has led to a continuing chemical revolution in the development and application of man-made organic polymers. Elements other than carbon also share to some extent this same ability, though only recently has their potential come to be appreciated. For example, phosphazene or phosphonitrile polymers are cyclic or linear molecules comprising a long chain of alternating phosphorus and nitrogen atoms with two substituent atoms or groups on each phosphorus atom. Shorter, cyclic versions of these molecules, the trimeric and tetrameric chlorophosphazenes (1), had long been known from the reaction of ammonium chloride and phosphorus pentachloride. These brittle, crystalline compounds could be converted by heating into the transparent plastic polymer polydichlorophosphazene, also known as inorganic rubber, an interesting but almost valueless material because it decomposes in air after

1

chlorophosphazene trimer, tetramer; polydichlorophosphazene

$$PCl_5 + NH_4Cl \xrightarrow[(-HCl)]{120^\circ C} (NPCl_2)_3 \text{ (trimer)} + (NPCl_2)_4 \text{ (tetramer)} \xrightarrow{250^\circ C} [-N=PCl_2-]_n$$

the hydrolysis of its P — Cl bonds by atmospheric moisture.

Phosphazene chemistry languished until recently, when Harry R. Allcock and co-workers at Pennsylvania State University succeeded in stabilizing the polymer by devising methods of synthesizing linear chains with nonhydrolyzable substituent groups. By a judicious choice of substituents, a variety of polymers with differing and useful properties was obtained (2). Since then, many other experimenters, including those at Firestone Tire & Rubber Co. and at Horizons, Inc., a Cleveland, Ohio, research firm, have prepared a selection of phosphazene polymers that can be used as rigid plastics, as rubbery elastomers, as films and coatings, and as expandable foams.

In many respects, some of the new phosphazenes far exceed the performance of a number of the older, widely used polymers in their water repellency, flame-retardant properties, and suppleness at extremely low temperatures. Scientists from the Army Materials and Mechanics Research Center in Watertown, Mass., announced their consideration of phosphazene polymers for fuel hoses to service military equipment that must operate in subzero Arctic climates. From their recent research at Penn State, Allcock and his colleagues reported that polyfluoroalkoxyphosphazenes can be used to waterproof cotton fabrics, and Celanese Corp. textile chemists demonstrated that other phosphazenes act as flame retardants for normally flammable polyester fibers.

Phosphazenes also may have many important biomedical applications. For example, Allcock suggested that phosphazenes with amino acid esters attached to their basic structures may be useful in reconstructive surgery; they should be biodegradable, eventually breaking down and allowing normal tissue to grow back and replace them.

Currently dampening the promise of the new and versatile phosphazenes is their price—about $100 a pound. Firestone, however, began producing one type of fluorinated phosphazene elastomer on a pilot-plant scale, and many chemists expect that growing demand and large-scale production will bring down the price of phosphazenes to levels that the average consumer can afford.

Biomedical applications of polyalkylsulfones. Another versatile class of polymers, prepared by reacting sulfur dioxide (SO_2) with a wide range of olefins (unsaturated hydrocarbons containing at least one pair of carbon atoms connected by a double bond), appears to have properties that make these materials especially useful in several biomedical applications, including artificial lung-support machines, contact lenses, and temporary artificial skin. Synthesis and ongoing studies of these copolymers, called polyalkylsulfones (PAS), were discussed extensively in late 1975 at a joint meeting of the American Chemical Society, the Asociación Farmacéutica Mexicana, the Chemical Institute of Canada, the Instituto Mexicano de Ingenieros Químicos, and the Sociedad Química de México in Mexico City.

According to Don N. Gray of Owens-Illinois, Inc., Toledo, Ohio, the properties of the PAS copolymers differ significantly depending on the chain length of the olefin with which the SO_2 is reacted. For example, when the olefin chain contains six carbon atoms, the end product (PAS-6) is hard and brittle. On the other hand, PAS-16, made from an olefin containing 16 carbon atoms, appears especially desirable for biomedical applications because of its great tensile strength, biocompatibility, and high permeability to oxygen, carbon dioxide, and other gases of physiological importance. Gray also suggested that the permeability of PAS-16 should make it ideal as well for use in protective dressings for surgical and burn patients.

PAS-16 holds great promise as an efficient, long-lasting membrane in experimental lung-support systems currently being tested by scientists at Arthur D. Little, Inc., Massachusetts General Hospital, and the National Heart and Lung Institute. The new polymer may help eliminate one of the major problems involved in using conventional lung-support systems, in which the patient's blood must be removed from his body and passed through a chamber where oxygen is fed in and carbon dioxide removed. After six to eight hours in

Photographs, courtesy, G. B. Street, H. Arnal, W. D. Gill, P. M. Grant, R. L. Greene, IBM Research Laboratories

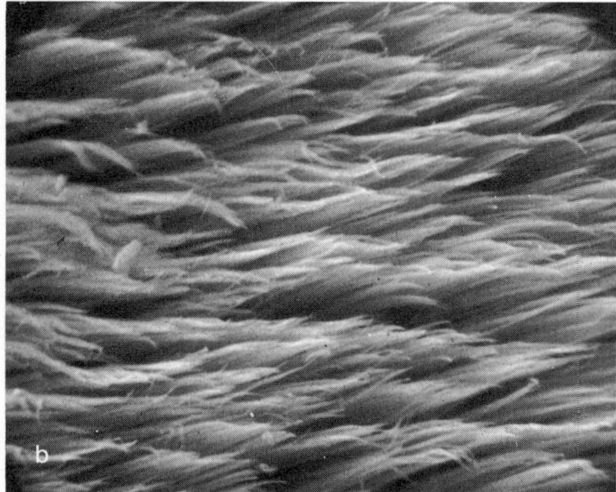

Magnified images of twinned crystals of polysulfur nitride or $(SN)_x$ (top) reveal constituent fibers of the crystals along their b axes meeting at the line of intersection (center) with fiber termination on crystal face at angle to b axis (bottom). An inorganic polymer composed of nonmetallic atoms, $(SN)_x$ has recently been shown to have metal-like electrical and optical properties and to be superconducting below 0.3° K.

the older machines, direct and prolonged exposure to oxygen can damage and even burst blood cells. By contrast, the experimental system prevents direct contact by separating blood cells and oxygen with a highly permeable membrane; using membrane material other than PAS-16, it has been tested on patients for as long as two weeks without causing cellular damage.

Compatibility tests of PAS-16 at Massachusetts General Hospital showed that rings of PAS-16 implanted in dogs in the vena cava portion of their hearts did not obstruct the oxygen flow into the heart through clot formation and functioned well for periods as long as two weeks. In addition, Warren Zapol of the hospital and John Ketteringham of Arthur D. Little found that PAS-16 membranes were about 7.5 times more permeable to oxygen and 6 times more permeable to carbon dioxide than the silicone and silicone-polycarbonate copolymer membranes currently used in many experimental lung-support machines.

When Ketteringham and Zapol built an experimental lung-support machine using a PAS-16 membrane and tested its performance on a lamb's blood supply, they reported that the animal survived the experience well. More important, the PAS-16 membrane was equal to the silicones in oxygen permeability and from six to ten times better in removing carbon dioxide from the blood. The improved performance could eventually lead to lung-support machines that use membranes with less surface area, reducing the potential hazard to the patient from blood-clot formation.

Recycled plastics. Responding generally to the environmental protection movement and to the growing shortage of petrochemicals from which most plastics and polymers are manufactured, a trend toward recycling of discarded plastics picked up momentum in 1976. In one approach, currently being tested on a pilot-plant scale at the University of Hamburg, West Germany, plastic wastes are pyrolyzed or broken down at high temperatures in an inert atmosphere into useful hydrocarbons and other products. Depending on the plastic starting material and the temperature at which it is pyrolyzed, the reaction products can vary widely. For example, Walter Kaminsky and his colleagues at the university's Institute for Inorganic and Applied Chemistry found that when polyethylene is heated to 740° C (1,365° F), it yields about 25% ethylene, by weight, 16% methane, 12% benzene, 10% propylene, and small amounts of carbon, hydrogen, and other products. Heating polyvinyl chloride produces about 56% hydrochloric acid, 2% ethylene, 22% assorted aromatics, and a collection of other products. An old rubber tire subjected to the same conditions

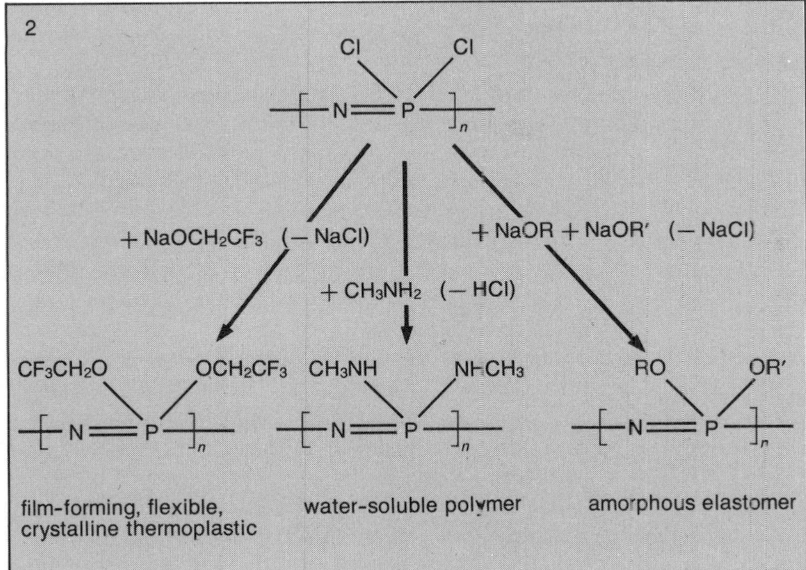

yields about 43% carbon, 10% methane, 18% aromatics, and smaller amounts of unsaturated and saturated hydrocarbons. Produced at the rate of 250,000 metric tons a year in West Germany alone, plastic wastes, if recycled efficiently, should relieve some of the strain on world oil resources and help solve an environmental problem by eliminating the vast amounts of plastic wastes accumulating upon the Earth.

In another line of attack, economy-minded chemical companies have increased their efforts to salvage usable plastics from discarded plastic products. The Monsanto Co. began recycling the nitrile resins used in plastic soft-drink bottles, and car manufacturers were developing techniques for recycling foamed polyurethanes from junked automobiles. In Piqua, Ohio, the Polymer Recovery Corp. built a plant that can salvage 24 million lb of plastic wastes per year, primarily polyvinyl chloride from furniture and old automobile upholstery. In Yakima, Wash., Western Foam Pack was recycling foamed polystyrene trays, while Mobil Oil Corp. was recovering polystyrene from old egg cartons, converting the waste into pellets, combining it with "virgin" polystyrene, and manufacturing a new generation of egg cartons.

Malodor counteractants. In 1968, while screening chemicals with human test subjects for possible new fragrances, Alfred A. Schleppnik of Monsanto Flavor/Essence, Inc., St. Louis, Mo., discovered several compounds that produced a "fresh air" smell. In contrast with the usual masking effect of conventional deodorizers, which create a stronger odor in order to smother the malodor, the new compounds produced what amounted to an almost instantaneous elimination of malodor perception. Such peculiar activity seemed to contradict then current theories of olfaction and went unappreciated for several years until officials at Monsanto realized the potential of the discovery and initiated a strong research and development program.

Most of the commonly encountered malodors are molecules of relatively low molecular weight that can either accept or donate a proton (a hydrogen atom minus its electron). These include many thiols and phenols, certain carboxylic acids found in perspiration and rancid foods, and a number of amines responsible for typical fishy odors. The mechanism by which these molecules react with receptors in the nose is a subject of much speculation among scientists. Schleppnik's own explanation incorporates popular theories concerning two contributing receptor systems—the olfactory system and the trigeminal system. The olfactory system apparently operates through the reception of odor molecules on specific submolecular regions or sites on protein molecules embedded in the membranes of olfactory receptor cells. Associated with these olfactory nerve endings are the endings of the trigeminus nerve branches, which are thought to signal the presence of odor molecules with a molecular weight of 150 or less by producing burning or tingling sensations. Schleppnik noted that the malodor counteractants all have molecular weights between 150 and 250, possess or can assume the same basic geometric shape, have similar distributions of electron density and polarity, and elicit stereospecific activity. He suggested that these properties allow the counteractant molecule to react with an allosteric submolecular region on the receptor molecule (a second site somewhat distant from the receptor region), causing the receptor molecule to assume a different geometrical conformation, one which shields the receptor region from incoming malodor molecules. The net effect is a selective loss of perception in the nose, an apparent lowering of the odor level in the air, and a tingling, fresh-air effect caused by direct stimulation of the trigeminal nerve endings.

The activity of the counteractant persists as long as a concentration of at least one part per million is maintained; otherwise, olfactory sensitivity to malodors returns within a few seconds. In tests, no residual effects either to periodic exposure or to high concentrations of the chemicals have appeared. Perhaps the only drawback of the counteractants noted thus far was their lack of complete selectivity. Not all proton-donor and proton-acceptor molecules are considered offensive; certain appealing food odors, *e.g.*, blue cheese, would also be eliminated by the counteractants.

Monsanto will market the chemicals—which it calls amals—for use in home air fresheners and in such products as depilatories, shampoos, and cosmetics, which often contain necessary but malodorous components. Other uses might be in personal deodorants, douches, and pet litter boxes. In microencapsulated forms, the counteractants could be processed into paper, cardboard, plastic sheets and tile, and other industrial products that often give off harmless but offensive odors. In purely scientific applications, they might contribute significantly to the elucidation of human and animal olfaction mechanisms and to the study of proton-transfer enzyme systems.

Hyperbaric hydrogen therapy. In an unexpected practical application of physical chemistry, three Texas scientists combined their knowledge of the kinetics of free-radical decay with theories concerning the biochemistry of cancer to invent what may eventually become a treatment for certain human carcinomas. During previous studies of the radiation chemistry of polyethylene, chemist Malcolm Dole of Baylor University, Waco, had observed that molecular hydrogen at high pressures had a significant catalyzing effect on the decay of alkyl free radicals, $-CH_2CHCH_2-$, that had been induced in solid polyethylene via irradiation. Free radicals, which are atoms or groups of atoms characterized by the presence of an unbonded electron (represented in the alkyl radical above by a black dot), also have been found to be of importance in many biochemical processes and are thought by some to take part in cancerous growth processes.

Making a logical connection from this information, Dole, in collaboration with physiologist William Fife of Texas A & M University, College Station, and biologist F. Ray Wilson of Baylor, devised an experiment in which hairless mice with ultraviolet-light-induced squamous cell carcinomas were sealed in a high-pressure chamber maintained at 8.3 atmospheres with a mixture of 2.5% oxygen and 97.5% hydrogen. The oxygen content of the mixture was well below the value that was required for the mixture to become explosive, and, under such high pressure, the partial pressure of oxygen was equivalent to that of a normal atmosphere.

After ten days under such conditions, the test mice showed dramatic improvement. Their tumors had turned black, some had fallen off, and some seemed in the process of being pinched off at their bases. In addition, the mice exhibited no undesirable side effects from the treatment. By contrast, a control group of cancerous mice sealed in a chamber in which helium was substituted for hydrogen showed no regression of their carcinomas. Despite the apparent success of the therapy, however, many questions remained to be resolved concerning the exact mechanism of the hyperbaric hydrogen effect, the relationship between exposure time and tumor regression, the permanency of the regression, the applicability of the therapy to other types of cancerous tumors, and, most important, its applicability to the treatment of cancer in human beings.

—John F. Henahan

Earth sciences

Earth scientists pursued their studies over a broad range of subjects during the past year. Meteorologists worked to develop theories to explain the long-term changes in the Earth's climate and, at the other end of the spectrum, also sought to improve daily weather forecasts. The function of the ozone layer in the stratosphere and its possible destruction by nitrogen oxides and chlorofluorocarbons was a matter of particular concern. Geologists continued to probe deep into the crust below the ocean in an effort to determine the driving force for seafloor spreading, and also devised a technique for determining the depth limits of hydrocarbons. Earthquakes and ocean pollution also received considerable attention.

Atmospheric sciences

During the past year the subject of climate continued to receive a great deal of attention from the scientific community and the public as well. The long drought in Africa made it dramatically evident that prolonged dry spells, particularly in regions overgrazed by animals, can have disastrous effects. Another drought in the Soviet Union during 1975 did serious damage to grain production; the wheat crop was said to be the worst in ten years, and the Soviets were forced to import large quantities of grain. Fortunately the weather over the U.S. during the growing season was quite favorable. If the grain belt had suffered a prolonged drought, premature frosts, or flood-producing storms, farm yields would have been reduced and the social costs would have been extreme.

Research on climate. The need to understand the factors governing seasonal weather and climatic changes stimulated research in these areas. The weather and climate depend on the quantity and distribution of solar energy reaching the Earth, the composition of the atmosphere, the characteristics of the solid and liquid parts of the Earth, the thermodynamic properties of the oceans, and the interactions of these and a variety of other factors.

The climate problem was being attacked on various fronts. Measurements of meteorological variables such as temperature go back only 100–200 years. This has made it necessary to use indirect measurements of the state of the atmosphere. Radioisotope techniques have been particularly useful. Cores have been drilled in Arctic and Antarctic ice and in the ocean bottom. By means of available procedures, estimates can be made of the period during which various carbonaceous substances were created and the approximate temperatures that prevailed at the time.

By measuring the widths of the rings of ancient trees, dendrochronologists can estimate the weather conditions over the period of the tree's growth. This requires the study of those tree species that develop easily identifiable rings and which have grown under environments in which changes of rainfall cause changes in available water for tree growth. A great deal of information has been gathered from bristlecone pines, which grow under semiarid conditions at high elevations

Diver drills into coral heads to obtain cores which exhibit growth bands indicating past climatic conditions. The dark, dense bands of 1969–70, 1963–64, 1957–58, and 1941–42 represent periods of unseasonably cold water caused by severe winters.

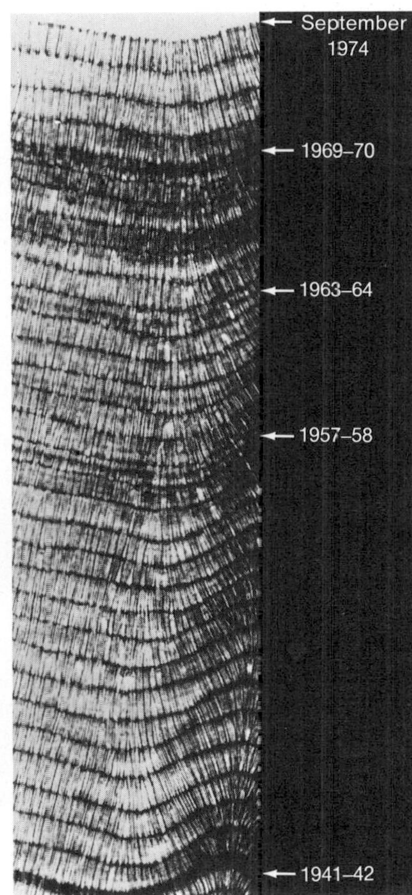

Photographs, courtesy, Eugene A. Shinn, U.S. Geological Survey

Courtesy, NASA Landsat

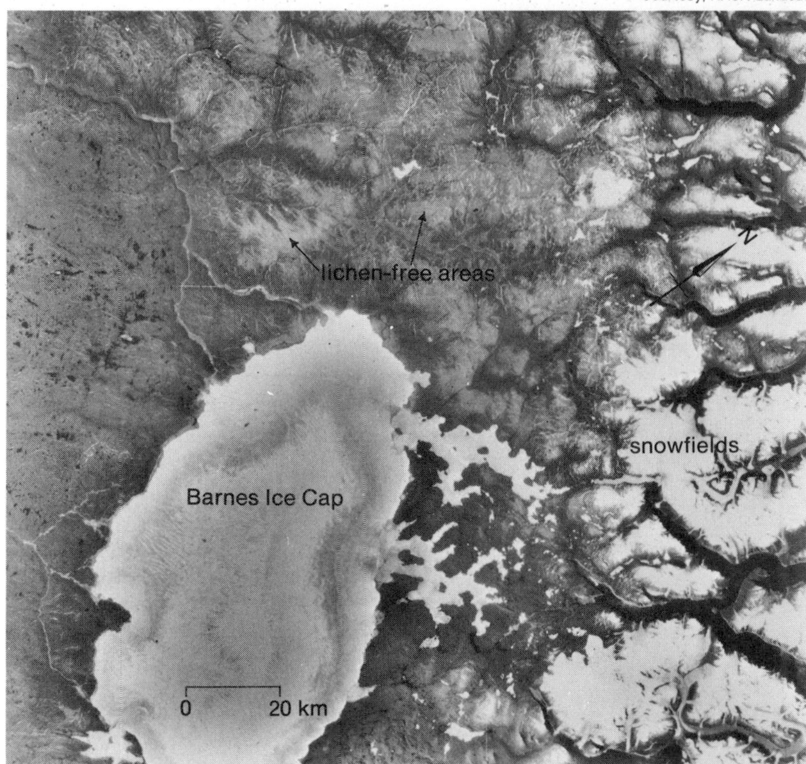

Satellite photograph of north-central Baffin Island in Canada provides evidence to support the idea that the great ice sheets of the Late Pleistocene were generated in this and similar areas by widespread lowering of the snow line. Light-toned lichen-free areas indicate places where the moss and lichen cover was killed by the thin permanent snowfields that developed during the "Little Ice Age" of the 17th through the 19th century.

in the western United States and can live for more than 4,000 years. The interpretation of tree-ring data in meteorological terms requires the use of statistical techniques for sorting out the meteorological from the other relevant factors. (See *1976 Yearbook of Science and the Future* Feature Article: DENDROCHRONOLOGY: HISTORY FROM TREE RINGS.)

It is of particular interest to learn about climate fluctuations over periods of decades, at least in part because it has been proposed that environmental pollution may be partly responsible for causing them. After a general warming of the low-level global atmosphere for the half century ending about 1940, there has been a cooling trend. Until the last few years it was believed that the temperatures were falling over most of the Earth. It has now been established that the cooling has occurred mostly in the Northern Hemisphere. Apparently the Southern Hemisphere, particularly at higher latitudes, has been subject to a warming trend over the last few decades.

The assertion that air pollutants in the form of carbon dioxide or particles of smoke and dust are responsible for global warming or cooling still has not been convincingly established. Nevertheless, the arguments that the effects of large quantities of atmospheric CO_2 (carbon dioxide) would lead to warming of the lower atmosphere are becoming more persuasive. As of 1976 the concentration of CO_2 in the atmosphere was about 330 parts per million and was increasing at slightly less than 1% per year. About half of the CO_2 emitted by the burning of fossil fuels remains in the atmosphere, while the rest goes into the oceans. It has been predicted that as the years go by, the fraction remaining in the atmosphere will increase.

Atmospheric scientists are attempting to derive a satisfactory theory of climate and to develop mathematical models that can take into account the many factors governing the circulation of the atmosphere. There are two general lines of attack. The first is being pursued in such laboratories as the Geophysical Fluid Dynamics Laboratory of the U.S. National Oceanic and Atmospheric Administration (NOAA), the National Center for Atmospheric Research (NCAR), and elsewhere. They have developed complex mathematical models of the general circulation of the Earth's atmosphere. The models attempt to incorporate all of the physical mechanisms believed to be important. Given an initial set of conditions describing the state of the air, water, and land and the properties of incoming solar radiation, scientists can calculate the properties of the atmosphere as a function of time. By means of a series of calculations, one may obtain statistics of atmospheric states that describe the climate.

A second approach to the development of a theory on climate is represented by the research of W. D. Sellers at the University of Arizona and M. I.

Budyko in the U.S.S.R. They started with simplified versions of the water and land configurations of the Earth, representing it as consisting of one large ocean and one large continent, each distributed by latitude essentially as on the real Earth. This simplified model accounts for the interactions of air, sea, and land, variations of snow and ice cover, and the latitudinal transport of energy. Sellers' calculations indicate that doubling the amount of CO_2 in the atmosphere would increase the global surface temperature by 1.3° C.

Recently, Jule Charney at Massachusetts Institute of Technology proposed that deserts may be intensified by a so-called biogeophysical feedback mechanism. His theory is as follows. When a desert is forming, the shortage of rainfall causes a reduction in vegetation. As a result, the soil reflects a greater fraction of the incoming solar radiation than it would if there were more vegetation. At the same time the desert emits large quantities of terrestrial radiation. Charney calculated that the consequence of this state of affairs is enhanced sinking of air over deserts, which intensifies desert conditions.

World Weather Program. The World Weather Program is an international effort to provide an understanding of the global atmosphere and to make available weather warnings, forecasts, and other weather services on a worldwide basis. In the U.S. the major components of the program are the World Weather Watch, the Global Atmospheric Research Program (GARP), and the System Design and Technological Development program. The last unit has been concerned with the development of polar and geostationary satellites and their associated sensor systems, carrier and constant-level balloons, remote sensing instruments, automatic weather stations, and data buoys. In mid-1975 the U.S. began nearly continuous (day and night) observations, by means of two geostationary satellites, of weather and storm systems over most of North and South America and the adjacent waters. The U.S.S.R., Japan, and the European Space Agency took steps to join the U.S. by 1977 in the operation of five geostationary satellites that would be stationed around the globe, keeping the entire atmosphere under surveillance.

During the summer of 1974, the GARP Atlantic Tropical Experiment was carried out over about a third of the Earth's tropical belt. A massive quantity of information was obtained and in 1976 was still being analyzed by meteorologists in many countries. Other major field programs of GARP are the Monsoon Experiment, to be carried out over the Indian Ocean in the late 1970s for the purpose of studying the mechanics of monsoon circulations, and the Polar Experiment, scheduled for 1976–77, dealing with the role of polar regions in global energetics. The lessons learned from these and other experiments are to be incorporated in planning the First GARP Global Experiment (FGGE), scheduled to be carried out during 1978.

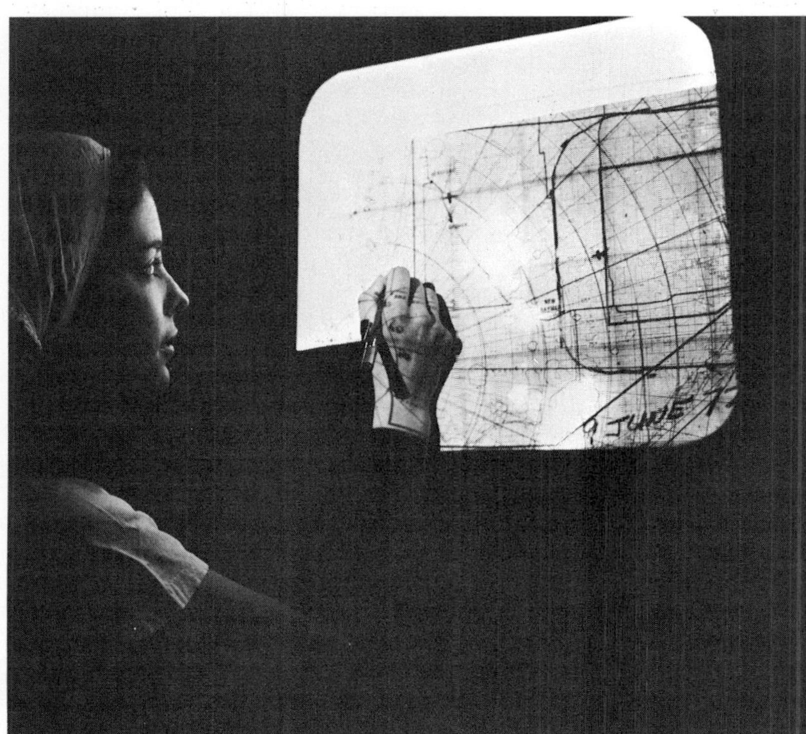

Radar data obtained by instruments such as the antenna being lowered onto its pedestal in Grover, Colorado (below), are plotted on a map of the field research area of the U.S. National Hail Research Experiment, a project aimed at testing the feasibility of modifying hail by cloud seeding.

Photographs, courtesy, NCAR

It will involve the collection of data on a global basis on a scale never before attempted.

The World Weather Watch is an international program for supplying better weather services for all countries. Since its inception in the early 1960s it has brought about significant improvements in observing the atmosphere, in the procedures for data handling and communication, and in the dissemination throughout the world of weather forecasts and warnings.

Violent weather. Unlike the preceding two years, 1975 was a relatively quiet one with regard to tornado occurrence. On the research front, impressive advances were made in the development of the pulsed-Doppler radar for use in detecting and tracking tornadoes. Because a Doppler radar measures the speed of reflecting targets toward or away from the radar, a high-velocity vortex can be identified by strong gradients of the radial wind over a small distance. Research at the National Severe Storms Laboratory in Norman, Okla., showed that of 25 identifiable vortexes observed in 1973 and 1974, 16 were associated with tornadoes and another 7 with strong wind and hail. No tornadoes were observed during the period of radar data collection that were not preceded by an observed vortex. An important conclusion of the analysis was that, on the average, the Doppler radar observed a vortex 34 minutes before the occurrence of a tornado.

Techniques for the display of Doppler radar data have improved greatly over the last few years. One particularly interesting innovation is the use of color to depict radar echo intensities and Doppler radar velocities. The color presentation makes it easy to observe regions where echoes are intense and where radial velocities are strong and variable. These two factors are helpful in identifying severe thunderstorms likely to produce hail and tornadoes. Research is continuing on the development and use of combinations of two or three Doppler radars in order to measure the three-dimensional air motions in storm systems containing rain or snow.

It is now recognized that in order to understand, predict, and possibly modify the major thunderstorm systems that sweep across the Great Plains and Middle West of the U.S., it is necessary to know more about the internal properties of the storms and the interactions of the storms with their environments. Scientists from NOAA and NCAR have been planning a program of research on these organized thunderstorm systems. The program, the Severe Environmental Storms and Mesoscale Experiment (SESAME), should move from the planning to the operation phase by the late 1970s.

Weather modification. Weather modification research encountered some disappointments in 1975. Beginning in 1970 the U.S. Bureau of Reclamation supported a research program over the San Juan Mountains of southwestern Colorado for the purpose of testing the extent to which a particular cloud-seeding procedure could increase snowfall. The project involved the introduction of silver iodide ice nuclei into suitable winter clouds. One of the purposes of the project was to check the results of earlier experiments conducted by scientists at Colorado State University. As of early 1976 the results had not been published, but available information indicated that the experiments did not prove that snowfall was increased by the cloud seeding.

The U.S. National Hail Research Experiment completed its third year of field observation and experimentation on the modification of hail by means of cloud seeding. The most recent analyses did not confirm persistent Soviet reports that seeding reduces the amount of hail that falls. Although the cloud-seeding techniques used in the U.S.S.R. and in Colorado are different, this does not appear to account entirely for the different results. Extensive studies of the structure of major Colorado hailstorms based on observations by radar, airplanes, and other instruments convinced the principal scientists in the National Hail Research Experiment that the major hail-producing storms cannot be modified by means of the Soviet seeding procedures.

The last major field effort of the scientific groups cooperating in the Metropolitan Meteorological Experiment, an investigation of the effects of a large industrialized urban area on the weather in its vicinity, took place in 1975. The results show that there are important effects. For example, in and just downwind of St. Louis, Mo., temperatures and summer rainfall are higher than in surrounding areas. (*See also* Feature Article: THE CLIMATE OF URBAN AREAS.)

Ozone. Although the total quantity of ozone (O_3) in the Earth's atmosphere is small, with maximum concentrations less than about ten parts per million of air, ozone plays a crucial role in the maintenance of life on the Earth. Ozone exists in the lowest 50 or 60 km (about 31 to 37 mi) of the atmosphere, but most of it is found at altitudes of 15–30 km (9–19 mi). It is formed by a complicated series of photochemical processes involving the absorption of solar ultraviolet radiation. If the quantity of ozone were to be reduced significantly, there would be an increase in the amount of ultraviolet radiation reaching the ground and this could lead to an increase in skin cancer and other biological damage. A 1975 report from the U.S.

Courtesy, F. S. Rowland, University of California, Irvine, photograph by Mike Jones

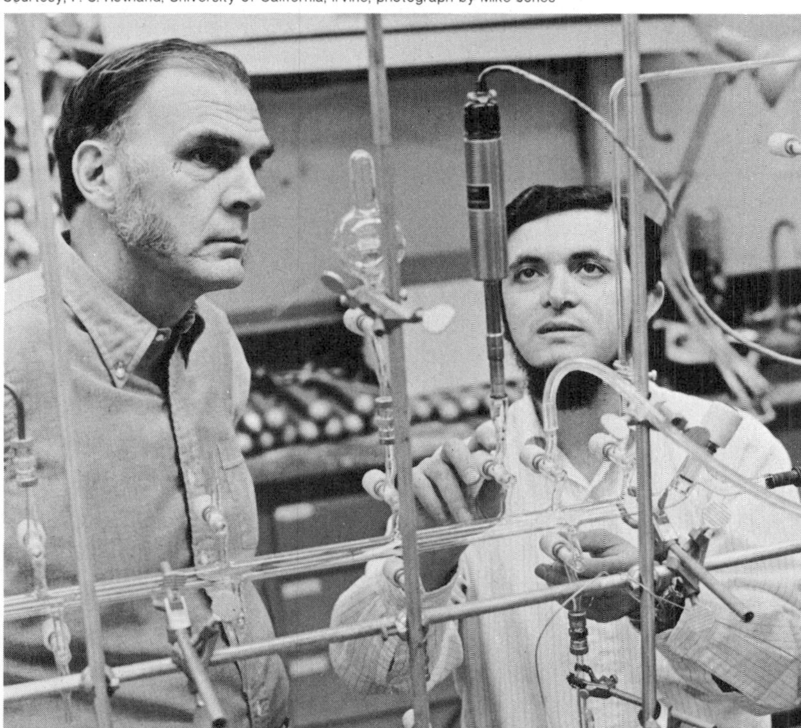

F. Sherwood Rowland (left) and Mario Molina, leaders in the research on ozone in the atmosphere, work with vacuum line used to measure the concentrations of fluorocarbons in a sample of stratospheric air.

National Academy of Sciences indicated that a 10% reduction of ozone could increase the incidence of skin cancer by 20%, or 80,000 cases per year in the U.S. The same report concluded that a fleet of 300–400 large supersonic transport (SST) airplanes could cause a 10% decrease in the ozone because nitrogen oxides emitted by the airplane engines would react with and destroy ozone.

The effects of SSTs on ozone gained significance during the year because of the request by the British and French governments that their new SST, the Concorde, be allowed to use certain U.S. airports. Objections to the request were raised on the grounds that the Concorde would create excessive noise and be the first of a future fleet of SSTs that would pose a threat to the ozone layer.

In 1975 a great amount of attention was focused on a second potential threat to the ozone layer, that presented by the use of chlorofluorocarbons and other similar compounds in aerosol spray cans and refrigerating systems. These substances are commonly referred to by the brand name Freons. Mostly on the basis of theoretical studies, it was stated that the fluorocarbon molecules, because of their stable chemical properties, remain in the atmosphere for a decade or more. During this period they slowly diffuse upward through the ozone layer, where they absorb ultraviolet radiation. This causes a dissociation of the molecules and the release of chlorine atoms, which can react with, and destroy, ozone molecules. Theoretical projections indicated that unless there is a marked reduction in the release of Freon-type material into the atmosphere, there will be a serious depletion of the ozone by the year 1990. This assertion was challenged by some scientists on the ground that it is based on unsupported assumptions about sources and sinks of chlorine and about the photochemical processes involved in ozone creation and destruction.

In an attempt to resolve this dispute, the U.S. government established a task force on Inadvertent Modification of the Stratosphere. It reported in June 1975 that a serious problem in the ozone layer did in fact exist and called for an aggressive program of research, a position taken also by the Panel on Atmospheric Chemistry of the U.S. National Academy of Sciences.

Scientists in various laboratories launched a number of relevant research programs. In June 1975 Arthur L. Schmeltekopf and his associates at NOAA's Environmental Research Laboratory sampled the air in the atmosphere at altitudes of 18 to 22 km by means of high-altitude ballons. They observed distributions of fluorocarbons, which supported the notion that the chlorine-bearing molecules diffuse through the lower atmosphere and are dissociated by solar ultraviolet as they pass through the region of high ozone concentration. Similar results were reported by scientists at the National Center for Atmospheric Research.

—Louis J. Battan

Geological sciences

Significant strides in international cooperation highlighted the year in the geological sciences. Among them, the International Geological Correlation Program had 281 national groups from 73 participating countries involved in 46 approved projects, supported by the International Union of Geological Sciences and UNESCO. The program was designed to strengthen ties of geologists in less developed countries with those in industrialized nations. The wide range of projects under way included: studies of worldwide sea-level fluctuations during the Holocene Epoch (the past 10,000 years), the genesis of kaolins (a group of clay minerals), accuracy in geologic time, and the identification of ore deposits separated by continental drift. Seabed assessment, a project of the International Decade of Ocean Exploration, included assembling mineral resource data on continental margins, on the relation of plate tectonics to the origin of metal deposits, and on manganese nodules. An important map series, including base, geologic, tectonic, mineral, and energy maps, was being compiled for the Pacific Basin and adjacent continental areas.

Geology and geochemistry. The U.S. deep-sea research vessel "Glomar Challenger" continued making headlines with important discoveries as it drilled in the North Atlantic, the Mediterranean, and the Black Sea. Remarkable symmetry in the opening of the North Atlantic was demonstrated with nearly identical rocks approximately 135 million years old cored on both the east and west flanks of the Mid-Atlantic Ridge. Two holes were made along the New England Seamount Chain to date the seamounts' volcanic aprons and test the hypothesis that plumes act as a driving force for seafloor spreading. Results were inconclusive because neither hole penetrated the original oceanic crust, but the possibility was raised that about 75 million years ago the entire seamount chain was affected by synchronous, episodic volcanism. Just south of the Grand Banks (shoals off the coast of Newfoundland) shallow-water reef carbonates about 125 million years old (Early Cretaceous) were recovered at a depth that shows that the reef has since subsided more than 4,100 m. This represented the most deeply subsided former island drilled in the ocean basins. The existence of this reef material also presents evidence of a possible continuous northward extension of the Early Cretaceous reef system documented off Florida and in the Bahamas.

The "Glomar Challenger" set a new record for deep drilling in a hole that reached bottom in upper Jurassic limestones dated at 140 million years old. Drilled on the rim of the Blake Plateau, it extended 1,412 m below the sea floor. Fauna and flora found in dolomitic marls of the eastern Mediterranean indicate that it was a lake receiving fresh or brackish waters about 10 million–12 million years ago. The Black Sea was presumably a playa lake, one formed in a flat-floored, undrained desert basin, for the past 2 million–3 million years, with only occasional marine inflows.

At the year's end the "Glomar Challenger" set out to penetrate and sample the ocean crust to a

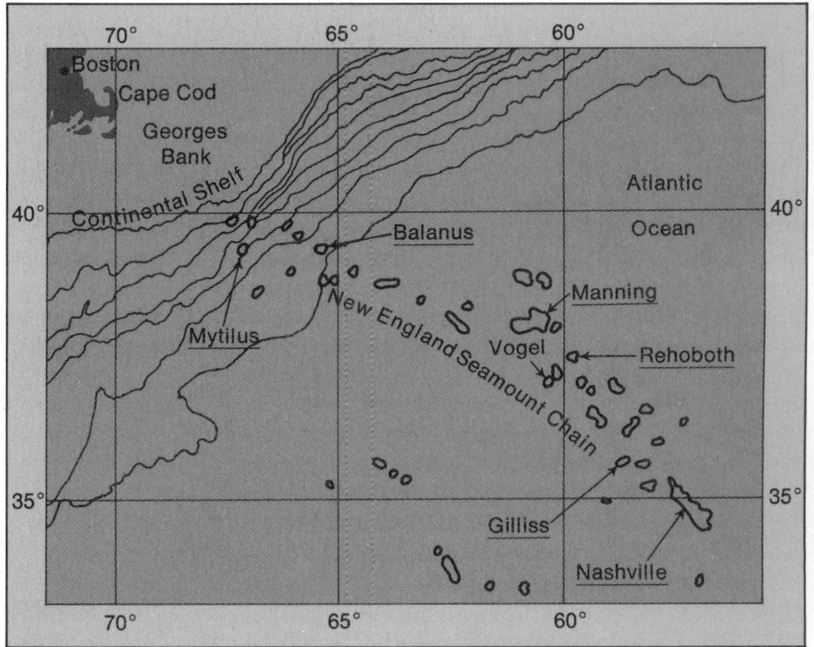

Specimens collected from a chain of submarine volcanoes indicate that the chain probably formed along a rift in the ocean's floor. Samples taken by the submarine "Alvin" from seamounts whose names are underlined show that Mytilus has subsided 10,000 feet since its formation. Vogel and Nashville were in eruption about 80 million years ago, according to samples taken by the "Glomar Challenger."

depth of several kilometers as part of the International Phase of Ocean Drilling. Drilling and coring of more sediments, especially along continental margins, were scheduled. Research on sediments was expected to prove fruitful in expanding knowledge on climatic shifts that the Earth has undergone over the past 100 million years. Dramatic changes in the configuration of the oceans and the distribution of continents appear as major factors in the cooling trend of the Earth's climate and the relatively recent establishment of polar ice caps. New insight into the potential hydrocarbon content of the continental margins was also anticipated; results may have a significant impact on projected petroleum reserves and substantially affect sites of future worldwide drilling programs.

Mineralogy and petrology. Much of the research in mineralogy during the year was concentrated on the crystal chemistry and thermochemistry of rock-forming silicates and on the fundamentals of chemical bonding in minerals. Such phenomena as lattice vibrations in crystal structures, transition-metal chemistry at high pressures, and quantitative estimates of strain energies in crystals were being explored. (A transition metal is one with an incomplete inner electron shell.) Application of the findings was to problems on a very large scale, chemical differentiation in the lithosphere (crust and upper mantle) and chemical reactions in zones where one crustal block descends beneath another being examples.

Experimental petrologists focused on the problem of calculating temperatures and pressures in the Earth's mantle more precisely. They believed that fragments of peridotite (an olivine-rich igneous rock) in volcanic eruptions, presumably derived from the mantle, might afford the best clues and yield insight into mantle convection and magma generation as well. Further studies on the distribution of calcium and aluminum among certain minerals in peridotite indicated that pressure-temperature histories of parts of the upper mantle could be reconstructed with increased confidence. The effects of carbon dioxide and water on rock melting were also being studied extensively.

A diamond-anvil pressure cell gave great promise for creating pressures and temperatures in the laboratory that simulate characteristics of the lower mantle. The device consists of two opposing diamonds, cut from gem-quality stones, pressed against each other by a piston of substantially greater area than that of the diamonds. Because the diamonds are transparent to visible and infrared light, X-rays, and gamma-rays, investigators can heat the experimental samples with laser beams and observe transformations directly. Pressures of 500 kb (kilobars), equivalent to those at a depth of about 1,000 km within the Earth (1 km = 0.62 mi), were obtained in laboratory studies.

Experiments with iron silicates and oxides in the cell indicated that previous estimates of temperature distribution in the Earth's mantle might require major revision. Compounds with ferrous iron may break down to yield native iron and ferric iron minerals. Changes such as this may mean that electrical conductivity in the lower mantle is substantially higher than previously estimated. Effects on thermal conductivity are less certain, but it is thought that they may result in significantly higher temperature estimates.

Geochemistry. Organic geochemistry provided one promising approach to defining the likelihood and depth limits of hydrocarbon concentration. The technique involved determining vitrinite reflectance of cored samples, that is, measuring the amount of light reflected from organic plant material. This reflectance reading indicates the degree to which organic matter in sediments is converted to commercial hydrocarbons. The extent of the conversion depends on two factors, temperature and duration of heating. Through temperature measurements and knowing the geologic age of the sediments involved, geochemists can improve the accuracy of their predictions of hydrocarbon occurrence. Temperature can be substituted for time in the conversion process. Thus, in areas of anomalously high temperature, such as the overpressured zones that occur in parts of the coast of the Gulf of Mexico, commercial quantities of hydrocarbons may occur in younger rocks and at shallower depths than otherwise anticipated. Likewise, factors including the temperatures and the amounts of H_2S and sulfur present control the maximum depth at which different types of hydrocarbons will be found. Oil gradually disappears between 4,250 and 7,600 m, and the odds are strong that at depths below 9,000 m no significant quantities of natural gas will exist.

Stable isotope geochemistry, particularly of oxygen, was applied by scientists to document the exchange of matter between magma and invaded country or host rock. Their findings had significant relevance in determining the genesis of ore deposits. In some deposits, the metals may have originated in the host rock rather than in the magma as previously assumed. Heat and related mobility created by the intrusions may cause movement and concentration of the metals.

Manganese nodules, which are extensive over portions of the seafloor, continued to attract attention. Manganese dioxide presumably acts as a catalyst in extracting metals, thus explaining the high concentrations of cobalt, copper, and nickel these nodules commonly contain. Experiments

Photographs, IN-BILD from Authenticated News International

Manganese nodules lying on the bottom of the Pacific Ocean (left) were photographed by the West German vessel "Valdivia" (right). It is estimated that the nodules of the Central Pacific contain about 17 billion tons of nickel and 12 billion tons of copper.

simulating Black Sea conditions substantiated this contention. They indicated that cobalt fixation by hydrous manganese dioxide is a function of pH (levels of acidity and alkalinity), surface area, and composition. Use of manganese dioxide may, therefore, prove effective in pollution control, with potential applications to industrial wastewaters and automobile exhaust systems.

The ability to achieve long-range prediction and ultimate control of worldwide climatic trends is of growing importance. Mounting evidence supports the idea that there are relatively rapid changes between glacial and interglacial periods. Thus the balance that governs whether the Earth will cool down or heat up may be a delicate one. If true, man's input of contaminants into the environment could be a critical factor. To establish causes and rates of climatic pattern shifts, geochronologists began dating tree rings and measuring carbon, deuterium, and oxygen isotopes contained in the cellulose. Once the rate of climate shifts and the delicacy of their balance are known, the effect of human involvement on them can be determined. The study of levels of carbon dioxide (CO_2) in the atmosphere exemplified the potential problem. The substantial CO_2 production generated by the consumption of fossil fuels in the next century or two could have a dramatic effect on the climate. Predictions are that the present cooling trend will then be reversed and possibly followed by an extended period of warmer temperatures.

Lunar and planetary geology. Continued analysis of lunar data led to refinements in the understanding of the Moon. Some lunar mare flows may be much younger than any volcanic rocks brought back to Earth by the Apollo astronauts, perhaps by hundreds of millions of years. Models of the Moon's thermal and chemical history indicate that the crust and mantle separated by means of early melting of a thick lunar surface. This supported data on chemical fractionation that dated the activity as occurring about 4.3 billion–4.4 billion years ago.

Mars appears to have an ancient blanket of thick (1 km or more), unconsolidated material covering much of its surface. Strata 1 to 6 km thick are indicated by topographic analysis of layered deposits exposed on slope faces. A lithosphere 100 to 200 km in thickness is suggested for Mars, based on correlations between gravity data and regional topography. Approximately 25% of Mercury has been photographed adequately for photogeologic interpretations. Mapping of Mercury indicates (in age sequence): very old intercrater plains, old craters and cratered terrains, mountainous and hilly lands, younger smooth plains, and very young craters.

Energy and minerals. Methods of in situ coal conversion (gasification and liquefaction) received growing support during the year as did efforts to gather basic data on the location and quality of U.S. coal reserves. Estimates indicated that coal deposits represented approximately two-thirds (on a BTU basis) of the U.S. fossil fuel reserves that are recoverable under present conditions. The need for effective exploration techniques to identify geothermal reserves was addressed. Belts of hot dry rock, geopressured systems with excessive pore-fluid pressure, and portions of the seafloor spreading zones were receiving special attention.

The possibility of tapping magma chambers at shallow depth was also under scrutiny.

Two new silver mines were developed in Idaho, and several new discoveries of metal deposits were made in both the U.S. and abroad. These included valuable base-metal deposits in Alaska's Ambler district; a large deposit of near-surface ore at the McDermitt mercury mine in Nevada; gold in the Philippines; manganese in Brazil; sulfides in the Northwest Territories of Canada; bauxite in the British Solomon Islands and India; and copper in Fiji, India, Indonesia, Mexico, Peru, and Western Australia. Nevertheless, a troubled economy, soaring development costs, and unsettled tax policies led to reduced exploration and development in many parts of the world.

Paleontology. Among vertebrate paleontologists the dispute as to whether dinosaurs were warm-blooded or cold-blooded received special attention. R. T. Bakker, a proponent that dinosaurs were warm-blooded animals, suggested placing dinosaurs, birds, and pterosaurs with their reptilian ancestors in the new class Archosauria. Under this scheme birds are viewed as highly specialized dinosaurs; therefore, some dinosaurs are not extinct but exist extensively and are warm-blooded with feathers.

A reorganized phylogenetic classification of living and fossil turtles using cranial characteristics rather than shell structure was proposed. The divergence of each line of descent is placed in a relative position through time and thus gives a more precise picture of the relationship between the various members of the group. This scheme was also being applied to other groups of animals.

Largest lunar metal sample found to date, over 5 mm in its longest dimension, consists of globules of nickel, cobalt, iron, and phosphorus.

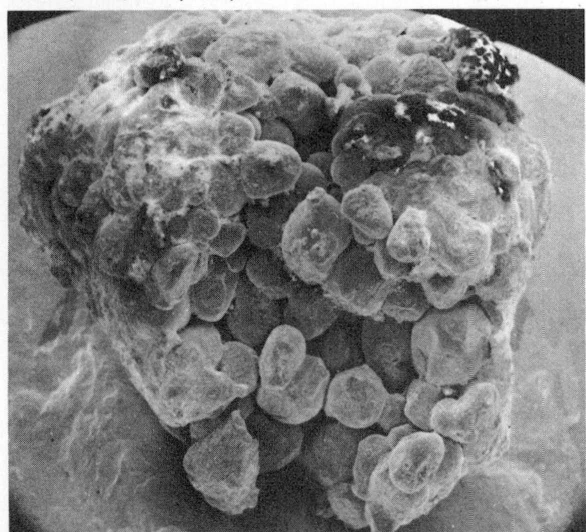

Courtesy, Joseph I. Goldstein, Lehigh University

New discoveries in vertebrate paleontology included remains (lower jaws and teeth) of hominids found by Mary Leakey at Laetolil in Tanzania and potassium-argon dated at 3,350,000–3,750,000 years old. The remains are assigned to the genus *Homo* on the basis of their resemblance to those of a skull found on the east side of Lake Rudolf in East Africa. Also found was a very primitive mammal from the Lower Cretaceous (100 million–120 million years ago) Cloverly Formation of Montana by Farish Jenkins.

Research on modeling of the fossil record and comparisons with the actual record continued to receive attention from paleontologists. These scientists hoped that refinements in modeling by using stochastic processes (programming of random events) might lead to identification and explanation of special circumstances (such as mass extinctions of species.) Work was being extended on the differences between geographic and phyletic (evolutionary) differentiation of species, and many scientists were speculating that geographic distribution may be the major factor in the origin of most species in the fossil record. Communities of invertebrates were being looked at carefully, especially in various marine environments. Nearshore environments were found to have low species diversity in contrast to outer zones, which had high diversity but also experienced more mass extinctions. The year also marked the first issues of *Paleobiology*, a new quarterly journal of the Paleontological Society.

Thermal color alteration of conodonts (Paleozoic fossils) provided promise as an index for assessing potential occurrence of oil and gas. Heating conodonts in the laboratory produces color differences, a feature previously observed in conodonts obtained from well samples. By using conodont color as a geothermometer, researchers can determine the level to which rocks in which they are found have been heated. This may define thresholds and cutoffs for the occurrence of oil and gas in the rocks.

—Robert E. Boyer

Geophysics. For most of its approximately 70-year history as a modern science, geophysics has been principally concerned with the accurate determination of the physical structure of the solid Earth. The definition of the petroleum-bearing and mineralized layers of the Earth's upper crust has been the primary focus of geophysical research for many years. But whether the structures of interest be petroleum-bearing sedimentary strata, the oceanic crust, or the Earth's core, the basic concern has almost always been the geometric configuration and material constitution of the structure itself rather than the physical processes

Wide World

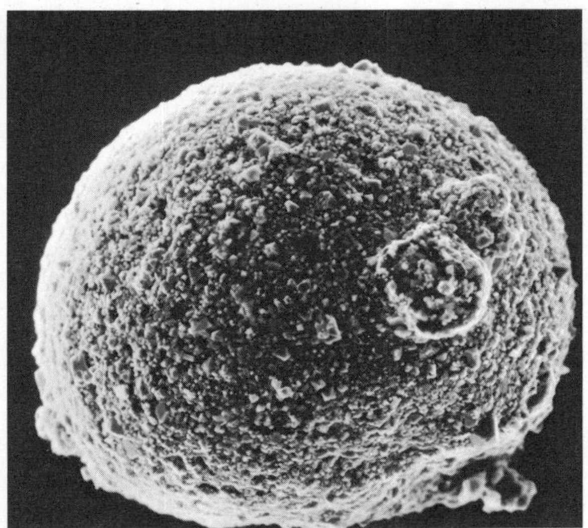

Sulfur droplet, magnified 460 times, was ejected in 1975 from a fumarole on Mt. Baker in Washington.

giving rise to it. In the past decade, however, this point of view has been fundamentally changed. Within geophysics and throughout the Earth sciences, there is considerable optimism that a preliminary understanding of the active geologic processes that shape the face of the Earth is emerging.

There are three principal reasons for this revitalized thinking. First, the advent and almost complete acceptance of plate tectonics has provided a global framework within which to place the causes and consequences of active geologic processes. Second, manned and unmanned explorations of the Moon, Mercury, Venus, and Mars have provided a comparative basis of knowledge with which to assess the long-term evolution of the Earth. Third, the location and development of ever-diminishing material resources, including energy sources derived from the solid Earth, require a more complete understanding of many active geological processes, not only those related to the origin and emplacement of these resources but also those that constitute a potential threat to man's efforts to develop and use these resources in an environmentally acceptable manner.

Nevertheless, it is plain that active geological processes, manifesting themselves in plate motions and mountain-building, fault zones and earthquakes, volcanoes and other geothermal systems, and in many forms of hydrothermal mineralization, are complex physical and chemical systems. Present knowledge of them is limited, in large part because scientists have access to them only through measurements made either at or close to the Earth's surface. *Continental Drilling*, published in June 1975 by the Carnegie Institution of Washington and the U.S. Geodynamics Committee, described a bold but as yet unfunded program to sample several of these active geological processes in situ at depths of about 10 km (6 mi). Federally sponsored research programs in geothermal energy and earthquake prediction are probing ever more deeply the responsible geological processes, and there was increasing realization that widespread, economically viable utilization of geothermal energy and reliable earthquake prediction will require a deeper understanding of the causative processes than is presently available.

Advances in earthquake prediction research were marked by a new appreciation for the complex processes operative in active crustal fault zones, the locus of most destructive earthquakes. Even in well-instrumented areas, there was increasing concern that a reliable earthquake prediction capability might take years if not decades to develop. Nevertheless, the destructive earthquake in the Yingkou-Haicheng areas of Liaoning Province, China (Feb. 4, 1975), was successfully predicted by Chinese scientists. As reported by Robin D. Adams of the Department of Scientific and Industrial Research of New Zealand, this successful prediction was a spectacular triumph in the history of earthquake seismology. No less impressive was the social value of this prediction. Evacuation plans in and around Haicheng were implemented and followed five hours before the earthquake.

As if to underscore the diversity of the many active geological processes, their many potential hazards to the works of man, and the many investigations seeking to discover these processes and mitigate their potential hazards, three other significant developments occurred in the past year. Rapid uplift of the Earth's surface relative to sea level over a broad region of southern California in the past 15 years has been documented by careful scrutiny of leveling data. While the cause of this uplift is as yet unknown, this sequence of events once again revealed the potential and real mobility of large crustal blocks. The damaging Oroville, Calif., earthquake of Aug. 1, 1975, occurred within 15 km (10 mi) of the Oroville Dam, the largest earth-filled dam in the U.S., and refocused attention on the possibility that large dam/reservoir systems can and have triggered damaging and locally destructive earthquakes in relatively aseismic areas prior to reservoir impoundment. Finally, publication of *Studies for Seismic Zonation of the San Francisco Bay Region* (U.S. Geological Survey Professional Paper 941-A) culminated a several-year effort in the development and documentation of geologic and geophysical techniques that

Courtesy, Sandia Laboratories

Laser strain seismometer consists of mile-long evacuated pipe with a granite pier at each end. A laser beam projected through the pipe can measure movements of the Earth between the piers as small as 0.000003 inch.

can, together with advances in earthquake-resistant construction and appropriate land-use policy, significantly reduce earthquake hazards, whether or not the responsible earthquake is successfully predicted.

The Palmdale uplift. Relative uplift or downwarping of the Earth's surface observed through a wide range of spatial and temporal dimensions reflects a variety of geological processes, some reasonably well known but others only poorly understood. The uplift of Fennoscandia (Finland, Sweden, and Norway) in the past 10,000 years, as measured by successively younger shorelines, is attributed to glacial rebound following the retreat of ice from northern Europe. The Colorado Plateau of the southwestern U.S. has been elevated by as much as several kilometers in the past several million years, but the cause of this uplift is not known. Geodetically observed oceanward tilting of the eastern coast of the island of Hokkaido in Japan is caused by elastic deformation of the lithosphere coupled to the Pacific plate downthrusting beneath the Japanese island arc; observation of this tilting was one element of the successful prediction of the Hokkaido earthquake, which occurred on June 17, 1973.

Examination of the record of repeated leveling surveys by Robert O. Castle, Michael R. Elliott, and Jack P. Church of the U.S. Geological Survey revealed that approximately 12,000 sq km (4,700 sq mi) of southern California have been uplifted by significant amounts since 1960. The uplifted area involves much of the Transverse Ranges and western Mojave Desert north of San Bernardino, Los Angeles, and Santa Barbara. It is elongated in an east-west direction with approximate symmetry about a north-south line passing through Palmdale, where the maximum uplift of 25 cm was observed.

Although the cause of this regional uplift remained unknown, the uplift is consistent with the long-term geologic history of this region, which has previously witnessed the uplift of the Transverse Ranges and western Mojave Desert relative to the Los Angeles Basin to the south. The startling aspect of the latest uplift, however, is the rate at which it took place and the implications for mobility of large crustal blocks. Palmdale rose about 20 cm between 1961 and 1962 alone, and an uplift of about 18 cm had spread northward to Mojave by 1965. Uplift also spread to the southeast. Between 1968 and 1973 the elevation at Cajon increased by about 15 cm.

Moreover, this uplift occurred without any noticeable increase of seismicity along the San Andreas Fault, which transects the uplift area from northwest to southeast, or along the Garlock Fault, to the north and northeast. Indeed, the uplifted area has been one of the least seismic regions of southern California since 1960.

The California Aqueduct, which brings water to the Los Angeles metropolitan area from the northern Sierra Nevada, traverses the uplifted area. Flow of water is achieved through the use of pumping stations and a precisely graded concrete channel. The channel grade in the western Mojave Desert is typically six centimeters per kilometer. While the changes in elevation gradient are as yet considerably less than the typical channel gradient, continued elevation changes in this region could adversely affect the flow of water through the aqueduct.

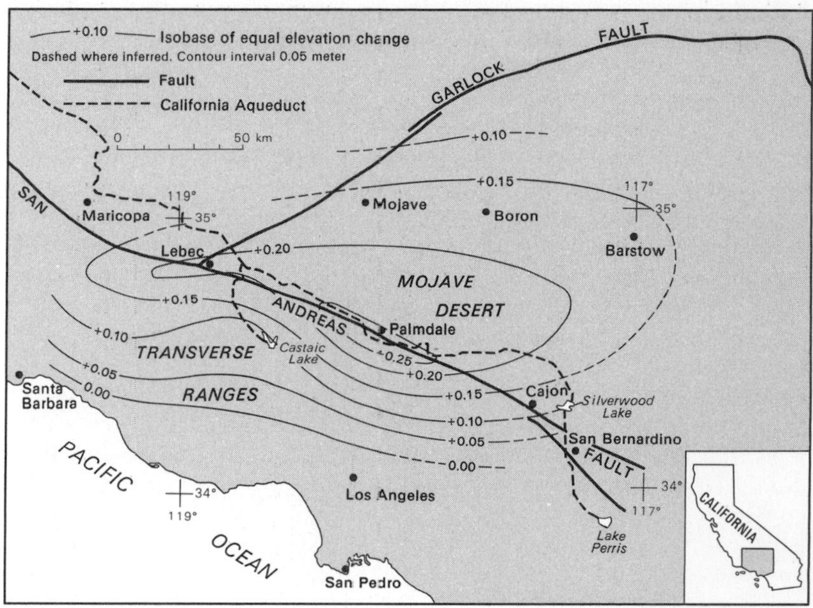

Area of 12,000 square kilometers (4,700 square miles) in southern California, centered on Palmdale, has been uplifted significantly since 1960. Figures indicate the amount of uplift of the Earth's surface in centimeters.

The Oroville earthquakes. The Oroville earthquake sequence, including foreshocks, main shock, and numerous aftershocks in the summer and fall of 1975, is a recent example of a damaging earthquake that perhaps was triggered by a nearby dam/reservoir system. Since Dean S. Carder first observed this phenomenon in association with the filling of Lake Mead behind Hoover Dam several decades ago, there have been many instances of markedly increased levels of seismicity associated with reservoir filling. Damaging earthquakes were closely related both in space and time to the filling of large reservoirs at Lake Kremasta in Greece; Lake Kariba in Rhodesia; Koyna in India; and the Hsinfengkiang Dam in China.

By California standards, the Oroville area in north-central California had been one of low seismic activity. Forty earthquakes of magnitude greater than 3.5 on the Richter scale had occurred within 100 km (60 mi) of Oroville between 1940 and 1974. Although earthquake locations in this region were not adequately determined until recently, none of these shocks occurred within 30 to 40 km of the area activated in the summer of 1975. The largest earthquake known to have affected this area prior to 1975 occurred in 1940 at a distance approximately 50 km north of Oroville and had a magnitude of 5.7.

The 1975 earthquake sequence began with a series of foreshocks on June 28; the largest of these had a magnitude of 3.5. Foreshocks continued at a low and intermittent level until August 1, when, five hours before the main shock, both the rates of occurrence and magnitudes of the foreshocks rapidly increased. The main shock (magnitude 5.7) occurred during the afternoon of August 1 and was followed by a large number of aftershocks. In the first ten days after the main shock, 12 aftershocks with magnitudes greater than 4 occurred. Aftershocks with magnitudes greater than 4 continued into November, although at a greatly diminished rate.

The two principal explanations for reservoir-induced seismicity are the increased stresses in the Earth's crust that are caused by the new load of the dam/reservoir system and the lowering of the shear strength of fault zones. The latter is caused by increased fluid pressure, provided that the reservoir water can gain access to the incipient rupture surface.

The first explanation seems unlikely in the case of Oroville, which is adjacent to the approximately north-south structural contact separating the crystalline rocks of the Sierra Nevada from the sedimentary rocks of the Great Valley. Motion on the fault surface, also approximately north-south, was such that the western block went down and to the west relative to the Sierra Nevada, a motion consistent with the long-term geologic history of the region. But the Oroville Dam and Reservoir lie entirely in the Sierra Nevada, and the excess load acts to retard the known sense of fault displacement rather than to abet it.

If increased fluid pressure along the incipient rupture surface is the triggering process, the six-year delay between reservoir filling and induced failure might be expected, according to the diffusion of fluid pressure into the Earth's crust. Earthquakes were inadvertently triggered near Denver, Colo., and controlled at Rangely, Colo., with fluid pressure variations artifically induced at a depth of several kilometers. Thus it would hardly be surprising if reservoir-induced earthquakes were related to reservoir-induced fluid pressure increases in the Earth's crust.

As of 1976, however, appropriate experiments had not been performed to test this hypothesis, either at Oroville or at any other locality. While the several instances of damaging earthquakes and many instances of increased levels of seismicity that are related to reservoir filling provide compelling circumstantial evidence that dam/reservoir systems can and have induced earthquakes, in no one instance has a direct causal relationship been established.

Earthquake hazards. Loss of life and property due to earthquake-related phenomena has occurred since the beginning of civilization. The principal hazard has been, and still is, damage to and failure of man-made structures. The principal agent of this damage and destruction is the primary passage of seismic radiation, which imparts dynamic forces to human dwellings and is caused by the rapid and extensive release of elastic strain energy in the course of a major earthquake. The loss of life consequent to failure of man-made structures induced by ground motion can be awesome; approximately 830,000 Chinese died in the Shensi Province earthquake of 1556 when widespread collapse of inadequately constructed family dwellings occurred. Early in 1976 thousands in Guatemala were victims of the same circumstances. Landslides induced by ground motion, and fires and flooding due to failure of engineering systems can also lead to significant losses. Of the nearly 100,000 people who died in the Tokyo earthquake of 1923, 38,000 perished in an uncontrolled fire. Approximately 20,000 people lost their lives when the town of Yungay, Peru, was obliterated by a landslide triggered by a major shock beneath the coast of Peru in 1970.

In any particular area earthquake-related hazards can be highly variable. An obvious variability is the proximity to active fault zones capable of producing major earthquakes. Both the regional and local geology can alter the amplitude and frequency content of surface ground motion and thus its influence on man-made structures. Soil failure, including landsliding and liquefaction, depends on a variety of local rock conditions, slope angle, and extent of water saturation in the ground.

Seismic zonation is the application of interdisciplinary investigations which seek to recognize, evaluate, and quantitatively analyze the behavior of natural systems in response to earthquakes on a regional and local basis. *Studies for Seismic Zonation of the San Francisco Bay Region* is a landmark effort in the development of methods, data, and application of seismic zonation on both a local and regional basis. Seismic zonation in the Bay area is developed with respect to the location of active faults, the magnitude and recurrence intervals of major earthquakes, estimation of bedrock motion for major earthquakes, amplification of bedrock motion by local geologic conditions, and potential for liquefaction and landsliding. The nature and extent of these effects are predicted for selected sites in the Bay area for a hypothetical, moderate earthquake (assumed magnitude of 6.5) occurring on the San Andreas Fault.

Seismic zonation, however, is only the first of several steps necessary to mitigate hazards related to earthquakes. The principal safeguards against these hazards in densely populated, earthquake-prone areas are the earthquake-resistant design and construction of engineering systems and structures and wise land-use planning. Thus, it is essential that the elements of seismic zonation be communicated not only to Earth scientists and engineers of several disciplines but to elected officials and the public as well. *Studies for Seismic Zonation of the San Francisco Bay Region* is written with a keen appreciation of this necessity and represents a rare attempt to communicate socially significant, scientific knowledge to a nontechnical audience.

—Thomas C. Hanks

Hydrologic sciences

Among the leading developments in hydrology and oceanography during the past year were advances in the use of recycled water, improvements in the pollution levels of Lake Erie and Lake Ontario, and the measurement of oceanographic conditions during a hurricane in the Gulf of Mexico. The International Decade of Ocean Exploration reached the five-year mark, and a major conference of scientists was held in France, the 16th General Assembly of the International Union of Geodesy and Geophysics.

Hydrology. During the past year hydrologists found themselves increasingly involved not just in applying hydrology to problems of water use but also in trying to apply it to decisions affecting social, economic, and political uses of water. For example, the U.S. Environmental Protection Agency (EPA) requires that an integrated approach must be employed in plans for resolving water pollution problems. Thus, in addition to the classical technical concerns are added analyses of alternative solutions involving overall cost; environmental, social, and economic values; and fiscal constraints at several levels of government.

Advances in research. New insight into the nature and occurrence of water came from several different lines of research. Astronomers identified ice in space as incubating a new star in the group of stars called T Tauri; scientists from the U.S. National Aeronautics and Space Administration found water vapor in the atmospheres of the planets Mars and Jupiter; the Argonne National Laboratory near Chicago developed a computer program that appears capable of analyzing the nature of water as a liquid; and work by a mathematician-hydrologist, done through cooperation between the U.S. Office of Water Research and Technology and the Desert Research Institute at the University of Nevada, resulted in a model able to provide a three-dimensional analysis of groundwater flow. The latter two modeling efforts were particularly significant to Earth scientists because they represent breakthroughs in the development of analytical techniques.

Application of the concept of mass transfer (movement of chemical constituents, and their interactions, through an aqueous medium) to problems of groundwater flow provided hydrologists at the U.S. Geological Survey and Princeton University with new insights into the relationship of groundwater to the chemical constituents it dissolves, carries, and precipitates. This work is particularly significant in its application to carbonate aquifers, whose flow characteristics are difficult to define in many areas throughout the world.

Multidisciplinary approaches to water problems offered new evidence regarding the intimate interplay between the physical and biological aspects of the environment. In many instances, the particular characteristics of individual biological species have led to their use in resolving water problems, at least on an experimental basis. For example, water hyacinths, whose rapid spread clogs navigation along many warm-water canals and inland waterways, have been found also to take up some pollutants (such as nickel, phenol, nitrogen, and phosphates) and also to be convertible to a locally usable fuel. A West German sewage treatment system, now being experimented with in Oklahoma, uses fish (such as carp, catfish, shad, and suckers) and crustaceans (such as crawfish, mussels, and clams) to clean up sewage treatment effluent. Shrimp, algae, and microbes were experimented with to clean up oil spills and industrial pollutants. In many instances the clean-up processes also result in "crops" of the creatures involved, and these were used for animal feed because the pollutants had been so altered by the

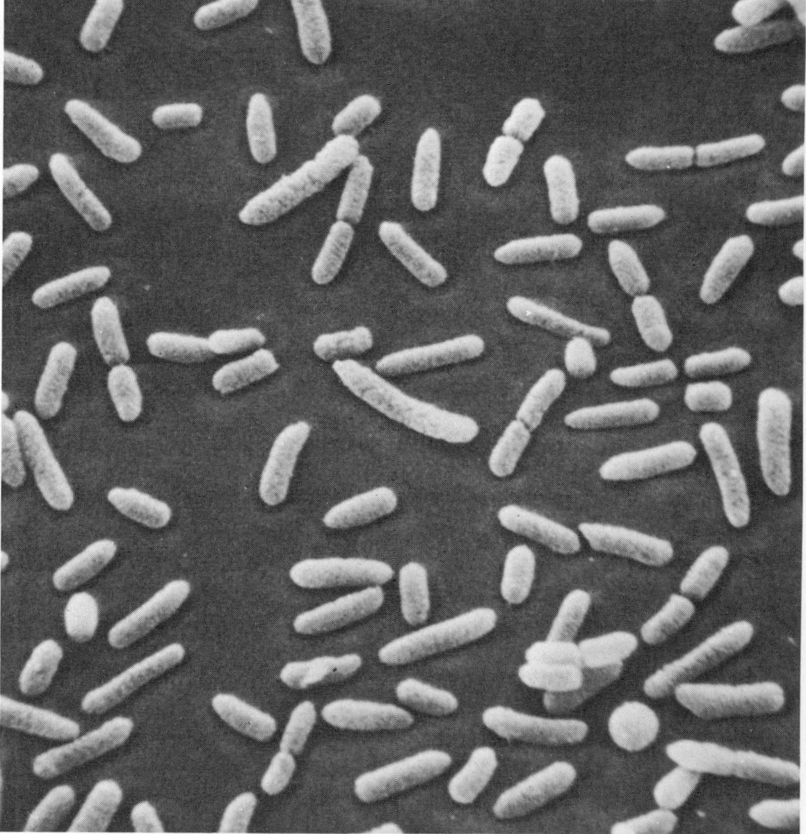

Microbes (right) created by Ananda Chakrabarty (above) were developed to clean up oil spills on waterways by digesting the petroleum and converting it to a form that marine animals can eat. The two flasks above contain petroleum and water; microbes have been added to the flask on the right and are beginning to digest the oil.

Photographs, courtesy, General Electric Research and Development Center digestive processes of the organisms that they were no longer at toxic levels.

Similarly, the concept of reducing water pollution by making useful products out of waste has been applied to municipal sewage and to barnyard and feedlot manure. Near Los Angeles solids settling out of sewage were being burned to form activated carbon, which was then used to purify the remaining wastewater; in addition, the gases from the burning process provide most of the fuel to fire the system. At Texas Tech University, beef cattle manure, a major source of water pollution in many localities, was being processed into ethylene, ethane, methane, and anhydrous ammonia for potentially profitable commercial use.

Advances in water use. The conscious use of recycled water has been generally accepted by agriculture but less so by industry and municipalities. Although some industries use recycled water to provide 90% of their demand, the practice is scattered. About 360 municipalities, mostly in the southwest, provide treated water for agricultural use, and some is used for industrial cooling, fish propagation, and recreation.

As of 1976 only one community in the U.S., Grand Canyon Village, Ariz., used recycled water in its municipal supply, but its use was restricted to non-potable purposes such as toilet flushing and lawn sprinkling. At the same time, some experts estimated that about one-third of the people in the U.S. unknowingly drink wastewater that had been discharged from municipal or industrial outlets only a short time earlier. In many instances such water is a far greater threat to public health than is the wastewater that has been purposely treated for public consumption.

The protection of coastal wetlands and inland swamplands was made the responsibility of the U.S. Army Corps of Engineers. These areas, which have been the sites of numerous recreational residence developments, were to be regulated by the Corps in compliance with EPA regulations and the advice of the U.S. Fish and Wildlife Service.

The groundwater reserves of the U.S., especially in the Colorado River Basin, received extensive attention because of their potential use in mining oil shale and coal, which require large quantities of water. Aquifers in the Colorado River Basin, in conjunction with surface water supplies, were found to be adequate to supply the anticipated demands because they contain approximately 115 million ac-ft of recoverable water in storage and are being replenished at the rate of about 4 million ac-ft per year.

Although groundwater supplies provide a huge potential national reserve, local overpumping creates problems. Land subsidence resulting from huge withdrawals of groundwater has affected many areas. In the San Joaquin Valley of California alone, 5,200 sq mi continue to sink, in some places dropping 10 m in 1975. Thousands of deep wells and miles of canals and ditches have been damaged. Near Houston, Texas, groundwater withdrawals threaten the possible inundation of large areas around Baytown Harbor.

The disposal of industrial wastes underground by means of deep injection wells received increasing attention during the past year. More than a million ac-ft of oil-field brine and liquid industrial waste were injected underground. A 4,000-ft well was begun near Raleigh, N.C., by the U.S. Geological Survey to test the storage potential of deep geologic basins along the east front of the Appalachian Mountains. The site was selected because the basins being tested are accessible to major industrial areas, appear to have no other mineral value, and are of great size.

Conditions in two of the most polluted of the Great Lakes, Erie and Ontario, provided encouraging news. Lake Erie, widely touted as "dead" a few years ago, provided a catch during the past year of no greater quantity but of better quality fish, including salmon and walleyed pike. Boating and swimming were again permitted where waste-control measures and devices had been established. In Gravenhurst Bay on Lake Ontario, a phosphorus improvement program begun in 1971 cleared the water and lowered the phosphorus content to a level better than any on record.

Sturgeon returned to the Hudson River to spawn, and hope was expressed for the rebirth of the caviar industry in the U.S. Atlantic salmon were back in the Connecticut River after a reported absence of more than 100 years. The return resulted from the release of young six-inch salmon in 1973, and an eight-year-long effort to clean up the river.

Acid rain, previously blamed entirely on the emission of pollutants into the atmosphere, was shown to be also a natural phenomenon. A study by University of Pittsburgh scientists cast doubt on the extent to which acid rain is composed of strong acids. The long record of increasing levels of acidity in rainfall in some parts of the eastern United States and Scandinavia, however, continued to be blamed for the failure of young fish to mature. An analogous debate continued regarding the feasibility of returning waste into U.S. rivers at a zero level of pollutant as required by the federal law. Zero discharge was claimed to be a physical, scientific, and social impossibility in that all streams are bound to receive runoff carrying pollutants from natural uncut areas as well as from agricultural and other open areas controlled or modified by man.

In western Maryland, drainage of acid from abandoned underground coal mines was shown to have been partly neutralized by natural processes without any help from man-made treatment. Approximately 70% of the acid was neutralized before it came to the surface, by means of contact with carbonate rocks as it percolated downward and to the surface. In some mines it was completely neutralized.

—L. A. Heindl

Oceanography. Oceanographers of all nations participated in a quadrennial review of achievements and progress in geophysics at the 16th General Assembly of the International Union of Geodesy and Geophysics (IUGG). The International Decade of Ocean Exploration reached the halfway mark, and both the achievements to date and plans for the next five years were summarized. Substantial progress was made in both the global and local oceanic measurement of pollution and its sources. The final reports on the Earth Resources Experiment Package of Skylab were written, and a group of scientists summarized the accomplishments of the program. Ways to measure oceanographic properties from space were being applied routinely and also developed for future spacecraft. The National Data Buoy Program scored a notable success by measuring meteorological and oceanographic conditions during a hurricane in the Gulf of Mexico.

IUGG. The 16th General Assembly of the IUGG was held from August 25 to September 6, 1975, in Grenoble, France. Much of the meeting was concerned with some aspect of oceanography and its interactions with other areas of the Earth sciences, such as vulcanism, plate tectonics, earthquake prediction, climatic changes, and weather prediction. The American Geophysical Union published a report summarizing the work of U.S. scientists for the IUGG.

One highlight of the assembly was a session describing the radar altimeter results from Skylab and GEOS-3 (geodynamics experimental ocean satellite) as they pertained to the further study of the geoid over the oceans, and the interaction of these results with other oceanographic problems. (The geoid is the surface around the Earth that is everywhere perpendicular to the direction of gravity and that coincides with mean sea level for the oceans.) Ways to study the geoid over a sea

Major U.S. IDOE Projects

Project	Number of institutions	Number of scientific investigators	Year initiated	Expected year of completion	Estimated total cost ($000,000)
Environmental forecasting					
Norpax	16	37	1971	1981	30.0
CLIMAP	5	18	1971	1979	7.0
MODE	16	45	1971	1974	8.0
ISOS	7	13	1974	1980	6.0
Polymode	10	30	1975	1979	10.0
Environmental quality					
Geosecs	14	28	1971	1978	18.5
Pollutant baselines	17	30	1971	1972	2.3
Pollutant transfer	9	10	1972	1980	10.0
Biological effects					
field (Cepex)	5	10	1973	1979	8.0
laboratory	6	8	1973	1981	10.0
Seabed assessment					
South Atlantic margins	2	15	1971	1975	4.0
Nazca Plate	3	40	1971	1977	6.0
Mid-Atlantic Ridge	4	8	1972	1975	1.0
Manganese nodules	10	18	1972	1978	4.0
Living resources					
CUEA	13	24	1972	1979	15.2
SES	10	11	1974	1981	7.0

mount both from deflection of the vertical measurements and from altimeter data were described by I. Fischer. B. V. Harmon and R. S. Mather described how GEOS-3 data would be used to study sea surface slopes off northeastern Australia.

Another session was concerned with air-sea interaction and the effect of anomalies in the temperature of the ocean surface as the cause of droughts. The severe drought in Africa was cited as an example by J. Namais. A paper by Klaus Wyrtki and W. C. Patzert developed and verified ways to predict when the anomalous warm "El Niño" current will occur off Peru.

IDOE. The International Decade of Ocean Exploration was proposed to the UN General Assembly in 1968 and began in 1970. The IDOE is concerned with four scientific areas: environmental quality, environmental forecasting, seabed assessment, and living resources. The scope of IDOE is shown in the accompanying table, where Norpax is the North Pacific Experiment; CLIMAP is Climate: Long-Range Investigation, Mapping, and Prediction; MODE is the Mid-Ocean Dynamics Experiment and Polymode is its extension; ISOS is the Integrated Southern Ocean Studies; Geosecs is the Geochemical Ocean Sections Study; Cepex is the Controlled Ecosystem Pollution Experiment; CUEA is the Coastal Upwelling Ecosystems Analysis; and SES is the Seagrass Ecosystems Study.

The IDOE is an international effort to explore the oceans, with 30 nations involved in it. One of its major accomplishments was the measurement of pollutants in the deep ocean.

Pollution. In the early 1970s the threat posed to various birds by DDT and its byproducts was recognized, and efforts were made to curtail its release into rivers, estuaries, and oceans. The results of those efforts as they concerned the brown pelican were reported during the year. According to D. W. Anderson and his colleagues, these birds nest on islands off the coast of southern California and Baja California. In 1969 the birds built 1,125 nests and raised only 4 young birds. In 1970, 727 nests produced 5 young; in 1971 there were 650 nests and 42 young; in 1972, 511 nests and 277 young; in 1973, 597 nests and 134 young; and in 1974 there were 1,286 nests and 1,185 young.

DDT-like compounds in the ocean are first absorbed in the plankton, which are eaten by the fish and concentrated. The pelicans eat the fish and further concentrate the compounds in their bodies. This causes an upset in their calcium metabolism and the pelicans lay eggs with very thin shells. The eggs are broken long before they can hatch, as the parents try to incubate the eggs.

A factory that manufactured DDT in Los Angeles was the major source of this pollutant in the ocean, and after 1970 this source was eliminated. In 1969 the anchovies on which the pelicans feed contained 4.27 parts per million of DDT-like compounds, and in 1974 this had decreased to 0.15 parts per million. In 1969 the pelican eggs had 1,204 ppm of DDT-like compounds in them, and in 1974 there were only 96.6 ppm. In 1969 the crushed eggshells were 0.288 mm thick, and in 1974 the shells were 0.378 to 0.482 mm thick.

A broad understanding of the total pollution problem in the New York Bight was evolving via the research programs of many organizations, many of which were supported by the Marine Ecosystems Analysis Program of the U.S. National Oceanic and Atmospheric Administration (NOAA). (A bight is an open bay formed by a bend in a coastline.) During a two-day symposium on the bight, held in New York City, papers by J. A. Mueller and two colleagues and by T. C. Malone provided new insights into the problem.

Before describing the two reports, some historical background is necessary. When it rains in New York City, both street runoff and sewage are discharged into the Hudson River, Upper New York Bay, and the East River. When it is not raining, some of the sewage is given secondary treatment in that the "solids" in it are removed and the effluent is chlorinated and discharged. Plans for giving all of the city's sewage secondary treatment when it is not raining will be realized in a few years.

The "solids" that are removed from the sewage are put in barges that are towed to the New York Bight and dumped in an area called the "sludge dumping" site. Other wastes from the city, mainly from industries and from harbor dredging, are also dumped into the bight. The sludge dumping became a major source of controversy in the New York City area with one scientist, W. H. Harris, maintaining that the sludge was moving onto the Long Island beaches.

The contribution of Mueller and his co-workers was to put the "sludge" into proper perspective with reference to the other sources of pollution in the bight. They found that the "sludge" is a minor part of the total amount of pollutants entering the bight. Dredge spoils are the biggest contributors to suspended solids, followed by river runoff. If sludge dumping were stopped tomorrow, approximately 90% of the pollution problem in the bight would still remain.

Only in excess can nitrates be thought of as pollutants. In reasonable amounts, nitrates and phosphates are the fertilizers on both the land and in the sea and along with carbon are essential for phytoplankton growth. Malone described what happens to the nitrates in the New York Bight. After entering it, mainly from the Hudson and East

Courtesy, NOAA

Data buoy in the Gulf of Mexico contains instruments that monitor such conditions as wind speed and direction, air temperature and humidity, and surface water temperature. This buoy was directly in the path of Hurricane Eloise in September 1975 and recorded important data about the storm.

rivers and Lower New York Bay, the nitrates, in sunlight, nourish the phytoplankton. The phytoplankton feed the zooplankton, which in turn feed the fish. The decrease in nitrates is almost linear with distance into the bight from the line separating the bight from Lower New York Bay.

The major problems of pollution in the bight are those of the heavy metals such as lead, mercury, cadmium, chromium, and zinc, and undesirable man-made toxic and carcinogenic chemicals. The latest of these are the PCBs, the polychlorinated biphenyls. These appear to be as serious a hazard as DDT and mercury. They concentrate in fish and are suspected as cancer-producing agents when ingested by people. Industries on the Hudson River discharge these chemicals into the bight.

Spacecraft studies. The study of the data obtained by Earth Resources Experiment Package of Skylab drew to a close during the year. More than 100 final reports on projects that used the data, many in the marine sciences, were submitted.

In one project, by the City University of New York and the University of Kansas, it was shown that each small elliptical area of the ocean surface, when scanned at a high angle of incidence (32° to 50°) by Skylab in the radiometer-scatterometer mode of S193, yielded data from which the wind speed could be computed with an accuracy at least as good as (and probably better than) that which could be provided by a weather ship near that area. (S193 was a multipurpose instrument that served as both a scanning pencil beam radiometer-scatterometer and an altimeter.) A future spacecraft, called SEASAT-A, will measure the wind speeds and directions near the surface of the ocean at 320,000 points per day.

Data buoys. An important project of NOAA during the year was the National Data Buoy Program. Large discus-shaped buoys, 40 ft in diameter, were instrumented to record the weather and various oceanographic data and radio the information to land. Two of these buoys were anchored in the Gulf of Mexico for many months, and in September 1975 they were almost directly in the path of Hurricane Eloise. The result was an entire volume of data for this hurricane. Waves about 27 ft high were recorded, some reaching 40 ft. Winds of 75–80 mph were detected by one of the buoys just before the eye of the hurricane passed.

—Willard J. Pierson

Energy

No single world economic issue commanded more attention during the year or had greater implications for future relations between the industrialized and less developed nations than did the continuing energy crisis. Despite much discussion and governmental prodding, the world's major consuming nations achieved little progress in energy conservation or in reducing their energy dependence on the small group of oil-producing and exporting nations known as the Organization of Petroleum Exporting Countries (OPEC).

World energy consumption in 1975 was estimated at about 120 million to 125 million barrels per day in terms of oil equivalent, down only slightly from 1974. The major industrialized countries (the U.S., Western Europe, Japan) consumed nearly 58% of this total. Oil supplied about 41% of total world energy. The industrial nations used nearly two-thirds of the world oil supply, and OPEC was the source of nearly three-fourths of their consumption.

Reduced oil demand varied among the major consumers and averaged nearly 9% less than during 1974. Most of this demand curtailment was due to the prolonged worldwide economic recession, which was itself largely caused by the massive OPEC oil price increases of late 1973 and early 1974. Efforts to develop large non-OPEC oil and other alternate energy supplies were not yet realized as of early 1976. U.S. oil and gas production continued to decline; only Mexico, China, and the North Sea were sites of large new oil discoveries, and their development was expected to take many years; the rate of development of nuclear energy slowed; coal development lagged; and efforts to develop new energy sources continued in the research and development stage.

On the other hand, the OPEC countries maintained production cutbacks to sustain their pricing policies, expanded their influence in other less developed countries, and emerged a more cohesive and stronger organization. On Oct. 1, 1975, OPEC unilaterally imposed an additional 10% price increase on world oil, bringing the FOB (free on board) price of Saudi Arabian light marker crude oil to $11.51 per barrel. Other OPEC nations did not fully apply the 10% increase, but increases of 5–8% were effected.

The financial and economic impact of the OPEC price increases continued to be felt throughout the world. Although 1975 OPEC oil exports declined to about 25 million–26 million barrels per day, or 20% below pre-embargo levels in 1973, OPEC oil income remained at about the same level in 1975 as in 1974, about $105 billion. Despite expenditures of $50 billion, surpluses of $50 billion to $60 billion accrued to the OPEC producers, with most going to Saudi Arabia, Iran, Iraq, and the United Arab Emirates. Thus, the consuming nations continued to amass large deficits to finance oil purchases. The non-oil-producing less developed countries had an annual deficit of about $20 billion. This was more than twice the economic assistance those countries received from all sources and caused sharp curtailments in their industrial development programs.

The producer and consumer nations alike took steps to seek an accommodation to the new world energy and economic developments. Acting through the 18-nation International Energy Agency, the major industrialized nations continued their efforts to lessen their dependence on oil imports by encouraging cooperative efforts to lower their consumption and accelerate the development of new energy supplies. The oil producers, aligned with 77 of the world's less developed countries, called for the establishment of a dialogue between the sides in a producer-consumer conference. A preparatory conference, held in April 1975 among producers and consumers, failed because of disputes over the issues of expanding talks from energy to other raw materials, aid to less developed countries, and financial and monetary considerations. As a result of subsequent French and U.S. initiatives, notably the U.S. proposals for world economic reform put forward by Secretary of State Henry Kissinger on September 1, the issues were resolved and agreement was reached to proceed with a producer-consumer conference.

On Dec. 16, 1975, the ministers of 27 nations, 8 from industrial countries and 19 from less developed countries, convened the first Conference on International Economic Cooperation. They agreed to establish four commissions to deal with problems in energy, raw materials, economic assistance, and financial and monetary affairs. Thus, a significant step toward resolution of energy and related economic problems was taken, although it remained to be seen how effective these commissions would be.

In the United States, Pres. Gerald Ford and the Congress carried on a national debate and sought legislative compromises that would permit agreement on a comprehensive national energy policy. There was no disagreement on the objectives of that policy, which were to develop adequate indigenous energy supplies and reduce consumption through conservation measures in order to achieve an adequate level of self-sufficiency by 1980–85. But there were serious differences in views on how best to accomplish an acceptable

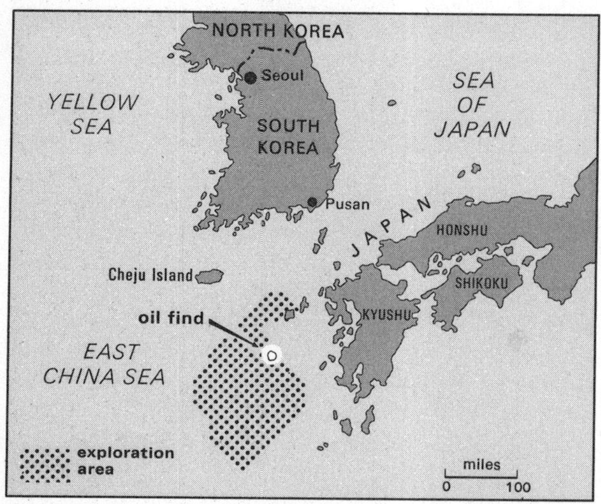

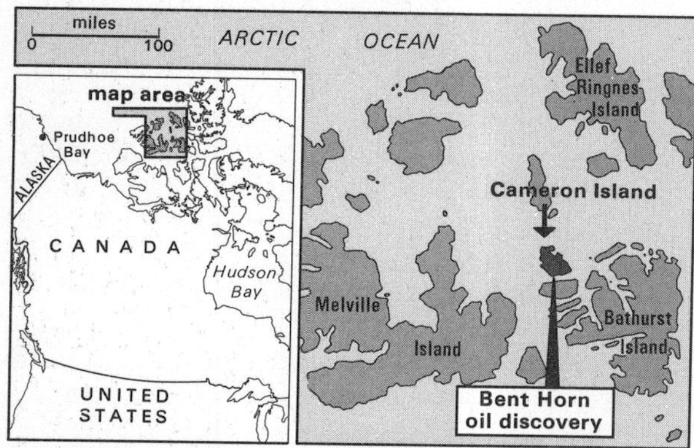

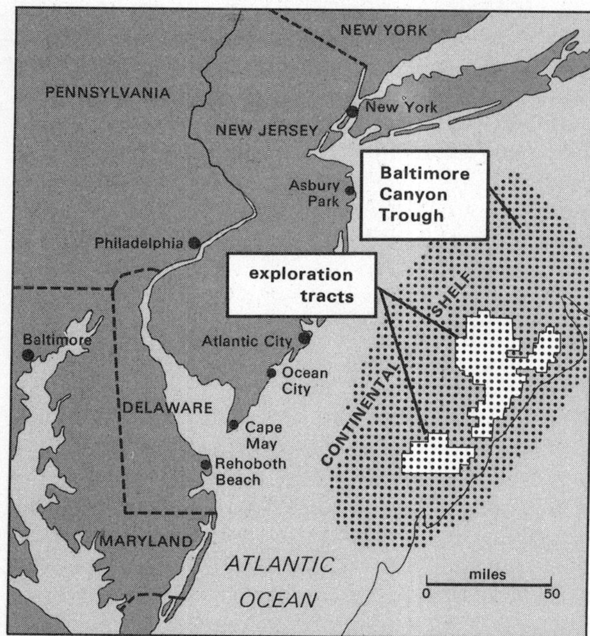

Potential new sources of petroleum include an area in the East China Sea (top), Cameron Island in Canada (center), and the continental shelf off the coast of New Jersey (bottom). Strikes have been made in the first two of these regions.

reduction of the dependence by the U.S. on oil imports. President Ford favored a "free market" solution that would remove current oil and gas price controls in a phased program having appropriate windfall tax provisions. It relied largely on the price mechanism to reduce demand and accelerate the development of new supply. Congress favored a slower relaxation of controls and the retention of price ceilings in order to lessen the impact of energy price increases on consumers and the possible adverse effects rapid energy price increases might have on recovery from the economic recession. Both agreed on the urgent necessity to establish a security storage reserve of crude oil and petroleum products to reduce the nation's vulnerability to an oil embargo.

President Ford imposed an import fee of $1 per barrel on crude oil on February 1 and then raised it to $2 on June 1 as a first step in implementing his program. In a series of confrontations between the president and Congress, Ford in September vetoed an act extending petroleum allocation and price control authority. In all, the president vetoed five energy-related bills that he believed had adverse long-range impacts on his energy goals; and none of the vetoes was overridden by Congress.

Finally, in November, the year-long deadlock between President Ford and Congress was broken. House-Senate conferees in Congress agreed on a bill that provided for an overall price ceiling on domestic crude oil of $7.66 per barrel, a phased decontrol of crude oil prices over a 40-month period that limited price increases to no more than 10% annually, a number of conservation measures, and an oil storage program. The effect of the bill was to roll back average crude oil prices from $8.75 to $7.66 per barrel. President Ford reluctantly signed the bill in December and at the same time rescinded the $2 import fee.

Petroleum and natural gas

The trend of annual increases of 6–7% in world oil demand, which had ended abruptly in 1974, was not resumed in 1975. World oil demand during the first nine months of 1975 averaged about 56 million barrels per day (bpd), about the same as in 1974. Oil demand in the non-Communist countries during this period was down nearly 9%, to about 42 million bpd. Despite signs of economic recovery in the latter half of 1975, U.S. demand through September was off a little more than 1% and averaged only 16.4 million bpd.

Crude oil production in non-Communist countries averaged 41.5 million bpd in the first nine months of 1975 but had increased to 45 million in September. The OPEC countries accounted for two-thirds of this output, with Middle East production averaging nearly 19.5 million bpd during this period. Production of crude oil and natural gas liquids in the U.S. declined by about 4% during this time, to an average of 10.2 million bpd. The rate of decline of U.S. crude oil production slowed during 1975, a slight reversal of a trend that began in 1971. Total crude oil and petroleum product imports were down by about 1% to 5.9 million bpd. Nevertheless, the cost of oil imports was expected to remain at about $26 billion during 1975.

The worldwide economic recession and continued high world oil prices were responsible for the severe drop in oil demand by non-Communist countries in 1975. Led by Saudi Arabia, Kuwait, and Iran, the OPEC countries curtailed their oil production and exports in order to maintain their price increases. Despite economic strains on some of its members, OPEC remained a cohes ve organization able not only to sustain its fourfold price increases of 1973–74 but to raise them another 10% as of Oct. 1, 1975. This unilateral action by OPEC demonstrated its leverage over the world oil market and its confidence in sustaining this control for many years.

This posture by OPEC resulted in greatly expanded efforts on the part of the major industrial consumers of oil to expand and diversify their oil supply sources. Oil and gas exploration and development activity in countries having developable resources reached record levels of activity during the year. Major strides were made in the development of the large oil and gas reserves of the North Sea. In the U.K. sector of the sea, the Argyll field went into operation in June to inaugurate oil production that is expected to lead to oil self-sufficiency for the U.K. by the early 1980s. Oil from the U.K. Forties field reached Britain on November 3 following completion of a 110-mi, 34-in.-diameter pipeline. Initial production was 40,000 bpd, expandable to 400,000 bpd by 1979.

In October production from Norway's huge Ekofisk field in the North Sea was increased to 300,000 bpd when the initial delivery was made through a

Wide World

Oil storage tanks under construction at Valdez, the southern terminal of the trans-Alaskan pipeline.

new 220-mi, 34-in.-diameter pipeline to a terminal at Teesside, England. Ultimately, up to one million bpd of Ekofisk crude oil were to be delivered through this pipeline to the terminal and then transshipped by tanker to Norway and other markets. Additional exploration drilling confirmed the Statfjord field in the Norwegian area as the North Sea's largest discovery, and 20 additional oil finds were reported from the North Sea area.

Mexico actively developed its large Reforma trend oil discoveries, raising production in the country by more than 200,000 bpd in 1975. China continued to develop its large new oil finds and increased its exports, mainly to Japan, to about 200,000 bpd. China was expected to expand exports greatly in the next few years, to as much as 800,000 bpd by 1980. A major new oil discovery was reported in Spain's Gulf of Valencia, and significant new finds were reported from Indonesia, Brazil, and India.

Despite this increased exploration activity, new discoveries for the non-Communist world proved only about 10 billion to 12 billion bbl of new reserves in 1975. By comparison, annual oil consumption was about 17 billion bbl. Obviously, these nations must reduce their demand and greatly accelerate production or be prepared for growing oil shortages in the not too distant future.

U.S. oil and gas well drilling activity continued at boom levels in 1975 despite lower oil company earnings, uncertainties over energy policies, and repeal of the 22% oil depletion tax allowance for major oil producers and companies having refining and marketing outlets exceeding 50,000 bpd. Active rotary rigs reached 1,740 in November, a 14-year peak. This record drilling activity was expected to continue into 1976 but could drop sharply because of reduced tax allowances and domestic oil price rollbacks.

U.S. oil company earnings in 1975 were approximately 30% lower than in 1974. Meanwhile, capital spending was up sharply. Compared with 1973, capital investment by the U.S. petroleum industry increased 56% in 1974 to $19.5 billion, and this level is believed to have been exceeded slightly in 1975. Refining capacity was expanded by approximately 5% in 1975, permitting a 10% drop in imports of oil products.

The first stretch of the 800-mi trans-Alaskan oil pipeline was installed in March 1975, and construction continued on schedule throughout the year. By the end of 1975 half of the pipeline was constructed. Its completion will open up the oil-rich North Slope of Alaska, where one-fourth of all U.S.

Liquefied natural gas from Algeria will by 1977 be brought ashore from ships through a tunnel (left) and stored in buffer tanks (right) at Cove Point, Maryland, on Chesapeake Bay.

Photographs, Grant Compton from "Business Week"

proved oil reserves are located, to markets in the other states.

In November the Federal Energy Administration (FEA) issued a final report on U.S. oil and gas reserves. The report was based upon a year-long survey of oil and gas fields made as an independent check on reserve estimates submitted by the industry. The FEA set proved reserves at 38 billion barrels of crude oil and 240 trillion cubic feet (Tcf) of natural gas as of Dec. 31, 1974. The report indicated that proved oil reserves were 11% higher than those published by the American Petroleum Institute and that gas reserves were 3% higher than those reported by the American Gas Association. The estimates, however, varied no more than might be expected when comparing such highly technical estimates from different sources.

Marketed natural gas consumption throughout the world remained at about 46 Tcf in 1975. Communist nations, led by the U.S.S.R., accounted for nearly 25% of the world total. Consumption in the non-Communist countries was down nearly 4%, largely because of a 7% decline in U.S. natural gas production. An 8% rise in Western European countries was offset by U.S. declines. As large new supplies from North Sea fields became available in 1975, The Netherlands supplied half of the Western European gas and the U.K. was approaching self-sufficiency.

Liquefied natural gas (LNG) projects had expanded in recent years but by 1975 accounted for only about 1% of total marketed sales. Total exports of natural gas, of 3.5 Tcf, accounted for only 8% of marketed consumption.

The U.S. was faced with growing natural gas shortages as both additions to reserves and producing rates continued to decline. Curtailments of deliveries to industrial customers were expected to cause supplies to fall 20% short of demand in the 1975–76 heating season. These curtailments were expected to affect users in 21 states. Industries with no access to alternate fuels faced the possibility of having to shut down for varying periods of time as the shortages peaked, but the situation was eased by a comparatively mild winter.

The U.S. Federal Power Commission (FPC) took several steps to ease shortages. It issued an order permitting industrial firms to contract directly with producers for intrastate gas supplies at going market prices for two years. It also took action to force producers of natural gas to deliver the full amounts of gas contracted for to their interstate pipeline customers. An FPC order permitting 180-day emergency natural gas sales in the winter of 1973–74, however, was held illegal by the U.S. Supreme Court.

The average price of natural gas rose to 65 cents per thousand cubic feet, up from 52 cents in 1974. Even at the new rate, natural gas on a thermal equivalent basis sold for only about one-fourth the price of alternate fuel oils. Congress was considering proposals to deregulate gas prices in order to stimulate the development of new supplies and reduce consumption.

Coal

Since the oil embargo of 1973 and the subsequent increases in oil prices, coal has assumed a more important role in the energy policies and plans of both developed and less developed countries. Coal consumption had been steadily declining during the years after World War II, but an upturn in demand was evident in 1975.

World production of bituminous coal and lignite was estimated at about 3.5 billion tons in 1975, a small increase over that of 1974. Most of the increase was in the Communist countries, which together accounted for nearly 58% of world consumption. The nine Western European countries

Scientist at the Lawrence Livermore Laboratory of the University of California views column filled with coal rubble down through which air is forced. The gray ash at the top is from burned coal, and the glow in the rubble is caused by burning. The flame at the right results from burning the gases produced in the column. The experiment is part of an effort to learn how to gasify coal deep underground.

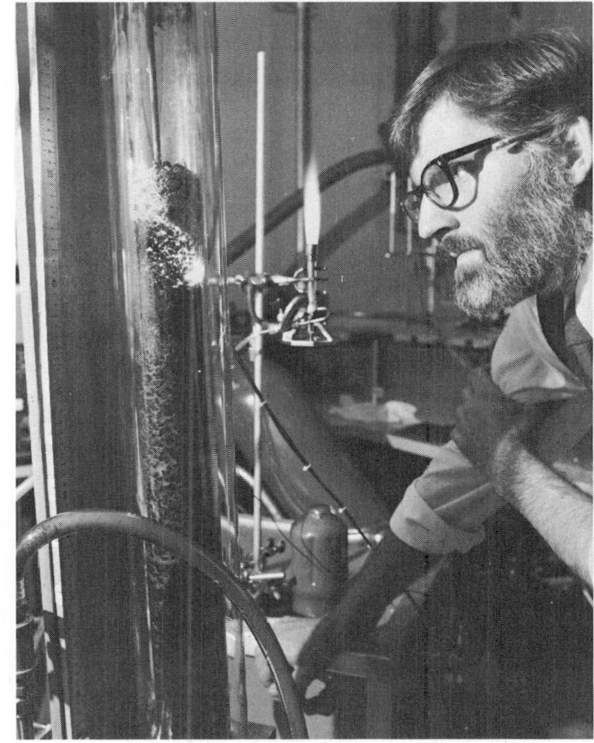

Courtesy, Lawrence Livermore Laboratory, University of California

of the European Economic Community (EEC) reported that bituminous coal output in the first nine months of 1975 climbed to 191 million tons; all the increase was in the U.K., where output in 1974 had been depressed by strikes of miners. The EEC reported that 345,000 miners were employed in 1975, the first increase in this labor force since 1958. U.S. coal production totaled 495 million tons during the first ten months of 1975 and was expected to be 635 million tons for the full year, about 5% more than in 1974.

World coal trade during 1975 was estimated at about 180 million tons, an increase of 5% over 1974. The U.S. continued to be the leading exporter, accounting for about one-third of the total. Other major exporters were Poland, the U.S.S.R., Australia, and West Germany. Japan was the leading importer, followed by France, Canada, Italy, West Germany, and Belgium.

U.S. coal exports were expected to reach 66 million tons in 1975, 10% more than in 1974. The value of these exports exceeded $3 billion and contributed favorably to the U.S. balance of trade. These increased exports and a buildup of coal stocks accounted for the increase in U.S. coal output, because domestic coal demand in 1975 remained essentially flat at 553 million tons.

Expansion of U.S. coal production is a prime requisite if the nation is to attain the degree of energy self-sufficiency that is required to hold oil imports to a manageable level. President Ford's energy program called for a doubling of production to 1.2 billion tons annually by 1985. This would require the opening of an estimated 200 Eastern and 100 Western mines. As of 1976 new mines were not being opened at anywhere near that rate. An early resolution of federal strip-mine and related environmental legislation, ending of

Harrison Station of the Allegheny Power System, a 1,950-megawatt coal-fired power plant completed in 1975, is located in West Virginia near the coal mines rather than the cities it serves. This has been made practicable by the development of extra-high-voltage transmission.

Courtesy, Dravo Corporation

the moratorium on the leasing of federal lands, major improvements in the railway systems, and a large expansion of the mine labor force were urgently needed to meet the expansion goal.

The U.S. Energy Research and Development Administration funded a great variety of coal conversion projects and awarded contracts for the construction of prototype plants for converting coal into synthetic gaseous and liquid fuels. The objective of these projects was the development of processes to demonstrate to industrial firms that the technical and economic risks involved in the building of commercial-sized conversion plants are acceptable.

As required by law, the U.S. Federal Energy Administration issued orders to 25 electric utility companies to cease using oil or natural gas at 74 power plants and to switch to the use of coal as their primary fuel for electricity generation. The U.S. Environmental Protection Agency must certify the date that each plant can burn coal and still meet air quality standards. Full conversion of these plants would save 64 million barrels of oil and 88 billion cu ft of gas each year.

Electric power

Depressed world economic activity, high fuel prices, and moderate weather in major consuming regions combined to cause the first substantial decline in electric power consumption in the post-World War II period. World electric power demand declined by an estimated 3–4%, to about 5.2 trillion kw-hr in 1975. Electricity generation by utilities in the United States in 1975 increased by about 1.5% to about 1.8 trillion kw-hr, according to preliminary estimates.

Fossil fuels continued as the primary energy source for electricity generation, but nuclear power plant capacity increased significantly in 1975 as higher oil prices and security concerns caused several industrialized countries to increase nuclear-energy development programs in 1975. Nuclear generation in 1975 increased its share of power output to nearly 5%, while hydroelectric power remained at 23% and fossil-fueled plants accounted for the remaining 72%. Nevertheless, previously established targets for nuclear capacity expansions were revised downward because of the economic recession and reduced energy demands, high inflation rates, tight money supplies, and challenges from environmentalists. In 1974 developed nations forecast nuclear capacity to increase from 58,000 Mw at the end of that year to 275,000 Mw in 1980. This target was later reduced sharply to the range of 175,000 to 205,000 Mw.

At the end of 1975 U.S. nuclear generating capacity was approximately 45,000 Mw, nearly 10.5% of the nation's total power generation. During the first half of 1975, electric utilities placed orders for 11 new reactors with a capacity of 13,700 Mw. This brought the total number of nuclear plants in operation, under construction, or on order in the U.S. to 218. Fears about safety and adverse impacts on the environment caused delays in some nuclear plant construction. A petition signed by 2,300 scientists called on the government to curtail plant construction drastically and suspend exports until safety and environmental issues could be resolved. The Nuclear Regulatory

Courtesy, General Electric Research and Development Center

Water-cooled gas turbine is designed to generate twice the power of those now in use and to operate at more efficient, higher temperatures.

Adapted from "Technology" in "Science News", vol. 109, no. 12, p. 185, March 20, 1976

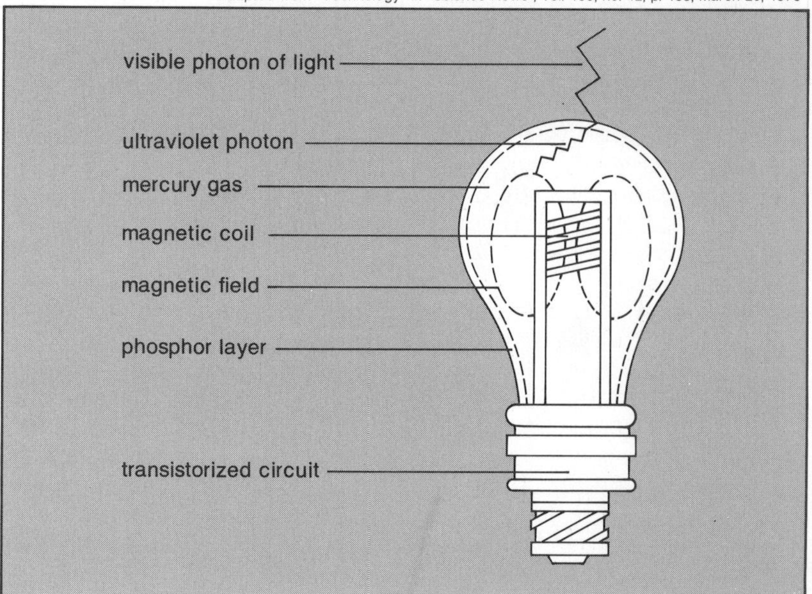

Newly developed "Litek" lamp is a fluorescent light in an ordinary screw-type bulb. A transistorized circuit in the base of the bulb supplies current to the coil of wire in the center of the glass. The electromagnetic field around the coil energizes atoms of mercury vapor, causing them to emit ultraviolet rays. These rays react with the phosphor layer coated on the inside of the bulb and cause it to glow. The Litek uses 70% less energy than ordinary incandescent bulbs and may last for ten years without burning out.

Commission refuted the charges, citing figures showing that nuclear plants have proved to be safe and reliable.

Enriched uranium required to fuel nuclear plants now in operation and under construction was in 1975 produced almost exclusively for the non-Communist world in U.S. government enrichment plants. Increasing demands required a vast expansion of this enrichment capacity. Two major projects were being developed in Europe, and the U.S. planned major expansions of its facilities. In a departure from previous national policy, President Ford asked for legislation that would allow all future U.S. enrichment plants to be built and operated by private industry.

Other energy sources

Solar and geothermal energy are potentially inexhaustible power sources, but as of 1976 no one had been able to develop the techniques required to convert either of them to widespread service. Economic reasons, more than technical ones, continued to limit the development of solar energy. Research programs were under way to change this by developing economical technologies that would convert solar energy to thermal and electric forms. The U.S. until recently lagged behind other countries in solar research, but the fiscal 1976 federal budget of $89 million exceeds the public expenditures of all other nations combined.

Almost-completed cooling tower of a nuclear power station rises alongside the Aare River in Gösgen, Switzerland.

Keystone

Solar energy is widely used in Israel, Australia, and Japan for heating water and other low-temperature applications; its total use, however, is less than 1% of overall energy consumption. Barring a major technological breakthrough, solar energy will remain restricted to limited areas and have local rather than national or international significance for many years.

In 1976 geothermal energy was being utilized in several countries for heating and for the generation of electricity; however, most scientists did not expect it to provide a significant amount of world energy until the next century. Low conversion efficiency and extraction problems in developing geothermal steam and hot-water resources continued to limit its use. Total world geothermal electric generating capacity in 1975 was estimated at 1.2 million kw, approximately equal to a single modern nuclear or coal power plant. The U.S. had about 400,000 kw, all located at the Geysers field in California. Italy had about the same capacity, largely at Larderello near Naples. New Zealand, Mexico, and Japan accounted for the remainder. Iceland was a major user of geothermal energy for space heating.

—James A. West

Environment

During the year a number of publications and events made it possible to evaluate the state of the environment and of environmental science more precisely than had been feasible in the past. Recent calculations suggest that the prognosis for the availability of food is poorer than ever before. New information on the dynamics of climate suggests that world population is already far too high, given the most likely long-term crop-growing conditions. Analysis of the effects of lower fuel supplies in early 1974 resulting from the actions of the Organization of Petroleum Exporting Countries (OPEC) revealed new insights. An important meeting near Houston, Texas, clarified the direction being taken by environmental science, as did several new books. New assessments of large-scale computer models as tools for environmental management (and other purposes) shed light on the reasons why they have had so little effect on policy. Certain recent governmental actions indicated new policy directions on environmental matters.

Population-resources imbalance. At the Worldwatch Institute in Washington, D.C., a "think tank" on emerging global problems, Lester R. Brown, the president, and his associates developed a number of revealing measures of the state of the world with respect to the balance between population and resources. One of these is "size of world grain reserves, expressed as days of world grain consumption at current rates." In his most recent publication, Brown indicates that this index dropped from 105 days in 1961 and 69 days in 1972 to a projected 31 days for 1976. The world is now perilously vulnerable to simultaneous bad crop years in the U.S.S.R., Canada, the U.S., France, Australia, Argentina, South Africa, and Thailand.

Indicating that disastrous crop failures in some of these areas are not unlikely, the 1975 Soviet grain harvest for example was only 60% of expectations. To put this in perspective, the shortfall of some 80 million tons was about twice a typical recent U.S. wheat harvest. Clearly, there was no way the Soviets could meet this shortage by importation, and they were forced to reduce their requirement for grain by slaughtering huge numbers of poultry and hogs. The grain they did have could be used five times more efficiently by feeding it directly to humans rather than by first processing it into meat and then feeding the meat to humans. This move away from a meat diet, which was expected to be a continuous process in all developed countries from now on, is an early-warning signal of world overpopulation.

Another index Brown computed is worldwide grain production per hectare. As of 1975–76, this measure had been dropping for three years. There are several explanations. The best agricultural land is always the first to be brought under cultivation in each country. Thus, the land recently added to the actively cultivated pool is likely to be the worst land for agriculture, and the yields from such land will depress national average yield figures. Also, the worldwide fuel shortage put a downward pressure on fertilizer inputs to agriculture. For example, in poor countries dung that might have been used for fertilizer was being used for fuel. Also, as fuel becomes short, there is more cutting of forests to obtain wood for burning. This leads to erosion, with resultant loss of soil fertility. Finally, a gradual deterioration of climate is occurring. Thus, the inherent crop-growing potential of the world is declining somewhat, and the capacity of the planet to support a large population is being reduced accordingly.

A shortage of food in certain countries, coupled with a surplus in others, would almost certainly intensify a process that had already begun: massive grain exports to pay for massive fuel imports.

Stephen Schneider and Clifford Mass shed more light on the kind of climate crop growers can expect in the future. They assembled a number of measures of climate going back many centuries and also developed a model to simulate the

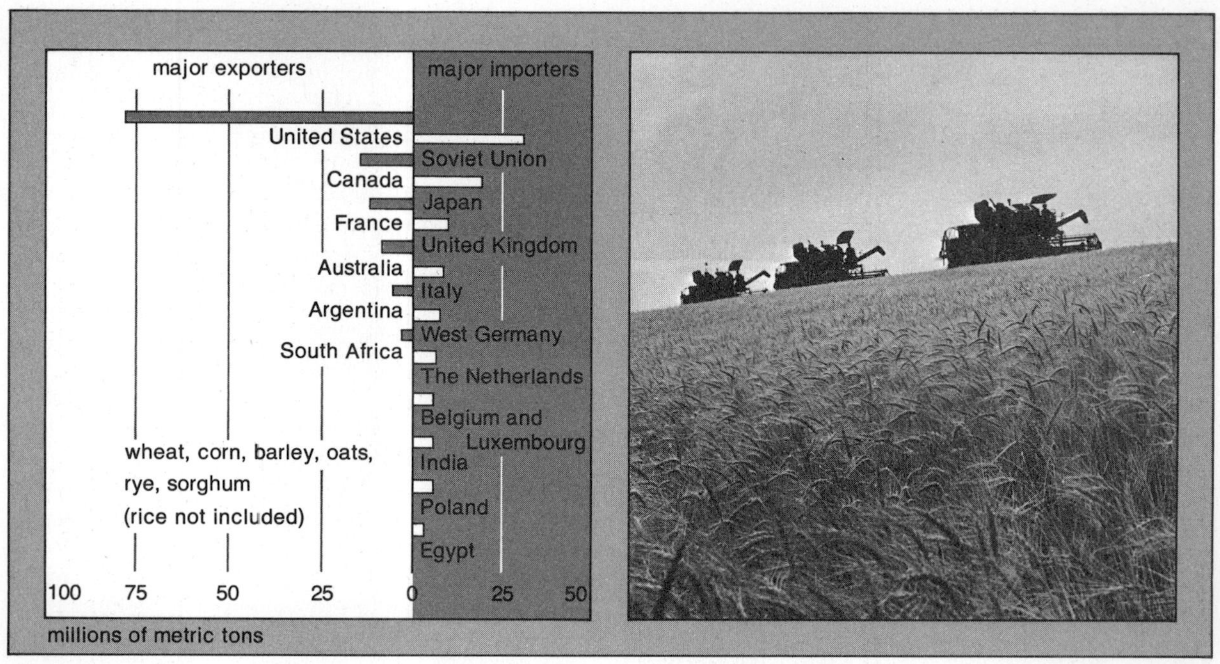

Combines harvest grain in the Soviet Union, which, as shown by the graph at left, was the largest importer of grain in the world in fiscal 1976.

(Left) Adapted from "U.S. Food Power, Ultimate Weapon in World Politics?" in "Business Week," p. 58, December 15, 1975; (right) NOVOSTI from Sovfoto

mechanism that determines world climate. Their research shows that the period 1930 to 1950 was unusually warm compared with the average for the last 800 years; as recently as 1650 to 1700 it was about 3° C colder. (Other studies have indicated a figure of 2° C.) Since a depression of about 1½° C in 1815–17 had a marked effect on world food production (grain prices doubled), a return to 1650–1700 weather would be a disaster now, when there are eight times as many mouths to feed. But the main point of the paper by Schneider and Mass is that planetary changes need not be invoked to explain these startling climatic shifts. Variations in the emission of energy from the Sun over the last four centuries can account for much of the variation in global temperature, and certainly for the great 1650–1700 change.

The implications of this research are extremely important. Even if man does not chill the planet significantly by polluting it in the next few decades, there could be a massive reduction in food-producing capacity resulting from solar radiation fluctuations, a natural process that is totally outside his control. Prudence dictates that man should adjust the planet's population to the level that can be supported by the average long-term energy emission from the Sun, not a climate that occurs over a few decades and represents the best conditions experienced in a millennium. This is one reason why predictions of much higher world populations are suspect; it is much more likely that the future world population will equilibrate at 500 million to 1 billion, one-eighth to one-fourth of the present level.

Meanwhile, the worldwide rate of population growth showed no sign of decreasing. The average annual rate of population increase between 1960 and 1965 was almost 2%, and more recently it climbed to 2.2%. However, this worldwide average pools very different rates for different regions. In some less developed countries, the rate of population increase dropped very rapidly in the last few years; in others there was no decrease. Of particular concern were India, where population was growing at 2.6% per year, and Indonesia, where the growth rate was 2.7% per year. By contrast, in South Korea the population growth rate had fallen from 2.8% between 1958 and 1963 to 1.7%.

The population growth rate of the U.S. was in a period of rapid change. The rate of natural increase (birthrate minus death rate) had been falling steadily, from 15.8 per 1,000 population in 1956 to 5.6 per 1,000 in 1973. Then the rate increased to 5.9 per 1,000 in 1974 and declined to 5.8 in 1975. However, a crude statistic such as this masks startling changes in specific components of the population. For example, because of the "baby boom" in the mid-1950s, the number of people in the reproductively most active ages, 20–24, will increase from 6.2% of the total population in 1960 to 9.5% in 1980. Therefore, tremendous declines in the age-specific fertility rate (births per 1,000 women) of young women would be required just to keep the number of births constant, and such drops were occurring. The annual fertility rate of women aged 15–44 was only 56% as large in 1975 as it was in 1955.

Moreover, there were reasons to think that the sharp downturn in birthrates would resume shortly. First, the effect on the birthrate of sharply increased unemployment among 20–24-year-olds, which began in 1975, would only begin to appear about September 1976. Provisional government statistics indicated that the 1975 birthrate was 1.3% below that for 1974. More strikingly, the birthrate for November 1975 was 8.7% below that of November 1974. Since no authorities were predicting a short-term drop in unemployment among the young, it was reasonable to expect a continuing decline in their birthrate. Another completely unexpected factor that could have a considerable effect on U.S. birthrates was the cost of malpractice insurance for obstetricians. Insurance rates were rising sharply, and local surveys indicated that the number of doctors prepared to deliver babies had fallen by as much as 90% during the year. In addition, the cost of delivery had escalated sharply. These factors seemed likely to depress birthrates still further.

As the rate of natural increase in the U.S. declines, another factor will become more and more controversial: illegal immigration. Currently the U.S. population was increasing by about 1,239,000 a year due to natural increase, 386,000 due to legal immigration, and between 650,000 and 1 million per year due to illegal immigration. The seriousness of this problem could hardly be overestimated. It was believed to be an important contributor to high U.S. unemployment rates and a drain on the welfare system, and it was a major undiscussed issue in the controversy surrounding the financial crisis of New York City. Illegal immigration was already of great concern to poor people, the unemployed, and such groups as farm workers' unions.

One further issue related to population is a potential time bomb for U.S. society. In 1950 the percentage of the population over 64 years of age was only 8.1, but by 1990 it would exceed 11. Social Security benefit payments, which amounted to 0.34% of the gross national product in 1950, reached almost 4% of GNP by 1973 and were rising rapidly. If one extrapolates present trends, the inescapable conclusion is that these payments will reach a minimum of 8.9% of GNP by 1990. At the least, this would impose a significant brake on future economic growth.

A paper by David Pimentel and associates gives a very precise set of estimates on the relationships between food production and the availability of land and energy for agriculture. World availability of arable land is 0.38 ha per capita. In the U.S., however, 0.62 ha of arable land per person plus a high input of fossil fuel energy have been necessary to provide the current diet. Thus the kind of diet now enjoyed in the U.S. is no longer a possibility for the existing population of the world as a whole, to say nothing of an increased population. If yield per acre were to be increased through the use of petroleum inputs to provide this diet worldwide, all known petroleum reserves would be gone in 13 years. Accordingly, the possibilities open to mankind are constrained by the limited availability of both land and energy.

Pimentel points out that one possibility is to increase food for humans by decreasing the use of meat. In the U.S. 91% of all cereal, legume, and vegetable protein suitable for human consumption is fed to livestock. It takes 24.6 million metric tons of this plant material to produce 5.3 million metric tons of animal material, so the transfer is only 22% efficient. Also, there are enormous differences in the number of kilocalories of fossil fuel energy required to produce one kilocalorie of food protein. For most U.S. food plants, it takes two to five energy units to produce one energy unit of food. In contrast, it takes 36 units to produce 1 of milk, 13 for eggs, 10 for rangeland beef, and 78 for feedlot beef. The conclusion seems inescapable: feedlot rearing of beef is soon going to be too expensive from an energy standpoint. Energy efficiency (and, soon, economic necessity) will force Americans closer to an all-plant diet. Meat is the food of populations in balance with their environment. Thus the first really compelling evidence of overpopulation will soon be affecting the U.S. population—evidence that was apparent in most other countries long ago.

The fuel crisis as a major experiment. Normally it is exceedingly difficult to ascertain the precise effect of air pollution on public health on a population-wide basis. The usual method is to determine the magnitude of the effect statistically, by comparing the morbidity and mortality rates of polluted and unpolluted urban areas. However, this is a task of great technical difficulty because urban areas differ from one another in such important respects as age distribution, climate, presence of natural noxious agents such as fungal spores, economic and educational status, meteorological conditions, and migration history. It is difficult to prove that observed differences in

Courtesy, NASA

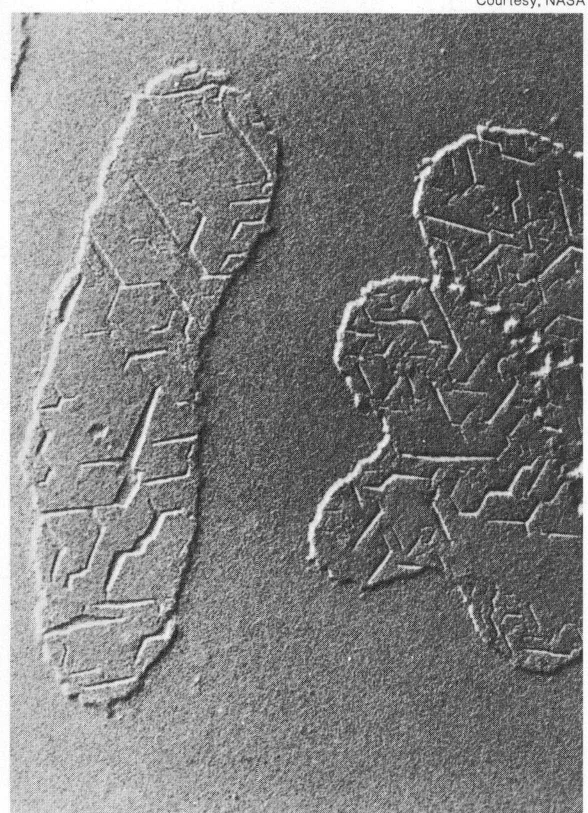

Electron micrograph of dried purple cell membrane surfaces in the bacterium Halobacterium halobium *shows cracks that indicate the presence of protein molecules in a crystalline lattice. This is the first indication of a photochemical mechanism other than chlorophyll-based photosynthesis in a living organism.*

mortality rates were due to differences in air pollution, to the exclusion of all other possible causes. The counterargument can always be made that some other cause not included in the analysis was the real culprit.

The only way this argument can be settled is by sharply decreasing air pollution in urban areas for extended periods and seeing what happens to the incidence of death and disease due to various causes. Although it was scarcely their intention, the OPEC countries had created the conditions for such an experiment to occur in the spring of 1974 by their cutback of petroleum shipments to the West, thus forcing curtailments in automobile use. According to an analysis by Stephen Brown, Michael Marmot, Susan Sacks, and Linda Kwok, in San Francisco and Alameda counties in California, deaths from chronic lung disease in the first quarter of 1974 were 33–38% below what they would normally be for the period. Deaths from all causes in the region ranged from 7.7 to 13.4% below the normal level.

The "OPEC experiment" also demonstrated a weakness in the current management of transportation in urban areas. In an effort to induce riders to shift from cars to mass transit, more mass transit facilities are being made available but disincentives to car use are not being applied. The "OPEC experiment" showed that in the San Francisco Bay area mass transit ridership could be increased roughly 23% by making it as difficult to get gasoline as it was in the spring of 1974. Clearly, if there is to be more efficient use of fuel in transportation systems, active measures must be taken to discourage car use at the same time that public transportation is expanded.

The limits to growth meeting. Perhaps the most important meeting on environmental matters during the year occurred in Woodlands, north of Houston, Oct. 19–21, 1975. The meeting was sponsored by the Club of Rome, the University of Houston, and Mitchell Energy and Development Corp. George P. Mitchell had become a major benefactor of this type of activity. He planned to sponsor five such conferences over the next ten years and had begun a program of awarding large cash prizes for outstanding essays on the consequences of declining economic growth. The essay contest for 1975 attracted 1,000 entries.

The Woodlands meeting was addressed by many of the world's best-known futurists, including Jay Forrester, Dennis and Donella Meadows, Lester Brown, E. F. Schumacher, Herman Daly, Herman Kahn, Robert Theobald, Yujiro Hayashi, and Edward Goldsmith. Besides the usual debate as to whether humanity will soon be bounded by various types of limits, it produced some important new ideas presented by Forrester. He stated his belief that the debate about limits to growth has focused too narrowly on physical limits, and that this has been counterproductive. "Technological optimists" can always argue that it is possible to circumvent physical limits on man's activities by means of technology, and this argument is basically unanswerable because it is difficult to know what technology might be discovered in the future. Forrester believes that social limits may constitute a better argument for limiting population growth and industrial activity.

This reasoning, which constitutes a major change in perception concerning the nature of limits on growth, is shared by many other global and national modelers. As Dennis Meadows pointed out at the meeting, technology with the potential to solve all the present problems of mankind has already been discovered, but the problems are still with us. Clearly, the lack of necessary technology is not the real difficulty. The true problem is institutional, political, economic, and social malfunctioning, which in turn leads to increased drug addiction, kidnapping, sabotage, terrorism, revolution, frustration, alienation, and disenchantment, as well as resource depletion and pollution.

Given the complexity of the problems that must be unraveled and the multitude of components involved, the only available methods of problem-solving would appear to be those advocated by such researchers as Forrester and Meadows: systems analysis using complex computer models. Systems thinking seems to be the natural and characteristic mode of thought in the environmental sciences. Consequently, a predictable development is that environmental science will place increasing reliance on the type of systems analysis conducted by Forrester, Meadows, Mesarovic, and others.

The $10,000 first prize in the Mitchell essay contest was awarded to Bruce M. Hannon of the University of Illinois at Urbana. Earlier in the year, Hannon published a penetrating essay on the relationships between wages, capital cost, and energy cost, in which he concluded that taxes on energy resources and rationing of energy should be used to make society more labor-intensive and less energy-intensive.

What is environmental science? With most traditional academic disciplines, there is no confusion about the nature of the subject; the discipline is defined in terms of a widely agreed-upon subject matter. But what is the subject matter of environmental science? Such popular texts as *Living in the Environment* by G. Tyler Miller, Jr., published in 1975, indicate that environmental science concerns itself with population, pollution, resources, soil and agriculture, energy, energy flow and ecosystems, food, forestry, parks, wilderness, wildlife, estuaries, cities and urbanization, minerals and mineral resources, water and water resources, economics, politics, ethics, religion, value systems, the culture, technology—in short, almost everything that goes on in the mind and world.

The inescapable conclusion is that environmental science cannot be defined meaningfully in terms of its subject matter. Rather, environmental science as now commonly understood is better defined in terms of the method it uses to analyze problems. The key characteristic of this method is a type of logic that perceives reality in terms of long, chainlike causal pathways that pass through different subject fields completely without regard to traditional boundaries. Furthermore, these chains can loop back on themselves. An enormously important implication of this mode of perceiving reality is that it can lead to conclusions completely opposite from those yielded by more conventional modes of analysis.

Thus, for example, a conventional analysis might conclude that cheap energy is desirable, since it implies low cost inputs into the production of goods and services; this, in turn, will have a dampening effect on inflation and lead to healthy growth in the real gross national product. The systems view, however, is a logical web rather than a straight line, and it takes into account side effects

UPI Compix

Diamond Express Lane (right) of the Santa Monica Freeway in Los Angeles is reserved for buses and cars with three or more passengers.

Courtesy, J. E. Cermak, Colorado State University

Scale model of Fort Wayne, Indiana, is studied at the Fluid Dynamics and Diffusion Laboratory at Colorado State University in an attempt to determine the effects of an urban heat island on the transport of air pollutants through the city.

and social and other variables that ordinarily might not be considered. Such an analysis might well show that cheap energy leads to the substitution of energy for labor (automation), resulting in unemployment and the undesirable social effects of unemployment, such as higher crime rates. Also, cheap energy encourages high consumption and, if imported oil is cheaper than domestic oil, the result is an undue dependence on imports and concomitant balance of payments difficulties. With world markets for manufactured goods approaching saturation, it may be necessary to pay for imported oil by exporting raw materials and agricultural products, such as grain, leading to shortages and higher prices for these commodities at home. By taking all these factors into account, such an analysis could thus lead to the conclusion—completely opposite to the conventional one—that cheap energy is undesirable.

It should also be noted that the system as perceived in this analysis tends to be ultimately self-correcting. The demand for commodity exports will lead to shortages of land suitable for agriculture, forestry, and mining; less land will be converted from these uses to urban use; cities will become more compact; higher densities and greater traffic volumes will force a return to greater use of public transportation; energy will be saved; and oil imports will decline.

Computer models for environmental management. As of 1968 many people believed that the complexities of modern society could only be understood by means of large-scale computer models, which would also be the most appropriate tool for making policy decisions. Consequently, the U.S. government, the Ford Foundation, and other institutions underwrote the development of many such models, the most publicized of which have been those of the Club of Rome. By 1975, however, it was clear that these models had not had the effect that many people had foreseen.

A survey commissioned by the U.S. National Science Foundation to analyze such activities with the aim of ascertaining their role in the policymaking process revealed that a great deal of money has been spent in this activity (the total bill, worldwide, is probably several hundred million dollars) and that the effect on policy has been marginal at best. The models are very diverse and in effect represent an elaborate statement of the preconceptions of the modelers. Bad predictions, particularly by national economic models, have cast doubt on the credibility of all large-scale models. Furthermore, it has turned out to be a remarkably difficult, time-consuming, and expensive task to communicate the results of the models to busy policymakers. But perhaps the most compelling reason for the low utilization of the models is that they did not tell politicians or the electorate what they wanted to hear: that the present socioeconomic system could continue to flourish and grow without any major changes. All the models suggest that major changes will be required if society is to continue without significant trauma. Clearly, a major job of public education on environmental and population matters is required.

Some steps have been taken by the U.S. government. A "foresight provision" went into effect in the 94th Congress requiring that each standing committee, except those concerned with budget and appropriations, undertake on a continuing basis future research and forecasting on matters within its jurisdiction. A specialist in future research became part of the Congressional Research Service, and the first conference on the future for Congress was held. Also, it appeared certain that there would be a new national solar energy research institute within the Division of Physical Research of the Energy Research and Development Administration.

—Kenneth E. F. Watt

Food and agriculture

Although the food-supply crisis that had loomed in 1974 receded somewhat in 1975, shortfalls in some important areas, especially in the Soviet Union, served as reminders that the problem had not been solved. Considerable attention centered on the need to conserve energy in agricultural production and on the use of alternate forms of animal feed to replace grain. Nutritionists continued to emphasize the need for education since, even where food supplies were adequate, a poor diet could cause malnutrition.

Agriculture

Several viewpoints concerning shortages and increasing prices for agricultural products were summarized by Fred H. Sanderson of the Brookings Institution. One was that events in the past few years may be the first signs of a fundamental change in the balance between food supply and food demand; we are rapidly approaching the point at which population growth and rising affluence will outrun the world's capacity to feed itself.

Another view foresees not imminent disaster but a reversal of past trends in agricultural technology aimed at reducing costs and the use of traditional fuels. In this view, the day of food surpluses is over and less developed countries will put increasing pressure on agricultural resources, bringing about a significant rise in real food prices. As a result it will become increasingly difficult to sustain present levels of food consumption. Still others believe that recent shortages can be traced to transient factors and that the possibilities of expanding food production in developed as well as less developed countries are far from exhausted. Some even suggest that this can be done without a dramatic increase in the real costs of production.

Over the last century agricultural production has increased manyfold in the United States and in several other countries. Over the last 25 years it has more than doubled in the U.S., so that while in 1930 one U.S. farmer fed 10 persons, today one farmer feeds 54 persons. Higher productivity, however, has been brought about largely by the increased use of energy. Until recently, energy supply has not been a constraint on human population as it has on other organisms. A further complicating factor is that a growing population tends to decrease the amount of land available for food production. According to recent U.S. Department of Agriculture (USDA) figures, there has been a 13% reduction in the land area grazed in the U.S. over the last two decades. Thus it is important to look at potentials for food and fiber production in terms of a balance between production technologies, energy supplies, and availability of natural resources, including land, water, and nutrients.

Food and agricultural research. Over the last century, the U.S. has been by far the largest contributor to agricultural research. More recently, however, other governments and many private organizations have intensified their efforts in this field. Research centers are now located in Southeast Asia, Latin America, Africa, and the Near East. Some of these efforts began as early as 1943, but the majority of them have been started within the last decade. China is a particularly interesting case. The Chinese have made extensive use of practices involving the production of several crops

Aqueduct 100 kilometers (60 miles) long carries water to irrigate rural areas of Hopeh Province in China.

Keystone

on the same land each year, either by growing a succession of crops on the same tract or by growing two or more crops together during a given time period. A full gamut of agricultural research, including studies of the use of fertilizers, plant protection, development of specific crop varieties for various climates, fruits, and animal production, is under way.

The U.S. Foreign Assistance Act of 1975 included a section proposing the establishment, in agriculturally less developed nations, of institutions similar to the U.S. land-grant colleges. These institutions would focus on developing the talents of the indigenous peoples and on initiating research appropriate to local needs rather than simply on the transfer of knowledge from the developed world. The program was to be organized through counterpart institutions in the U.S.

Energy and agriculture. It has been calculated that in the developed nations 29% of the total energy cost of food is accounted for by the growing process. Another 29% is used in processing procedures, 17% in transportation, and the remaining 25% in storing and cooking. In an effort to conserve fossil fuels, scientists in the USDA's Agricultural Research Service and several state agricultural experiment stations, including those at the University of Maryland, the University of Arizona, and Michigan State University, were concentrating on the use of solar energy to heat the buildings used in livestock production and other aspects of agriculture. Knowledge derived from studies of solar heating for houses could also have agricultural applications.

Peanut flaking was developed by scientists to increase the usability of peanuts in the diet.

Courtesy, Clemson University, Clemson, South Carolina

An interesting technological innovation emerged from research conducted at Purdue University, Lafayette, Ind., in which shredded rubber from worn-out tires was used as a garden mulch for ornamental plants; it was found to work as well as or, in some cases, better than stone, plastic, or bark chips. Scientists at the USDA, the University of Connecticut, and elsewhere addressed the question of high food prices in the northeastern U.S., which result, in large measure, from high transportation costs. Growing and storing food closer to the market should decrease the amount of fuel used and hence the cost of such food items.

Rangelands as a food resource. Rangelands comprise almost 50% of the total land surface of the Earth. They represent about 55% of the land area in the U.S. and about 70% of the land in the western states. With proper management, one pound of beef can be obtained for every ten pounds of growing range forage consumed by a steer. Research scientists, particularly in the western states and in Australia, were addressing questions of range improvement, not only to provide an important source of animal feed but also as an energy conservation measure. Unimproved rangelands cost nothing in terms of fuel or other energy inputs, and improved rangelands require only a minimal amount.

Alternate forms of animal feed. Researchers in Missouri found that, in addition to plants, ruminant animals can use urea as a source of nonprotein nitrogen. A new product, called ammonia super slurper, was based on the observation that when urea is mixed with starchy products the ammonia released inside the animal is detoxified; thus a ruminant animal can use this nonprotein nitrogen source instead of eating grains that could be consumed directly by humans. Scientists at the universities of Wisconsin and Maine, as well as at the USDA, had also worked on the problem. The potential for the product in the beef, dairy, and sheep industries was impressive.

A significant amount of the energy needed in the production of food could be obtained by recycling waste products and by-products. Approximately 150 lb of cellulose are produced per day for every one of the Earth's four billion people. The annual production of crop and animal wastes in the U.S. exceeds 800 million tons. Rations for poultry and livestock can be made that contain up to 15% animal waste products. Corn silage, with certain alkalis and acids added and mixed with cattle manure, cornstalks, low-quality straw and hays, tea leaves, sterilized poultry wastes, and apple pumice, appears to be useful in supplying nutrition to beef animals. Such waste products have been used to provide 70% of the total winter feed for cows at Pennsylvania State University, and scientists at Auburn (Ala.) University found that these techniques improve feed efficiency by approximately 25%. The ensilaged material has no offensive odor.

Food quality. Two-thirds of the protein intake in the average diet in the U.S. and Canada is made up of animal protein; by contrast, 97% of the protein consumed in South Asia comes from plant sources. Clearly, in developing new crop varieties for less developed areas as much attention must be paid to the quality and quantity of the protein as to the caloric value. Scientists have used a number of criteria, including resistance to disease and insects, length of time required to produce a crop, temperature, moisture and soil conditions required, protein content, the distribution of amino acids in the protein (protein quality), and caloric value. Varieties that have been developed in recent years include high-lysine corn and sorghum; in many parts of the world the indigenous varieties of these two grains were deficient in lysine, an amino acid vital to good nutrition.

During 1975 a number of significant new crop varieties were released, including a variety of sorghum from Texas A & M University that has improved digestibility. Wheat varieties with significant characteristics were released by the USDA and several state agricultural experiment stations. "Double Crop," released by the University of Arkansas, matures four to five days earlier than other varieties, while "Lindon," released by Colorado State University, has increased production on the U.S. High Plains by approximately 5%, although it is not as resistant to environmental stress as some other varieties. The high yield recorded for "Ruler," a new variety released by Ohio scientists, was an impressive 92.1 bu per ac.

A new high-density potato, released during 1976, is particularly well adapted to baking and has consistent size and quality. Scientists in South Carolina developed a peanut flaking technique that can be used as the basis for meat analogues,

Courtesy, USDA

Urban elm, a cross between an elm from The Netherlands and a Siberian elm, is believed to be resistant to Dutch elm disease. A limited supply of the trees was expected to be available by 1978.

sandwich spreads, snack dips, casseroles, and desserts. In Rhode Island a closed aquaculture system was developed for salmon. The system offers rapid growth, excellent protein conversion ratios, and high production per unit volume of water (13 lb of salmon produced per cubic foot of water per year).

Spoilage and pest control. A considerable amount of agricultural research in 1975 was concerned with reducing losses during the harvesting, processing, and retailing of food. Researchers in Idaho developed a system involving temperature, humidity, and air distribution characteristics that reportedly can decrease the loss from shrinkage and spoilage in stored potatoes from 25% to about 5%. Scientists in Nebraska were testing a promising vaccine against calf scours, a major cause of loss in calves born on the range.

A promising method for controlling blight in chestnut trees involves inoculation of less virulent strains of the blight fungus at the edge of cankers on the tree. If the canker has not yet girded the entire tree, this treatment appears to stop further growth of the pathogen and allows the canker to heal. A new variety of elm, said to be resistant to Dutch elm disease, was expected to be available in limited supply by about 1978. Called Urban elm, it is a cross between an elm from The Netherlands and a Siberian elm. Its shape is somewhat different from that of the American elm and it grows more slowly, but it attains a height of 60 to 80 ft when fully mature.

—John Patrick Jordan

Nutrition

Problems of food supply, assessment of nutritional status, immediate and long-term anomalies of malnutrition, and efforts to restore health after it has been damaged by imbalances of nutrients continued to challenge society during 1975–76. Following U.S. Secretary of State Henry Kissinger's announcement at the World Food Conference, held in Rome in late 1974, that the U.S. was prepared to provide financing for the conquest of nutritional diseases such as vitamin A-related blindness and iron deficiency anemia, U.S. Pres. Gerald Ford suggested that U.S. expertise be used to increase food production and alleviate malnutrition throughout the world. To implement this policy, an effort was being made to coordinate the work of U.S. departments and institutes doing research in nutrition and allied areas. Congress was to study the effectiveness of National Institutes of Health training and research grants in encouraging greater interest in nutrition; for example, grants for cancer research could focus attention on the importance of dietary factors in the genesis and treatment of malignancy.

A recurring awareness on the part of scientists that nutrition is disregarded in the medical school curriculum and in the practice of medicine led to the establishment of the National Nutrition Consortium. The aims of the consortium were to remove impediments to the practice of clinical nutrition and to gain acceptance for it within the medical community. The American Institute of Nutrition (AIN) and the American Society for Clinical Nutrition (ASCN) pledged their cooperation. The need for such efforts was pointed up by Bruce R. Stillings of Nabisco, Inc., editor of the AIN *Nutrition Notes*, who asked whether the U.S. nutrition community is sufficiently coordinated to meet the challenges facing it within the U.S., to say nothing of other countries. The general public is searching for nutrition information, but too often it gets false answers from pseudo-nutritionists through the news media. Congress, as well as students in the health and medical areas, is too often misled by those who are enthusiastic but untrained.

Legislation. Howard A. Schneider of the AIN public affairs office reported on problems of legislative action as illustrated by the effort to pass a Senate bill regulating high-dosage formulations of vitamins. When it seemed that an acceptable bill, based on reliable professional testimony, was certain to pass, a senator who was strongly supported by "health food" advocates secured passage by means of a parliamentary maneuver, of an amendment barring regulation of vitamin preparations. Only two senators voted against the amendment, and eight were absent or failed to vote. The desirability of self-chosen megadosage of vitamins was not the issue in the final vote; rather, it was "freedom of choice in the market place."

The fate of this bill led Oral L. Kline, secretary of AIN, to stress that nutritionists must redouble their efforts to teach people that concentrated nutrient preparations may be detrimental, since the constituents may not be properly balanced and excessive amounts of an ingredient that is healthful or necessary in small quantities may be harmful. To aid in informing the public, the National Nutrition Consortium published a paperback book entitled *Nutrition Labeling—How It Can Work for You.*

Professional organizations such as the American Dietetic Association are concerned about legislation not as lobbyists but as sources of information on legislative matters of interest to their members. An urgent need is to provide professional testimony to members and committees of Congress and other governmental groups that will assist them in formulating legislation and policies. The purposes of such testimony are to clarify

National Institutes of Health, Bethesda, Md.

Two rats of the same age show the importance of proteins; the smaller rat had a protein-deficient diet.

issues under consideration, provide current research results, specify the competencies of professionals for given tasks, and provide names of qualified persons to serve as referees and consultants.

Food supplies and population. Continuing concern over food supplies was indicated by the symposium on "World Food Needs and Resources" on the first day of the tenth International Congress of Nutrition, held in August 1975 at Kyoto, Japan. C. den Hartog, then president of the International Union of Nutritional Sciences, stressed that worldwide shortages can be solved only by cooperation among nations. A USDA Economic Research Service publication entitled *The World Food Situation and Prospects to 1985* concludes that the factors causing the current food shortage are transitory and can be corrected. Nonetheless, malnutrition will continue among the low-income groups in less developed countries, necessitating special national and international programs to help those most seriously threatened.

Charles Weitz of New York, a representative to the UN Food and Agriculture Organization, told U.S. farm leaders that current world food reserves are equivalent to only a 30-day supply and that 500 million people are in permanent hunger. With the world population growing by 273,000 a day, the situation can only get worse. Weitz stressed that countries unable to produce sufficient food to meet their needs must earn enough foreign exchange to enable them to import food from countries that can produce it.

News media refer to the "race between stork and plow" as today's critical issue in nutrition. In addressing this problem, agriculturists stress that, of the world land area, 90% has vegetation but only 11% is utilized as permanent cropland, while 52% is in pastures and meadows or covered with forests. Investments in reclamation and irrigation would make it possible to convert some pastures and meadows to cropland as climate permits.

Nutrition and the individual. A major concern within the nutrition community was exemplified by the theme of the tenth International Congress of Nutrition: "Through science and nutrition to human well-being." Although evaluation of the nutritional status of individuals and small groups had been dealt with at previous congresses, at this meeting more emphasis was placed on nutritional monitoring and surveillance to provide information for government health policy. For the first time several sessions were devoted to the social and cultural components of food acceptance, dietary practices, and nutrition education.

Repeatedly, malnutrition has been shown to occur even where available food supplies are adequate. People fail to select the food required for health. Each newborn individual must learn to accept food that will supply adequate nutrients for development and maintenance. Traditionally, the mother determined the family's eating patterns—it has even been suggested that food can be defined as that which the mother designates as edible. As women depart from their traditional role, however, they tend to delegate these duties to other people or to rely on devices and nutrient concentrates. The mother who gives her baby a vitamin-mineral supplement may feel less urgency about teaching her child to accept a variety of foods. Doctors have contributed to this false security by prescribing fortified formulas while neglecting to instruct mothers in proper infant and child feeding.

Over 55% of women think their major nutrition problem is overweight, which they combat by a faulty, unsuccessful "starve-stuff" schedule. Few families eat more than one meal together daily, even though most family members have six to seven food contacts. Trends are for families to spend 30 to 50% of their food budgets on food eaten outside the home, where choice is influenced by the availability of ready-to-eat items, by peer groups, and by advertising and clever merchandising. Of course, some facilities that provide

Courtesy, Swift & Co. and Estech Inc., Chicago

Nutrition information is provided by the "Food for Life" exhibit at Chicago's Museum of Science and Industry. Computer programs furnish personalized data to visitors through television monitors.

food away from home, such as day-care centers, school food services, and services providing meals for the elderly, are motivated to supply nutritionally balanced meals, often under government regulation.

The availability of prefabricated food and meals further affects nutrient intake. Such foods often furnish excessive or unbalanced amounts of nutrients. For example, a pre-cooked 6½-oz Salisbury steak with mushroom sauce contains 1,050 mg of sodium, according to data supplied by a reputable company. On the other hand, 6½ oz of cooked lean ground beef with natural gravy has 182 mg, when no sodium-containing seasonings such as salt or monosodium glutamate are added. Excessive use of prepared foods may result in daily intakes of over 9,000 mg of sodium even when no salt is added. Since the normal daily intake should be about 2,400 mg, these data imply that use of prepared food be limited, and the diet contain some plain garden and animal products.

Nutrition and disease. *Bone loss in the aged.* One of the major debilitating disorders of the aged is skeletal and alveolar bone loss, according to Anthony A. Albanese, director of the Nutrition and Metabolic Research Division, Burke Rehabilitation Center, White Plains, N.Y. About six million spontaneous fractures occurred in 1974 in persons over 45 years old. The Food and Nutrition Board of the National Research Council reaffirmed that the prevalence of osteoporosis in the elderly can be attributed to long-term or lifelong suboptimal intake of calcium and other nutrients. The 1974 Recommended Dietary Allowance of 800 mg of calcium per day is for the "normal healthy" aged person. However, recovery from bone loss requires intakes of about 1,100 mg per day.

Albanese's study demonstrated that subnormal bone density occurs in 10 to 15% of "healthy" people, starting in females at age 25. To achieve reversal of bone loss in aged females required 750 mg of calcium and 375 units of vitamin D_2 per day for 12 to 48 months. A nonsupplemented control group of females continued to experience a 2 to 3% bone loss annually. Albanese also found that weight-reducing diets cause a decrease not only in soft tissue but also in bone mass, with impaired bone formation during growth. When a previous deficit existed, diets and supplements prescribed for pregnant women did not prevent bone loss. Uncompensated calcium losses also occurred during breast feeding. These data point up the debilitating effects of marginal and deficient diets on health in later years.

Cancer. A summary of 93 references on "Nutrition, Diet and Cancer," published in *Dairy Council Digest* for September-October 1975 by the National Dairy Council, provides an excellent review of this complicated topic. The research is based primarily on results with experimental animals and has limited application to humans, but a few general concepts seem to be indicated. Excess calorie consumption resulting in obesity makes a person more susceptible, as does "high intake" of polyunsaturated fatty acids, high fat diets, and a few other nutrient anomalies. The traditional "adequate diet" containing all nutrients in the recommended amounts is assumed to be as anticarcinogenic as any, but it in no way assures freedom from the probability of a malignancy. Good nutritional status relates favorably to a patient's ability to survive the therapy used in cancer control. An adequate to high protein intake reduces tissue deterioration and negative nitrogen balances caused by surgery, radiation, and chemotherapy.

—Mina W. Lamb

Information sciences

The launching of new communications satellites with double the capacity of their predecessors, the introduction of increasingly sophisticated and also less expensive microprocessors, and the entry of the Arpanet computer network into its operational phase were among the major developments in the information sciences during the past year.

Computers

The rate of change in computer science and technology in the mid-1970s is no longer as great as in the 1960s, but in 1976 it was still too great to permit a complete annual review in an article of this size. This discussion will, therefore, focus on three important developments; namely, the rapid spread of computer technology, an area at the forefront of research in computer science, and a computer application that may affect almost everyone in the next few years.

Networks. A computer scientist arrives home in suburban Los Angeles. Before dinner he checks the mail, not by going to a mailbox but by sitting down at his computer terminal, typing a few characters, and then waiting while the terminal types (or displays) messages to him from colleagues throughout the world transmitted from similar terminals over telephone networks, satellites, and other communications facilities.

A colleague needs to contact a U.S. computer expert who is in Brazil. After failing to reach him by telephone because of problems with the international telephone system, he finally succeeds in doing so by sending a message over a computer network from a terminal in the United States to one in Brazil.

Both of these are actual incidents, the former occurring almost daily and the latter an actual incident of 1975. They were made possible by networks of computers, which have been in existence for several years but which in 1976 came into clearer focus in terms of their massive future impact on technology and society.

Most computer networks fall into one of the following classes: star networks, in which a number of remote terminals and small remote computers, usually in a relatively restricted geographical area, are all connected to a large central computer; small heterogeneous networks, which may contain a few large central computers of different types, usually in a relatively restricted geographic area, to which are connected a variety of terminals and smaller computers; and homogeneous networks, which may cover a wide geographical area and include many large central computers that are all the same or similar. The leading commercial and governmental networks fall generally in this third category.

But the most important, interesting, and significant network falls into none of the above three categories. This is Arpanet, developed by the Advanced Research Projects Agency (ARPA) of the U.S. Department of Defense (now DARPA, the Defense Advanced Research Projects Agency) in collaboration with a number of universities and research laboratories. Arpanet is probably much more representative of future computer networks than the three types described above. It is the most far-flung of all computer networks, extending from Hawaii to London, and also the most diverse,

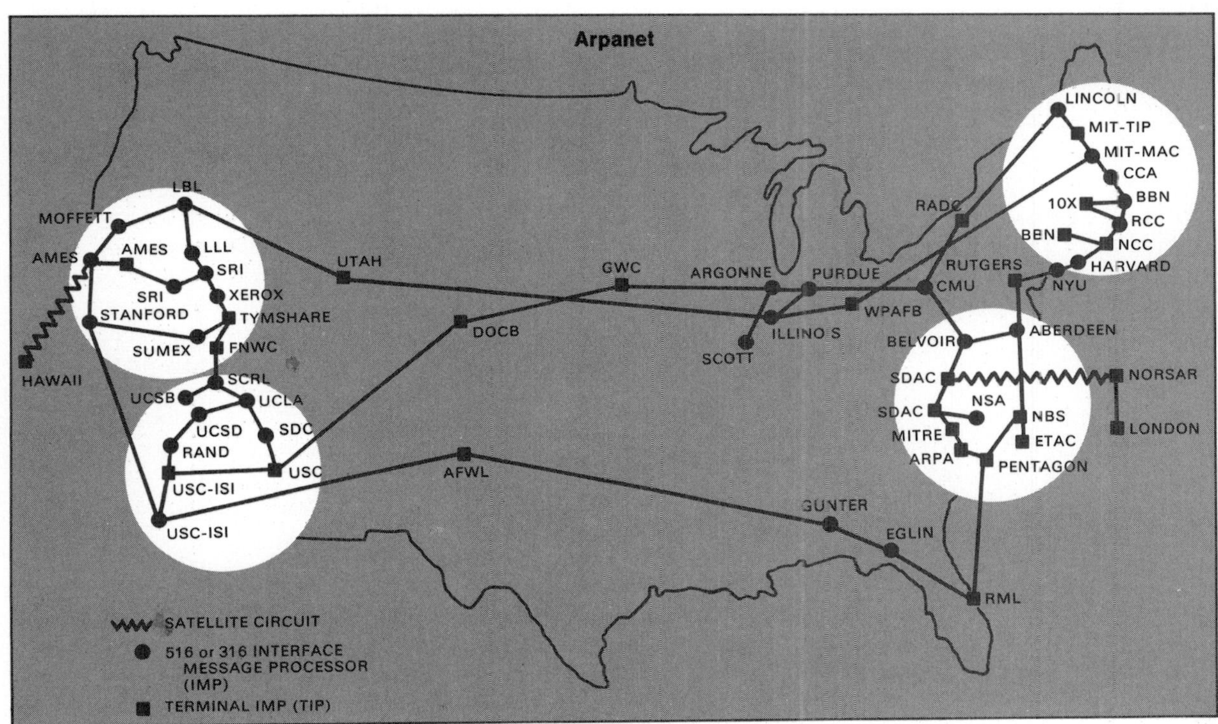

Kaku Kurita from Gamma/Liaison

Shopper in computerized Japanese supermarket orders the goods she wants by pressing buttons next to windows in which the merchandise is displayed.

containing an extensive array of major and different central computers. Both of the examples at the start of this section refer to the use of Arpanet.

Arpanet consists of several groups of components. These include host computers, which range from the largest and most powerful in existence (ILLIAC IV) to various minicomputers. In some instances, minicomputers or relatively small computers serve as "front ends" to larger computers with the smaller of the duo handling various communications and administrative tasks so that the larger can be kept busy processing data. Also part of Arpanet are the interface message processors (IMPs), which connect each host computer to the network and which are then connected to each other. Finally, there are the terminal IMPs (TIPs), which allow the connection of both host computers and smaller terminals to the network.

Arpanet pioneered resource sharing among computers in which not only programs and data but also special hardware (such as computer-controlled laboratory equipment) can be shared among many users. Particularly worth mentioning in this context was the development of software protocols that allow computers with different operating systems to communicate with each other.

But the most significant aspect of Arpanet is perhaps its communications facilities. IMPs and TIPs, which form the "nodes" of the network, are connected by 50 kilobit/second communication lines, with satellite communication used between Hawaii and California and between Norway and the eastern United States. Except for its satellite links the network is multiconnected in that there is more than one path between any two nodes. Therefore, a hardware failure at any one node puts only that node out of business.

The two main forms of communication in networks are circuit switching, in which a dedicated circuit (one used only for the transmission for which it was set up) is established from source to destination before the start of transmission, and message switching, whereby a message proceeds from node to node, as a circuit is available, until it arrives at its destination. Arpanet uses a form of message switching called packet switching, in which each message is subdivided into packets (of 1,024 bits each) that proceed from source to destination independently of each other by the shortest available path at the time of transmission. Because more than one path exists between each pair of nodes, different packets may follow different paths, depending upon the message traffic at the time each is sent; it is even possible for packets to arrive at their destination IMP in a different order from that in which they were sent. One function of the IMP is to reassemble the packets into their correct order before sending them to the destination host. Packet switching is one of the major contributions of Arpanet; in 1975 the first commercial packet switching services were instituted as a direct outgrowth of Arpanet technology.

In 1975 Arpanet passed from a research and development phase to an operational status when it was transferred from DARPA to the U.S. Defense Communications Agency (DCA). Because various aspects of the network remained subsidized, it was not clear as of 1976 whether the services provided by Arpanet are economically viable. But there can be no doubt that the decreasing cost of computer and communications technology will make networks such as Arpanet economical at some point in the not too distant future. Arpanet is undoubtedly the precursor of vast computer-communications networks which within 10–20 years will begin connecting individual homes to computer networks.

Computational complexity. Even before the existence of computers, mathematicians were interested in more efficient ways of solving problems. One of the most familiar examples of this concerns the evaluation of the polynomial

$$p_n(x) = a_nx^n + a_{n-1}x^{n-1} + \ldots + a_1x + a_0.$$

The "brute force" algorithm (method) requires one to compute $x^2, x^3, \ldots, x^n$ and then compute $p \leftarrow a_0$, $p \leftarrow a_i x^i + p$, and $i = 1, \ldots, n$. This method needs $2n - 1$ multiplications and n additions, and for that reason it is better to use Horner's method, which requires one to compute $p \leftarrow a_n$, $p \leftarrow px + a_{i-1}$, and $i = n, n-1, \ldots, 1$, which needs n multiplications and additions.

This example illustrates that there is usually more than one algorithm for solving a problem. Computational complexity is concerned with the inherent difficulty of problem solving. It attempts to find the best algorithm for the solution where "best" may mean fastest or requiring the least storage. At first glance it is paradoxical that in 1976 this should be one of the most active and rapidly growing areas of research in computer science, at a time when the speed of computers per unit cost is increasing rapidly and the cost of a unit of storage is decreasing rapidly. But an important effect of the increasing power of computers has been to allow consideration of problems that heretofore could not practically have been considered for computer solution. The more that computational horizons have widened the more important it has become to know what is possible.

A general but somewhat simplified formulation of the problem of computational complexity is as follows: Let A be the class of algorithms to solve a problem X and let $T(\alpha, X)$ be a measure of the time (or work or storage) required to solve X with algorithm $\alpha \epsilon A$. Such questions then arise as: For all $\alpha \epsilon A$ what is the minimum of $T(\alpha, X)$? Or what $\alpha \epsilon A$ achieves this minimum? The brief exposition below illustrates the ideas of computational complexity with two examples.

The first example deals with the evaluation of polynomials. Is Horner's method the best way to evaluate $p_n(x)$? If one has no a priori knowledge of the coefficients a_i and must write a computer program to evaluate $p_n(x)$ for any value of x and any set of coefficients, then Horner's method is best in that no smaller number of multiplications or additions is possible. But if the a_i are known and $p_n(x)$ must be computed for many values of x, then another method is better. For example, to evaluate any fourth-degree polynomial

$$p_4(x) = a_4x^4 + a_3x^3 + a_2x^2 + a_1x + a_0$$

one can precompute (compute once and for all, independent of x)

$$\alpha_0 = \tfrac{1}{2}(a_3/a_4 - 1) \qquad \beta = a_2/a_4 - \alpha_0(\alpha_0 + 1)$$
$$\alpha_1 = a_1/a_4 - \alpha_0\beta$$
$$\alpha_2 = \beta - 2\alpha_1$$
$$\alpha_3 = a_0/a_4 - \alpha_1(\alpha_1 + \alpha_2)$$
$$\alpha_4 = a_4$$

and then for any x compute

$$y = (x + \alpha_0)x + \alpha_1$$
$$p_4(x) = [(y + x + \alpha_2)y + \alpha_3]\alpha_4.$$

This method requires three multiplications and five additions, a slight but useful improvement because addition is faster than multiplication.

More generally, it is known that any method of evaluating a polynomial of degree n must require at least $\lfloor n/2 \rfloor + 1$ multiplications and n additions where $\lfloor \ldots \rfloor$ denotes "the integral part of," that is, dropping the fractional part so that $\lfloor 2.64 \rfloor = 2$, $\lfloor 47.235 \rfloor = 47$. The result above for fourth-degree polynomials is, therefore, the best possible for multiplications but not for additions. Indeed, it can be shown that no method for fourth-degree polynomials exists that requires both the minimum of three multiplications and the minimum of four additions. Many questions remain open concerning the best methods for polynomials of particular degrees.

A second problem is that of multiplying two n-digit numbers. The usual method requires n^2 elementary multiplications, that is, multiplications of one digit by another resulting in a digit and perhaps a carry to the column immediately to the left. For example,

```
       4,683
       9,274
    --------
       18732
      32781
      9366
    42147
    --------
    43430142
```

requires 16 elementary multiplications. Is it possible to do better? Surprisingly, perhaps, the answer is yes. Suppose u and v both have $2m$ digits. One can then write $u = 10^mU_1 + U_0$ and $v = 10^mV_1 + V_0$, where U_0, U_1, V_0, and V_1 are m-digit integers. Then $uv = (10^{2m} + 10^m)U_1V_1 + 10^m(U_1 - U_0)\ (V_0 - V_1) + (10^m + 1)U_0V_0$. This requires three multiplications of m-digit numbers plus some additions and shifts (multiplication by powers of 10). When this is applied to the example above with $u = 4{,}683$ and $v = 9{,}274$, the results are $U_1 = 46$, $U_0 = 83$, $V_1 = 92$, $V_0 = 74$, and $U_1V_1 = 46 \times 92 = 4{,}232$; $(U_1 - U_0)(V_0 - V_1) = -37 \times (-18) = 666$; and $U_0V_0 = 83 \times 74 = 6{,}142$.

From these results one can determine

```
uv = 42,320,000
        423,200
         66,600
        614,200
          6,142
     ----------
     43,430,142
```

This procedure used only 12 elementary multiplications. Moreover, this process can be applied recursively; for example, $46 \times 92 = (10^2 + 10)36 + 10 \times 14 + (10 + 1)12$. Therefore, whereas one would expect $T(n)$, the time required to multiply two n-digit numbers, to be proportional to n^2, it can be shown by using the above procedure that $T(n)$ can be made proportional to $n^{\log_2 3} \approx n^{1.57}$ in the sense that there exists a constant c such that $T(n) < cn^{\log_2 3}$.

Can the exponent of n be reduced as low as 1 so that multiplication of two n-digit numbers could be accomplished with elementary multiplications proportional to n, a procedure analogous to adding two n-digit numbers with n elementary additions? This cannot be done, but a recent result shows that there exists an algorithm such that for any $\epsilon > 0$, $T(n) < c(\epsilon)n^{1+\epsilon}$ and that no better inequality is possible. As might be expected, the constant c, which depends upon ϵ, gets very large as ϵ gets small so that only for large values of n is there any improvement over the standard multiplication method.

Although both of the above examples have practical applications (the latter to multiple-precision arithmetic algorithms), a more important aspect of the study of computational complexity is the insights it provides into the limits of computational algorithms and into the equivalencies that can be shown between the inherent complexities of—on the surface, at least—very different problems. For these reasons computational complexity was in the mid-1970s perhaps the most rapidly growing area of research in computer science.

Electronic funds transfer systems. The cashless society is not just around the corner, but in 1976 the first major steps toward it may be taken. Technological progress, particularly in the area of point-of-sale terminals and systems, made it possible to consider replacing paper transfer of funds with electronic transfers. The financial and social implications of so-called electronic funds transfer systems led in 1974 to the establishment in the U.S. of the National Commission on Electronic Funds Transfer.

Electronic funds transfer refers to all means by which financial transactions are handled electronically. Banks have, for example, used electronic means to handle large interbank transfers for years. But the term electronic funds transfer systems (EFTS) is normally used to refer to systems in which electronic means replace the traditional paper transactions by which individual customers interact with banks. In 1976 the EFTS on which most attention was focused was the use of customer bank communications terminals (CBCTs), located remotely from a bank, through which a bank customer may carry on most normal banking transactions. Such terminals may be operated by someone other than the customer, or they may be automated teller machines operated entirely by the customer.

The use of CBCTs raised various questions of public and social policy. One of these was whether such terminals are branch banks of the bank that places them. This question was expected to be settled by the courts in 1976 for federally regulated banks. The answer will have important implications for the future of banking in the U.S., particularly the relationship between the large, powerful banks, which might place CBCTs in many locations, and smaller, local banks. A second policy question is whether CBCTs portend a centralization of financial accounting and an increase of record keeping on personal financial transactions that would further decrease personal privacy. Related to this question are auxiliary issues concerning the possible need for governmental in-

Customer bank communications terminal (CBCT) allows many banking transactions to be carried out at a location away from the bank.

Courtesy, Continental Illinois National Bank and Trust Company of Chicago volvement in EFTS and the potential dangers of such involvement.

The advent of CBCTs suggests the possible proliferation of EFTS services and functions, which in turn raises new questions. For example, the system in a large retail chain such as Sears, Roebuck and Co., in which point-of-sale terminals in retail outlets are connected to the retailer's central computer for accounting and credit-checking purposes, could be extended via computer-communications networks to debit automatically the customer's bank account and transfer the funds to the retailer. How safe would such systems be from tampering? What might be the effect of hardware failures? Would knowledge of such individual transactions allow surveillance of individual movement and activities that would constitute an invasion of privacy and a potential instrument of social control?

These are some of the questions and issues concerning EFTS that the National Commission was expected to study. The legislation establishing the commission specified 12 of its 26 members; the other 14 presidentially appointed representatives were not named until October 1975. Therefore, the report of the commission, originally scheduled for late 1976, was postponed until 1977. Unfortunately, the commission contained no member with specific expertise in computers and data processing, despite the fact that the policy issues raised by EFTS are inextricably intertwined with information processing technology.

—Anthony Ralston

Communications systems

Leading developments in communications systems during the past year included the introduction of new models of microprocessors and the application of these devices to scientific and engineering instruments, the demonstration of a branch telephone exchange containing a miniaturized computer with its own memory, and the testing of a new underground transmission system that uses nitrogen in a hollow steel pipe.

Microprocessors. One of the most exciting new products to enter the industrial market is the microprocessor. It is the result of a series of technological advances that began with the introduction of the transistor and culminated in large-scale integrated circuits. Using this technology, engineers and scientists are capable of fabricating complex circuits consisting of thousands of transistors and components of a highly specialized nature on a thin silicon wafer that occupies an area typically one-quarter-inch square. The result is "intelligence" on a silicon chip. Despite the complexity of this device the number of new models introduced in 1975 equaled those produced in previous years so that there were almost 40 different microprocessors from which to choose.

Courtesy, Fairchild Camera and Instrument Corporation

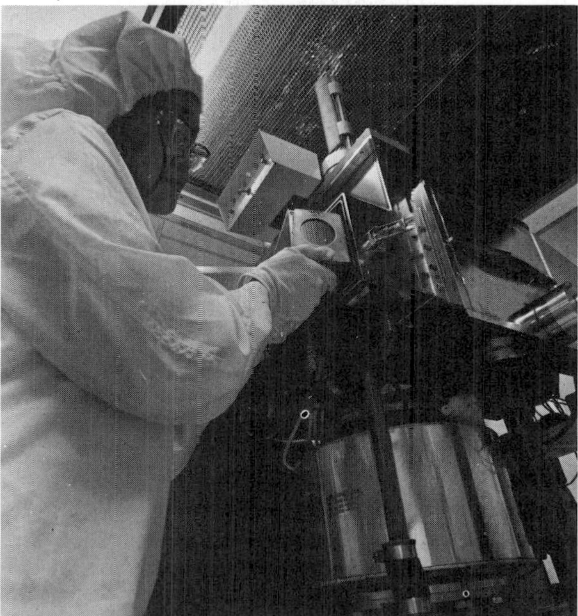

Silicon wafers are placed in an ion implantation machine during fabrication of large-scale integrated circuits used in microprocessor memories.

It is rare that a technological advance produces an order-of-magnitude change upon the culture in which it is introduced. (An order of magnitude is a tenfold increase.) The printing press produced such a change, and devices such as the vacuum tube and the transistor resulted in substantial changes in living patterns. Of equal significance for the present day is the microprocessor.

A microprocessor is equivalent to the central processing unit of a computer. It is in this unit that decisions are made. Of course, these decisions are predetermined by the designer, who ultimately places the constraints on the system.

Two other elements are required to make up a microprocessing system. The first of these are the memories; they consist of the permanent type, containing unalterable instructions, and the volatile type for storing data. These memories are fabricated by using large-scale integration techniques. The second element required to make up the microprocessor system is the capability to communicate with the outside world. This function is performed by the input-output devices.

Because a microprocessor is a central processor it has a repertoire of instructions it can perform. A program for a microprocessor is written using sequential logic, and execution is controlled by a regulated clock that keeps events occurring in the proper relationship. The first order is a "fetch," and it is followed by an "execute." The instruction may be arithmetic, such as addition or subtraction in a register called the accumulator; logical, such as comparing the accumulator with another register; or rotational, during which data are shifted.

There are two major areas of application for the microprocessor. The first is the microcomputer, a small device with great ability to perform complex operations. One major introduction into the microcomputer market in 1975 was from MITS, Inc., which offered the Altair 8800. This was one of the first microcomputers to be available to the general public with high-level programming languages at an affordable price.

The other major area of use for the microprocessor is that of "dedicated control," and the range of applications is almost without limit. One of the first control devices using the microprocessor was the automatic traffic light. Many considerations must be included in the design of a traffic-light control system such as local and state jurisdiction, traffic patterns and densities, intersection layout, time of day, and climatic conditions. Under microprocessor control it is possible to sense traffic conditions, including weather, and adjust signal lights to obtain the optimum flow of traffic. Because the system is intelligent the optimization is done on a continuous basis. It is also possible to monitor the system to find out if all the sensors and displays are operating properly and to notify a central agency when a malfunction occurs.

Other dedicated control applications include automated test equipment, machine control, medical systems, radio navigation equipment, telecommunications, automatic time clock and payroll systems, point-of-sale terminals, inventory control, crime information networks, and entertainment devices. Automotive applications are expected to demand a considerable share of the microprocessor market for use in improving engine efficiency, controlling emissions, and for non-skid braking.

In scientific instrumentation one of the most useful tools is the oscilloscope. Instruments introduced during the year contained microprocessors and thus have the ability to determine peak-to-peak amplitude, frequency, phase, power, and signal average along with calculator functions.

When microprocessors are incorporated into digital instruments, the meter can select the position that provides the most significant digits. True power values, voltages, currents, resistances, and ratios can be measured, and at the same time the meter can perform calibrations on itself so as to indicate to the user that it is operating correctly and producing the best possible output. With a microprocessor it is possible to satisfy a customer's special problem by replacing one of the system's memory circuits with one suited to his particular needs. Outputs are not only visual but can be structured so that they are directly available for processing under control of a central computer.

Telephones. American Telephone and Telegraph Co. (AT&T) continued to demonstrate Dimension PBX, a private branch exchange that featured a miniaturized computer with its own memory. The exchange could provide automatic

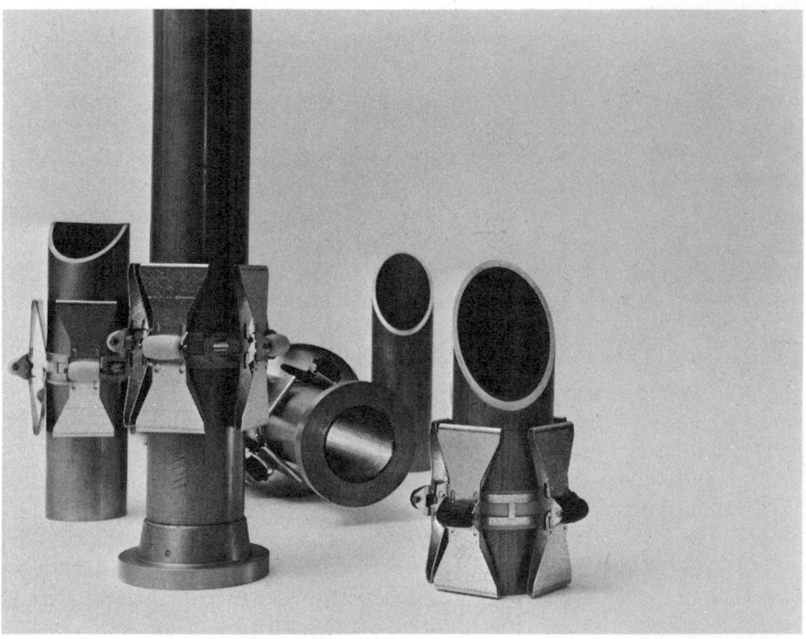

Hollow steel pipe lined with copper and polyethylene was tested by the Bell System for use in an underground telephone communications network.

Courtesy, Bell Laboratories

Courtesy, Motorola Inc. Semiconductor Group, Phoenix, Ariz.

Plastic sheets with microcomputer circuits imprinted in various colors are used to check the accuracy of a process in which the circuits are photoengraved on a silicon chip.

intraoffice calling when two lines are free, automatic route selection to achieve placement of calls at the most economic rate, and a three-way conference transfer that would permit privacy between two parties of a conference call.

Scientists at the Bell System during the year tested a 2.5-in.-diameter hollow steel pipe lined with copper and polyethylene and filled with nitrogen, to be used in an underground transmission system. They believed that it would be capable of carrying up to 230,000 telephone conversations at one time, more than twice the capacity of the largest buried coaxial cable used by Bell in 1975.

In April 1975 the General Telephone and Electronics Corp. obtained a $500 million contract to update and expand the telephone system of Iran. The project included installation of high-speed, computer-controlled switching systems in more than 500 telephone exchanges throughout the nation; these systems were expected to provide Iran with an additional 950,000 telephone lines.

Cables. The last section of an undersea cable from Sydney, Australia, to Auckland, N.Z., was completed in September. The cable provided 640 telephone links between the two countries.

Nine firms from various parts of the world announced in August 1975 that they planned to cooperate in building two undersea cables connecting the U.S. with the Caribbean and South America. One was to provide 640 telephone links between Venezuela and the Virgin Islands, and the other, 3,000 links between the Virgin Islands and the U.S. Both cables were scheduled to be completed in 1977.

—John Paul Froehlich

Satellite systems

Applications satellites are those Earth-orbiting spacecraft which use their unique vantage point in space for a variety of services of benefit to man. The great majority of such satellites are launched by the United States and the Soviet Union because of their developed space boosters. However, France, West Germany, Japan, Great Britain, Canada, and China have launched precursors of national or regional operational systems. The U.S. National Aeronautics and Space Administration (NASA) has launched some of these on a cost-reimbursable basis.

Communications satellites. The year 1975 marked the 15th anniversary of the first, passive, communications satellite, Echo. During those 15 years many active repeater satellites that conduct global communications were placed into orbit around the Earth. At the orbital altitude of 22,350 mi (36,000 km) over the Equator the angular velocity of a satellite matches that of the surface of the Earth. Thus, such satellites remain fixed, relatively, above the same point over the Earth's surface, and are described as synchronous.

In 1975 the International Telecommunications Satellite Consortium (Intelsat), comprised of approximately 90 nations, launched the last of the series of Intelsat 4 satellites (May 22) and the first of the advanced 4A satellites (September 25). Weighing about 3,300 lb (1,500 kg), each Intelsat 4A has an average assigned use of 6,250 circuits and 2 color television channels, or it may have various combinations of facsimile, telegraph, or data transmission. This is nearly double the capacity of an Intelsat 4.

This launching brought the total of operational transoceanic communications satellites of Intelsat to eight; four were over the Atlantic, and two each were over the Pacific and Indian oceans. From 1964 to 1974 transoceanic telephone traffic grew from 3 million calls to more than 50 million, a steady increase of 20% annually. During the same period the cost of such telephone service was reduced more than 50%, and quality was improved.

In August the U.S. applications technology satellite 6 (ATS 6) began a year-long program for the Indian government by transmitting instructional TV to thousands of remote villages, each having a communal color television set. The programs, developed by India, included family planning,

NOVOSTI from Sovfoto

TASS from Sovfoto

Orbita communications system of the Soviet Union comprises ground stations such as that at the left, which receive and relay television signals from Molniya 1 (right) and other satellites.

health and hygiene, child education, and agricultural and other vocational information. For the large majority of the audience of millions, the impact was great because none had ever seen a television program. One effect was a substantial increase in attendance by children at school.

In December 1975 the Soviet Union filed plans to establish a global network of seven synchronous Statsionar communications satellites in the period 1978–80. This plan, if effected, would duplicate the coverage of Intelsat. Three earlier-launched Statsionar satellites would provide a new communications network for the U.S.S.R. As of 1976 a system of multiple non-synchronous satellites, known as Orbita, provided national coverage.

The past year was one of much activity in the U.S. in the field of national communications satellites. A ruling of the Federal Communications Commission (FCC) in December opened the way for several giant communications firms to launch communications satellite systems for domestic use. The two Western Union satellites (Westar) launched in 1974 continued to function successfully. Monthly lease of a private-line hookup between Los Angeles and New York was offered by Western Union at less than half the cost of similar service by conventional land link via ground cable or microwave transmission.

In December RCA Corp. launched Satcom 1, first of a series of three 24-channel satellites planned by 1977. Significant was the relatively small size of Satcom, even though it had double the capacity of Westar. This permitted the use of an uprated Delta launch vehicle rather than the much more costly Atlas-Centaur used for the Intelsat 4 satellites.

The largest U.S. utility, AT&T, was at work on developing a joint satellite communications system with the country's second largest utility, General Telephone and Electronics Corp. Comsat General Corp., a subsidiary of Comsat (Communications Satellite Corp.), which conducts operations for Intelsat, contracted with NASA for the launch of two Comstar satellites in 1976. They were to be leased to AT&T.

International regional communications satellite systems moved ahead with the second successful launch of a Franco-West German Symphonie satellite. The European Space Agency (ESA), comprised of 11 European member nations, participated in the programs undertaken by the Canadian/NASA communications technology satellite (CTS). Launched January 17, the CTS had many innovations. Although of much smaller volume at launch than the ATS 6, it carried a 30 ft (9 m) reflector antenna and a powerful (200 w) transmitter. This permitted reception of television signals using relatively small ground antennas.

In 1976 Comsat planned to launch two communications satellites for specialized maritime use. Designated Marisat, one each would be positioned over the Atlantic and Pacific oceans. Both the U.S. Navy and commercial maritime shipping firms were expected to make heavy use of this specialized communications network. Another potentially valuable satellite system is known as Aerosat. Comsat and ESA were jointly exploring such a satellite system, which would serve aircraft flying heavily traveled transoceanic routes.

Earth observation satellites. This category of satellites has three major forms: weather, Earth resources, and military reconnaissance.

Weather satellites. As with communications satellites, the first weather (or meteorological) satellite, Tiros 1, was launched in 1960. By contrast, the satellites of 1975 and 1976 were capable of much improved televised scanning of the Earth's surface and cloud cover, and could also produce cloud photographs at night, cloud temperatures and altitudes, vertical soundings of the temperature of the atmosphere, measurements of rainfall over oceans, soil moisture and temperatures, and ice conditions over polar regions. Collection and transmission of data from unattended buoys and remote ground stations are some of the routine tasks performed by weather satellites circling the Earth in polar orbits.

Two synchronous meteorological satellites, SMS 1 and 2, launched in 1974–75, monitored U.S. cloud cover on a 24-hour basis. In geostationary orbit, 60° apart, their information on cloud pattern movement across the U.S. is used regularly on evening television news broadcasts.

The SMS satellites are forerunners of a global weather observation network. The first operational geostationary operational environmental satellite, GOES A, was launched in October 1975. GOES A, with a lifetime of five years, replaced SMS 1, which went into standby status. By 1979 GOES B and C will have been launched to join one similar satellite each over Japan, the U.S.S.R., and Europe in an international cooperative program. Each GOES will have the capability of collecting meteorological data from thousands of unmanned data collection platforms daily. The collected data will be relayed to ground stations for correlation, computer analysis, and production of more accurate weather maps and forecasts.

Earth resources satellites. The Earth resources technology satellite Landsat 2 was launched in January 1975 to join Landsat 1 (formerly ERTS 1). Although the lifetime expectancy of Landsat 1 operation had been only one year, it was still functioning in 1976 after more than three years. These Earth survey satellites in near-polar orbit record and transmit to ground stations pictures of every spot on the Earth's surface between 82° north and south latitudes. With the arrival of Landsat 2, separated from Landsat 1 by 180° in orbital position,

Courtesy, NASA

Satcom 1 is launched for the RCA Corp. by a Delta booster rocket (above). Final adjustments are made to the first U.S. geostationary operational environmental satellite, GOES A (below).

Courtesy, Aeronutronic Ford Corp., Newport Beach, Calif.

Courtesy, EROS Data Center, U.S. Geological Survey, Sioux Falls, South Dakota

Coast of California from just north of San Francisco to the southern end of Monterey Bay is clearly revealed in photograph taken from a Landsat satellite approximately 560 miles above the Earth's surface.

the time between repeat flyovers of the same area was halved from 18 days to 9 days.

Photography from the Landsat satellites is available to all countries from the U.S. Geological Survey. The multispectral (four-color) photographs proved immensely valuable for aiding nations to explore the Earth's limited natural resources and to detect threats to the environment.

Among the applications of the data provided by the Landsats are measurements of crop acreage, detection of blight, and estimation of yield; mapping of mountain snow to forecast spring runoff for irrigation, water supply, and generation of hydroelectric power; detection of sewage and industrial pollution of river and offshore waters; monitoring of environmental effects of strip mining; mapping of floods to evaluate damage and plan relief; surveying of forest and rangeland resources; detection of potential earthquake zones as an aid to planning future construction; and determination of likely locations for deposits of oil and minerals.

Military reconnaissance satellites. As of 1976 the U.S. and the U.S.S.R. were the only nations with both the technological capability and wealth to utilize satellites for photographic and electronic monitoring. Such programs have been conducted by both nations since about 1959. In fact, reconnaissance is the oldest of the applications satellites. Although there has never been official disclosure of such programs by either country, the use of reconnaissance satellites continues by tacit agreement. Such satellites are credited with being the main technological underpinning in nuclear-arms control under the terms of the 1972 strategic arms limitation talks (SALT).

—F. C. Durant III

Information systems

Statesmen and planners require specialized information to help them cope with national and international crises and with the normal day-to-day needs of our complex society. To meet these needs, new services are being created and research continues to explore better ways of collecting, organizing, and disseminating information. The goal of information systems can be described as providing the right information to the right people at the right time.

Yet, a word of caution is in order, for as computer-based record-keeping systems become ever more common there is a concomitant need to safeguard the privacy of the individuals whose names and records may be included in them. The U.S. National Bureau of Standards (NBS) is responsible for devising standards of data security, protecting the rights of individuals, and guarding against the misuse of such data. Periodically, the NBS sponsors meetings where groups can meet, share views, and make recommendations concerning needed legislative and administrative procedures for safeguarding data confidentiality. While much more research and education is needed before these difficult problems can be solved, four fundamental principles have evolved for protecting the privacy of individuals. These are: (1) individuals should have access to information about themselves in record-keeping systems, and there should be some procedure for finding out how and by whom this information is being used; (2) there should be some way for an individual to correct or amend an inaccurate record; (3) an individual should be able to prevent information from being improperly disclosed or used for other than authorized purposes without his or her consent, unless required by law; and (4) the custodian of data files containing sensitive information must take reasonable precautions to ensure that the data are reliable and not misused.

International information systems. To encourage research in the less developed countries and to ensure effective dissemination of available information about development, an International Development Research Center was established. The director of this agency, John Woolston of Canada, proposed that an international information system for economic and social aspects of development be established. The system would have the following objectives: to provide improved access to economic and social information for individuals and institutions in less developed countries and to all those involved in the formulation and implementation of development activities; to foster the building of national and international resources needed to meet this goal; and to improve coordination among development information facilities. The center was to be a voluntary organization that would function within the UN.

UNESCO continued to support the organization of science information services in countries throughout the world. Financial aid was given to Bulgaria, Colombia, Guinea, India, and Sudan in order to enable them to establish national information systems (Natis) and/or to improve existing services. At the Bulgarian Automated Center for Scientific and Technical Information in Sofia, UNESCO funds were to be used for computerized cataloging of Bulgarian works and translations of foreign material into that language. Colombia was to continue the development of its national information system by linking existing documentation centers with university and special library collections. On the African continent the projects in Guinea were to concentrate on acquiring books and reference materials, while Sudan was expected to begin establishing a National Documentation Center for Science and Technology as the central unit of its Natis.

The Food and Agriculture Organization of the UN was asked by the World Food Conference to plan for the establishment of a global "early warning system" on food and agriculture. Participating governments would voluntarily furnish current information and forecasts on the production status of basic food products. It would then be possible to monitor the world food supply and to identify countries and regions where there is likely to be an imminent danger of food shortages and malnutrition. Information supplied by the early warning system would alert the governments involved and thereby enable them to take appropriate measures in order to reduce the harmful effects of agricultural shortages.

An International Oil Marketing Information System was being developed and operated on a permanent basis by the secretariat of the Organization for Economic Cooperation and Development. The system was designed to monitor fluctuations in the international oil market and the activities of oil companies. Additionally, in times of emergency, the system would prepare measures for allocating oil during a shortage. The first of its two data banks was to contain general information relating to oil companies, such as their organizational structure, finances, sources of crude oil, rates of production, and allocation policies. The second data bank, concerned with emergency measures, would include information on oil reserves, transportation facilities, and projected levels of international supply and demand. All information in the system would be confidential so as not to harm competitive interests, but the information would be available to alleviate the harmful

Courtesy, British Post Office

Troposcatter antennas near Aberdeen, Scotland, permit communication with oil platforms in the North Sea by bouncing radio waves over the horizon, using the scattering of the waves that takes place in the troposphere 1–2 kilometers (0.6–1.2 miles) above the Earth.

effects of precipitous reductions in oil supply.

Information on the environment was being exchanged during the year by means of an International Referral System (IRS) developed by the UN Environmental Program. IRS operated through a decentralized network of regional focal points, each of which was responsible for compiling a directory of potential sources of environmental information together with details that describe the fields of activity and capabilities of those sources. This information is compiled into a *National Directory of Environmental Sources*, available at all local focal points. When a question is received, the *National Directory* is consulted first; if one or more appropriate sources are found, these are provided to the questioner, who may contact them directly. If there is no adequate local source, the IRS *International Directory* is consulted and the user notified of these sources.

National information systems and services. The U.S. spent approximately $11.8 billion in 1975 on scientific and technical information activities, according to figures released by the National Science Foundation. The figures covered all costs associated with identifiable scientific and technical information processes. These included costs for preparation and recording of information by researchers, up to and including manuscript preparation, $2.4 billion; distribution of information, including all functions associated with publication, distribution, storage, and retrieval of information, $6.1 billion; and assimilation of information by scientists and engineers, mainly salary costs attributable to browsing, searching for, and reading information, $3.3 billion.

Although the expenditures were large, they were not keeping pace with the growth rates for the funding of scientific research and development projects. In order to continue providing an adequate level of information support, information scientists must make the dissemination of information more efficient through the use of improved technology and the development of more effective information retrieval systems. Progress and the solution of critical national problems, such as energy, medical services, education, and safety, depend to a considerable degree on the availability and free flow of information.

A computer program called HELPS (Handicapped Education Learner's Planning System) was developed in Ohio to provide teachers of handicapped students with comprehensive information on various resources that can be used in designing individualized educational programs for those with learning problems. In a demonstration of the system, a teacher seated at a typewriter-like computer terminal typed a request for educational materials about job training skills suitable for 13-year-old students with a mental age of 9 and a second- to third-grade reading ability. In response, the computer was able to provide a listing of teaching materials, classroom activities, performance measures, and references to books and filmstrips. To retrieve this information, the teacher spent about 15 minutes at the console conducting the search and used approximately 30 seconds of computer time.

The city of Boston offered the public a free health information service called Tel-Med. Anyone needing health information could dial the Tel-Med

Courtesy, Department of Health and Hospitals, Boston

Tel-Med operator in Boston receives calls from people wanting health information. She selects the relevant tape cartridge from among those stored in the case at the rear and inserts it into a playback unit.

number and ask the operator to play a tape relevant to the problem area. The operator inserts the proper cartridge into the Tel-Med playback unit, and the tape begins automatically. When the message is ended, the tape ejects itself and disconnects the phone line. In 1976 the Tel-Med tape library contained almost 200 recorded health messages, 3–7 minutes long, many of which were available in both English and Spanish. The facility was receiving thousands of phone calls requesting health information.

Information research and future developments. Most existing information retrieval systems do not retrieve information per se; they retrieve bibliographic citations and possibly abstracts. In 1976 a research project was under way at Lehigh University, Bethlehem, Pa., to explore and evaluate the underlying technology required for enabling a user to obtain specific answers to a question concerning a research report. In the proposed fact retrieval system, for example, the prospective user should be able to ask for and obtain descriptions of hypotheses that were tested, instrumentation that was used, and the results that were obtained.

Before such a true information retrieval system can become operational many problems will have to be solved. It will be necessary to design computer programs that will analyze questions and determine the kind of output (or answer) that will be most appropriate; organize heterogeneous data sources into a suitable file structure; retrieve and evaluate passages from text; and assemble the retrieved pieces of information into an appropriate form for computer output.

Another approach to improving information systems was being investigated at the Massachusetts Institute of Technology. Researchers there were studying ways of "humanizing" computer-aided instruction systems, which they decided to achieve by having the computer deliver its responses in good-quality synthetic speech rather than in the form of a printed page. This project included the building of a prototype system that will convert text input to digitally stored speech that can then be produced in the form of messages that sound natural.

—Harold Borko

Life sciences

During the year, investigators in the life sciences uncovered valuable information on the mechanisms by which disease-producing bacteria adhere to the surfaces of human tissue and suggested a new diagnostic technique that can determine genetic deficiencies in unborn fetuses. Other significant events included the development of improved methods for ascertaining the exact sequence of nucleotides that comprise a specific gene, information concerning the reproductive system of the coelacanth, a primitive lobe-finned fish, and a unique insight into the humanness of Neanderthal man.

Botany

Botanists continued to study plants on all levels of organization. For example, much research was done on the cellular level to uncover information

Courtesy, Argonne National Laboratory, Argonne, Ill.

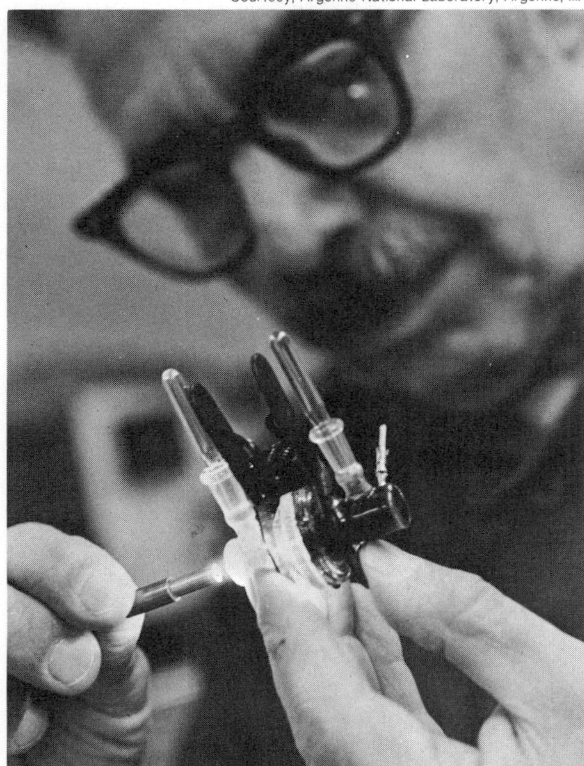

"Synthetic leaf" developed by scientists at Argonne National Laboratory employs a plastic membrane impregnated with a chlorophyll-water adduct to mimic the photosynthetic processes of green plants. When connected to electronic monitoring circuitry and exposed to light, the glass-encased cell generates voltages and currents with an efficiency comparable to the best organic photovoltaic cells known.

about the structure and functions of plant cells. On the organismic and populational levels, plant-animal relationships were investigated and forestry problems brought closer to solution. Extinct plants offered some insights into Neanderthal culture, and endangered plant species became a subject of national debate.

Photoperiodism, phytochromes, and red light. Certain developmental processes, including flowering, of many plants are known to be photoperiodic; that is, they seem to depend upon a certain day length for inducement. Some plants are short-day plants, blooming in spring or fall in temperate regions, and others are long-day plants, blooming in summer. Under certain conditions, the use of artificial lighting will induce a long-day plant to bloom like a short-day plant. If a long-day plant is illuminated in the middle of a long night, it will bloom even when the days are short. Monochromatic light from the red part of the visible-light spectrum will produce the same effect.

The results of such experiments indicate that it is actually the length of night that determines flowering. Short-day plants are more correctly long-night plants; if the night is broken by an interval of light, these plants will not respond normally. On the other hand, when a long-day plant is grown during a short-day period with its long night, the plant can be induced to flower if the long night is broken by a few minutes of light.

Such effects of light, and especially of its red component, on flowering and on such phenomena as seed germination inspired plant physiologists to look for the responsible mechanism. Some years ago a pigment called phytochrome was discovered that is sensitive to red light and, in some manner, is involved in photoperiodic plant responses. Although investigators subsequently have characterized the chemical changes in phytochrome that occur during exposure to red light, they are only on the verge of understanding why these changes should promote plant growth.

In general terms, periods of darkness are necessary to sensitize phytochrome to red light. Although there is no red light at times of darkness, there is far-red light, which is not part of visible light. Far-red light is able to change phytochrome to its red-sensitive form at night while there is virtually no red light to counteract the process. During the day, red light changes phytochrome to its far-red-sensitive form, thus making the process photoreversible. Current research efforts are attempting to uncover the mechanism by which red-light-stimulated phytochrome actually promotes growth during flowering and seed germination.

In one series of investigations, Richard W. Jones and Robert W. Sheard at the University of Guelph, Ont., showed that phytochrome is involved in the production of nitrate reductase, an enzyme important to the synthesis of amino acids and, therefore, of proteins. Because protein synthesis is also necessary for growth, it is possible that phytochrome acts at least through this enzymatic route to elicit some of the observed growth responses in plants. The specific chemical relationship of phytochrome to nitrate reductase, however, is not clear. Verna R. Lawson and Robert L. Weintraub, two researchers at George Washington University, Washington, D.C., studied the effects of red light on wheat and barley coleoptiles, the sheathlike structures covering the first shoot of grass seedlings, and proposed a mechanism by which phytochrome actually enters the growth process. Red light produces a small change in the phytochrome that enables it to combine with a receptor site of an enzyme; this allows the enzyme to initiate a synthetic pathway resulting in the assembly of microtubules in coleoptile cells. As cells grow,

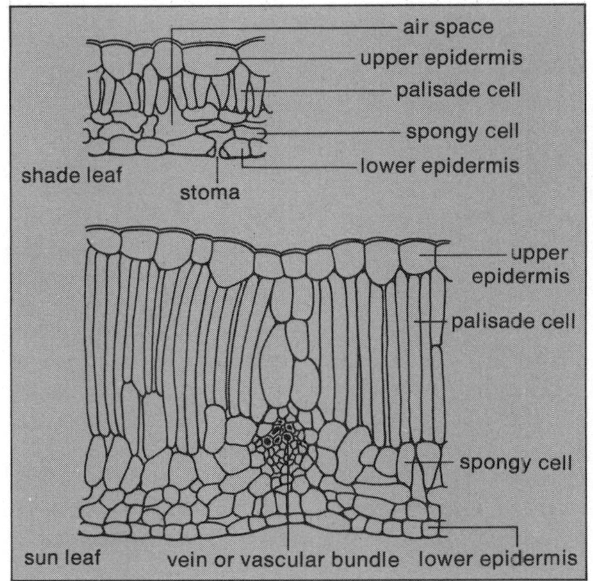

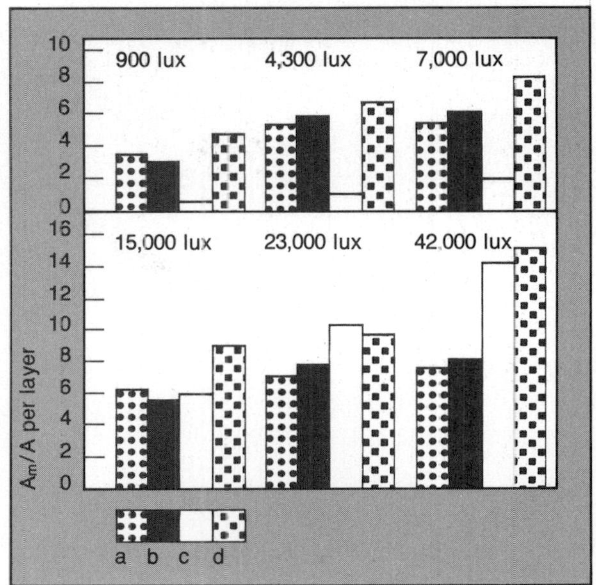

(Left) Cross sections of shade and leaves of Norway maple, Acer platanoides. *Graph (right) depicts contributions of individual layers of palisade and spongy mesophyll to the total surface area of mesophyll per unit leaf area (A_m/A) of* Plectranthus parviflorus *grown under six different levels of illumination. Bars denote contributions of (a) uppermost palisade layer, (b) next palisade layer, (c) lower palisade layers, and (d) spongy mesophyll.*

microtubules are assembled, which in turn regulate the pattern of microfibril orientation necessary for expansion of the cell wall.

Leaf size and sunlight. It is an easily verified observation that tree leaves growing in the shade (shade leaves) tend to be broader and thinner than those growing in more direct and abundant sunlight (sun leaves). Because the difference may be seen on the same plant, the phenomenon must be considered a response to environmental causes rather than to genetic ones. When leaves begin to sprout from bursting buds, the process of growth involves cell enlargement and specialization rather than multiplication, primarily because most of the cells have been produced as part of the bud formation. Therefore, exposure to light exerts its influence on the size and thickness of leaves after cell formation and during the period of cell specialization.

Microscopic examination of leaf cross sections reveals a structure comprising four tissue layers. Two of these are the upper and lower epidermis, each one cell thick. The lower epidermis usually bears stomata, which open and close to regulate gas exchange. The other two layers, called mesophyll, are the tissues that are modified by environmental conditions during their formation. In a shade leaf, the upper portion of mesophyll may consist of one palisade layer, characterized by rectangular cells packed together in an upright position and occupied by numerous chloroplasts. The lower portion of mesophyll, called spongy mesophyll, consists of rounder, loosely arranged cells that leave large air spaces connecting with stomata; these cells also contain fewer chloroplasts than palisade cells. This arrangement maximizes for each shade leaf the interception and utilization of light. In sun leaves, both palisade and spongy layers are thicker. The palisade cells either form multiple rows or grow longer and remain fairly tightly packed. The spongy cells form more but smaller air spaces, an arrangement that produces an increase in the surface area of spongy cells in contact with air.

Park S. Nobel and others at the University of California, Los Angeles, established that sun leaves have a higher photosynthetic rate than shade leaves, the difference being related to the surface-area exposure of the spongy mesophyll. Enlarging the surface area facilitates a greater carbon dioxide flow, necessary for increased photosynthesis.

Plant-animal interrelationships. Each passing year of scientific investigation brings more reports of specific ways in which plants and animals interact. For example, recent observations indicated that the reproductive behavior of certain animals can be affected by plant products in their diets. In one controversial study, Lawrence B. Hendry and his co-workers at Pennsylvania State University

(Left) C. E. Mohr from National Audubon Society, (below) Devereux Butcher

Surviving only in cultivated gardens since the early 1800s, the franklinia, a native U.S. tree (flower shown below), is officially listed as extinct by the Smithsonian Institution. The royal palm (right) is a common sight along streets and roads in southern Florida, but endangered wild groves of the species are protected in state and national parks.

reported evidence that certain chemicals originating in oak leaves eaten by larvae of the oak leaf roller moth persist in the female until adulthood, at which time they are employed as pheromones or sex attractants. Adult males recognize only these specific chemicals because, as larvae, they also consumed oak leaves containing the same chemicals and were impressed with the information necessary to discriminate among a variety of chemical attractants secreted by female insects with differing diets. These observations appeared to contradict the widely accepted idea that individual pheromone systems remain unique to each species because they are synthesized by the female, regardless of diet.

Hendry's findings, however, were challenged by Wendell L. Roelofs and colleagues of the State University of New York, who failed to confirm the presence of certain oak-leaf chemicals that Hendry had reported in the oak leaf roller moth pheromone system. In addition, when Roelofs' team raised a related species of moth on a variety of synthetic diets, the moths showed no change in the identities or ratios of components in their distinctive pheromone system.

Another report of the influence of plant chemicals on animal reproduction was presented by a group of California scientists studying the California quail, *Lophortyx californicus*. They noticed that, during a relatively dry year (1972) when vegetation was scant, quail reproduction was very low. During a year with generous rainfall (1973) when vegetation was abundant, quail reproduction was vigorous. Examination of the vegetation produced in each year showed that the relative abundance of phytoestrogens was much higher in the sparse plants of 1972 than in the abundant plants of 1973. Phytoestrogens are plant hormones, *e.g.*, formononetin, biochanin A, and daidzein, that seem to inhibit reproduction. It was assumed that quail, eating the sparse vegetation, experienced reduced reproduction. Conversely, quail that ate the more luxuriant plants consumed smaller amounts of phytoestrogens and did not experience such inhibition. This relationship between plants and quail reproduction can be interpreted in terms of a reproductive strategy. Only when there is abundant food will many young be produced.

Tussock moths and DDT. In early 1974 hearings were held by the U.S. Environmental Protection Agency (EPA) to determine whether a widespread application of the banned pesticide DDT should be used to curb an epidemic of the Douglas-fir tussock moth, *Orgyia pseudotsugata*, in the forests of the northwestern U.S. In the three years preceding the hearings, the moth, whose larvae consume the needles and buds of Douglas fir and true firs, had defoliated at least 800,000 ac of forest in Washington, Oregon, and Idaho; the U.S. Forest Service predicted the loss of an additional 650,000 ac during 1974. Other damage was expected from the loss of timber, increased fire incidence, watershed disruption, and diminished recreational use. Even after salvage of logs, estimates of loss ranged between \$28 million and \$67 million. Although tussock-moth outbreaks usually had lasted about three years because of a natural control through a virus in the moth population, the latest one appeared more serious. Reluctantly, EPA administrator Russell Train permitted controlled spraying with DDT during 1974.

Preliminary reports from the Forest Service to the EPA in 1975 substantially supported the effectiveness of DDT in controlling the epidemic. Any adverse effects of DDT use had not yet emerged. In addition, both the ongoing efforts of the Forest Service and the safeguards imposed by the conditional use of DDT produced some encouraging information about tussock-moth control that, except for the crisis of the epidemic, might not have been available. Among the more recently discovered chemical alternatives to DDT is a substance called Sevin-4-oil; it may not be as harmful as DDT in the ecosystem but is not yet fully tested. More promise is held for biological controls, foremost of which are a virus and a bacterium. These may hold moth populations at low levels when artificially introduced, but again they have not been adequately tested. A program of the U.S. Department of Agriculture planned to continue to study the tussock moth for two years.

Meanwhile, reforestation efforts were continuing; Boise Cascade Corp., for example, planted about 200,000 ponderosa pine and Douglas fir seedlings in affected forests. Interestingly, these seedlings had become available one year earlier than normal because geothermal energy was used in the nurseries to accelerate growth.

Neanderthals and flowers. Probably the earliest known evidence of the inclusion of flowers in burial was reported during the year by Arlette Leroi-Gourhan of the Musée de l'Homme, Paris, and R. S. Solecki of Columbia University, New York City. Excavations at Shanidar cave in northern Iraq uncovered several Neanderthal skeletons which were accompanied by abundant flower pollen, an indication that the flowers had been buried with the remains of the Neanderthals. This is an especially exciting discovery, not only because the pollen was preserved even when conditions were poor for skeleton preservation but also because most of the plant genera represented have herbal or medicinal qualities. Speculation is now possible concerning Neanderthal ability to associate various plants with their medicinal qualities and, further, about the very act of placing flowers at the burial site. Solecki believed that this find is evidence of Neanderthal man's "humanness" and, as such, that it will affect certain theories of the origin and development of man.

The list of plants identified at the site comprises genera still found in Iraq, many in the vicinity of Shanidar, although species are different. Members of the composite (aster) family belong to the following genera: *Achillea*, widely used in herbal medicine; *Centaurea*, a thistlelike plant; and *Senecio*. A grape hyacinth, *Muscari*, represents the lily family and *Althaea* the mallow family. Gymnosperms are represented by *Ephedra*, a member of the family *Gnetaceae*. Though it is only inferred that Neanderthals actually employed the plants for medicinal purposes, the nearly exclusive choice of flowers from medicinal plants makes the inference rather strong.

Endangered plants. Although it is true that gardening and nature preservation groups have sought to preserve rare and endangered plants for many years, only recently have preservation efforts secured national attention. The U.S. Endangered Species Act of 1973 instructed the Smithsonian Institution, Washington, D.C., to prepare a list of endangered and threatened plant species. Methods of preservation in nature were also to be recommended. In a 200-page report submitted to Congress in January 1975, the Smithsonian Institution labeled 2,099 species, or about 10% of U.S. flora, as extinct, endangered, threatened, or commercially exploited.

The report recommended preservation of habitats of endangered species rather than propagation of the plants in conservatories and botanical gardens. This precipitated much controversy. Many endangered plants are found in such specialized habitats as ravines and bogs, which are part of larger regions targeted for dam construction, logging, and other projects. To many people, it does not make sense to stop such beneficial large-scale activity in order to save a species of small plant. To most botanists and conservationists, however, it makes much sense, because diversity of plants and animals seems to be important for the maintenance of a stable biota. As species disappear, one by one, diversity decreases and the effects of such decrease are difficult to predict.

The Endangered Species Act prohibits any project that threatens an endangered plant or animal species. Although this appears to be rather final, the act also requires the U.S. departments of Interior and Commerce to verify that the listed plants are actually endangered before they are officially published in the *Federal Register*. The process of verification is extremely slow because the habitats of most of the species are hard to reach, adequate manpower is lacking, and prediction of the effects of exploitation of the habitat is difficult. Nevertheless, the Department of the Interior's Office of Endangered Species had its intent to accept the entire Smithsonian list read into the *Federal Register* in July 1975 and soon afterward began its verification tasks.

—Albert J. Smith

Microbiology

Interest in the use of microorganisms to solve problems concerning the environment, energy production, and waste disposal remained high during the past year. These three world concerns are interrelated because microorganisms can be used to convert waste materials into fuels and into single-cell protein. Although much of the work currently under way is experimental in nature, it nevertheless portrays what may become reality in the future. That microorganisms are useful in the solution of these problems is well established. However, it is still necessary to develop more efficient bioconversion processes. One approach has been through the selection of microorganisms that break down the materials more rapidly and show increased yields of products.

General, environmental, and applied microbiology. Research efforts continued to be directed toward the utilization of bacteria to produce methane by the bioconversion of waste and low-cost materials, especially of those that contain cellulose. Cellulosic materials are readily available as agricultural wastes or as discarded newspapers. For example, more than 200 million tons of agricultural cellulosic wastes are produced each year in the U.S. alone.

Methane production by bacteria is an anaerobic process; *i.e.*, carried out in the absence of air. Bioconversion in general and anaerobic digestion in particular are slow processes. There are at least four potentially rate-limiting steps in the bioconversion of materials to methane. First, in the case of cellulosic materials, the insoluble cellulose must be converted into its soluble sugar components by microbial enzymes called cellulases. Second, microorganisms that are not methane producers must convert the sugars to acidic products, chiefly acetic and propionic acids. Third, the methane-producing bacteria then must convert these acids to methane and carbon dioxide. Fourth, the dissolved gases must undergo transfer from the liquid to the gas phase.

Photographs, courtesy, Jerome J. Paulin, University of Georgia

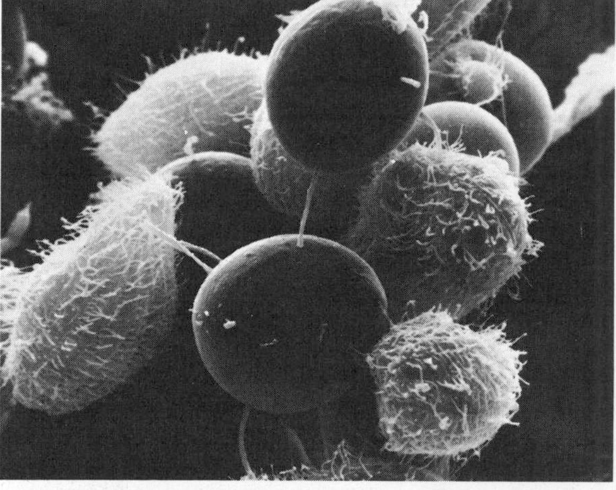

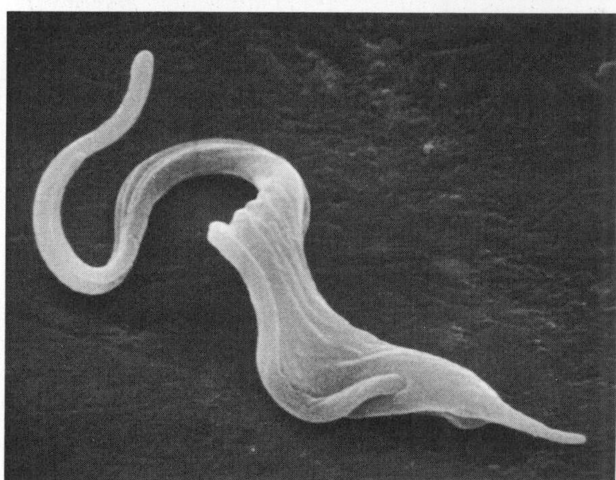

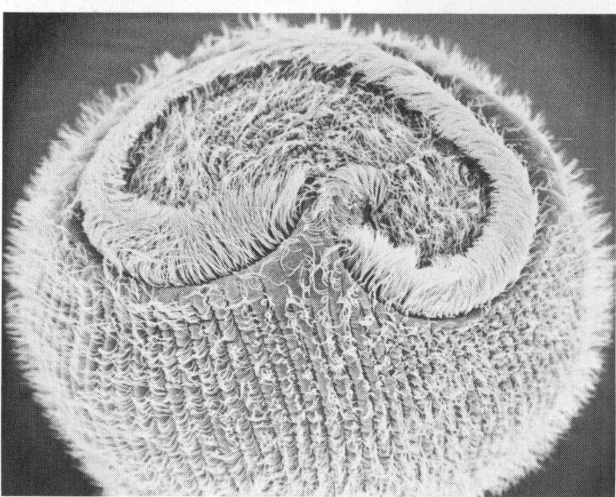

The beauty and eeriness of the microscopic world are revealed in scanning electron micrographs of (top to bottom) smooth, spherical suctorian Discophrya *feeding on captured* Tetrahymena *species; parasitic flagellate* Trypanosoma equiperdum, *a close relative of the causative agent of African sleeping sickness; and protozoan* Stentor, *whose band of cilia creates vortices of water currents that propel food into its mouth.*

One school of thought holds that the transfer to the gas phase is actually the crucial rate-limiting step. Moreover, there is evidence that, at high nutrient concentrations, gas bubbles surround the methane-producing bacteria and interfere with the diffusion of nutrients into the bacterial cells, thus inhibiting their normal metabolic activity. Clifton D. Finney and Robert S. Evans II of Drake University, Des Moines, Iowa, recently reported that this rate-limiting factor may be overcome by microbial digestion in a low-pressure environment with vigorous agitation and elevated temperatures.

An approach to speed up the rate of methane production through a process involving microbial fermentation at elevated temperatures was under study by P. J. Weimer and J. G. Zeikus of the University of Wisconsin. One experimental fermentation system employed thermophilic bacteria, which thrive only at elevated temperatures, and operated at 140° F (60° C). The system utilized two different thermophilic bacteria: one that converts cellulose to hydrogen, carbon dioxide, ethyl alcohol, and organic acids; and one that converts hydrogen and certain of the organic acids to methane. These bacteria, moreover, are able to degrade a variety of cellulosic waste materials, including newspaper, cotton, weeds, and wood-pulping wastes. As expected, the methane yield of the new system was significantly higher than that of lower temperature systems. The thermophilic system offers an added advantage because its higher operating temperature prevents growth of many undesirable microbial contaminants.

Still in the experimental stage are processes to produce methane from animal wastes. It has been estimated that U.S. farm animals produce more than three million tons of waste per day. The major source of such wastes is cattle, and beef cattle produce about 60% of the total amount. Past experiments indicated that the rate of bioconversion of animal wastes to methane is economically poor. The latest approach, taken by V. H. Varel and coworkers from the University of Illinois, to produce methane economically from animal wastes also involves the digestion of wastes by thermophilic bacteria at 140° F. The process degrades waste material to 40–50% of its original dry weight. The remaining material, which contains microbial protein and nitrogen in the form of ammonia, could be refed to cattle or other ruminants as a protein source or could be used as fertilizer.

Because of the high energy consumption necessary for the industrial synthesis of nitrogen fertilizers, there has been considerable economic interest in both microbial nitrogen fixation and microbial action on nitrogen compounds introduced into the soil. These considerations are also important from a humanitarian viewpoint. Crop yields must be increased in order to feed an increasing world population, and available nitrogen compounds in the soil have a direct bearing on these yields.

In 1974 a Brazilian scientist reported the discovery of a nitrogen-fixing bacterium growing in association with the roots of a number of tropical grasses. Nitrogen fixation by root-nodule bacteria previously had been thought to occur only in such leguminous plants as soybeans. More recently, Joachim F. W. Von Bülow and Johanna Döbereiner of the Federal Rural University of Rio de Janeiro, Brazil, noted that certain strains of corn have roots with the capacity to fix nitrogen. This was attributed to the presence in the roots of the same nitrogen-fixing bacterium previously discovered in roots of tropical grasses. To date, efficient nitrogen fixation by this bacterium has been noted only in tropical climates where the soil temperatures are between 88° and 104° F; little nitrogen fixation occurs below 77° F. Preliminary work by Robert Burris of the University of Wisconsin, however, indicated that corn growing in a temperate climate can be infected with the nitrogen-fixing bacterium, although it was too early to determine whether this discovery would increase crop yields. It is unknown whether this microorganism or some other as yet undiscovered bacterium could be made to infect and fix nitrogen in the roots of such cereals as wheat. Nevertheless, an international effort is in progress to investigate that possibility.

The fate of nitrogen compounds in the soil is also of importance. Plants use nitrogen in the oxidized form as nitrate, and soil microorganisms do likewise. For maximal efficiency, most of the nitrogen compounds in fertilizers must be available to crops without depletion because of microbial activity. That microorganisms have a considerable appetite for nitrate was demonstrated on arid land in California. When nitrogen fertilizer and water were added to the soil, over 95% of the nitrogen disappeared from the soil almost immediately, presumably due to microbial action.

Medical microbiology. Diabetes is a disease in which the metabolism of carbohydrates is impaired while the metabolism of fats and proteins is enhanced. It is estimated that there are five million diabetics in the United States, and the number is increasing. Scientists had long considered the possibility that viruses cause some types of diabetes, although proof had been lacking. Recently, however, three different viruses were found that produce a diabetes-like disease in animals. All three are members of a family of small, commonly found RNA-containing viruses called picornaviruses, some of which produce upper respiratory infections and a pattern of symptoms similar to poliomyelitis except that there is no paralysis. (The poliomyelitis virus itself belongs to the picornavirus family.) Although these observations do not constitute a definitive proof for the linking of viruses and diabetes, they are strongly suggestive of a relationship. Should this prove to be true, the prevention of diabetes through immunization becomes a possibility.

Venereal diseases continued to attract attention. An interesting finding was that *Neisseria gonorrhoeae*, the causative agent of gonorrhea, produces an enzyme that cleaves the A1 subclass of the human immunoglobulin A (IgA). IgA is a type of secretory antibody involved in immunity to various bacterial diseases. Thus, the elaboration of an enzyme that cleaves IgA would seem to be of clinical importance. To date, however, such promise does not seem to be forthcoming because, paradoxically, IgA does not appear to afford protection against gonorrhea.

Cells of *N. gonorrhoeae* have long, hairlike strands of protein material projecting from their surfaces. Possessed by certain other bacteria as well, these structures are known as pili. It is believed that bacterial pili enable bacteria to adhere to surfaces, and adherence of *N. gonorrhoeae* to cell membranes in the human urethra is considered to be an important part of the infectious process. Charles C. Brinton of the University of Pittsburgh, Pa., recently isolated pili from *Neisseria gonorrhoeae* for use as a vaccine. Injection of the purified pilic material, which cannot cause disease, into animals and human volunteers induced the production of antibodies to gonorrhea. The level of protection, or immunity, against infection by *N. gonorrhoeae* in the human volunteers was found to be about 60 times higher than in unprotected subjects.

Scanning electron micrograph of the bacterium Vibrio cholerae *adhering to villi in the small intestine of an infant rabbit 1.5 hours after inoculation.*

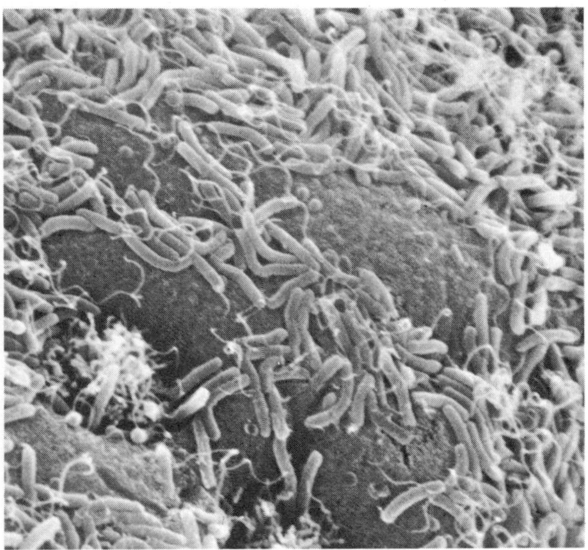

Courtesy, E. Nelson, J. Cements, R. A. Finkelstein, University of Texas

The use of pilic material to cause the body to produce protective antibodies is a novel approach to immunization. Should it prove successful in protecting against gonorrhea, the same approach might be useful for the control of other diseases caused by piliated bacteria.

Adherence of other disease-producing bacteria to tissues also received increasing attention. It appears that many disease-producing bacteria possess specific surface proteins and pili that enable them to adhere to a particular tissue in the body. Examples are the adherence of *Vibrio cholerae*, the causative agent of Asiatic cholera, to the intestinal membrane, and the attachment of particular streptococci to tooth surfaces, a precondition to the formation of dental caries. Understanding the mechanisms that enable disease-producing bacteria to adhere to specific surfaces is expected to shed light on disease-producing mechanisms and on methods to prevent such attachments.

Since the advent of antibiotics, drug resistance in microorganisms has been a topic of concern. Microorganisms that are initially susceptible to drugs frequently acquire resistance through mutation or through acquisition of genes containing genetic information for drug resistance from other drug-resistant microorganisms. Two major mechanisms of drug resistance have been recognized for several years: an alteration in the chemical target of the antibiotic within the microorganism so that the antibiotic becomes ineffectual, and the chemical alteration or destruction of the antibiotic itself by means of specific enzymes produced by the resistant microorganism. Recently, a third mechanism of antibiotic resistance was observed: the cell envelope of the microorganism becomes impermeable to the antibiotic and thus acts as a barrier to prevent the antibiotic from reaching its target. This mechanism is receiving much attention and seems destined to become an important subject of future research.

—Robert G. Eagon

Molecular biology

The past year was one of paradox for those trying to understand the workings of genes in detail. An astonishing array of procedures was developed, some of which will be described below, to determine the precise sequence of nucleotides in particular stretches of DNA molecules. The stretches whose sequences already have been determined include the critical parts of several genes that determine the rate at which the genes will be expressed. It was still not possible, however, to specify exactly what it is that permits one stretch of nucleotides in a DNA molecule to bind repressors, the protein molecules that turn particular genes on or off, and another nearby stretch to bind RNA polymerase, the enzyme responsible for transcribing the nucleotide sequence of the gene into messenger RNA. The solution may have to wait for the determination of the three-dimensional structure of crystalline repressor-operator complexes by X-ray diffraction methods. Nevertheless, recently acquired nucleotide-sequence data provided a sufficient number of tantalizing clues to provoke extensive speculation concerning the basis of repressor-operator recognition. Before proceeding, it will be useful to describe the basic molecular features of gene expression and some of the physiology and genetics of the two systems whose thorough investigation provided much of what is known about those features: the bacterial virus lambda and the genes controlling metabolism of β-galactosides in the bacterium *Escherichia coli.*

Features of gene expression. A gene consists essentially of a sequence of nucleotides along a DNA molecule. The DNA molecule contains two chains of nucleotides wound around each other in the form of a double helix; the two chains are held together by hydrogen bonds between individual nucleotides. The hydrogen bonds specify the pairing of nucleotides on each of the two chains. Thus, wherever adenine (A) occurs on one chain, it is linked by hydrogen bonds to thymine (T) on the other chain. Similarly, guanine (G) is always paired with cytosine (C). These nucleotide pairing rules govern both DNA replication, which assures the faithful transmission of genes from generation to generation, and the transcription of RNA. The latter is accomplished by an enzyme called RNA polymerase, which first must bind to a part of the gene called the promoter and then must transcribe the nucleotide sequence of one of the DNA chains, using the pairing rules, into the nucleotide sequence of an RNA molecule called messenger RNA. The message then attaches to the protein-synthesizing machinery called ribosomes, on which the nucleotide sequence in the RNA is translated, three nucleotides at a time, into the amino-acid sequence of a protein.

The rate at which a gene is expressed appears to be determined mainly by the frequency with which RNA polymerase binds to the promoter sequence to initiate transcription of the gene. Mutations in the promoters of particular genes have been shown to correspond to changes in single nucleotide pairs; these mutations can either increase or decrease the rate of transcription. In addition to this control, there exist special cases in which genes can be completely shut off by proteins called repressors, which bind firmly to short

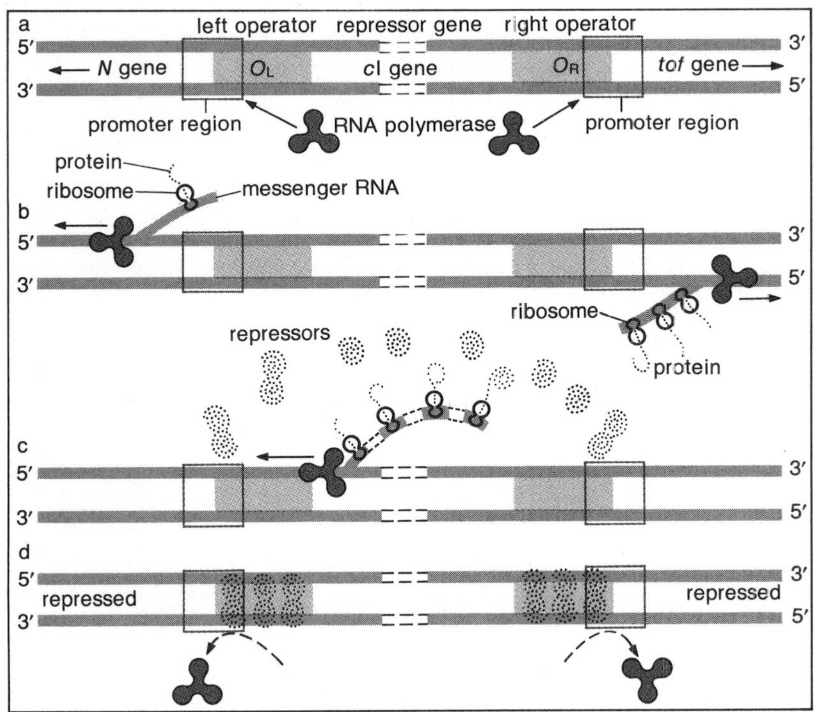

Adapted from "A DNA Operator-Repressor System" by Tom Maniatis and Mark Ptashne,

Under conditions that favor reproduction of the bacterial virus lambda, normal transcription of its genetic DNA into messenger RNA is effected by the enzyme RNA polymerase, which binds to a stretch of gene called the promoter located on either side of the cI repressor gene (a). The two molecules of RNA polymerase proceed in opposite directions along strands of DNA, using the nucleotide base pairing rules to assemble chains of messenger RNA that contain coded instructions for protein synthesis by viral ribosomes (b). Under conditions of viral dormancy, the cI repressor gene is transcribed by RNA polymerase and translated by ribosomes into repressor proteins (c) that migrate to binding sites in the two operator regions and block access of RNA polymerase to the promoters (d).

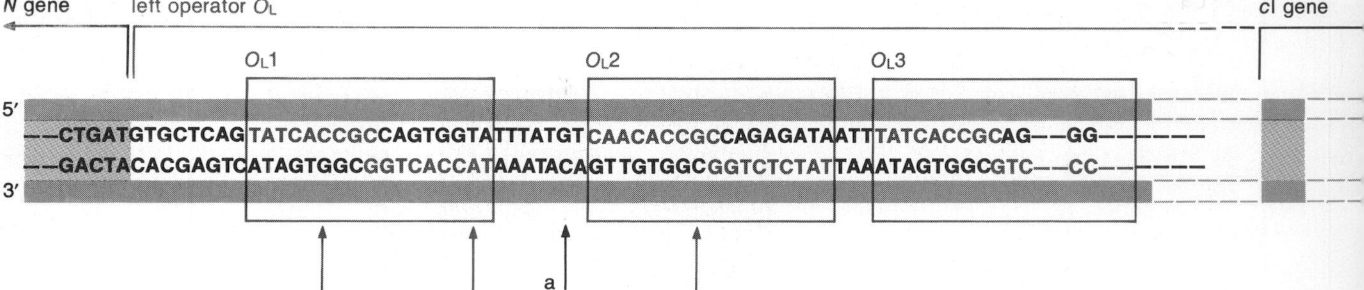

Sequences of bases for about 60 nucleotide pairs in the left operator of virus lambda DNA and about 70 pairs in the right operator. Sequence of O_L3 is not completely known. See text.

nucleotide sequences, called operators, located at the beginnings of certain genes. Several of these operator sequences, together with their adjacent promoters, have been determined by researchers during the past year.

Structure of the *lac* operon. In the bacterium *E. coli*, transcription of three linked genes coding for enzymes involved in the metabolism of β-galactosides is under control of a single operator-promoter. Under normal circumstances, the set of genes is turned off or repressed by a specific repressor protein (the product of another gene controlled independently) that binds to the operator. When a β-galactoside such as lactose is added to the cells, the galactoside or a galactoside metabolite binds to the repressor, causing it to fall off the operator. The system is then open for transcription.

Several kinds of mutations can render the system permanently open. For example, a mutation in the repressor gene can lead to production of faulty repressor, which fails to bind to the operator. Alternatively, certain mutations in the operator also prevent the repressor from binding. In either case, transcription is no longer dependent upon the presence of lactose; it is regulated only by the promoter's ability to bind RNA polymerase. (French scientists François Jacob and Jacques Monod received the Nobel Prize for Physiology or Medicine in 1965 for their discovery of the relationships between repressor, operator, and gene regulation.)

The repressor protein for this so-called *lac* operon was first purified by Benno Müller-Hill and Walter Gilbert at Harvard University in 1966. Since then, Gilbert's laboratory has been engaged in determining the sequence of nucleotides of the operator and the promoter. One of the first approaches was to purify a DNA fragment containing the operator sequence by protecting that sequence with repressor and then digesting away all of the nucleotides outside the protected segment with deoxyribonuclease, a rather nonspecific degradative enzyme that cleaves DNA randomly, that is, without regard to nucleotide sequence. The protected DNA segment, containing the operator, was isolated by first binding the repressor-operator complex to a nitrocellulose filter and then releasing the DNA with a lactose analogue that causes the repressor to let go. The sequence of nucleotides in the released fragment could subsequently be determined.

Until recently, all of the methods available for sequence determination had been difficult to apply successfully. Sequencing methods depend entirely upon the ability to tag the DNA, or a faithful RNA transcript of the DNA, with radioactive phosphorus atoms; the amounts of material available for analysis are too small for any other method of detection, and phosphorus-32 is the only radioactive atom suitable for such experiments in terms of cost and half-life. Unfortunately, bacteria grown on media containing ^{32}P yield *lac* operator DNA with radioactivity barely sufficient for conventional sequencing methods.

What revolutionized the nucleotide sequencing procedure was the introduction of a new approach developed by Frederick Sanger and his colleagues at the Medical Research Council (MRC) Laboratory of Molecular Biology in Cambridge, England, and extended by students of Gilbert and of Mark Ptashne at Harvard. First, a ^{32}P atom is added in vitro to one end of each strand of an otherwise unlabeled DNA chain, which is then separated into two strands, each with one end labeled. Next, the fragments are digested with exonuclease, an enzyme that removes one nucleotide at a time from the unlabeled end. Stopping the exonuclease reaction at various points generates a population of partially digested molecules, all labeled and identical at one end, differing from each other only by the number of nucleotides removed from the unlabeled end. This mixture is then subjected to a two-dimensional fractionation procedure.

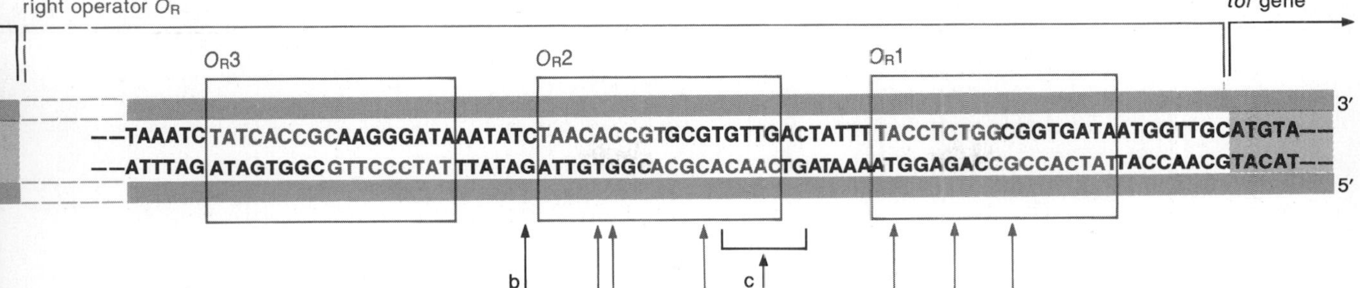

The first dimension is an electrophoresis that separates molecules on the basis of charge, the second a chromatographic separation based on size. The second step allows identification of fragments containing from one to n nucleotides. Comparing the electrophoretic mobility, in the first dimension, of the fragment containing n nucleotides with that containing $n-1$ nucleotides, one can deduce whether it was an A, a G, a C, or a T that had been removed in the conversion of n to $n-1$ nucleotides. The ^{32}P atom added in vitro can have extremely high specific activity; it appears that DNA fragments up to 40 nucleotide pairs long can be sequenced readily by this method.

There were, however, a few shortcomings to be eliminated from the fragment isolation techniques in the initial stage of the determination. First, the operator fragment isolated by deoxyribonuclease digestion of repressor-protected DNA has somewhat ragged ends due to the random nature of the nuclease cleavage. Second, it is preferable to have a longer fragment, one that includes the promoter, in order to determine the relationship between the operator and the promoter. Such fragments became available during the past year by using host-restriction endonucleases. These enzymes have the remarkable property of recognizing and cleaving particular sequences in DNA. They have been of crucial importance in the cleavage and splicing reactions that make possible genetic engineering in bacteria.

Operator-promoter complexes in the bacterial virus lambda. Host-restriction endonucleases are widespread in bacteria. As the name implies, they are enzymes that appear to recognize "foreign" DNA and cleave it at particular sites. One such enzyme, for example, will cut a DNA molecule wherever it finds the sequence CCGG (paired with GGCC on the other strand); another cuts at the sequence AATT; another at the sequence CTTGAC. With a suitable collection of these enzymes, it is possible to cut a DNA molecule first, then to use repressor to bind a particular fragment, and then either to sequence the fragment or to cut it further with other of the restriction endonucleases.

The power of these methods has made it possible to dissect the structure of the operator-promoter regions of the bacterial virus lambda. It had been known for some years that the lambda chromosome contained two operators, each controlling the transcription of several genes. The same repressor recognizes both operators. Several years ago at Harvard, Ptashne and Vincenzo Pirrotta showed that repressor at high concentrations protects an 85-nucleotide-pair fragment from deoxyribonuclease; more recent work from the laboratories of Ptashne at Harvard and of Pirrotta at the Biocenter of the University of Basel (Switz.) showed that each 85-nucleotide fragment contains three repressor binding sites. During the past year, Ptashne's laboratory reported the complete sequence of both operators, each containing three slightly different versions of a 17-nucleotide-pair sequence separated by short stretches of nearly pure AT pairs.

The figure above shows the sequence of the two lambda operators, as determined at Harvard and in Basel, with a speculative interpretation by Ptashne and his associate Tom Maniatis. The actual repressor binding sites are suggested to be the almost symmetrical 17-nucleotide-pair segments enclosed in boxes. The nonlettered arrows indicate nucleotide pairs at which mutation completely abolishes repressor binding to that particular site. Furthermore, Pirrotta found that RNA polymerase protects a 45-nucleotide-pair segment that includes most of O_R1 and some 20 nucleotide pairs of the gene to the right. Thus, repressor and RNA polymerase must compete for the same site. Put another way, the promoter includes the operator. The lettered arrows indicate nucleotide pairs altered in genetically identified promoter mutations. Since they are outside the region protected by RNA polymerase, the mechanism by which they affect the frequency of transcription initiation is a mystery.

That repressors act by direct competition with

RNA polymerase for the same or overlapping sites on DNA was also shown last year by Charles Yanofsky and co-workers at Stanford University, who studied the genes controlling biosynthesis of the amino acid tryptophan. Students of Gilbert at Harvard, correcting an earlier report from another laboratory, showed that the same competition is seen between the *lac* repressor and RNA polymerase. Thus every system examined so far has produced the same result, leading to the same paradox: the operators are recognized uniquely by individual repressors (*e.g.*, the *lac* repressor does not bind to lambda operators), but they are also a major part of the promoters, all of which are recognized by the same RNA polymerase molecule.

Chromosome organization. Molecular biologists in increasing numbers have been turning to studies of cells more complicated than bacteria. One neglected subject that has returned to center stage is the structure of chromatin—the complex of DNA, histones, and other proteins found in cell nuclei during interphase (the stage between successive cell divisions). These studies had their origin in a provocative article by A. L. and D. E. Olins, of the Oak Ridge (Tenn.) National Laboratory, published in 1974, which showed electron micrographs of chromatin that consisted of nu bodies, particles appearing as beads on a string of DNA. Subsequent biochemical work by Roger Kornberg, Markus Noll, and others at the MRC Laboratory of Molecular Biology, Pierre Chambon and co-workers in Strasbourg, France, and many other laboratories confirmed and extended the Olins' views.

From this research the following picture has emerged: Chromatin consists of tightly packed beads 125 angstroms in diameter (1 angstrom = 10^{-8} cm), connected by a thread of DNA probably coated with histone H1. Each bead (nucleosome or nu body) contains two molecules each of histone H2A, H2B, H3, and H4 and about 200 nucleotide pairs of DNA. The four latter histones form an octameric structure in the absence of DNA; the octamer appears to consist of a tetramer "core" of H3 and H4, to which two pairs of H2A-H2B are joined. Chromosomal DNA is probably wrapped around the outside of this structure. From data based on digestion with nucleases, about half of the DNA in chromatin was found to be associated with nu bodies and the remainder with the threads connecting them; recent data, however, suggest that more, perhaps over 70%, of the chromosome DNA is associated with nu bodies. The evidence for the revised estimate is quite straightforward. When one isolates, from a brief partial nuclease digest, chromatin fragments containing one, two, or three nu bodies, the fragments are found to contain DNA segments about 180, 360, and 540 nucleotide pairs long. When the nuclease digestion is more extensive, the chromatin fragments containing one, two, or three nu bodies yield DNA segments about 140, 320, or 500 nucleotide pairs long. It can be concluded, therefore, that each nu body contains approximately 140 nucleotide pairs, and that two bodies are interconnected by a 40-nucleotide-pair segment of DNA.

—Robert Haselkorn

Zoology

Evidence for the cause of a form of mental retardation and promise of a new diagnostic technique for the early detection of genetically related birth defects highlighted zoological research on the cellular level. The discovery of several animal pheromones was reported, including one that enabled male crayfish to differentiate between sexes. An unusual experiment to investigate altruism in the mountain bluebird lent support to theories that animal behavior is basically selfish.

Cellular zoology. A gene is a specific sequence of nucleotides along the double-stranded DNA molecule. The genes that constitute the chromosomes in the cell or cells of a living organism individually determine the features and functions of that organism, generally through control over the synthesis of enzymes. Eye color is thought to be controlled by a single pair of genes, and body size and shape or skin color by many interacting genes. Certain diseases such as Turner's syndrome are caused by the absence of genes and entire chromosomes. Conditions such as sexual dimorphism result from the presence of extra genes and chromosomes.

Gene mapping, the determination of the specific order and sequence of nucleotides in an individual gene, and chromosome mapping, the assignment of genes to chromosomes in proper order, have become subjects of intensive study for many cellular investigators. Working with bacteria, molecular biologists have mapped most of the 1,000 genes on the relatively few chromosomes these organisms possess. *Drosophila melanogaster*, the fruit fly, has approximately 5,000 genes, of which 2,000 have been mapped. Humans are thought to carry 100,000 to 200,000 genes on their chromosomes. Approximately 100 human genes have been mapped to date using cell-fusion techniques, in which human cells are fused with mouse cells in tissue culture. The fusion process yields hybrid cells possessing a small fraction of functional human chromosome material; thus, any enzyme that a hybrid cell produces only when it possesses a specific human chromosomal fragment can be

Adapted from "Putting Human Genes on the Map" by Joan Arehart-Treichel, "Science News," vol. 108, no. 15, p. 235, October 11, 1975

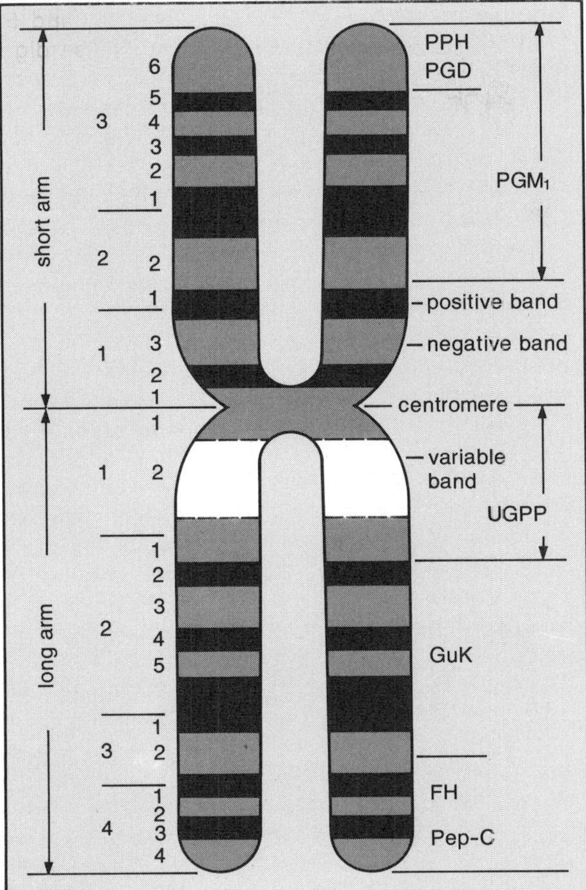

To date the human chromosome most thoroughly mapped is chromosome number one. General regions of gene location are denoted by abbreviations of protein products that the genes express.

assigned to that fragment and becomes an indicator of the general location of the gene responsible for its production.

Such mapping remained crude and highly theoretical, and techniques were not available to give it practical value in the prevention of genetic diseases. For several years, however, a harmless technique called amniocentesis has been employed to withdraw samples of amniotic fluid containing sloughed skin cells from human fetuses still in the mother's womb. Because every cell possesses all of the genetic information of its parent organism, fusion of a fetal skin cell and a mouse cell in culture could trigger the hybrid cell to produce enzymes the fetal cell would not normally express. This concept could be of great benefit in diagnosing certain genetic deficiencies in the fetus before birth. For instance, if the hybrid cell could be induced to synthesize a particular brain-cell enzyme expressed by a gene that is characteristically absent in some kinds of mental retardation, the fetus would be expected to be normal because the gene must be present. Genetic detection of diseases is perhaps five to ten years in the future, but by then, researchers probably will have mapped a thousand or more human genes.

In other reports of advances in cellular research, Stephen S. Wachtel of the Sloan-Kettering Institute for Cancer Research in New York City assigned a gene to the human Y chromosome; with one possible exception, this was believed to be the first gene so assigned. Norman Davidson and Madeline Wu of the California Institute of Technology described an electron-microscopic technique for visualizing those regions on DNA molecules that serve as genes. They combined, or duplexed, a certain DNA molecule with an RNA molecule and mixed the duplex with another protein that binds only to single-stranded segments of DNA. The proteins attach to all parts of the DNA molecule except those sections duplexed with RNA. The protein aggregates serve as markers between genes (the duplexed DNA-RNA sections). The technique is useful in estimating the number of nucleotides in each visualized gene.

Other discoveries within the cell included that by Michael Locke and P. Huie, of the University of Western Ontario, of new components found only in certain granular bodies called Golgi complexes. The components appear as rings of beads 10–12 nanometers in diameter (1 nanometer $= 10^{-9}$ m) and are thought to be related to some of the basic functions of the Golgi complex, which assists in membrane construction, protein and lipid storage, and particle transport across the cell membrane. Dominick Purpora of the Albert Einstein College of Medicine, New York City, used the Golgi method—the hundred-year-old technique involving the staining of nerve tissue in silver nitrate—on tissue from brains of children with severe mental retardation and discovered that certain brain cells had a significantly decreased number of dendrites (the minute branching extensions found only on nerve cells) or an altered dendritic structure. Because nerve cells receive impulses via dendrites, the decrease could be evidence of a diminished communication among brain cells that manifests itself as retardation. Purpora also discovered large structures with spinelike processes, called meganeurites, between the cell body and the axon of certain nerve cells of victims of Tay-Sachs disease. Some of the meganeurite processes might interfere with normal impulse transmission between nerve cells, thereby causing retardation.

Actin and myosin, the specialized proteins involved in muscle contraction, had been suspected of contributing as well to the movement of chromosomes along spindle fibers during cell division. Joseph Sanger of the University of Pennsylvania School of Medicine settled part of the issue when, using a new fluorescent technique, he found patterns of actin on the mitotic spindle fibers during all stages of mitosis. He also found actin in the nucleoli of cells, a discovery which may help explain their role in spindle formation. Current tissue-preparation techniques apparently destroy any myosin present, but investigators predicted that myosin will also be confirmed as an agent in the movement of chromosomes.

General zoology. Among social insects, swarming serves to establish new colonies or to move a colony to a new site. As has been known about bees for many years, swarming requires communication among individuals. In fact, many investigators believe that communication among bees, whether by dance language or by odor, is second in complexity and abstraction only to human language. Wasps are regarded as more primitively social than bees, and their swarming mechanisms are not well understood. In the case of certain tropical wasps frequently preyed upon by army ants, Martin G. Naumann of the University of Connecticut reported that these wasps respond to the random disruption, brood loss, and nest loss caused by ant invasions by escaping and immediately relocating the colony. Such swarming does appear to be facilitated by both communication within the swarm and by orientation to sites along the trail marked with odoriferous substances.

Ruth Chadab and Carl W. Rettenmeyer of the University of Connecticut reported the discovery of a pheromone (a chemical substance liberated by animals to attract, warn, or otherwise communicate with others of their own species) used by scout army ants to recruit other ants and direct them to a newly discovered source of food. The scout forages as far as 100 m and lays a chemical trail back to the nest. Then it touches other ants with its antenna and body, releasing another chemical that excites them and sends them off along the trail in formidable numbers.

Among many pheromones discovered during the year were a sea-anemone alarm pheromone, a crayfish sex-recognition pheromone, and a crayfish pheromone that attracts young crayfish to their own mothers. The sea-anemone pheromone was isolated and its structure elucidated by Nathan R. Howe and Younus M. Sheikh in California using ion-exchange chromatography and spectroscopic methods. Previously E. O. Wilson of Harvard University had noted that, unlike airborne pheromones which must be relatively small molecules to be volatile, molecules of waterborne pheromones need not be small to be water soluble. Christopher Ameyaw-Akumfi and Brian Hazlett of the University of Michigan discovered pheromones responsible for aggression in male crayfish *Procambarus clarkii* toward other males and submission and courtship toward females. These chemicals, thought to be carbohydrates, are sensed by receptors on the antennules of the male, telling him the sex of other crayfish and triggering his behavioral response. The maternal attractant pheromone was discovered by Edward Little of Florida State University; released by brooding female crayfish, it helps the immature crayfish return to their mothers until they can fend for themselves.

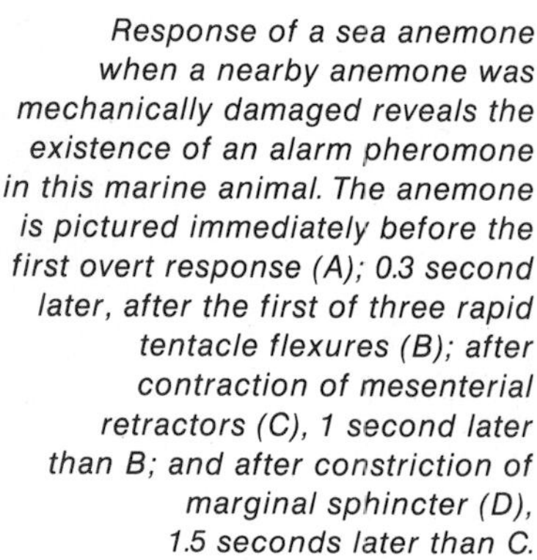

Response of a sea anemone when a nearby anemone was mechanically damaged reveals the existence of an alarm pheromone in this marine animal. The anemone is pictured immediately before the first overt response (A); 0.3 second later, after the first of three rapid tentacle flexures (B); after contraction of mesenterial retractors (C), 1 second later than B; and after constriction of marginal sphincter (D), 1.5 seconds later than C.

Courtesy, New York Zoological Society

With only 10–20 breeding females remaining, the American crocodile was recently afforded protection as an endangered species under the U.S. Endangered Species Act of 1973.

A team of U.S. scientists at the Hebrew University at Elat, Israel, reported that the flashlight fish, *Photoblepharon palpebratus,* uses its bioluminescent organs with more versatility than any other organism yet described. Anglerfish attract prey with their luminescent organs, others such as the sea squid escape predators by squirting a bioluminescent substance into the water, and fireflies use luminescence to attract mates. These investigators reported that flashlight fish can do all three with luminescent organs located just below its eyes, suggesting "a flashlight, whose owner can exercise options in its use."

Although the total destruction caused by in-flight encounters between birds and aircraft was unknown, the U.S. Air Force estimated that in 1973 300–400 collisions occurred between its aircraft and birds, resulting in two pilot deaths and a cost of $32,000 per collision. In an attempt to alleviate this serious hazard, a team of scientists from Rockefeller University and the State University of New York, Stony Brook, studied the behavior of birds that suddenly find themselves in the flight path of oncoming aircraft. They observed that night migratory birds, which comprise a large fraction of American bird species, quickly shift to evasive flight maneuvers in order to avoid collision and discovered that strong lights allow even large flocks to react with a swiftness sufficient to avoid oncoming planes. Their report recommended that planes be equipped with "appropriately designed and programmed lights" to shine in the "zone of collision hazard."

Mountain bluebirds were used to study altruism and to add more information to a controversy among zoologists concerning the basic nature of animal behavior. Many textbooks of biology, it was alleged, led readers to believe animals are basically altruistic; *i.e.,* they will contribute to the reproductive success of others in a situation of no direct benefit to themselves. The difficulty with theories of altruism, however, is that they simply do not describe the character of many animals. Darwin's theory of natural selection, the Hardy-Weinberg law of genetic equilibrium within populations, and other useful generalizations based upon actual observations indicate a kind of evident selfishness in animals, especially with respect to reproductive behavior and in matters relating to care of offspring. Harry Power of the University of Michigan reported that one way to measure the frequency of true altruism was to give bluebirds the choice of behaving altruistically or selfishly toward the offspring of other bluebirds. He removed one male or female adult from each of a number of nests with fledglings and allowed the missing birds to be replaced naturally by courting birds, which he called consorts. He found that all male consorts were reproductively selfish; none fed the young, cleaned the nests, sounded the alarm, or protected the young from danger as he would have done for his own. Yet the consorts did not kill the nestlings, as some mammals would have, nor did they interfere with the natural mothers' services to the young. The one female consort that was observed successfully in his study finally gave care after five days but was not considered truly altruistic because of her delay in accepting the duties of motherhood.

Ecology. Ecology is the study of the relationships between populations of organisms and the environment and the effects produced by changes in the balance of nature by such global forces as pollution and the extinction of species. Although

Courtesy (bottom), The American Museum of Natural History; photo (top) Russ Kinne from Photo Researchers

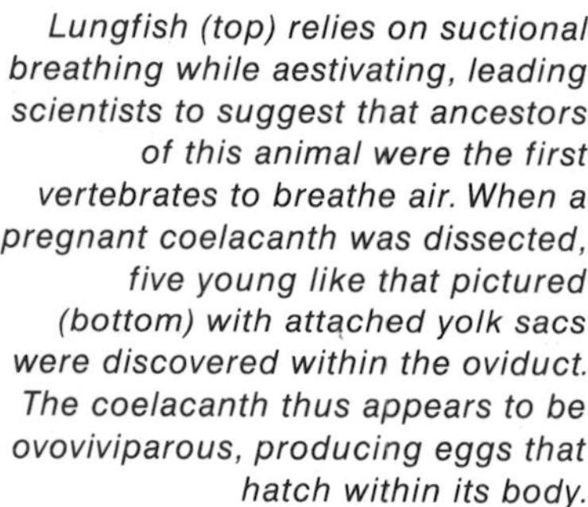

Lungfish (top) relies on suctional breathing while aestivating, leading scientists to suggest that ancestors of this animal were the first vertebrates to breathe air. When a pregnant coelacanth was dissected, five young like that pictured (bottom) with attached yolk sacs were discovered within the oviduct. The coelacanth thus appears to be ovoviviparous, producing eggs that hatch within its body.

concern for the quality of the environment and the preservation of endangered life has become a dominant issue in recent years, many ecologists continued to pursue traditional studies of the behavior and relationships among organisms. Such research was exemplified by that of Neal Griffith Smith of the Smithsonian Tropical Research Institute in the Canal Zone, who reported that New World songbirds are attracted to the sound "spsh" because it mimics sounds made by certain tropical birds with which the songbirds socialize during their annual southward migrations. The songbirds' normal diet of insects is apparently scarce in the tropical forests they visit; instead they must rely on the ability of tropical birds to find nectar and fruit. In another traditional kind of report, Robert E. Otis, W. C. Johnson, and D. F. Cowan of Ripon College in Wisconsin observed that overcrowding geese leads to a breakdown in their pecking order and increases their aggression. A return to normal population size restores the normal ranking.

Ecological studies often contribute valuable information toward the evaluation of proposed pest-control practices and public-health measures. One recent example was the calculated elimination of over 100,000 giant snails originally introduced into Florida by a child who brought some home as souvenirs from Hawaii. The snails were hermaphrodites, reproduced quickly, and, in their search for food in an alien habitat, produced slimy trails everywhere and ate house paint to meet nutritional needs. The snails were also vectors of rat lungworm to which humans are susceptible. Another potential health hazard came to light after a report by F. Marilyn Bozeman and co-workers of the U.S. Food and Drug Administration revealed that flying squirrels in Virginia and Florida are hosts to typhus. Their study indicated a reservoir of this disease can exist in a wild rodent population without outbreaks of the disease. Team members reported that the ticks, fleas, and mites that convey the disease-causing organisms from squirrels to humans must also be studied before any preventive action can be taken.

A principal tenet in the theory of evolution states that survival through evolution occurs in a large population because of its large pool of genes. Conversely, a small population becomes extinct because of the limited number and variation of genes. Little work has been done on the accumulative effects of random changes in inherited traits in populations—the so-called genetic drift or Sewall Wright effect. Genetic drift is basic to species survival, and when populations dwindle to 20–30 individuals the group no longer has in its gene pool the entire reservoir of genetic variability that once permitted large populations of the species to adapt to changes in the environment. Unfortunately, there are many species with dangerously few living members. A few encouraging reports of increases in populations of some very endangered species were received during the year. For instance, the population of the white-tailed kite, represented by only one known pair in 1913, was reported to have risen to 70 pairs because of the decrease in killing by man; another Mauritius kestrel was found, bringing their total number to 9; 35 Manchurian cranes and 4 Japanese crested ibis, previously believed extinct, were found in Korea; and peregrine falcons and immature bald eagles increased in North America because of man's efforts on their behalf.

However, as of early 1976, those species of animals officially listed as endangered by the U.S. Department of the Interior numbered only about 100; an additional 9 species were classified as threatened. Intended to offer protection to species under the Endangered Species Act of 1973, this timid list drew criticism sufficient to encourage the federal bureaucracy to accelerate its listing process. Consequently, ten more species were proposed for the official list and eight were accepted: the American crocodile, including about 10–20 remaining breeding females; the Cedros Island mule deer, which number about a dozen; the peninsular pronghorn antelope; the Hawaiian honeycreeper (a bird); the Scioto madtom (a small Ohio fish); a Hawaiian bird called the po'o uli; the gray bat; and the Mexican wolf.

The U.S. Fish and Wildlife Service nominated 74 species of plants and 171 species of animals for addition to the list, including, on a worldwide basis, the Asian elephant, grizzly bear, peregrine falcon, leopard, jaguar, and proboscis monkey. The Audubon Society and other organizations listed a number of birds possibly destined for extinction, including Newell's Manx shearwater, the canvasback duck, the purple martin, the reddish egret, the mountain quail, the upland plover, the common nighthawk, Lewis's woodpecker, and the lesser goldfinch. Approximately 25,000 species have been slated for review, but since the problem is an international one, it often has been complicated by political and economic issues.

Evolution. It is believed that life began about three billion years ago. Lois Nagy of the University of Arizona found very simple microfossils in rocks 2.6 billion years old and more complex and diversified microfossils in rocks 2.3 billion years old. Accordingly, the indications are that life on Earth did begin at least 2.6 billion years ago and that evolution to more complex and diversified forms occurred during the 300-million-year period immediately following.

Because fossils and artifacts are the major sources of data concerning the history of life on Earth, it would be expected that the protection of fossil-rich sites would be welcomed by all. But paleontologists, like ecologists, occasionally find themselves in serious competition with the necessities of modern man. Near Darmstadt, West Germany, lies the Messel Pit, an Eocene lake bed abounding in fossils of the animals which roamed there 50 million years ago. Very recent excavations included skeletons of an eohippus (an Eocene ancestor of the horse), a large bat with two smaller bats attached at its breasts, and snakes and crocodiles. The rock structure is such that it is even possible to observe shadows of soft organs, which do not fossilize. The local museum has had insufficient funds to sponsor work in the pit, and the best finds have often found their way into private and amateur collections. A greater problem concerns the use of the pit as a garbage dump; local officials calculated that the pit could be filled with 25 million cubic meters of trash by the year 2000. There are no equivalent fossil sites; other European sites mainly reflect life as it was in the Miocene, some 7 million to 25 million years ago. In response to this problem, concerned scientists issued a plea for a mobilization of world opinion to save the pit as a fossil site.

Lungfish and the coelacanth *Latimeria* are considered living fossils, and many zoologists study them for information about primitive systems. J. P. Lomholt and K. Johansen of the University of Aarhus in Denmark and G. M. O. Maloiy of the University of Nairobi, Kenya, studied lungfish during aestivation (hibernation during hot, waterless conditions) and reported experimental data supporting the theory that suctional breathing first developed in the ancestors of these fish while aestivating. By contrast, extinct amphibians had always been considered the first group of vertebrates to breathe air. The lobe-finned fish *Latimeria chalumnae* is closely related to the extinct rhipidistian fishes, which apparently were the ancestors of land vertebrates. Coelacanths were thought to be extinct since the Late Cretaceous period, about 60 million years ago, but since 1938, 80 have been caught off the coast of Africa. Recently C. Lavett Smith, Bobb Schaeffer, and James Atz of the American Museum of Natural History, New York City, and Charles S. Rand of Long Island University dissected a gravid (pregnant) coelacanth and found five advanced young averaging 318 millimeters in length. Each had a large yolk sac with no apparent connection to the mother, and it was concluded that the fish is ovoviviparous; the female produces eggs that develop and hatch within her body. However, because many modern fish (including lungfish) are ovoviviparous, it would be dangerous to impute the conclusions of this investigation to the reproduction of rhipidistians or primitive tetrapods.

—John G. Lepp

Materials sciences

Research in the field of materials sciences during the year was highlighted by new developments in the fabrication and testing of high-temperature ceramics, in slag chemistry, and in gaining an understanding of the kinetics of high-temperature gas-liquid reactions.

Ceramics

Many outstanding developments in the investigation of ceramic materials during 1975 centered on their fabrication and on the improvement of their mechanical properties for turbine engine applications, in which their resistance to very high temperatures and corrosive environments offered significant advantages over other materials.

Sintering studies. Most ceramics traditionally have been fabricated from fine metal oxide powders by heating them at high temperatures to eliminate porosity and thus obtain dense, strong components. For use in emerging turbine engine applications, however, the most interesting high-temperature ceramics are nitrides and carbides, materials whose atomic structures are more nearly covalently bonded than metal oxides. Many studies on these materials, and on such other covalently bonded materials as silicon, germanium, and boron, have shown that, unlike oxides, they do not densify, or sinter, easily at high temperatures. They have therefore resisted attempts to fabricate them into inexpensive, complex shapes.

Structure of boron carbide fiber composite is shown in microscopic cross section. Outstanding thermal resistance and high directional strength make the material attractive for gas-turbine applications.

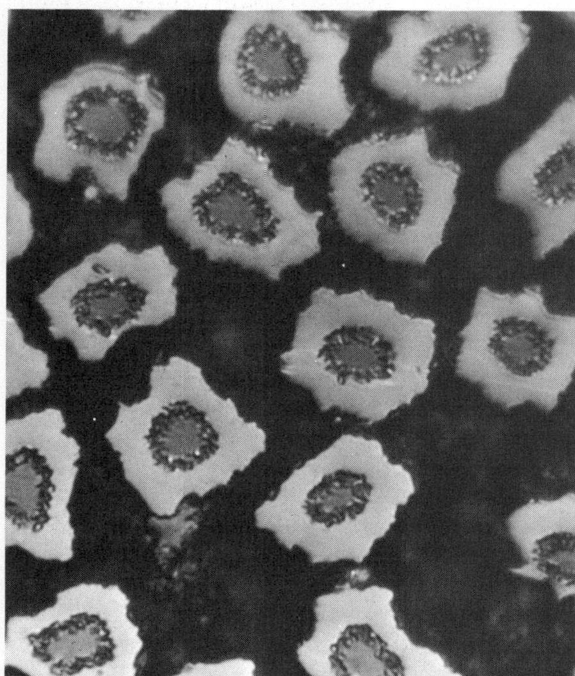

Courtesy, James Economy, W. D. Smith and Ruey Y. Lin, Carborundum Company, Research and Development Division

One of the major breakthroughs in non-oxide ceramic technology occurred in 1974, when scientists at the General Electric Research and Development Center reported that silicon carbide can be sintered with the addition of small amounts of carbon and boron. The reason for this behavior, however, was not clear. Recent studies at General Electric suggest that, in pure silicon carbide, rearrangement of atoms by surface diffusion, which reduces surface area and the driving force for sintering without actually producing any beneficial densification, occurs at intermediate temperatures before useful high-temperature, volume-diffusion processes can proceed. Precisely the same kind of behavior was reported during the year by scientists at Sandia Laboratories in theoretical and experimental studies on boron. Further clarification of these concepts and their extension to the sintering of silicon nitride, one of the most promising materials for turbine engine construction, are expected to lead to important improvements in materials and processing.

Silicon nitride. Until ways are found to sinter silicon nitride, dense, strong parts can be made from it only by hot-pressing techniques, which are expensive and limited to the production of relatively simple shapes. Silicon nitride has been so hard to densify that, even when using hot-pressing methods, additives are required to reduce the prevailing temperatures, times, and pressures. Several years ago, magnesium oxide was discovered to be an effective additive, but recently its use has been shown to limit seriously the service temperature of the components so fabricated. It has been suggested that the magnesium oxide and the presence of various impurities contribute to the formation of a glassy phase at silicon nitride grain boundaries, and that it is this glassy phase that reduces the high-temperature strength and creep resistance of silicon nitride.

Many studies in the past year focused on the search for new additives with less harmful side effects than magnesium oxide. For example, George E. Gazza of the U.S. Army Materials and Mechanics Research Center, Watertown, Mass., and A. Tsuge and co-workers of the Tokyo Shibaura Electric Co.

Photos, courtesy, George E. Gazza, Army Materials and Mechanics Research Center

Grain structure of hot-pressed silicon nitride is revealed in electron micrographs just before densification (left) and after fully densified (right). Yttrium oxide was added to help achieve the strong fibrous structure of the densified state.

recently discussed the behavior of yttrium oxide when used as an additive. The grain boundary phase formed in this case apparently transforms into a crystalline yttrium-silicon oxynitride, probably $Y_2Si_3O_3N_4$, with much more advantageous high-temperature mechanical properties than the glassy phases found in commercially available hot-pressed silicon nitride. Unfortunately, there were also some indications that the crystalline $Y_2Si_3O_3N_4$ phase may contribute to rapid degradation of the ceramic material in air at intermediate temperatures.

Another promising approach to the improvement of silicon nitride was the announcement by Roy W. Rice and W. J. McDonough of the U.S. Naval Research Laboratory of the effectiveness of zirconium oxide, zirconium silicate, zirconium nitride, and zirconium carbide as densification additives. They found that these compounds promote adequate hot-pressability, high strength, unusually high fracture surface energies, and excellent oxidation and creep resistances.

Mechanical properties. Prompted by the potential of high-temperature ceramics to increase turbine engine efficiencies and decrease their noxious emissions by allowing higher operating temperatures, ceramists also sought to improve their understanding of the mechanical properties of these brittle materials. A controlled surface-flaw technique, reported by J. J. Petrovic and co-workers at the U.S. Air Force Aerospace Research Laboratories at Wright-Patterson Air Force Base, Dayton, Ohio, was shown to be a useful way of reducing the statistical scatter in strengths normally measured in testing brittle ceramics and to be a convenient way of evaluating the critical stress-intensity factors, or fracture toughnesses, of these materials under various environmental conditions.

Unfortunately, the very high strengths and low fracture toughnesses of these brittle high-temperature ceramics imply that failures can be expected to originate at flaws as small as 10–50 μm (10^{-6} m) in effective length. Until nondestructive testing techniques are developed that will reliably find such small defects, ways are needed to eliminate components that will fail unacceptably in service and to predict the safe service lives of critical parts. One such approach undertaken by A. G. Evans of Rockwell International Science Center, S. M. Wiederhorn and co-workers of the U.S. National Bureau of Standards, R. W. Davidge of the Atomic Energy Research Establishment at Harwell, England, and others involves the use of proof testing and fracture mechanics. Their studies demonstrated that proof testing can be used to limit the size of the largest flaw initially present and that, together with measured subcritical crack growth rates, it can be used to predict minimum times to failure. A recent study by D. P. H. Hasselman and E. P. Chen at Lehigh (Pa.) University and by C. L. Ammann and colleagues at Pratt and Whitney Aircraft showed that a fracture-mechanics treatment of slow crack growth in these materials also can be used to predict their lifetimes under thermal fatigue, which will be a limiting factor in their use in many turbine engine applications.

Another concern about the use of ceramics in

Courtesy, Kawecki Berylco Industries, Inc.

Material about to undergo hot isostatic pressing is lowered into a high-temperature, inert-gas autoclave. Use of the HIP technique for the consolidation of ceramic and metal-alloy powders permits fabrication of relatively complex shapes that are close to their final configurations.

turbine engine applications has been their poor resistance to impact. Ceramic components can be damaged easily through the introduction of minute, but dangerous, microcracks during rough handling, and entire stages of components can be lost in service by cascading impact failures caused by the ingestion of even small foreign objects into the engine. It had been known, at least qualitatively, that the poor impact resistance of ceramics is a result of the relatively small amount of energy absorbed in their brittle failure. Until recently, however, little had been done to measure and better understand the impact resistance of ceramics under the temperatures and impact velocities characteristic of the turbine engine environment or to improve upon the impact resistances of currently available ceramics.

In a series of carefully instrumented drop-weight and ballistic impact tests conducted at the U.S. Air Force Materials Laboratory, I. Bransky and his colleagues measured the impact resistance of reaction bonded and hot-pressed silicon nitride and silicon carbide ceramics over a wide range of temperatures and impact velocities. They showed that, while the impact strengths (the dynamic strengths at failure) of these materials varied by a factor of two, their impact resistances (the amounts of elastic energy stored at failure) varied by a factor of five. Therefore, in order to increase impact resistances significantly, ceramists will need to achieve much higher dynamic strengths in these materials or find techniques of increasing their energy-absorption characteristics, by modifying either their failure mechanisms or their component designs.

One of the most important new approaches to improved impact resistance has been the use of energy-absorbing layers, as described by Henry P. Kirchner and J. Seretsky of the Ceramic Finishing Co., State College, Pa., and by scientists of the General Electric Research and Development Center. They showed that such layers, containing intentionally introduced porosity or microcracks, could provide significant increases in impact resistance under testing conditions where failure originates from the formation of Hertzian (small cone-shaped) cracks at the point of impact.

Ceramic processing. Hot isostatic pressing (HIP), the use of a high-temperature, inert-gas autoclave for the densification of powders, has been one of the most important general processing developments of recent years and is already finding rapid and widespread application in the compaction of metal powders. A study published in 1975 by K. H. Hardtl of the Philips Research Laboratory described the application of the HIP process to ceramic powders without the need for soft metal or glass molds. In this new technique, the ceramic powders first are sintered to about 95% of their theoretical density by a conventional firing step. At this point, the remaining porosity has become isolated in the form of small, closed voids within the interior of the ceramic component and is no longer in any way connected with the surface. The component then is hot isostatically pressed in an inert gas at pressures of only several hundred atmospheres to produce fully dense parts of complex shape, without the need to apply a mold that might react with and harm the component. Hardtl has applied this inexpensive method to the fabrication of several important ceramics, including alumina, some ferroelectric oxides, and some ferromagnetic oxides.

—Norman M. Tallan

Metallurgy

An understanding of the mechanisms of the chemical reactions that occur between slag, metal, and gaseous phases in pyrometallurgical smelting and refining processes has been a goal of metallurgists for several decades. Within the past year, significant advances have been made in the fields of slag chemistry and in the kinetics of high-temperature gas-liquid reactions.

The constitution of metallurgical slags. Attainment of an understanding of slag chemistry has been hindered by the complexity of metallurgical slags, which can contain up to 20 elements with concentrations each exceeding 1%. The fundamental questions of interest are: what determines the rates at which slag-metal refining reactions occur; to what extent do the reactions proceed before thermodynamic equilibrium is attained; and what determines this equilibrium state? Answers have been sought through determination of the structure of slags on the atomic scale.

The natures of the ionic complexes present in slags are determined primarily by an acid-base equilibrium established in the melt. Basic oxide components such as calcium oxide (CaO), magnesium oxide (MgO), manganese oxide (MnO), and iron(II) oxide (FeO) provide positively charged ions (cations) and free oxygen ions in the melt, and acidic oxides such as silica (SiO_2) and phosphorus pentoxide (P_2O_5) remove free oxygen ions to form negatively charged ions (anions) of varying complexity. This equilibrium determines both the nature of the complexes and the nature of the interactions that occur among the various ions present. Thus, the actual structure of the slag melt and, hence, its physicochemical properties are determined by the concentrations and acidities or basicities of its components.

A better understanding of these relationships has been provided by application of the principles of classical polymer theory to binary systems composed of one basic oxide and silica as the acidic oxide. On the assumption that such melts contain a range of sizes of linear-chain and branching-chain polysilicate anions, application of the statistical methods of polymer theory has permitted the most probable size distribution of the polysilicate ions to be calculated as a function of the degree of polymerization in the melt. This has led to the development of a thermodynamic and structural model of binary silicate melts in which the size distribution of ions and the thermodynamic activities

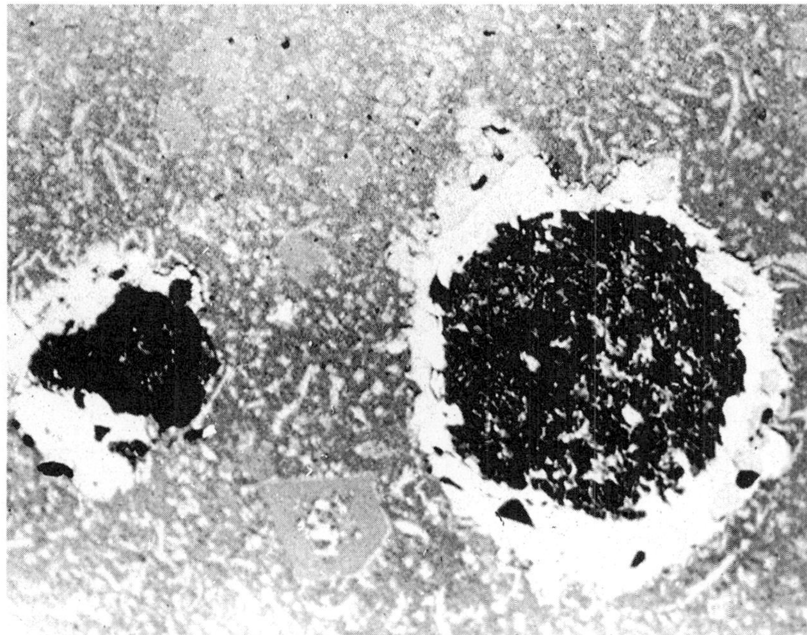

Courtesy, Carlos A. Villachica, Colorado School of Mines Research Institute

Microphotograph of cooled slag produced in the smelting of lead ore depicts two metallic lead globules surrounded by magnetite in a matrix of silicate glass. In the furnace, lead and iron dissolved in the molten slag as Pb^{2+} and Fe^{2+} together with oxygen form a uniform silicate phase. Tapping of slag from the furnace disrupts this gas-metal-slag equilibrium, inducing a phase separation according to the reaction $2Fe^{2+} + Pb^{2+} \rightarrow Pb$ (metal) $+ 2Fe^{3+}$ (in magnetite).

of components of the melt can be calculated as functions of composition and the relative basicity of the basic oxide present. Excellent agreement between theoretical prediction and experimentally measured thermodynamic properties has been obtained for melts between the basic metal oxide and orthosilicate compositions. Deviation of the theoretical predictions from actual behavior between the orthosilicate and metasilicate compositions suggests that, in this range, significant concentrations of ring silicate anions form in the melts. At the present state of knowledge, however, polymer theory cannot treat quantitatively the probability of ring-ion formation.

Any prediction of structure with respect to kinetic processes must be examined experimentally, and research toward this end is currently in progress. The experimental method involves extracting silicate ions from quenched slags as their trimethylsilyl derivatives; *e.g.*, the SiO_4^{4-} monomer and $Si_2O_7^{6-}$ dimer ions are extracted as $SiO_4[Si(CH_3)_3]_4$ and $Si_2O_7[Si(CH_3)_3]_6$, respectively. These derivatives are sufficiently volatile that they can be separated and quantitatively determined by gas-liquid chromatography and can be identified by mass spectrometry. Examination of quenched lead silicates and cobalt silicates has shown the presence of SiO_4^{4-}, $Si_2O_7^{6-}$, the linear trimer $Si_3O_{10}^{8-}$, and the cyclic tetramer $Si_4O_{12}^{8-}$ in both systems. The ratio of monomer to dimer measured in these systems differed from the theoretical prediction, which was expected in view of the existence of a ring ion and the mathematically necessary exclusion of such ions in the model.

The complexity of actual metallurgical slags has been demonstrated by the results of trimethylsilylations of samples of slags from iron blast furnaces and open-hearth steelmaking. The extracts from these slags were found to contain 26 types of discrete silicate and silicophosphate anions in the form of linear-chain ions up to the tetramer; simple trimeric, tetrameric, and pentameric ring ions; tetrameric rings with one and two internal bridges; and pentameric rings with one, two, three, and four internal bridges. For each ion type, both silicate and silicophosphate ions (in which one silicon atom is replaced by a phosphorus atom) were found. However, in the open-hearth slags, linear-chain and the cyclic-tetramer silicate ions accounted for 86% of the total anion population, while for iron-blast-furnace slags, this figure rose as high as 95%. In the open-hearth slags, 8% of the ions occurred as linear silicophosphates, and in accord with theoretical prediction, the relative proportions of the linear-chain ions present decreased with increasing polymer size.

Recently, this research has been extended to include examination of the influence of fluorides on the constitutions of slags. Preliminary measurements in the system $PbO–PbF_2–SiO_2$ show the occurrence of singly fluorinated linear-chain ions up to the tetramer, doubly fluorinated chain ions from the dimer to the tetramer, the triply fluorinated linear trimer, and the ring ions $Si_4O_{11}F^{7-}$, $Si_5O_{14}F^{9-}$, and $Si_7O_{18}F^{9-}$. This work has particular significance with respect to the current problem of finding a substitute for fluorspar (CaF_2) as a slag flux and fluidizer, which for years has greatly facilitated steelmaking operations. However, rising costs and concern for the ecological effects of fluorine emissions from steelmaking furnaces have prompted steelmakers to seek a substitute for CaF_2. Determination of the effect of fluorides on the ionic constitution of silicate melts is expected to provide the type of information necessary for the selection of such an alternative.

Kinetics of a decarburization reaction. In recent years an increased awareness of the importance of the concepts of mass transport on a macroscopic scale as employed in traditional chemical engineering has aided metallurgists in their approach to the experimental study of the kinetics of high-temperature reactions. One high-temperature gas-liquid reaction of considerable industrial importance is the decarburization of liquid iron-carbon alloys by gaseous oxidants, a reaction whose experimental study has produced many differing opinions as to the nature of the reaction rate-controlling mechanism. In 1975 metallurgists analyzed the previous investigations of one type of this reaction, in which carbon dioxide (CO_2) gas introduced to the molten-alloy surface results in decarburization via the overall reaction: CO_2 (gas) + C (dissolved in the liquid iron) → 2CO (gas). They determined that in most cases the measured rates were limited solely by mass transport of gaseous CO_2; *i.e.*, that the rate at which CO_2 gas was introduced to the metal surface was slower than the inherent chemically controlled rate at which the reaction was capable of proceeding. Consequently, a novel experimental arrangement was designed to eliminate the influence of gaseous mass transport by providing a sufficient flow of gaseous CO_2 to the melt surface. Under such conditions, it was found that the rate of decarburization was proportional to the partial pressure of CO_2 in the reactant gas and, for carbon concentrations of at least 1%, was independent of carbon concentration in the melt. These observations suggested that chemisorption of CO_2 from the gaseous phase onto the surface of the liquid alloy was the reaction rate-controlling step. However, it also was found that differing rates of decarburization were measured with iron-carbon melts prepared from nominally pure irons obtained from different commercial suppliers. Further investigation attributed this variation to differences in the sulfur contents of the nominally pure irons; *e.g.*, at 1,450° C (2,640° F), the occurrence of 20 ppm (parts per million) of sulfur in the melt decreased the rate constant for the decarburization reaction by a factor of two below that obtained with an essentially sulfur-free melt.

Analysis of the dependency of the decarburization rate on the sulfur concentration showed that the effect could be quantitatively explained in terms of ideal Langmuir adsorption of sulfur on the surface of the liquid iron, which, by decreasing the number of surface sites available for the adsorption of CO_2 from the gaseous phase, decreased the decarburization reaction rate. Hence, in addition to being proportional to the partial pressure of CO_2 in the reactant gas, the reaction rate was proportional to the fraction of the iron surface not already occupied by adsorbed sulfur atoms. Quantification of this effect showed very good agreement with earlier measurements of the influence of adsorbed sulfur on the surface tension of liquid iron.

—David R. Gaskell

Mathematics

Progress in mathematics results from creative tension between theory and application. As the language of science, mathematics grows both as new ideas are expressed in mathematical terms and as these ideas find employment in the various sciences. Research in the mathematical sciences ranges from the most abstract forms of pure thought to the pragmatic demands of scientific computation. While traditional applications of mathematics have depended largely on such classical tools as calculus, algebra, and probability theory, in the mid-1970s mathematical research has reflected a renaissance of interest in applications made possible by theoretical advances of recent years.

Typical of these advances are three that will be examined in some detail: catastrophe theory, computational complexity, and nonstandard analysis. Catastrophe theory arose from an application of topology, the study of geometric properties of surfaces as they are bent and distorted. The analysis of computational complexity is part of the larger field of combinatorics—the study of the patterns in which things can be arranged. And nonstandard

"I tend to agree with you—especially since $6 \cdot 10^{-9} \sqrt{t_c}$ is my lucky number."

analysis, a revolutionary new approach to classical calculus, is deeply rooted in mathematical logic, the study of the reasoning processes of mathematics itself.

Catastrophe theory. The premier achievement of mathematics is the astonishingly effective synthesis of the calculus of smooth change—a purely abstract construct—with the behavior of a mechanical system—a purely physical event. This synthesis, articulated first by Newton in his *Principia* and refined later by such geniuses as Euler, Gauss, and Einstein, established the relation between calculus and physics as a paradigm for all scientific investigation.

Despite this general success, Newtonian theory and its successors do not really provide adequate models for such physical processes as earthquakes and cell differentiation, in which continuous causes produce sudden, discontinuous consequences. Recent work by the French topologist René Thom has resulted in a new geometric theory of such processes, called catastrophes. While scientific application of Thom's theory remains in the future, a small coterie of followers led principally by E. Christopher Zeeman of the University of Warwick, England, has applied catastrophe theory to a wide range of natural and social processes and discovered it to be an unusually apt metaphor for scientific investigations.

A good example of one common catastrophe—the cusp catastrophe—can be exhibited with a piece of paper gently folded to form a pleat. If points on the pleated sheet of paper represent states of the stock market as determined by investor demand and speculation, then a crash of the stock market can be explained as the result of a demand-speculation structure that forces the market out over the cliff edge on the upper sheet of the pleat. Exact prediction of stock market crashes based on the cusp catastrophe model would require analysis of large amounts of data; current research efforts are focused primarily on testing the new model rather than on using it as a forecasting device.

It is in developmental biology that catastrophe theory thus far has enjoyed its greatest scientific success. Zeeman and his colleagues have used catastrophe models to study such diverse phenomena as the firing of a neuron in controlling heartbeat and the division of a cell. Like many other basic life processes, each of these events is a discontinuous consequence of smoothly varying causes. The growth of an embryo is motivated by the continual ingestion of food, yet its behavior is a succession of discontinuities (cell division and differentiation) separated by gradual change in cell size and position. Catastrophe theory provides a model that makes it possible to explain how a cell divides and to predict when it will divide. Perhaps more importantly, catastrophe theory also explains why cells divide. According to Thom's central theorem, a cell must divide because the geometry of the chemicals within the cell makes it impossible for it not to do so; it is riding on the surface of a catastrophe and at some point must fall over the edge.

The heart of the theory of catastrophes is a surprising and very difficult theorem proved by Thom in the late 1960s. It assigns every physical discontinuity to one of seven elementary types. Specifically, in a system restricted to four control variables (for instance, the three dimensions of space and one of time) all possible catastrophes must follow one of seven specific paradigms. The cusp catastrophe is one of the most elementary paradigms; others are more complex, with exotic names like the swallowtail catastrophe and the hyperbolic umbilic catastrophe. Each arises from unfolding a singularity on a surface, just as the pleat on the cusp catastrophe surface unfolds the singularity that occurs where the single surface splits into three sheets.

Thom's classification theorem is immensely powerful, for it says that discontinuous processes that result from continuous causes are forced, by the logical nature of geometry rather than by the empirical facts of science, to fit one of seven elementary molds. Catastrophe theory thus opens up an entirely new approach to mathematical modeling of scientific phenomena, one that many believe will do for the inexact sciences of the 20th century what Newton's calculus did three centuries ago for the fledgling sciences of physics and astronomy.

Computational complexity. A great many practical problems concerning scheduling—for instance, of school classes, of assembly-line processes, of airline flights—involve so many combinations of events that even the fastest computers cannot cope with them. Many problems grow exponentially with the number of variables; adding one more variable is likely to double the amount of work involved in achieving a solution, and adding ten variables can increase the work a thousandfold.

A good example is the common task confronting telephone companies when they determine charges for corporate customers who rent entire networks to link their various offices. Charges are based on the network of least total length that could be used to link the given offices, rather than on the network actually used by the telephone company, which may be much longer than actually necessary. All of the known methods for solving

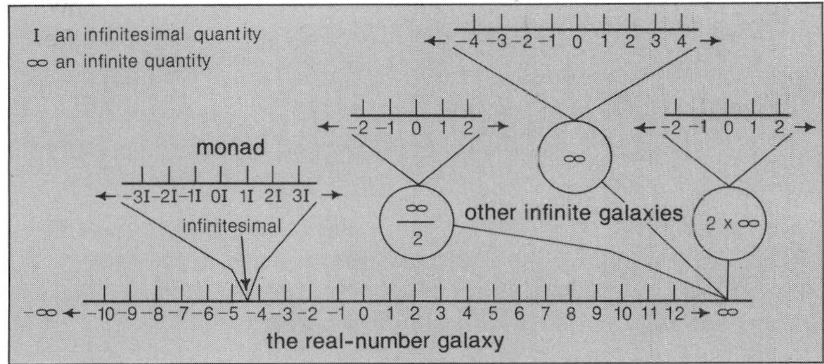

Nonstandard numbering involves expansion of the real number sequence at the left. New number monads and galaxies are generated by consideration of infinitesimally small and infinitely large quantities.

this problem grow so rapidly with the size of the network that, practically speaking, it is impossible to find exact solutions for large networks linking 50–100 cities.

Typically, mathematicians and computer scientists use various algorithms—that is, step-by-step solution procedures—to attack combinatorial problems. Ideally, before a proposed algorithm for a particular problem is programmed on a computer to carry out a solution, it is first analyzed theoretically to determine if it will actually yield a solution. For most complex scheduling problems, however, mathematicians have been confronted by a frustrating Catch-22 barrier: those algorithms known to work in theory were too inefficient to use on a computer, whereas those that actually could be used were not guaranteed to succeed.

Several years ago Stephen Cook of the University of Toronto, Ontario, and Richard Karp of the University of California at Berkeley found a way to isolate and identify certain combinatorial problems that were essentially equivalent in the sense that an algorithm which would solve one of them efficiently could be adapted to solve them all. This class of problems, termed NP complete, is currently the subject of intensive investigation to determine how extensive it is and whether or not it is possible to solve any (and hence all) NP-complete problems by any efficient algorithm. In late 1975 Ronald Graham and his colleagues at Bell Laboratories succeeded in showing what had been suspected; namely, that the minimal telephone network problem is NP complete.

The major conjecture, as yet unproved, is that no efficient algorithms exist that can provide exact solutions to NP-complete problems. Accordingly, many researchers have abandoned hope for theoretically perfect algorithms and are seeking good approximate ones. Others are attempting to determine whether these problems indeed cannot have efficient algorithms.

Nonstandard analysis. Tough combinatorial problems arise because of the difficulty of dealing with large but still finite numbers of events. Other mathematical models require analysis of infinite numbers, and recent applications of mathematical logic have provided surprising new insights into this esoteric and nearly unimaginable realm. Over a decade ago, Abraham Robinson of Yale University discovered that there exist, in addition to the standard system of so-called real numbers (which include positive and negative whole numbers, fractions, and irrational numbers), certain larger systems that contain both infinitely large and infinitesimally small numbers. At a conference at Yale in May 1975, on the first anniversary of Robinson's death, logicians, mathematicians, physicists, and economists gathered to explore ways in which this nonstandard analysis of infinitely large and infinitesimally small numbers could provide new approaches to outstanding problems in pure and applied mathematics.

Just as the solution to a combinatorial problem requires a precisely stated algorithm, so too the description of an infinitely large or infinitely small number requires a precise definition setting forth its properties. For the last hundred years, mathematicians have worked quite successfully with a definition of number that glossed over certain fine points. Robinson explored the relation between the language in which mathematics is expressed and the objects described by that language to produce a totally new model of numbers that contains "galaxies" of infinite numbers and "monads" of infinitesimally small numbers.

Economists and physicists find these new models both useful and exciting because, for the first time, they can extend directly to infinite numbers the intuition and reason that have been honed on finite numbers. Scientists often need to analyze infinite models because such models are good approximations of excessively complex situations involving a large finite number of events. Formerly, when approaching an infinite model, they had to develop ad hoc devices with little scientific reality simply to avoid fatal contradictions in the mathematical model. Nonstandard methods permit scientists a direct transition from well-known finite models to little-known infinite ones.

In economics, for instance, many theories assume as a hypothesis that individual traders have only an insignificant or infinitesimal influence on the market as a whole. Market conditions are determined, instead, by the aggregate behavior of all individuals acting simultaneously. This very natural principle of economic theory is awkward to express in conventional Newtonian calculus. But it has proved quite amenable to nonstandard analysis through a direct translation from finite to infinite computations.

—Lynn A. Steen

Mechanical engineering

Steady state as an engineering description implies a use of energy and materials such that the output of both equals the input. An oil-burning steam power plant operates for much of its life at steady state. Its outpouring of electrical and waste energy and of material from its exhaust stack and condenser equals the energy and material supplied to it in the forms of coal, air for combustion, and seawater to cool the condenser. An automobile engine is operating at steady state as it propels a car at a constant speed along a level road. Steady state is a useful mode of operation for many machines and also has relevance to societal affairs.

Some societies in times past and some less developed countries today have approached a nearly steady-state mode of material, energy, and population existence. Water and land management yielded food and material supplies that were balanced against societal wastes. Nature would then provide, through sunlight, air-water-earth chemistry, and links with other living systems, a restoration of materials and energy. The community thus maintained or approached a steady-state material balance.

Mechanical engineering has played a major role in dislocating Western civilizations from this primitive equilibrium. There were good reasons for doing this, the most important being that whims of nature in the forms of unexpected cold or hot weather, droughts, floods, fires, and disease continuously tormented man and often demanded heavy toil and suffering from him. To forestall such disasters engineers in modern times have used energy and materials accumulated in the Earth's crust over centuries to achieve a metamorphosis in the standard of living of people in the industrialized nations. There is an inclination on the part of many to preserve the status quo; that is, to stay in this metamorphic or nonsteady state. The current and projected stresses imposed on the resources and environment of the Earth by such an outlook demand, however, that it be rejected. One alternative view would be a return to a primitive existence, but this would result in mass privation and starvation.

Contemporary scientific and engineering knowledge in conjunction with enlightened leadership can make available another choice. This is to restore steady-state operation on a worldwide basis so as to come again into balance with all other living things but on a far more comfortable and controlled plateau than any past societies have been able to achieve.

Mechanical engineering activities are tending in this direction. Many mechanical engineers believe

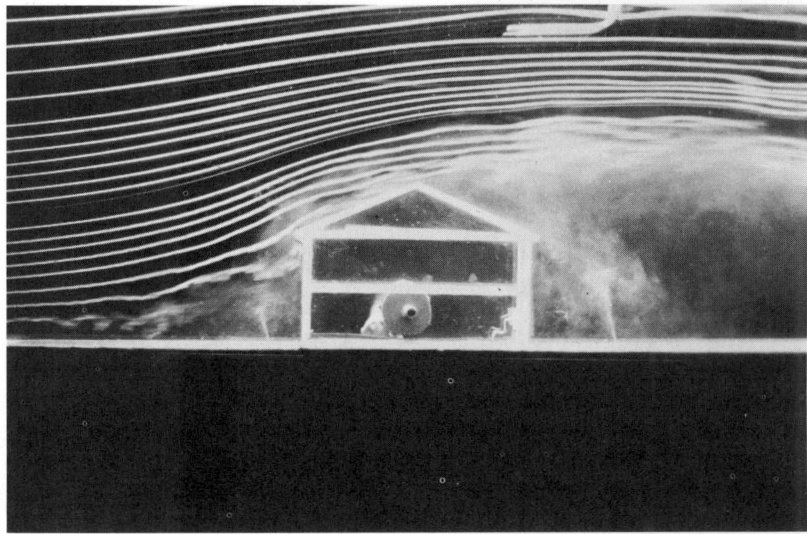

Wind-flow patterns around a house are visualized by placing a model of a New Jersey townhouse in a wind tunnel that has smoke seeped into it. The experiment is part of a study of the effects of wind speed on energy consumption in a house.

Courtesy, G. E. Mattingly, Princeton University, Princeton, N.J.

Courtesy, Richard K. Pefley

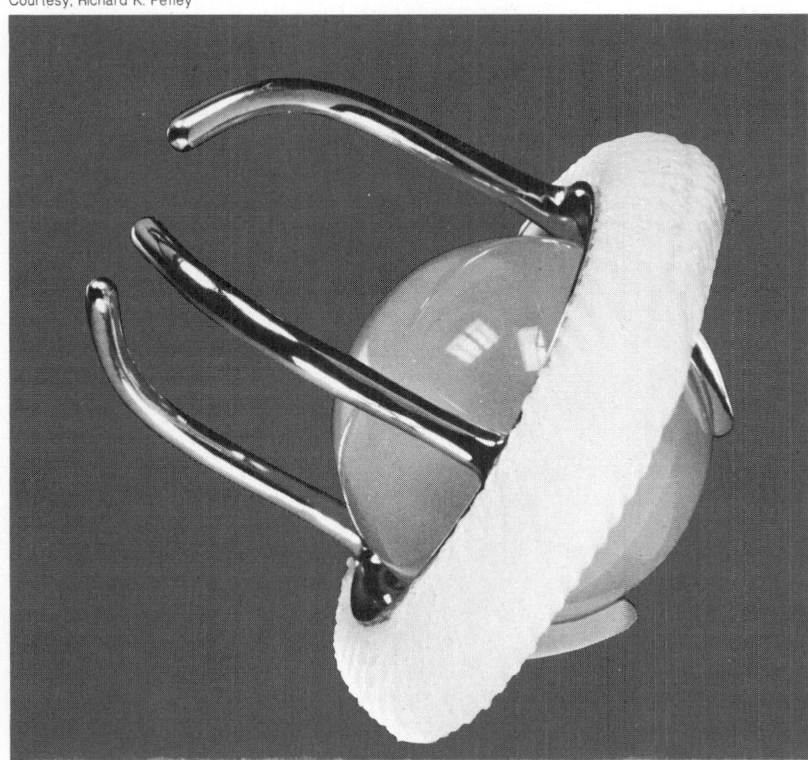

Prosthetic heart valve made of plastic was developed jointly by mechanical engineers and cardiovascular surgeons in California.

that their main task is to strike a better balance between supplying society with energy and products and maintaining the quality of the landscape and atmosphere. In this regard, some possibilities of the near future can be considered.

Coal, once a major partner in the supply of energy, may return to a dominant position. But, since the time of coal's previous dominance, concern has grown for the quality of the environment. How can the use of larger quantities of coal be resumed without spoiling the landscape and fouling the atmosphere? The equipment and mechanical engineering skills exist to achieve such goals. Establishment of demonstration projects could restore the confidence of the public that these concerns can be met.

The coal, instead of being burned and producing a dirty exhaust, can be processed into a sulfur-free clean gas known as "synthesis gas." This can be used in place of natural gas. When burned correctly in home heaters, steam power plants, or gas turbines, it can significantly lower the oxides of nitrogen, which are a worrisome exhaust pollutant. Alternatively, the synthesis gas can be processed into a liquid fuel known as methanol. This fuel is an alcohol sometimes referred to as "wood alcohol."

Methanol is attracting engineering and scientific attention for many reasons. Being a liquid, it is easy to store and transport. It can be made from any cellulose fiber, such as municipal, agricultural, or logging wastes. This leads to an appealing way to harvest the Sun's energy: Grow plants, process them into methanol, return the minerals to the soil, and store the energy for subsequent use. Methanol thus offers a means for changing from the use of the rapidly shrinking capital energy resources, such as petroleum and natural gas, to coal and, subsequently, to the direct harvest of solar energy.

Research investigations indicate that methanol can maintain an automobile's performance, improve the engine's efficiency, and reduce its exhaust emissions. It is also a satisfactory fuel for gas turbines and steam power plants, which have been operating on gas and oil. Methanol is safer than gasoline from a fire hazard viewpoint; water easily extinguishes a methanol flame because the fuel is soluble in the water. Massive land and water spills of methanol are more biodegradable than are oil spills.

Engineering support for medicine has grown rapidly since the advent of plastic materials. This area provides an excellent example of how modern technology can achieve a steady-state material balance that is an improvement in the quality of life in comparison to the steady-state existence of times past. As a case in point, the Smeloff-Cutter prosthetic heart valve was a joint venture involving faculty from the department of mechanical engineering at California State University, Sacramento, and cardiovascular surgeons from the Sutter Hospital's Medical Research Foundation. The group worked together as a team to develop an improved prosthetic heart valve. Many aspects of mechanical engineering were involved. Once the basic concept of the caged ball prosthesis was established, a heart simulator had to be designed and constructed so that hemodynamic flow studies could be done on the valve. Material compatibility studies, stress analysis of critical parts, and fatigue testing of the entire structure were conducted. Valves were thoroughly tested in animals before clinical implantation was undertaken. Finally, manufacturing methods and quality control standards had to be established.

The first clinical implantation of the valve was in 1963. Modifications were made in 1966, and the valve has remained unchanged since that time. More than 30,000 valves of this design have been implanted, and as of 1976 it was the most widely used ball-valve prosthesis in the world.

It has been suggested that improvement of the quality and precision of performance of consumer products would be superior to continually increasing the quantity of material possessions. Mechanical engineers have tried to achieve that goal, as is evident if one examines the evolution of many products. An example that can be cited is the supplying and retrieving of information to the storage disk file of a computer.

Inductive magnetic recording in disk files is accomplished by the relative motion between a stationary read/write element and a high-speed rotating disk. In general, the read/write element is embedded in a spring-loaded slider bearing that is separated from the moving disk by an air film that is a few millionths of an inch thick. This is necessary for the attainment of high recording density and transcription accuracy. Maximum density and accuracy are obtained if the read/write element and the disk are in contact. This results, however, in wear of both the read/write element and the disk. Thus, to satisfy the requirements of both high recording density and reliability, the slider containing the read/write element is designed to fly at the smallest possible spacing at which material wear is negligible.

This flying task can be visualized by imagining a football field (the slider) flying over the surface of the Earth (the recording disk) with a separation of one inch (the few millionths of an inch clearance between the slider and the disk) at a speed of more than 100 mph. After considerable research mechanical engineers succeeded in designing a slider that fulfilled this requirement.

—Richard K. Pefley

Medical sciences

Developments in medicine during the past year ranged from large increases in the cost of malpractice insurance for physicians to the introduction of a new drug that relieves pain caused by arthritis. Orthopedic surgeons made advances in the treatment of low back pain and in the replacement of joints, and dental researchers found that the use of plastic sealant materials reduces the rate of tooth decay in children. In the United Kingdom an investigative commission published a report that included recommendations for the long-term development of veterinary medicine.

General medicine

"There is a spirit of solidarity and activism throughout the profession," said the chairman of the American Medical Association's Board of Trustees in 1975. He spoke of the new militancy among U.S. practitioners of medicine, who were reacting with strikes, slowdowns, and lawsuits to years of criticism, rising malpractice insurance rates, and fresh governmental regulations. Although the fight against disease and death went on, accompanied by the usual setbacks and victories, the fighting U.S. physician was at the center of much recent medical news.

Malpractice and related issues. Malpractice, a term seldom heard until the early 1970s, was a major concern. More than 40,000 claims were filed in 1975, an increase of over 60% since 1970 according to AMA estimates. There is no discernible pattern to either the problem or proposed solutions. When the immediate past president of AMA, California surgeon Malcolm Todd, announced that he could no longer afford annual liability insurance payments of over $31,000, he typified the frustration of many doctors, whose policy rates were expected to increase 170–500% in 1976. Todd planned to continue to practice, but without insurance. Many physicians, including up to 30% of those in Alaska, were doing the same.

Others chose to protest in stronger ways. The first large-scale doctors' strike in U.S. history took place in March 1975, affecting 22 New York City hospitals. The primary issue was working conditions of the house staff (interns and residents). With the support of organized medicine, more than two dozen similar "job actions" followed, primarily protests against high malpractice insurance rates. A lengthy slowdown virtually halted elective surgery and other routine procedures in more than 100 southern California hospitals.

Some 400 legislative bills were introduced in virtually every state legislature in attempts to solve

Russell Reif from Pictorial Parade

Striking interns and residents picket outside the Klingenstein Clinical Center of New York City's Mount Sinai Hospital in an effort to achieve better working conditions.

the malpractice insurance problem. Most of the bills that set claims limits were struck down by the courts. The administration of U.S. Pres. Gerald Ford maintained that federal solutions cannot be invoked until the states have tried and failed.

National malpractice statistics began to emerge during the year. Surgeons, anesthesiologists, obstetrician-gynecologists, and cardiologists were the most likely to be sued, according to the National Association of Insurance Commissioners. Preliminary statistics indicated that judgments tended to be higher against physicians who lack certificates of advanced training (known as board-certification). A five-year study by the American College of Surgeons and other national medical groups showed, meanwhile, that perhaps as many as 30,000 doctors perform surgical procedures who should not be doing so. The report recommended restricting most surgery to either board-certified clinicians or those in training.

Professional competence throughout the medical profession became another storm center. Todd and other officials estimated that 5–6% of the 320,000 physicians in the U.S. may be incompetent to practice. The causes vary from psychiatric illness and drug or alcohol addiction to simple failure to keep abreast of medical advances. A Gallup Poll of the general public indicated that 85% favored stricter policing of medical competency. The responses of medical leaders were mixed, although most agreed that current regulatory measures are inadequate.

Two proposals made during the year were opposed by most medical leaders. The U.S. Federal Trade Commission asked for an end to the traditional professional ban on public advertising, and the Food and Drug Administration began to develop "package inserts" in lay language to inform patients of the ingredients, side effects, and proper uses of drugs. The first of these inserts, for oral contraceptives, went on the market in 1975. In the past, package inserts had been solely for the information of physicians and pharmacists and were written in scientific language.

The best publicized medical controversies took place in the law courts. Renewing the "death with dignity" debate, the parents of Karen Ann Quinlan, 21, sued to remove their foster daughter from a respirator that had kept her alive since she entered

"Sorry I'm late, but it took them about a week to determine I was legally dead."

Photos, courtesy, Children's Hospital of Philadelphia

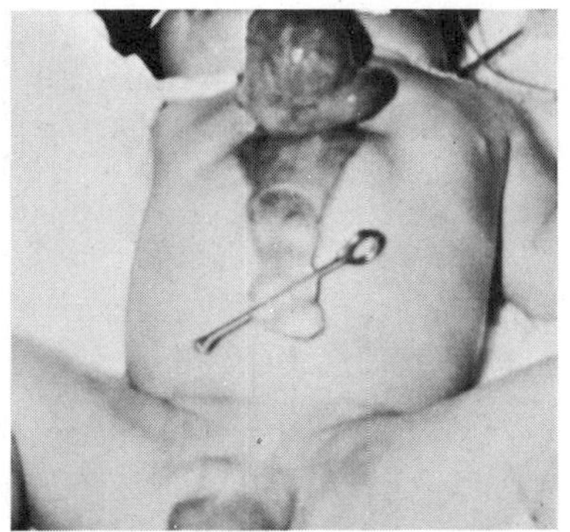

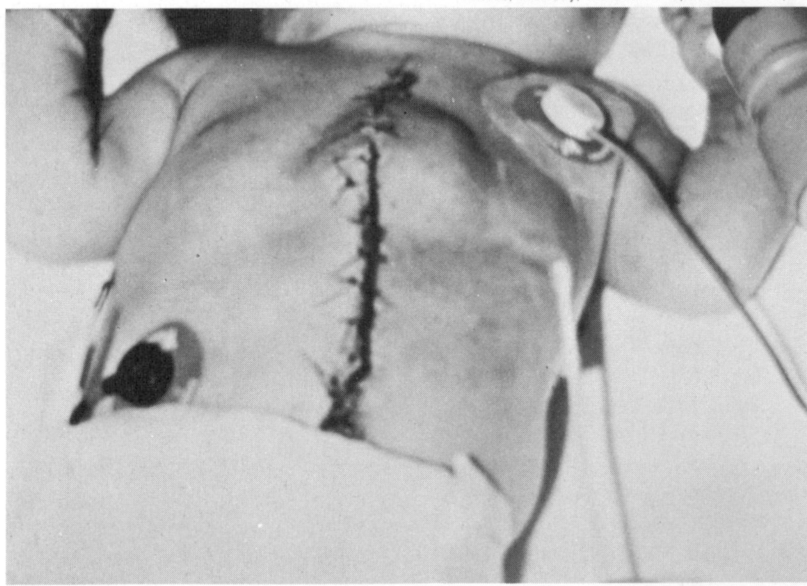

Baby born with his heart entirely outside his chest (above) became the first child with this defect to be saved when a Philadelphia surgeon closed skin flaps over the organ.

a coma in April 1975. The New Jersey Supreme Court ruled in March 1976 that the respirator could be disconnected if the attending physicians and a panel of hospital officials agreed that there was "no reasonable possibility" of recovery. Another verdict involved Boston physician Kenneth Edelin. A jury convicted him of manslaughter two years after he performed an abortion. The decision led to a sharp decline in the number of abortions done in the U.S. Edelin was sentenced to probation for one year.

Advances in treatment and diagnosis. Medicine was not all controversy. Successes were scored in several clinical areas. "The Heimlich maneuver," named for the Cincinnati surgeon who developed the technique, was able to save hundreds of the estimated 4,000 individuals who die each year from food or other particulate matter that becomes lodged in their throats. The choking victim is embraced from behind, just under the rib cage, and the rescuer's closed fist is thrust into the abdomen and upward. During the first nine months following public descriptions of the technique, Henry Heimlich received reports of 162 instances where it had been employed with success.

A procedure that reverses vasectomies with the aid of high-powered operating microscopes was developed by a team of urologic surgeons in San Francisco and Sydney, Australia. Some four million vasectomies have been performed in the U.S. since 1970, according to the Association for Voluntary Sterilization. Ordinarily, efforts to restore procreative function are successful in about 25% of cases. The microsurgery technique, done primarily on men who have remarried, was successful in more than 75% of the patients.

A rare but spectacular birth defect, ectopia cordis, was corrected for the first time by surgeons at the Children's Hospital of Philadelphia. Christopher Wall, Jr., was born with his heart outside his chest. Only about 200 such cases have been recorded in five centuries, but the surgeons were able to place the heart close to its natural resting place by following directions in a theoretical paper published just two years ago.

Heart transplants, a surgical "miracle" of the 1960s, continued to be done, cautiously. About 50 of the 286 who had received hearts since 1967 survived in January 1976. Following the death in 1975 of Charles Perrin Johnston of Canada at the age of 61, the longest-term survivor was a 56-year-old Wisconsin woman, who received her transplant in 1968. Johnston, who died of lung disease, had a "perfect heart" nearly seven years after his operation. Fighting the body's natural tendency to reject foreign tissue continued to be the principal problem, said Norman Shumway of Stanford University, who had operated on over a third of all of the heart transplants.

Plastic and reconstructive surgeons reported success in replacing large gaps in bone and cartilage with individually formed trays of bone chips. Cartilage then fills in the defects, growing through the bed of chips. The technique was used for accident victims who had lost parts of their skulls, lower jaws, and ears.

Plastic, steel, and rubber were successfully formed into artificial knuckles and toe joints to replace those painfully deformed by arthritis. The artificial knuckles enable patients to regain almost full use of their hands, including the ability to grip and pinch, about six weeks after the operation.

Greatly improved strength and movement, with resulting ability to walk distances, was obtained with the toe implants.

A new drug relieving arthritic pain, with fewer bad effects than aspirin, was among the 37 drugs introduced in the past two years. Ibuprofen, as it is known generically, does not usually cause the gastrointestinal bleeding or stomach upsets many arthritics experience when using aspirin. Other new drugs were aimed primarily against infections, including those caused by previously resistant bacteria and fungi.

Disease detection continued to become more sophisticated. Radiologists at Wayne State University, Detroit, reported that soft-tissue X-rays of the breast, called mammography, can be used to detect those women who are most likely to develop that type of cancer. Previously, the technique had been thought effective only in detecting cancer already present in the breast, but the Detroit team was able to detect precancerous tissue changes on the films.

New York immunologists reported finding a substance in the serum of the blood that seems associated with the presence of cancer in the ovaries. If further research confirms this antigen, more specific and early detection will be possible. Examinations of the blood also can be used to identify carriers of hemophilia, again by detecting the presence of a specific antigen. The technique confirmed the genetically transmitted characteristic in 53 of 56 known carriers.

Space-age technology resulted in U.S. Air Force development of a portable blood-pressure recorder that buzzes when patients reach dangerous levels of hypertension. The device uses a sensitive, tiny microphone taped to the skin over the brachial artery. In Cincinnati, Ohio, physicians at Jewish Hospital were working with former U.S. astronaut Neil Armstrong to adapt for general use the thermal fluid pump that kept his body temperature stable as he walked on the Moon. The pump, smaller and more efficient than those currently in operating rooms, would be used to circulate blood during heart-lung surgery.

Severe psoriasis, spread over major portions of the body, was treated with a combination of drugs and ultraviolet light by dermatologists at Harvard University and the University of Vienna. Of 300 patients, 294 achieved at least 90% clearing of this skin disorder for more than a year after undergoing the treatment.

Corns and calluses are relieved by the injection of a small amount of liquid silicone between the growth and bone, a Los Angeles physician reported. The cushion prevents pressure and pain. Silicone injections also are useful in smoothing sunken scars, facial wrinkles, and other deep lines, according to a New York dermatologist. Liquid forms of silicone, however, continued to be banned from large-scale injections, such as those for augmenting breasts.

For the third consecutive year, public health organizations and the U.S. Public Health Service Center for Disease Control (CDC) scheduled October as Immunization Action Month. But despite their efforts an estimated 5.3 million U.S. children, mainly of preschool age, remained vulnerable or incompletely immunized to common diseases for which vaccines exist, such as poliomyelitis, mumps, rubella (German measles), and rubeola (red measles). As a result, the incidence of these

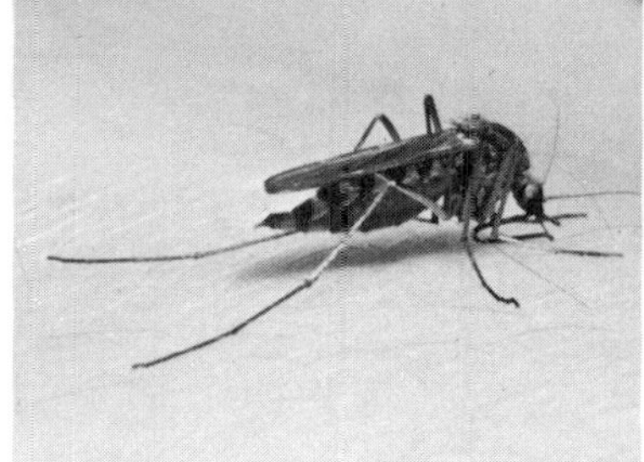

(Left) UPI Compix; (right) Russ Kinne from Photo Researchers

Helicopter sprays malathion insecticide on a forest preserve near Chicago in an effort to kill the Culex pipiens *mosquito (above), responsible for a severe outbreak of encephalitis in the U.S. in 1975.*

diseases was beginning to edge upward. In 1975, for example, 16,209 rubella cases were reported, compared with 11,845 in 1974.

The "flu season" in the U.S. was comparatively mild in 1975, despite large epidemics of type A influenza elsewhere in the world. The nation was not so fortunate with encephalitis. The mosquito-borne illness appeared in 26 states and the District of Columbia during the late summer and fall. Some 3,000 victims were reported, with those in the Ohio and Mississippi river valleys having the St. Louis strain and those in the Red River valley area (Minnesota, North Dakota) contracting the so-called Western equine type. Winter killed both the epidemic and the mosquitoes.

Scabies, food poisoning, and venereal disease, all diseases preventable by humans, not vaccines, also increased. Although the number of cases could not be estimated, the CDC reported that scabies has been increasing since 1963 and that by 1975 all socioeconomic groups were involved. The skin disorder is caused by a tiny mite, spread by close physical contact. While the nationwide incidence of syphilis remained about the same as in 1974, nearly a million reports of gonorrhea were submitted to the CDC, close to 100,000 more than in 1974. The sharpest increase was among males.

Often mistaken for the "24-hour bug," food poisoning affects more than two million persons each year. The primary agent, transmitted by faulty food handling or processing, is salmonella. This organism is found most frequently in raw meats, poultry, and egg products.

The chronic use of marijuana can result in the impairment of lung function, a study at the University of California at Los Angeles indicated. The impaired function was detected after subjects smoked an average of 5.2 marijuana cigarettes per day for 11 days.

Hepatitis struck more than 55,000 persons in 1975, a "typical" year for the liver-damaging disease. The U.S. government proposed that blood taken from paid donors be clearly labeled. "Skid row" blood is thought to be a principal source of type B hepatitis. Illinois has required such labeling since 1973 and has had an estimated reduction of 60% in the disease. California passed a similar law in 1975, over the opposition of commercial blood banks and many physicians.

The viruses believed to cause both type B hepatitis and the more common type A variety were identified during the year by several medical centers. At least two kinds of type B vaccine were being tested in chimpanzees. Studies at Rush-Presbyterian-St. Luke's Medical Center in Chicago indicated, however, that oral or sexual contact, in addition to transfusions, can transmit the type B disease. This would make control by vaccine more difficult. In addition, a New York Blood Center report indicated that there may be a third, type C, hepatitis virus. Approximately 8,000 hepatitis cases could not be identified as either type A or type B in 1975.

Sexual contact also transmits a herpes-type virus associated with enhanced risk of cervical cancer, scientists at Emory University, Atlanta, Ga., confirmed. Antibodies to the specific antigen of this virus were being used as a way to detect this type of cancer.

Several studies, suggesting an association between cancer of the mucous membrane of the uterus (endometrium) and the use of estrogen to relieve symptoms of menopause, led to advice for cautious further use of hormonal therapy. In some studies the incidence of endometrial cancer was seven times higher in estrogen users than in non-users, but scientists could not exclude the possibility that some other factor might be responsible. Obesity, hypertension, diabetes, and parity (number of children) all have been linked with an increased uterine cancer risk.

Sexual dysfunction, including the inability to have orgasms, may be related to biochemical factors, according to studies at the State University of New York at Stony Brook. Higher levels of glucose and insulin, as well as lower levels of nitrogen, phosphorus, cholesterol, and calcium, were found in the blood of "nonorgasmic women." Whether or not such factors affect brain chemistry as it relates to sexual response was being investigated.

Maternal death rates continued to decline in the U.S., a ten-year survey revealed. The most marked decreases took place among non-white women. Improved prenatal care, reducing the risk of such serious complications of pregnancy as toxemia, was credited with a major role in the decline, which in 1975 stood at fewer than 19 deaths per 100,000 live births.

Perhaps the most feared disease worldwide, smallpox, had been declared eliminated by the World Health Organization. It was, however, found to be still alive in 1975 when nine new cases were discovered in Bangladesh. Smallpox caused three million deaths as recently as 1967. WHO officials believed that 1976 would mark the end of a fight against the disease that began with Edward Jenner's introduction of smallpox vaccine in 1796.

Physician-patient relations. A nationwide survey by the U.S. National Center for Health Statistics was released, estimating that the average person visits a physician 3.1 times a year. The survey, to be made annually, found that 60% of the visits were by females and that more than half were of 15 minutes or less.

Family practitioners, formerly known as "GPs" (for general practitioners), were the clinicians consulted in some 40%, or about 258 million, of those patient visits. This illustrates the continued pressure on a branch of medicine where a continued shortage of personnel has been acknowledged. The number of medical schools teaching family practice doubled from 1973 to 1975, and 58% of newly graduated doctors were entering this specialty in 1975, according to AMA statistics. An American Academy of Family Physicians survey, however, indicated that nearly one-half of the new family practitioners chose to work in communities of 30,000 or more. The most serious shortages of doctors exist in smaller communities. The federal government allotted almost $40 million for training family practitioners, but any plans to correct physician maldistribution by legislative action were opposed by medical organizations. The U.S. House of Representatives voted down a bill that would have regulated the number and type of residency programs, but approved a medical-student loan program that would allow repayment by service in doctor-poor areas.

A visit to the physician results in drug therapy almost 60% of the time, a national survey found. Additional millions attempt to cope with their illnesses by self-medication with old prescriptions or over-the-counter (nonprescription) products. The U.S. National Institute on Drug Abuse announced survey results indicating that one in three Americans during 1975 took a drug that would affect his mental function. Perhaps one-fourth of these did so without physician advice, and many of them developed difficulties in coping with others or even suffered overdoses from the drugs.

Health expenditures. The nation's "health bill" in 1975 reached a record $547 per person in fiscal year 1975, according to U.S. government statistics. This amounted to 8.3% of the gross national product, or $118.5 billion. The federal government spent $38.8 billion of this total, the remainder coming from private insurance and personal funds. The $14.5 billion jump in the cost of medical care for fiscal 1975 was the largest on record. By comparison, it exceeded the total national investment in health care for 1950 ($12 billion).

National health insurance, a federal program that would substantially increase health expenditures in the U.S., was ruled out as both politically and economically impractical in the climate of the recent recession. U.S. Pres. Gerald Ford proposed a program of coverage of "catastrophes," but a national "cradle-to-grave" system of coverage, previously proposed by both political parties, seemed to be indefinitely shelved. Ironically, the National Institute of Mental Health reported that the unemployment generated by the recession was associated with an increase in mental illness. At the same time, use of hospital beds declined for the first time in 15 years, due to patients delaying "elective" operations that, in a better economy, they might have allowed.

—Byron T. Scott

Courtesy, National Institutes of Health

Virologist uses laminar flow hood to perform tissue culture studies with the herpes simplex virus. This virus causes cold sores and is believed to be implicated in the development of certain forms of cancer.

Orthopedic surgery

An orthopedic surgeon may be defined as a physician who specializes in caring for the abnormalities of the musculoskeletal system that are due to injury, birth defects, disease, or degenerative influences. Thus, he treats problems that arise with the bones, joints, ligaments, muscles, and nerves of the human body. Orthopedics also might be said to consist of straightening or correcting skeletal abnormalities whether they are caused by nature or man.

The word orthopedic is derived from the Greek, meaning, freely translated, to straighten the crippled child. Orthopedic surgery has advanced quietly but spectacularly in the years since World War II. Before that war there were fewer than 1,000 orthopedic surgeons in the U.S., whereas in 1975 the number had grown to about 9,000. This great increase is partly due to the growth of the nation's population, of course, but it also has resulted from growth of knowledge and the availability of new procedures.

Before World War II orthopedic surgeons were sometimes derisively known as "brace and buckle" men. Their chief concern was care of the results of poliomyelitis, osteomyelitis, congenital deformities, and postural problems. They had a small repertoire of surgical procedures, but these were dangerous because of the great risk of infection in bones and joints. The advent of antibiotics made new and more imaginative operations possible. The other single great advance was the availability of precisely matched blood, a necessity in many procedures.

Other beneficial postwar developments also occurred: operating rooms became far more sterile; no visitors were allowed; special ventilation systems were established; and disposable gowns and drapes were used. Discoveries in metallurgy and technology were most important, affecting as they did implants, instruments, and even power tools.

Important advances have been made in the last decade with regard to a better approach to some of the previous "unsolved problems" of orthopedic surgery. The most significant of the new devices and concepts are the following: (1) new developments in lumbar disc surgery, including the use of the disc-dissolving enzyme chymopapain; (2) joint replacement devices for destroyed joints, especially the hip and the knee; (3) internal fixation "hardware," which has greatly improved the treatment of fractures; (4) improved braces and artificial limbs; (5) improved equipment for prevention of sports injuries; (6) improved X-ray techniques; and (7) new methods for treating curvature of the spine.

Low back pain. An affliction that affects an estimated 80% of the adult population of the U.S., pain in the lower back has many and varied causes. In recent years many patients who had been judged to be curable only by surgery were able to obtain relief by the use of chymopapain, an enzyme injected into the lumbar disc causing the pain. The U.S. Federal Drug Administration (FDA), however, withdrew its approval of the drug in late 1975. The reasons given were unclear, and orthopedic surgeons generally hoped that permission to use this method would soon be granted again.

To understand chymopapain treatment one must first ask "What is a disc?" A disc is a fibrocartilaginous, shock-absorbing material between the vertebrae. The lumbar region contains five vertebrae in the lower part of the back and is the area of most trouble.

The disc serves as a cushion between the vertebral bodies. It is usually about one-quarter inch thick and is as wide as the vertebral body, which in the lower lumbar region of the normal adult is about one and three-quarter inches at its widest point. The outer layer of the disc consists of a firm and tough layer called the annulus fibrosus. The center of the disc is soft and pulpy and called the nucleus pulposus. As a result of normal aging, the annulus fibrosus becomes brittle and the nucleus becomes drier. This in itself can be painful and is called disc degeneration. When the annulus cracks and fissures, however, the soft nucleus pulposus in the center oozes out through these layers and may produce a bulge that presses on the nerve as it exits from the vertebral column. This causes low back pain and pain down the leg because this nerve becomes the sciatic nerve in the lower extremity. The pain in the leg is called sciatica.

A disc with a bulge caused by nucleus pulposus is called a herniated disc. Sometimes the disc ruptures through the containing ligaments and out into the spinal canal where the nerves exit. This is called an extruded disc. Extruded discs were not affected by the chymopapain injections, but herniated discs were.

Chymopapain itself is an enzyme that acts selectively on the nucleus pulposus. It is extracted from the common papaya fruit. If the chymopapain treatment works perfectly, the nucleus will be absorbed and replaced by fibrous tissue. In that respect it accomplishes the same thing as surgery.

The treatment of lumbar disc disease by chymopapain injection, a technique developed principally by Lyman Smith of Elgin, Ill., was begun in 1963. There was a gradual, slow acceptance of the method, and the FDA approved 35 orthopedic surgeons to investigate the drug. It was never released for general use in the U.S., but has been used in Canada, Sweden, Italy, and The Netherlands. About 15,000 cases were treated with chymopapain from 1965 to 1975, and the results were gratifying in that surgery was not necessary in about 50% of the cases.

Ordinary X-rays of the spine do not diagnose a herniated disc. Myelography is an X-ray procedure that does permit visualization of the location and extent of the problem, however. A radiopaque dye is injected into the spinal canal and the site of herniation can readily be found. This procedure has been used since World War II by utilizing Pantopaque, an iodine-based oil that must be removed from the patient. The method may cause pain and is somewhat lengthy. Other procedures include electromyography, the use of electrodes to provide a graphic record of nerve involvement, and discography, involving the use of water-soluble dye injected into the disc itself.

Surgery on the disc is necessary if the following conditions are present: severe and incapacitating sciatica unrelieved by at least three weeks of bed rest, progressive evidence of nerve damage, and frequent recurrences of incapacitating sciatica. In the past several years the incidence of unsatisfactory results of surgery has been decreased by recognition of hidden anatomical features, improved treatment of complications, and better preoperative diagnostic methods. For example, the spinal canal itself may be too narrow and thus cause pain. If this "spinal stenosis" is recognized and treated by wide excision of bone, other failures may be prevented. Adhesions have been partially attacked by using a resorbable membrane around the nerves and linings of the spinal canal.

By no means are all backaches treated by surgery or enzyme injection. In the majority of patients with low back pain, with or without sciatica, the cause is unknown. Results of recent studies indicated that both chemical and mechanical factors are probably of importance. It seems safe to say that simple bed rest, the use of salicylates and mechanical support devices, and following the dictates of "ergonomics" can be curative in 90% of people with backaches. Ergonomics is a term popularized by Alf Nachimson of Sweden, who used it to describe a discipline of work habits to prevent the onset of back pain. This includes certain isometric exercises and postural habits in work and at rest.

Total joint replacement. Perhaps the most dramatic development in orthopedics in the past 20 years is total joint replacement. Preparation for this breakthrough was long and laborious. There have been many blind alleys, but the road is now clear for successful treatment of at least two of the major joint problems, those involving the hip and the knee.

Rheumatoid and degenerative disease of the hip and knee have always plagued humankind. The first great discovery in their treatment was that of M. N. Smith-Petersen of Boston in the 1930s. He devised a "cup" for the hip joint which acted quite satisfactorily but required long postoperative treatment and was successful only in cooperative patients who had access to competent rehabilitation centers. In contrast, the total hip replacement is a procedure that relieves pain and disability almost overnight and does not require long hospitalization.

The total hip operation was developed by John Charnley of Manchester, England, about 1960. His key to his success was the use of polymethylmethacrylate cement and high-density polyethylene, which can articulate with stainless steel. Thus, in the hip joint the new artificial socket is made of high-density polyethylene and is cemented into the pelvis. The other side of the joint, the head and the neck of the femur, is made of stainless steel or cobalt-chromium alloys. This also is heavily cemented to the main bone or the femur. Both segments are then fitted tightly together. Because there are no nerve endings in the new ball or socket, pain is eliminated and the low-frictional quality causes the joint to function smoothly.

The knee is somewhat more complicated than the hip. Although the knee's motions might seem to be no more than those of a hinge, they are actually a complex combination of sliding, spinning, and rolling, in which the muscle forces exerted in walking can be more than 400 lb. Arthritic knees are more common than damaged hips, and attempts at surgical solution are more than 100 years old. Since the mid-1950s, both in the U.S. and in Europe, intense efforts at designing a new knee have been made, and several different designs have been successful. In all cases, however, the ligament structures must be intact to support the new joint.

The first total knee replacement in the U.S. resulted from the efforts of a committee composed of five orthopedic surgeons. The committee in turn received comments and advice from other orthopedic surgeons. Later, Mark Coventry of the Mayo Clinic reported on 307 knees replaced by himself and a group of surgeons since March 1971. The average age of the patients was 61, but some were over 80. After the operations the patients were all able to walk without crutches and cane, climb stairs, and arise from a chair or bed.

In total knee replacement the roughened cartilage of the joint is removed, as is done in the hip.

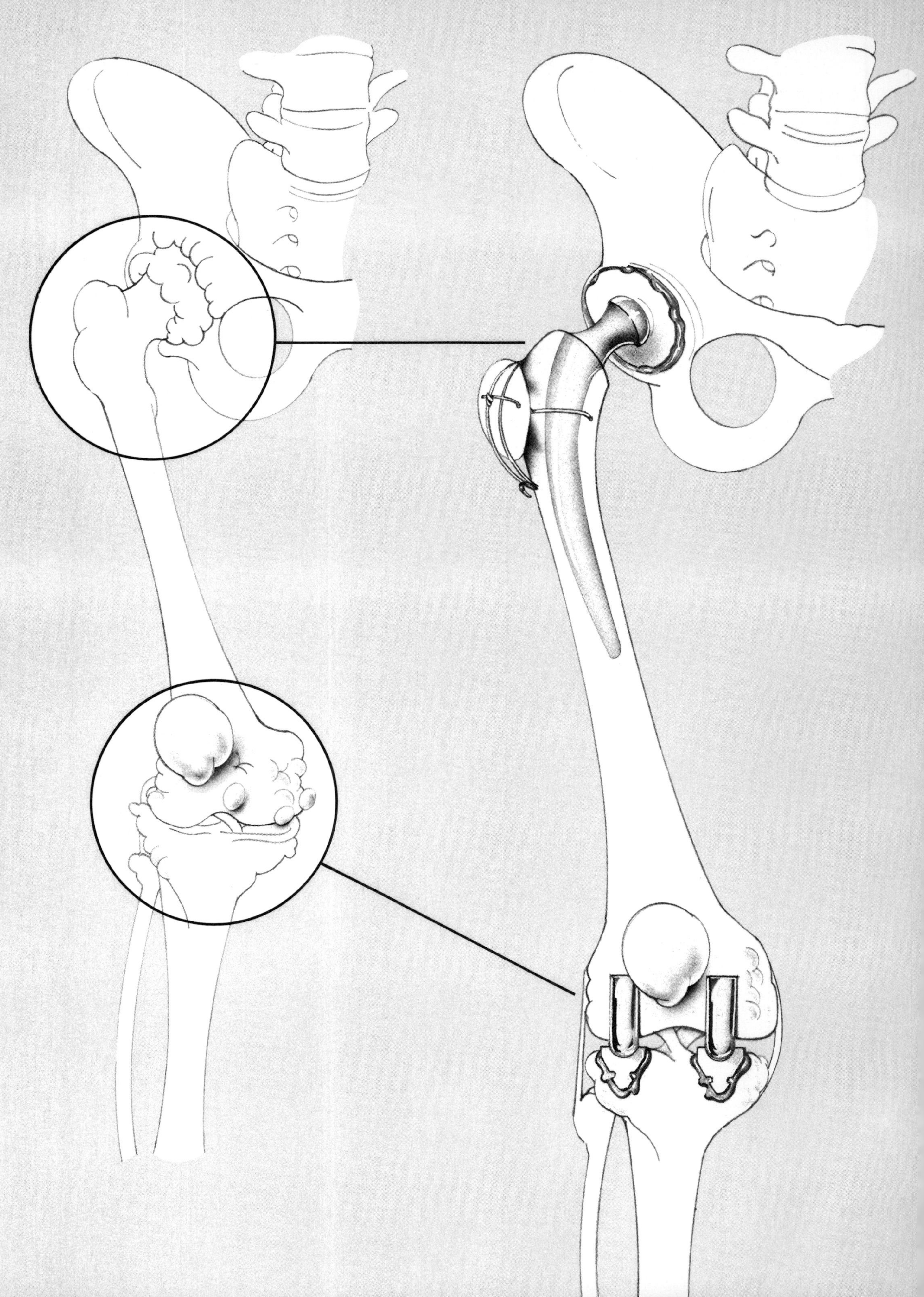

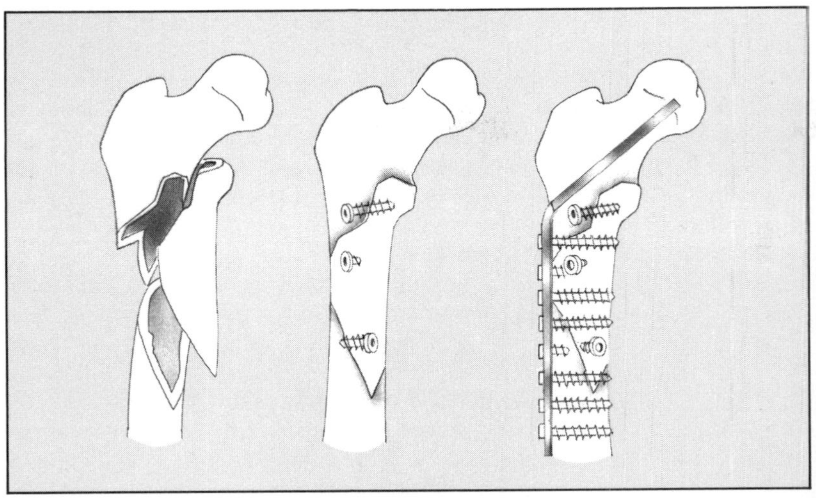

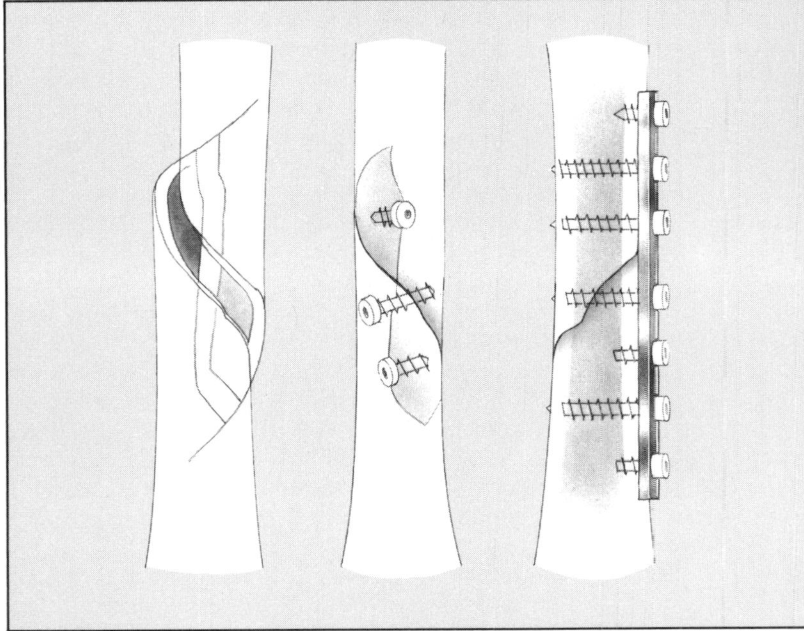

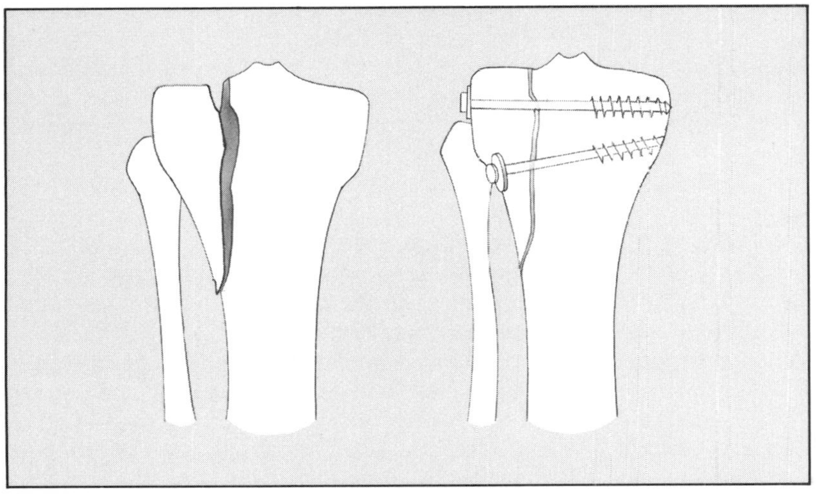

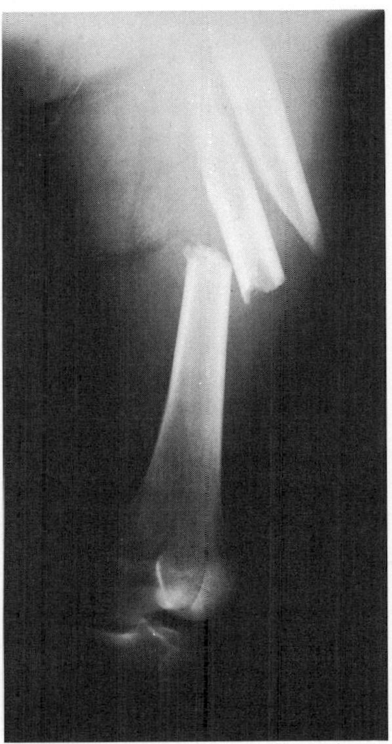

Total replacements of the knee and hip are shown on the facing page. Both joints at the far left have been affected by severe osteoarthritis, involving the loss of cartilage and roughening and deformation of bone surfaces, which restricts joint motion. Depicted are the Charnley-type total hip and polycentric-type total knee prostheses.

Severe compound fractures of the femur such as that shown above can be set and repaired by a variety of internal fixation techniques. Pinning of three different kinds of fractures is shown at left. Additionally, a rod may be inserted along the longitudinal axis of the bone in certain cases.

The femoral end is replaced by metal runners, which work on a polyethylene base. Both components are cemented firmly. The patient can be out of bed in two to three days and able to move the knee in five days in the uncomplicated cases.

Other major joints that could be replaced in the mid-1970s were the elbow, shoulder, and ankle. The finger joints have been replaced by a material called silastic. This type of replacement has been helpful for those suffering from rheumatoid arthritis. It seems likely that a multitude of artificial devices will become available in the future that will make possible the replacement of most, if not all, of the joints of the body.

The situation is not entirely rosy—complications are possible in any type of major surgery. Infections occur in a small percentage of cases and can result in catastrophic failure of the operation. Preoperative, intraoperative, and postoperative antibiotics of the broad-spectrum types have been used in recent years and have helped reduce the incidence of infection.

The most frequent complications of hip and the knee replacements are thrombophlebitis, or clotting in the veins, and its possible sequel, pulmonary embolism, a clot in the principal veins of the lung. It has been said that 25% of all patients have non-symptomatic thrombophlebitis as they leave the operating room. The best treatment is preventive, and to that end recent research has indicated that small doses of anticoagulant drugs both before and after the operation are effective. Surprisingly, it is now evident that common aspirin given before surgery and during the patient's hospital stay has been as successful as some other more dangerous anticoagulant drugs such as Coumadin and Heparin. Other means of preventing thrombophlebitis and pulmonary embolism are the use of elastic stockings, elevation of the foot of the bed, and, more important, early ambulation; the patient should be made to get out of bed the day after surgery if possible.

Joint replacements are not the ultimate cure for the wear, tear, and aging of the human body. It is correct to describe them as a great slow step forward to alleviate pain if the step is taken very carefully. They are best thought of as devices for replacement of the joint surfaces, which do not include the ligaments and other soft tissues.

Fractures. A fracture is a broken bone. Any bone in the body can be broken, and the treatment of fractures is as old as the history of humankind. Immobilization by such means as splints and traction has been used for centuries. Plaster-of-paris casts are relatively recent, having first been applied in the 1850s. The value of tried and tested methods of fracture treatment is unquestioned, and the use of surgical treatment is usually reserved for complicated cases in adults only. Screws, pins, bolts, and plates have been the traditional tools of surgery, but a method called AO, which utilizes new principles and designs, has become increasingly effective since its introduction in the early 1960s. This technique was developed by the Swiss Association for the Study of Internal Fixation.

The chief aim in fracture treatment is the return of the injured limb to full activity. In order to prevent loss of position, joint stiffness, and soft tissue damage resulting from circulatory disturbances, surgeons hope to achieve such rigid internal fixation that a long period in plaster is no longer necessary and early joint movement is possible. To achieve this goal the AO method uses compression techniques (screws and plates) and intramedullary nailing (rods and pins in the marrow canal of long bones).

A major value of the AO program has been the comprehensiveness of its organization. Standard instrument sets were developed, a documentation center was founded, and orthopedic surgeons throughout the world were taught how to use the method in Davos, Switz.

Orthotics. A relatively new branch of orthopedics termed orthotics is emerging. This is the science of controlling the motion of body segments by the use of external devices. It is a dynamic concept in that braces are used not only to support a limb but also to help in its movement and control. In the U.S. there are an estimated three million patients who need orthotic appliances. These devices are used in the treatment of lower limb fractures and congenital and acquired deformities.

New materials have had great impact on the design and development of orthotic devices. Thermosetting plastics are lightweight and can be combined with lightweight metals. There has been an upgrading of the suppliers of these devices, who now must pass examinations and become registered as orthotic practitioners.

Effects of sports. Sports injuries constitute an ever increasing problem now that more and more people are participating in sports. Millions of people play tennis and golf, go skiing and snowmobiling, and some even use skateboards and hang-gliders. The school population and amateur sport enthusiasts constitute the majority who fall victim to various types of injury.

Professional athletes often receive considerable publicity when hurt, and this may have helped call attention to the problem. An organized study of sports medicine was the result, and the American Orthopedic Society for Sports Medicine has been extremely active in this regard. The thrust is that of increased cooperation among physicians, trainers, and suppliers of sports equipment to heighten the awareness of the participating athlete with regard to possible injuries and how to prevent them.

Spinal curvature. Scoliosis is a condition that involves lateral curvature of the spine in growing children. For centuries man has been baffled by the cause and management of this condition, which is serious not only because of its effect on a person's appearance but also because serious cases can threaten the life of the victim. Hippocrates was the first to coin the term "scoliosis" and to use crude methods of treatment. No real progress was made until the first spine fusion was performed in 1911. A breakthrough occurred with the design of the Milwaukee Brace in 1946. Though other improvements have been made since then, the cause of scoliosis remains unknown and the treatment remains that of attacking the curvature after it has developed.

The Milwaukee Brace and subsequent design improvements on it have been successful in halting progression of curvature in perhaps 70% of cases. The brace is generally worn 23 hours a day for a number of years until the child is mature. In severe scoliosis, "halo" traction, a recent development, may be used. The halo screws into the skull with surprisingly little discomfort. Pins are placed in the thigh bones, and both ends are connected. Gradual pull is then exerted until the curve is corrected. Surgery may also be done, stiffening the vertebrae to obtain permanent correction. Another method recently developed is that of the "Harrington" rod, an adjustable stainless steel strut that is inserted into the back between a lower and an upper vertebra to straighten the curvature.

Future outlook. At the beginning of the 20th century the physician treating orthopedic problems had nothing to give but himself. Though offering a comforting hand and heart, he sometimes did more harm than good. Difficult fractures were treated by amputation; infection was a constant hazard; X-ray was in its infancy; and surgery was extremely hazardous. Between the two world wars orthopedic surgery struggled to establish itself as a specialty, and there were very few physicians to shoulder the burden of the problems of the bone and joint patient.

It has been said that more progress in orthopedic surgery has been made in the 30 years since 1945 than had been achieved in all preceding eras. The rapid pace of discovery and utilization of new techniques has not abated, and in the next few years continued advances should be made in the relief of pain and deformity.

—Samuel H. Fraerman

Dentistry

The continuing debate over national health insurance, ramifications of a new health manpower law, and a renewed effort to improve the dental health of the nation's children were among key legislative developments of concern to the U.S. dental profession during 1975. Testifying before congressional committees, officials of the American Dental Association reiterated the ADA's "fundamental conviction" that a "national health insurance program—including its dental component—be based upon the existing system of delivery."

Congress also heard testimony by Robert B. Shira of Boston, president of the ADA, on the proposed Health Manpower Act. Shira called on Congress to recognize the federal government's obligation to support the health professions' educational institutions in a meaningful way without attaching burdensome conditions such as mandatory enrollment increases. He also voiced strong opposition to proposed federal standards for licensure and relicensure of dentists and physicians, as well as to federally supported expanded-

Plastic braces are bonded directly to the teeth before starting tooth movement (top) and are bonded to upper teeth with an orthodontic wire in place for tooth movement (bottom).

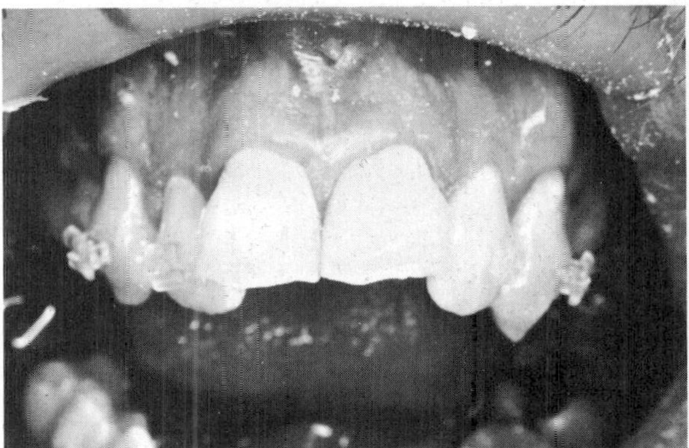

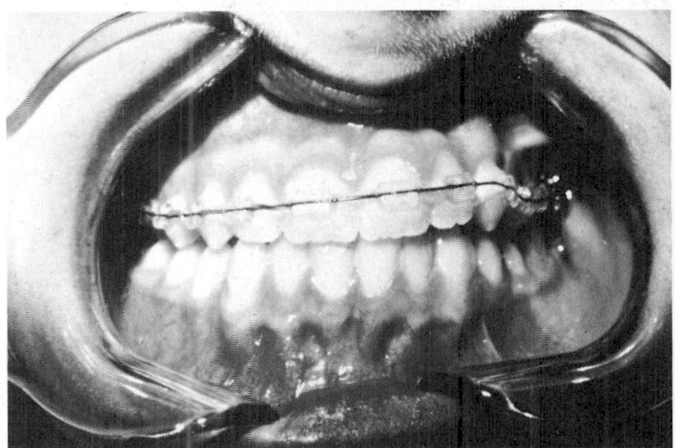

Photographs, courtesy, Norman W. Garn, Northwestern University duty dental auxiliary training programs that are in conflict with existing state laws.

For the second time in five years, a bill developed in cooperation with the ADA—the Children's Dental Health Act—was introduced into Congress. The bill calls for federal project grants for early and periodic screening, diagnosis, and treatment programs for Medicaid-eligible children. It would also earmark funds for school dental health education programs and one-time federal grants to states and communities that are planning fluoridation programs.

Many adults who had been afraid to smile because of protruding or crooked, unsightly teeth were discovering that orthodontic treatment could benefit them. A nationwide survey by the American Association of Orthodontists revealed that at least one out of every ten orthodontic patients was an adult. Isabel W. Grayson, a New York City orthodontist, said that frequently an adult will not realize that his teeth can be straightened until he meets someone his age wearing braces. Speaking at the ADA annual session held in October 1975 in Chicago, she described some of the faulty bite or malocclusion problems found in adults and how they can be corrected through simple orthodontic treatment. "Plastic orthodontic attachments which are clear and almost invisible can be glued or bonded directly to each tooth, thus eliminating the need for unsightly metal orthodontic bands and attachments formerly necessary to fix orthodontic arches or wires directly to the teeth." Also ideal for adult orthodontic treatment are removable appliances utilizing simple wires and an elastic system to move teeth. A barely visible system of tiny wire springs attached to a plastic removable appliance can help correct single or multiple tooth units into what is described as a reverse or cross bite.

A psychologist-dentist team at Case Western Reserve University in Cleveland, Ohio, launched a study aimed at helping children overcome anxiety in dental treatment. Children undergoing dental treatment under the Head Start program would be shown a dental health film, after which dentists would use various specific behavioral approaches to them. By comparing anxiety levels in children who had been approached with praise, verbal admonishment, or a neutral reaction to their behavior, the scientists hoped to determine which attitude was best.

A study at the University of Connecticut's dental clinic in Farmington was expected to shed some light on whether a child's age and need for treatment on his first dental visit affects anxiety and behavior during subsequent visits. The reaction of more than 200 children aged two to five was to be filmed by closed-circuit television. Physiological measures would also be used, together with a self-report in which the child selects a cartoon character that best matches his own feelings. By comparing the videotapes, the physiological measures, and the self-reports, it was hoped that the best time for the child's first visit to the dentist could be determined.

Courtesy, Boys' Clubs of America

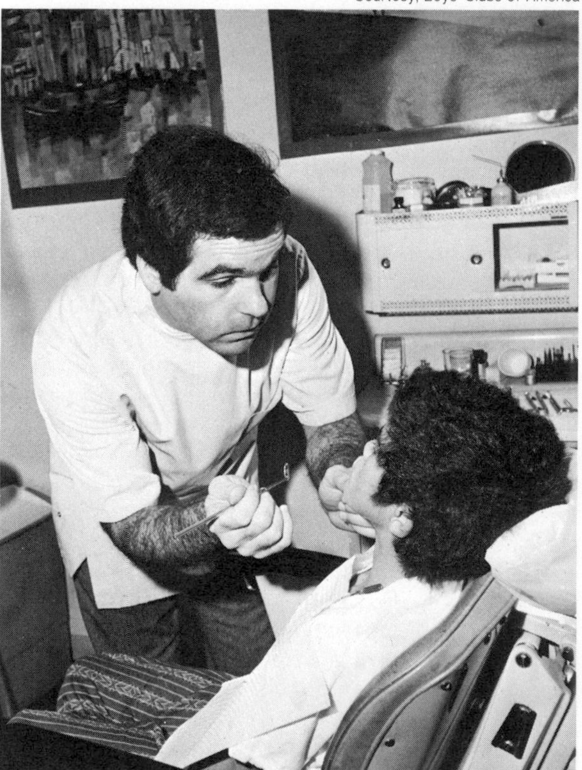

Dentist uses friendly persuasion in an attempt to calm the fears of an eight-year-old boy during the child's initial visit to a dentist's office.

Public health researchers at Harvard University found that children pattern their dental behavior (oral hygiene practice and dental visits) closely after that of their parents. Reporting at the annual session of the American Association for Dental Research in New York City in April 1975, the researchers concluded that "there appears to be little value in attempting to change the dental practice of children if there is no way to modify parents' behaviors as well."

A New York University dental scientist suggested that pregnant women and mothers should be made aware of the dental advantages of breast feeding their babies. "Most cow milk formulae for infant feedings contain added carbohydrates in the form of lactose or corn sugar, which makes it inherently sweeter than human milk," according to Julius Ozick, research assistant professor at the university's department of pedodontics. "Babies who are bottle fed . . . are being fed a sweet tasting, high carbohydrate diet from birth." These infants grow into toddlers who may desire sweets because of their previous feeding experience, and by the time a child reaches three or four, a sugar-consuming preference has been strongly ingrained in his eating habits.

Thorkild Karring of Skanderborg, Denmark, reported a new surgical technique that may speed the regeneration of lost bone because of periodontal or gum disease. The bone supporting the teeth of patients suffering from gum disease frequently resorbs, and if the missing bone is not restored the teeth may loosen and eventually fall out. The new method involves the grafting of soft tissue taken from other parts of the mouth to cover the repair sites. Under conventional surgery procedures, bone grafts used to rebuild the missing bone are covered with a soft tissue flap, but the epithelium or surface layer of the tissue flap frequently grows into the host site and thus may interfere with the regeneration of new bone. In the "free tissue" graft technique the epithelium of the tissue graft is shed, possibly because the donor tissue is less vital since it has been severed from any blood supply. Consequently, there is no migration of the epithelium to the repair site, and extensive new tissue attachment of the graft is achieved.

Studies in various parts of the world indicated that the use of plastic sealant materials can greatly reduce the rate of tooth decay in children according to Leon M. Silverstone of London. Although statistical data vary, the overall picture reveals that pit and fissure sealants applied to the chewing or occlusal surfaces of teeth are "highly effective" in the prevention of decay, and that the retention of such materials is relatively long-lasting. Silverstone reported that a review of studies in the U.S. and U.K., some conducted over a period as long as four years, showed a decay rate reduction as high as 90%. Especially high reduction rates were noted in communities with fluoridated water supplies.

A National Institute of Dental Research team reported that tooth decay in newly weaned rats infected with an organism that causes decay in humans is reduced more effectively when the animals drink water containing sodium fluoride than when they eat the same amount in food. In the experiments, decay decreased as the concentration of fluoride was increased in the food or drink of matched groups of rats. All animals were fed an identical high-sugar diet, and bacterial sampling showed that the same human strain of *Streptococcus mutans* continued to infect them throughout the experiment.

Versatile new dental materials called composite resins were receiving wide acclaim for their value in restoring accidentally fractured teeth in children, according to David R. Avery of Indianapolis, Ind. The substances are reinforced acrylic plastic materials that can match the color of natural teeth. They are best suited for use in teeth in the front of the mouth, where they can restore the function and appearance of teeth damaged by decay or accidental fracture. Avery noted, however, that they are not recommended for use in the rear of the mouth because they are less resistant than metals to the tremendous wearing forces that are encountered there.

An unexplained injury to a child's mouth or face or exceptionally stoic behavior by a young patient should make a dentist suspect that he may be faced with a child abuse case, according to Stanley Schwartz of Tufts University, Medford, Mass. The dentist's suspicion should be aroused whenever he sees a child "who is uncommunicative, unemotional, unresponsive with lethargic-automatic reflexes, lacking fear or inquisitivity who just drops into the dental chair." This psychotic withdrawal may be manifested by focusing head and eyes on the floor, shuffling gait, no fight or flight reaction—in fact, no response to the dental environment at all. Schwartz recommended that once this suspicion is formed the dentist should search for physical signs of possible child abuse. Dentists have the same responsibility as physicians to report suspected child abuse cases to the proper authorities.

—Lou Joseph

Veterinary medicine

An event of major significance to the veterinary profession in the United Kingdom was the publication, in 1975, of the report of the Committee of Inquiry into the Veterinary Profession, better known as the Swann Report (after its chairman, Sir Michael Swann). It presented some 60 recommendations as a guide to long-term development of veterinary medicine in the U.K., and many of its points were applicable to other developed countries. As specified by Sir Michael, the six most important of these were:

1. Greater emphasis on preventive veterinary medicine on farms, including the giving of advice on animal husbandry and management and collaboration with agricultural advisers.

The part-time employment of rural veterinarians by agricultural departments was expected to cost up to £850,000 per year initially in Britain, increasing to a maximum of £8.5 million in 10 to 15 years.

2. Closer links between animal and public health, including supervision of slaughter and of

Courtesy, USDA

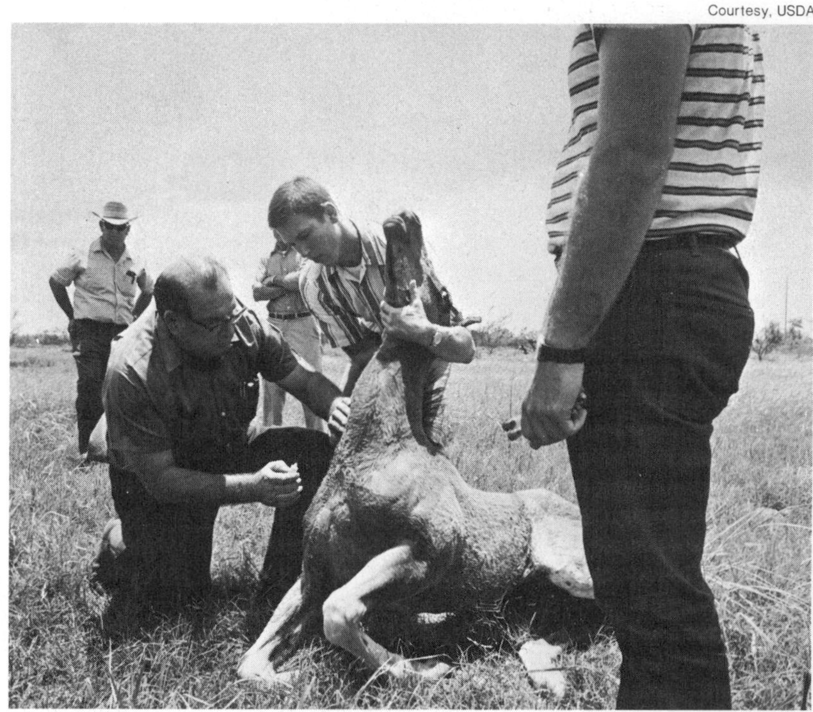

Veterinarian with the U.S. Department of Agriculture Animal Health Division obtains blood sample from colt suspected of being diseased.

meat inspection, with systematic monitoring to assess animal disease problems.

Most of the meat produced in the U.S. is under federal supervision, and the U.S. Department of Agriculture employs many full- and part-time veterinarians in fieldwork on animal disease. Working within the Food and Drug Administration, the Bureau of Veterinary Medicine is charged with determining the safety and efficacy of animal health products. On the state level, veterinary extension services provide a direct link with animal owners in addition to dealing with statewide animal disease problems.

3. Greater use of lay personnel to extend the knowledge and skills of veterinarians to more clients and animals, in both large-animal as well as small-animal practices.

In both the U.K. and the U.S., lay assistants are widely employed in animal hospitals, but their use in the field has been deterred to some extent by laws requiring direct supervision of their work. Legislation is required to encourage flexibility in the employment of assistants, but with safeguards against unauthorized practice. The Royal College of Veterinary Surgeons in Great Britain and the several state veterinary boards in the U.S. should work toward establishing mechanisms for certifying and registering veterinary assistants, especially in farm practice.

4. More research in the veterinary field, especially with regard to farm animals, though more also is needed on small-animal problems.

The work being done on epidemiology, preventive medicine, and economics of disease control should be expanded to provide a foundation for judgments on priorities. The veterinary profession should also contribute more to research on the conservation and control of wildlife and to multidisciplinary studies in human medicine and biology in general.

5. Changes in education, with emphasis on preventive medicine and epidemiology, and the provision of more clinical experience at the undergraduate level to better prepare students for the realities of practice.

The schools should keep in close touch with private practitioners and provide opportunities for students to become familiar with the actual practice of veterinary medicine. In the U.S. a number of the veterinary schools offer or require periods of preceptorship, during which students work under the direct supervision of veterinarians in private practice.

6. A better relationship of supply to demand in the profession, with the recommendation that the six schools in Britain produce an additional 1,625 veterinarians. This would provide an active workforce of about 7,200 by 1990, an increase of about 30% over current numbers.

In the U.S., the National Academy of Sciences had projected a need for 42,000 veterinarians by 1980, compared with 28,100 active in 1970. In December 1975 the number was estimated to be 37,600, which with a projected loss of some 2,200 during the next five years would require an additional 6,600 to meet 1980 needs. Even with the new schools in Louisiana (1974), Florida (1976), and Tennessee (1976), it was conceded that this number could not be supplied.

In Britain the recommendations of the Swann Committee were received with cautious approval by most individuals and professional groups, though "it will take many years before all the report's recommendations are tackled." In 1969 the same committee had reported on its study of animal feed additives and recommended that no antibiotic used for treatment of human disease be fed to animals unless prescribed by a veterinarian; in 1971 the feeding of three commonly used antibiotics was banned in the United Kingdom and this led to rigorous inquiry into their use in the United States.

After an intensive 12-year eradication program, Texas became, in 1974, the last of the 50 U.S. states to be declared free of hog cholera. In mid-1975, however, the disease was discovered in two Texas herds, and a 100-sq mi area was placed under federal quarantine. These 2 herds, together with 23 others in Texas and 6 in Oklahoma which may have been exposed, were destroyed at a cost of $350,000 paid to the owners in indemnities. Before 1962, when the eradication program was initiated, the cost to producers for vaccination and losses from the disease amounted to tens of millions of dollars annually.

An outbreak of infectious disease, even when minor, usually attracts national attention. Thus an outbreak of anthrax that claimed some 200 animals in Texas and Oklahoma in 1974 was widely publicized, whereas much larger disasters often pass with little notice. For example, flooding in the state of Washington in 1975 resulted in the drowning of 9,000 dairy cows, accounting for 15% of the dairy cattle population in two counties.

After nearly a year of investigation, the cause of a "mysterious" illness of livestock and poultry in Michigan was attributed to accidental contamination of feed with polybrominated biphenyl (PBB), a fire retardant. Although there were few deaths from poisoning by the chemical, affected animals became completely unproductive, and they could not be used for meat because of tissue residues. Altogether, about 30,000 livestock, mostly dairy cattle, and 1.6 million poultry on 507 premises were destroyed and disposed of by burial in a remote 60-ac site. The state spent nearly $1 million in handling the outbreak; the total cost to the responsible private companies was estimated at more than $25 million, with unsettled claims running to several times this amount.

The factors responsible for the malpractice insurance crisis in human medicine had not affected liability insurance rates for veterinarians during the 13 years that this coverage had been available through the American Veterinary Medical Association. The average cost for medical practitioners had risen more than 1,000% during the past 15 years, to $30,000 or more for some specialists ($1 million coverage), whereas the premium for veterinarians in 1976 was $63, up from $62.25 in 1962. One reason for this was that claims were limited to the actual value of animals, though this had nearly doubled during the past decade.

The major differences between medical and veterinary malpractice insurance include enrollment of nearly all veterinarians in one program, the trustees for which are all veterinarians who also review all claims. The insurance company vigorously defends its subscribers, and claims are settled only on merit, not for convenience or economic reasons. The Insurance Trust also

Michigan farmer readies for burial dairy cattle that he shot because they had accidentally been infected with polybrominated biphenyl (PBB).

UPI Compix

sponsors educational projects to reduce malpractice claims.

During the year the scope and capacity of the computer education system (PLATO) at the University of Illinois College of Veterinary Medicine were more than doubled. More than 100 lessons were available, and student-contact hours exceeded 10,000. Through telephone-line or microwave communication, the system would be an economically feasible tool in veterinary education nationwide. The computer could support 1,000 terminals, only 33 of which were currently required at Illinois, and it was anticipated that the ever widening student-faculty ratio would make the program attractive at other schools.

In late 1975 investigators at the Harvard School of Public Health reported finding significant evidence of a natural body defense system in cats that destroys cancer cells before tumors develop. In a study of 51 leukemic cats from one household, the levels of specific antibodies triggered by leukemia cells induced by leukemia virus were ten times higher in cats that did not develop leukemia than in those that did. It was believed that similar mechanisms against other forms of cancer might be operative under natural conditions in humans as well as other animals.

—J. F. Smithcors

Optical engineering

Advances in optical communication and in applications of the laser to an ever increasing number of scientific and technological problems provided the most exciting developments in optical engineering during the past year.

Optical communication. The impetus for the commercial development of fiber optics communication systems has been the urgent need for a compact, high-capacity system to replace the current, much bulkier network of electrical cables and microwaves. Already in use by the military because of their compactness and relative immunity to interference, fiber optics systems show great promise for increasing the capacity of the presently clogged telephone lines in central urban regions, where the cost of tearing up city streets to install additional lines is prohibitive. (See *1975 Yearbook of Science and the Future* Feature Article: FIBER OPTICS: COMMUNICATIONS SYSTEM OF THE FUTURE?)

Field tests of fiber optics systems for telephoning with beams of light were conducted during the year in the U.S., by both Bell Telephone and General Telephone and Electronics, and in Japan, Great Britain, France, and West Germany. Recent tests performed in an actual operating environment at a Bell Laboratories facility in Atlanta, Ga., emphasized equipment that had been packaged for compatibility with the conventional electronics used in central office equipment. In one set of experiments, light-emitting diodes were used as a light source. A second set of experiments made use of an aluminum-gallium arsenide laser source comparable in size to a grain of salt.

Even as testing of prototype commercial systems proceeded, new research sought to develop still more sophisticated systems. One necessary component in any future commercial system appears to be some form of optical repeater capable of boosting the power of a signal at intervals of perhaps 5–10 km. In present systems, the light signal is converted to an electrical signal, amplified, and then reconverted to light. What is needed is a device for the direct amplification of light, *i.e.*, an optical transistor. Such a device was first demonstrated at Bell Laboratories in 1975, but in its initial state it was much too massive (more than 20 kg) and too easily mistuned to be useful in practical applications. In addition, it employed an optical cell filled with sodium vapor, which slowly formed an undesirable deposit on the mirror ends of the cell, rendering them ineffective. Nevertheless, the principle of direct optical amplification was successfully demonstrated, and it is probably only a matter of time before solid-state optical transistors will be coupled to the thin glass fibers of existing optical communications systems.

The ultimate foreseeable goal of optical communications technology is the fabrication of most of the devices needed to manipulate light within a single crystalline wafer of gallium arsenide. Laser, waveguide, modulator, and detector would all be incorporated into an integrated optical circuit. As of 1976 no such integrated circuit had been produced, but a distributed feedback aluminum-gallium arsenide laser developed by Hitachi Ltd. of Japan represented an important advance. With distributed feedback, the laser frequency is determined by an optical grating created on the substrate of the gallium arsenide wafer, and is compatible with existing integrated-circuit fabrication techniques. Although conventional solid-state lasers can be made with dimensions sufficiently small to be compatible with the hairlike waveguides to which they must be coupled, they utilize cleaved surfaces acting as mirrors to achieve optical resonance. The manufacturing techniques required to achieve these surfaces are not suited to integrated-circuit fabrication.

Holography and speckled patterns. By contrast with photography, which records a two-dimensional image on a photographic emulsion, holography records all of the information available in

Adapted from "First Optical Transistor Demonstrated," "Science News," vol. 108, no. 17, p. 262, October 25, 1975

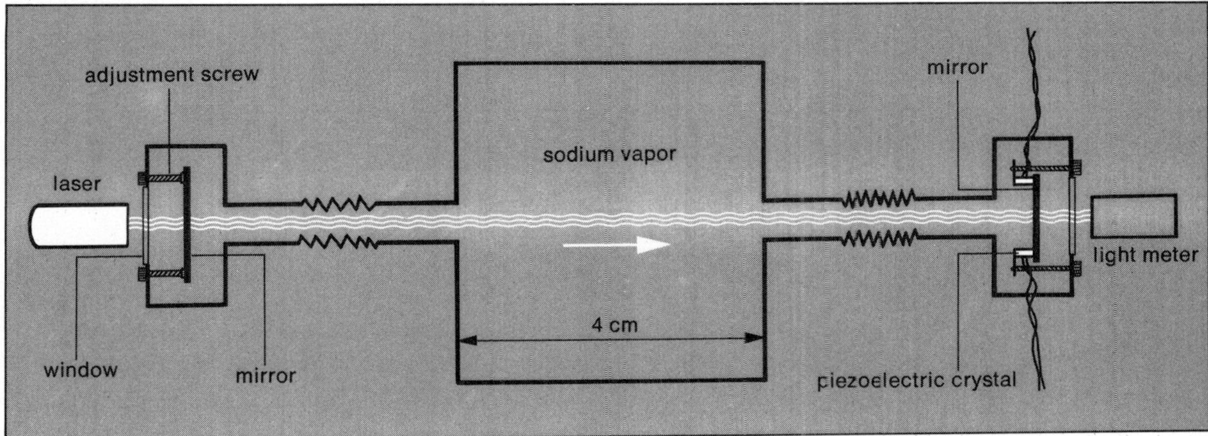

Bell Laboratories' optical transistor consists of a chamber filled with sodium vapor and terminated at the ends with two parallel, partially reflecting mirrors, spaced apart by an exact number of wavelengths. Manipulation of the mirrors or the contained gas controls a beam of laser light passing through the chamber in ways analogous to the control of an electric current by a transistor.

the light reflected from an object. A holographic recording permits one to reconstruct the original waves reflected from the object and, thus, to view the object in three dimensions as if it were the original. Practical holography has been made possible only by the invention of the laser. To produce the recording itself, called a hologram, a beam of laser light is split into two components. One component is directed toward the object, from which it is reflected, scattered, and diffracted, and is then allowed to strike a photographic emulsion. The second component, or reference beam, is reflected with mirrors directly onto the emulsion. At their point of convergence, the scattered waves and reference waves form an interference pattern that is recorded on the emulsion as a complex pattern of stripes and whorls. The recorded interference pattern is the hologram and contains all of the information needed to reconstruct a three-dimensional image of the object. If the hologram is illuminated with a coherent beam of light, the detailed pattern on the film acts as a diffraction grating to bend the beam in an exact reversal of the lighting conditions that created the hologram. To the viewer, this reconstructed wave appears as a three-dimensional image of the original object.

Since its initial success, holography has enjoyed a variety of applications. In 1975 Westinghouse Research Laboratories announced yet another: the successful use of holography in the design of turbine blades. The method produces a hologram using a double exposure. Air is forced through arrays of turbine blades and an exposure is made with light from a ruby laser. A second exposure on the same emulsion is made with no airflow. The resultant interference pattern provides a view of shock waves, boundary layer effects, and other fluid-flow phenomena not ordinarily visible, and it does so without perturbing the flow pattern.

Although holography is important and new applications will continue to be found, it has not developed as quickly as was predicted in the early 1960s. A derivative of holography, called speckle interferometry, currently holds much promise, however. Speckle is the granular appearance of laser light that has been reflected from or transmitted through a rough surface. It arises because the rough surface imparts a random phase to the light scattered from each of the many microscopic surface regions. The interference pattern produced by the interaction of all the reflected waves appears as a speckle pattern.

Dennis Gabor, the originator of holography (1948), commented in 1971 that "the novelty in holography is speckle noise. . . . It is not really noise but unwanted information. When we put a sheet of white paper into laser light, the reflected light conveys to us information on the roughness of the paper in which we are not at all interested." But, of course, modern technology is deeply interested in the roughness of things, and Gabor knew this when he made his tongue-in-cheek comment. Though speckle may be unwanted in most applications of lasers, particularly in holography, it finds use in surface-roughness measurements simply because it is directly caused by surface roughness. During the past year, considerable attention was given to the application of speckle methods to surface finish measurements, vibration studies, and three-dimensional strain analysis in transparent materials. A particularly important application, and one of the exciting recent developments in astronomy, has been the effort to better determine surface features and other properties of

Courtesy, Westinghouse Research Laboratories

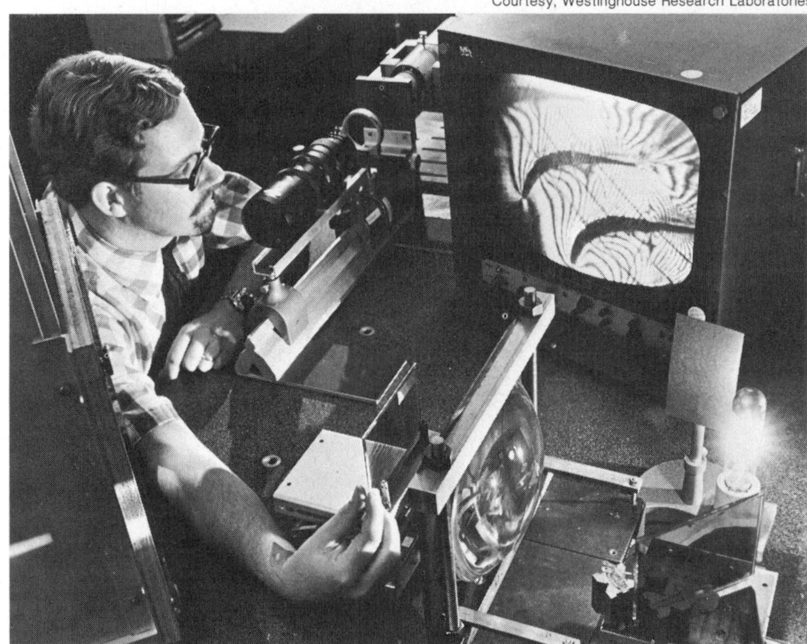

Successful use of holography in the design of turbine blades was announced in 1975 by scientists at Westinghouse Research Laboratories. Image on TV screen, a reconstructed holographic interference pattern, reveals the air flow around blade sections.

astronomical objects by reducing the interference caused by atmospheric turbulence.

Carbon dioxide lasers. Industrial applications of lasers for drilling, cutting, and welding advanced markedly during the past year. The increasingly widespread use of lasers for these purposes has been made possible primarily by the development of carbon dioxide (CO_2) lasers capable of operating continuously at power levels as high as 20 kw, with efficiencies of 15% or more, and during time intervals sufficient to be useful in manufacturing operations. Metering holes in aerosol valves are now drilled quickly and accurately using 50-w CO_2 lasers. Holes as small as 0.01 cm in diameter can be drilled in times ranging from 0.01 to 0.1 second. The interest in techniques to drill holes of small size has been accentuated by the reaction of manufacturers to the concern expressed by the U.S. Food and Drug Administration over the use of certain aerosol propellants.

Higher power CO_2 lasers are being used by Ford Motor Co. for metal cutting and welding of automobile underbodies. A 6-kw CO_2 laser can lap-weld two sheets of 0.035-in.-thick metal at the rate of ten meters per minute. An advantage of laser welding is that the power can be delivered to a spot as small as 0.01 cm in diameter. Thus, the net heat input is small and large amounts of heat are not dissipated in the surrounding material.

Spacecraft navigation. The two U.S. Viking spacecraft launched in August 1975 represent the most sophisticated of unmanned space explorations to date. Though the missions are directed primarily toward the search for evidence of life or its potential, they also herald an important advance in navigation technology—the first use of optical navigation. In the past, spacecraft tracking and guidance had been accomplished almost entirely from Earth-based stations, using the properties of radio signals sent to and from the spacecraft. While ascertaining distance with this method is quite accurate, determining the orientation of the spacecraft is difficult. One solution would be direct on-board measurement and control, and a step in this direction has been the installation of instruments aboard the Viking spacecraft for the acquisition of optical data to use in their navigation. Pictures of both the target and a reference star are transmitted to Earth, for instance, when an analysis of a sequence of such pictures will permit determination of the orientation and trajectory of the spacecraft. Of course, the time delay in transmission would prohibit use of the method on probes scheduled for close approaches to the most distant planets. Eventually, it seems likely that on-board computers will be used to make decisions without the need to transmit back to Earth for instructions.

—Frederick Wooten

Physics

Research in physics during the year produced significant experimental support for a modified form of the quark model and the existence of the quantum property called charm. A number of theories concerning the structure and behavior of the atomic nucleus were confirmed by experiments, and synchrotron radiation, once a mere by-product of physical research, received considerable attention as a valuable experimental tool in its own right.

High-energy physics

The study of physics at the highest obtainable energies has been directed toward the understanding of elementary particles and their interactions. In the past 25 years the number of so-called elementary particle states identified has mushroomed to more than 200. However, regularities in particle properties also have been discovered, and this in turn has led to their organization into families according to symmetry groups. For a detailed description of elementary particles and their quantum properties, *see* Feature Article: PARTICLE PHYSICS: A REALM OF CHARM, STRANGENESS, AND SYMMETRY.

The significance of the psi particle. In November 1974 the discovery of a qualitatively new kind of particle, announced by two groups that had conducted two very different experiments, stimulated the world of theoretical and experimental particle physics more than any other event in the past decade. The new particle, called the psi or the J particle (the nomenclature is still unsettled), was found by a Massachusetts Institute of Technology-Brookhaven team at the Brookhaven (N.Y.) National Laboratory as a sharp peak at 3.1 billion electron volts (GeV) in the invariant mass spectrum of electron-positron pairs produced by protons of 30 GeV striking a target of beryllium. The data were analyzed as if a massive particle had been produced that promptly decayed into the observed electron and positron. The second group, comprising scientists from Stanford University and the University of California, Berkeley, detected the new particle at the Stanford Linear Accelerator Center (SLAC) from a sharp rise in hadron production during electron-positron collisions having a total energy of 3.1 GeV. Again, the data were interpreted as the production of a short-lived particle that decayed rapidly into the observed hadrons.

Many elementary particle states had been discovered by such studies of the invariant mass combinations of the particles into which they decayed. In fact, of the many particle states identified to date, only the proton, electron, photon, and the two kinds of neutrinos are stable; all others decay to these particles, either directly or indirectly. All of the known heavier particles are hadrons, *i.e.*, particles that interact through the strong interaction, and these decay typically with lifetimes of only about 10^{-23} second. The surprising feature of the psi particle is that its lifetime is 1,000 times greater than had been expected for so massive a state. The most attractive explanation

Circular structure at center is the electron synchrotron accelerator of DESY at Hamburg, West Germany. White rectangular building to the left rests above its electron-positron storage rings.

Keystone

for this anomaly involves the assumption of a new quantum number, or property, of matter.

The quark model. It was mentioned above that an organization of the myriads of elementary particles according to their quantum properties had been accomplished using concepts of symmetry groups. Quark theory, a tangible but still unproven physical model for this organization, postulated at its outset only three kinds of truly fundamental particles called quarks, which are usually designated by *u*, *d*, and *s* (plus their corresponding antiparticles $\bar{u}$, $\bar{d}$, and $\bar{s}$). Baryons—the proton, neutron, and the other hadrons that decay into protons or neutrons—are presumed to be made up of three quarks, and each particular quark combination together with its distinguishing quantum properties corresponds to a particular baryon "elementary particle." Likewise, mesons—the hadrons that are not baryons—are presumed made of a quark-antiquark pair, and again the particular combination determines the meson's properties.

It had previously been established that two quantum properties of hadronic matter—strangeness and isotopic spin—must be conserved in interactions or decays mediated by the strong interaction. Particles with strangeness may only decay to nonstrange particles through the weak interaction, a process about 10^{13} times more feeble than the strong interaction; in such a strangeness-changing reaction, an *s* quark (the only quark that possesses strangeness) is transformed into a *u* or *d* quark. Furthermore, although quark-antiquark pairs may materialize from free space where energy and momentum conservation permit, it seems that such states as $s\bar{s}$ (the phi meson) have anomalously long lifetimes, as if the *s* quark and its antiquark are reluctant to annihilate.

From its decay into electrons and positrons in the Brookhaven experiment and its materialization from them in the SLAC experiments, the psi particle can be expected to have largely the same quantum numbers as the photon or gamma ray. Yet, if it were any combination of known quark-antiquark pairs, there would be no way of understanding its long lifetime. From this dilemma has arisen the need for a new quantum number.

Charm and charmonium. As it happened, such a new quantum number had been proposed in the mid-1960s for a variety of theoretical reasons. Even before the discovery of psi, the new property, called charm, had experienced renewed interest as a means of reconciling certain conflicts of the quark model with experimental observation. Hence, the revised quark model contains a charmed quark, *c*, carrying the property of charm. Like strangeness, charm also would be conserved in strong interactions. Given this revision, the psi particle can be considered a $c\bar{c}$ compound, just as the phi meson is interpreted as an $s\bar{s}$ compound. The long lifetime is then plausible if there are no mesons with charm that are less than half the mass of the psi particle.

If this model is taken literally, however, the observed psi should only be one energy state of the $c\bar{c}$ system; other combinations of *c* and $\bar{c}$ spins and angular momenta should presumably lead to different, equally long-lived states, just as the different relative states of a proton and an electron are responsible for the spectrum of energy levels of the hydrogen atom in atomic physics. Soon after the discovery of psi, a more massive (3.7-GeV) state of the $c\bar{c}$ system, the psi-prime, was discovered at SLAC. Subsequently, several other states have been found there and at a similar facility of the Deutsches Elektronen Synchrotron (DESY) in Hamburg, West Germany. These excited psi states seem to correspond qualitatively to the theoretically predicted states of "charmonium," as the general $c\bar{c}$ system has been called. If this interpretation of the system is correct, physicists must accept four discrete quarks and their antiquarks as constituents of hadrons and identify a new quantum number, charm, as a characteristic of matter.

New particle searches. Granted the existence of charm and its quark, in addition to the charmonium states discovered recently, there should be an entire new class of hadrons carrying a charm quantum number, about which certain predictions could be made. First, because the psi particle is so much more massive than the phi meson (by about a factor of three), charmed mesons and baryons should be generally more massive than most of the known hadrons. This would presumably be a consequence of the greater mass of the charmed quark and would conveniently explain why they have not been discovered earlier. The cross section for the production of particles falls as their mass requirements become greater. Second, like known hadrons with strangeness, charmed hadrons would be relatively slow to decay due to the nature of the weak interaction. However, their lifetimes would be considerably shortened if their masses are as large as expected; one guess might be on the order of 10^{-13} second. Third, a significant fraction of the decays should involve such other weakly interacting particles as mu mesons (muons), as has been observed in the decay of the strange kappa meson.

Several deliberate searches based on each of these properties have been carried out within the past year. A Japanese group working with very-high-energy proton interactions at the Fermi National Accelerator Laboratory (Fermilab), near Batavia, Ill., observed several unusual events that

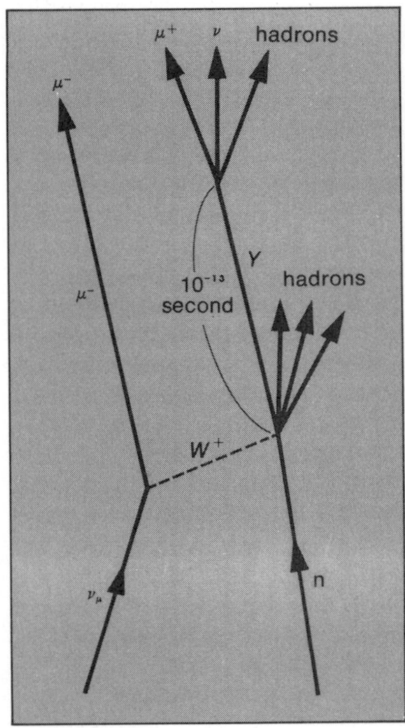

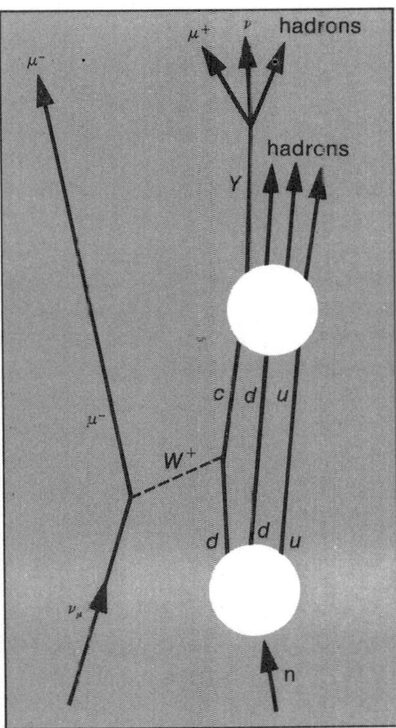

Results of neutrino-beam experiments at Fermilab involving the weak interaction of a muon neutrino ν_μ with a neutron n suggest the fleeting existence of a member of a new class of hadrons carrying a charm quantum number. As shown in the diagram at the far left, the new hadron, designated Y, *is produced during the exchange of an intermediate vector boson* W^+ *and subsequently decays in* 10^{-13} *second or less into a muon* μ^-, *a neutrino* ν, *and a spectrum of known hadrons. Interpreted in terms of quark theory (left), the neutron is composed of three quarks, labeled* d, d, *and* u. *Through the* W^+ *exchange, the neutrino* ν_μ *converts one of the* d *quarks into a new quark,* c, *with the quantum number of charm. The quarks then interact to produce several hadrons including at least one charmed hadron* Y.

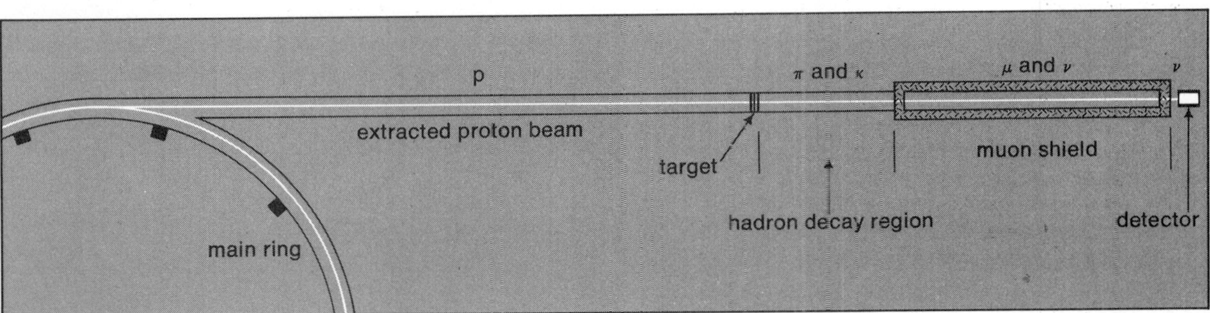

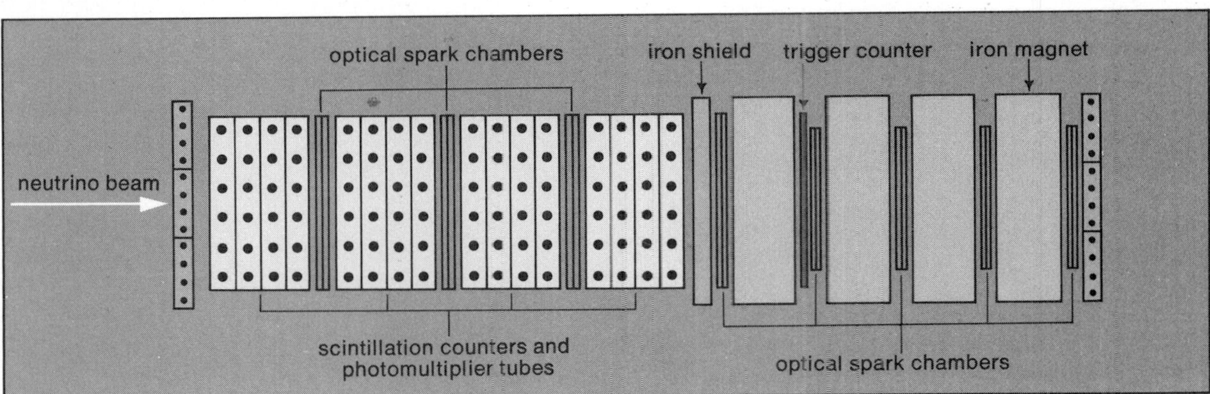

Neutrino beams at Fermilab are generated from a proton beam extracted from the main ring of the laboratory's 400-GeV proton synchrotron (center). Extracted protons p *collide with a metal target to produce pions* π *and kaons* κ, *which decay primarily into muons* μ, *neutrinos, and antineutrinos* ν. *A one-kilometer-long mound of earth filters out the muons, leaving a beam composed only of neutrinos and antineutrinos. Placed in the path of the beam is a detector (bottom) that, by virtue of its counters, shielding, and iron magnets, is sensitive only to events producing two muons in the final state, such as those for which the involvement of a new class of charmed hadron has been postulated.*

seemed to involve a decay in flight of a particle with a lifetime of the order of 10^{-13} second. In at least one case, it appears that two such particles were produced in the same proton interaction. Other interpretations of these observations cannot be ruled out, and further data are needed.

A group of physicists from three U.S. universities and Fermilab, working with neutrino interactions at Fermilab, observed events during which a muon was produced in a manner best understood as the decay of a massive hadron into the observed muon plus several other charged hadrons. While no firm lifetime estimates could be made from this experiment, the muons could come from the decay of a newly created charmed hadron. Subsequently, another group at Fermilab, studying the interaction of neutrinos in a large bubble chamber filled with liquid hydrogen, found four events in which a positron and a neutral kaon are produced together. Again, the experimenters believe that the most plausible interpretation of their observations is the production and subsequent decay into the observed particles of a charmed hadron with a mass estimated as roughly twice that of a proton and a lifetime of perhaps less than 10^{-13} second.

In addition to direct searches for quarks and charmed particles, physicists have continued to seek evidence for unusual predicted or unanticipated fundamental particles. One particle allowed by the laws of quantum electrodynamics is the magnetic monopole, a free magnetic particle analogous to a free electric charge. In 1975 a group from Berkeley and the University of Houston, Texas, announced evidence for such a monopole in a cosmic-ray experiment that had been designed to study the flux of heavy nuclei in the upper atmosphere. Other physicists who interpreted the data believe that the event was produced by a sequence of interactions of an ordinary heavy nucleus in the detector—an unlikely but possible occurrence. Given the latter explanation, there is now general skepticism that the monopole interpretation is correct.

In another experiment at SLAC involving positron-electron collisions, a significant number of events were observed in which a two-particle collision produced in its final state one electron, one muon, and some missing energy that presumably was lost in the form of neutrinos. These events are not easy to interpret in terms of conventional concepts of particle physics, even if charm is included. The most likely possibility is that the reaction proceeds through the production of a pair of new particles, perhaps massive leptonic relatives of the electron, each of which decays into a muon or an electron and some neutrinos.

—Lawrence W. Jones

Nuclear physics

The past year was marked by a dramatic increase in national commitments to nuclear science outside the U.S. In France a major, double-cyclotron, heavy-ion research laboratory with the acronym GANIL was under construction at Caen. New, very large electrostatic accelerators have been purchased from U.S. manufacturers by the Italian government for installation at Catania and at Legnaro and by the Japanese government for installation at the University of Tsukuba and the Japan Atomic Energy Research Institute. The Canadian government funded construction of the first large superconducting cyclotron for installation at Chalk River, Ont., and the governments of Denmark, Norway, and Sweden have encouraged their scientists to proceed vigorously toward acquisition of a tandem electrostatic accelerator in the 30-million-volt (Mv) range.

In addition to these new projects, the UNILAC, a very large heavy-ion linear accelerator destined for some time to be the most powerful in the world recently became operational at Darmstadt, West Germany. A 14-Mv, U.S.-built tandem accelerator has begun operations at Canberra, Australia, and a similar machine is currently being installed at the Weizmann Institute in Israel. Construction is well along on the 30-Mv tandem accelerator being built locally at the Daresbury Laboratory in the United Kingdom. At higher energies, the SIN cyclotron has begun full operation in Zürich, Switz., as has the TRIUMF cyclotron in Vancouver, B.C.

Such activity demonstrated a major expansion in international interest in the nuclear sciences with particular emphasis on heavy-ion and medium-energy research, both of which had been developed and brought to initial maturity in the U.S. By contrast, progress in the U.S. has been markedly slow. Only a single new major project, the National Heavy Ion Laboratory at Oak Ridge, Tenn., centering on a 25-Mv tandem accelerator, has been initiated in the past decade, and to date only part of this laboratory has been authorized. Despite this handicap, the U.S. scientific community does have a number of very powerful facilities, initiated prior to 1965 but maintained at frontier status by continued upgrading and modification. In particular, the Clinton P. Anderson Meson Physics Facility at Los Alamos, N.M., has just completed modifications that have increased its beam current by more than a factor of ten and allowed it to remain the world's most powerful medium-energy nuclear-science facility.

Nevertheless, unless adequate support can be provided within the U.S., the preeminence that U.S. nuclear science has enjoyed for the past three

Adapted from "France Boosts Nuclear Physics" by Roger Woodham, "Nature," vol. 257, p. 351, October 2, 1975

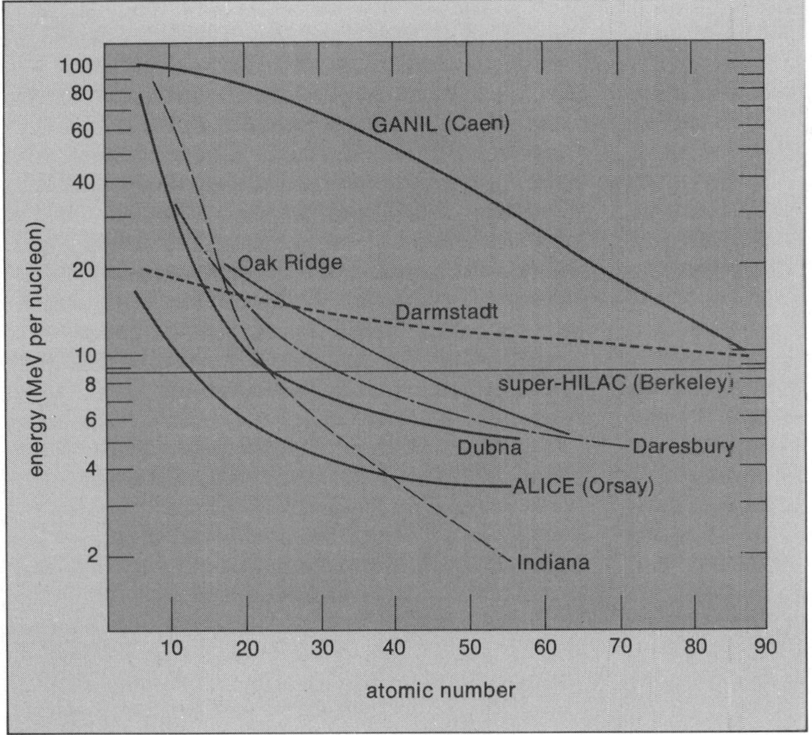

A comparison of the energy per nucleon available for operational accelerator facilities or expected for planned facilities.

decades—and the potential for continued major contributions to energy research, medicine, and defense—will be dissipated and lost. The U.S. nuclear-science community faces a critical period during the next few years: Either a posture of adequate financial support for new and existing facilities must be evolved, or the U.S. must become resigned to relinquishing the leadership role that it has enjoyed throughout the entire effective history of nuclear science.

Nuclear rotations. The existence in nuclei of well-defined bands of rotational quantum levels has been recognized for more than 20 years. These were interpreted to reflect the collective rotation of an intrinsic, nonspherical nuclear shape shared by all the band members. The award of the 1975 Nobel Prize for Physics to Aage Bohr and Ben Mottelson of the Niels Bohr Institute in Copenhagen and to L. James Rainwater of Columbia University, New York City, memorialized this recognition. (*See* Year in Review: SCIENTISTS OF THE YEAR.)

The fact that the moment of inertia of the rotating nucleus was measured to be substantially less than would be the case were it to rotate as a rigid body suggested that the neutrons and protons constituting the nucleus and involved in the rotation must respond to it in a very special way—as if highly correlated in a superfluid or superconducting state much like that of electrons in a metal cooled to temperatures approaching absolute zero. Pursuing this analogy, Mottelson and other physicists questioned whether there might not also be a phase change in nuclei, similar to the quenching of superconductivity in metals as an applied magnetic field exceeds some critical value beyond which the special electron coupling involved in superconductivity is destroyed. They suggested that in nuclei this coupling would be destroyed by the Coriolis forces involved in very rapid rotation.

When precise heavy-ion beams that were capable of subjecting nuclear matter to great centrifugal stresses became available, it finally was possible to test this very important suggestion—important not only to nuclear science but to all science where many-body problems play central roles. It was found that, as spin is increased, a new kind of nuclear rotation appears wherein one, or at most a few, of the nucleons carrying individual large angular momenta decouple themselves from the overall nuclear rotation and simply line up their spins with the axis of overall nuclear rotation, a phenomenon that was both observed and explained by Frank Stephens and his co-workers at the University of California, Berkeley. This behavior implies that in nuclei the phase change does not occur all at once; rather the individual nucleons uncouple themselves one by one from the parent nucleus. There are some indications that the effective moment of inertia first increases gradually with increasing spin simply because the nonspherical nucleus stretches centrifugally; then, as superfluidity is systematically destroyed, the moment of inertia increases rapidly with spin until it reaches the rigid body value; thereafter, it continues a slow increase with further centrifugal stretching. Nevertheless, the question of whether there are situations in which the nucleus can undergo a complete, essentially instantaneous phase change remains open, and much work is being devoted to it.

Nuclei far from stability. Of the 1,600 nuclear species known to science, only 300 are stable in their ground states; over the years some 1,300 additional radioisotopes have been produced artificially for study. With the increasing availability of new heavy-ion and higher energy projectiles, this number is being extended very rapidly as more new radioisotopes with numbers of neutrons and protons very different from those characterizing the stable isotopes of the element in question are produced and identified. A particularly powerful technique has been introduced at the CERN synchrotron in Geneva, Switz., utilizing the ISOLDE on-line mass separator to study very short-lived nuclear species produced by 24-GeV (billion-electron-volt) proton bombardment of various targets. C. Thibault and co-workers have recently examined the sodium (Na) isotopes in this fashion and have found that all isotopes from ^{26}Na to ^{32}Na appeared. Whereas naturally occuring sodium ^{23}Na has 12 neutrons, ^{32}Na has 21 neutrons or 9 more than normal. Among the other interesting results of these particular measurements was the discovery that a closed shell of 20 neutrons in the nucleus did not impart stability in this part of the nuclear chart. As yet there has been no explanation for this apparent breakdown of shell-model behavior, though it serves to illustrate the point that all extrapolation of nuclear behavior from the narrow range of nuclear stability to regions far removed from it is hazardous. Within the next few years the ISOLDE group at CERN, Vadim V. Volkov and his associates at Dubna in the U.S.S.R., Arthur Poskanzer and his associates at Berkeley, and others are expected to map large regions of this unknown nuclear domain.

Strange nuclei. The neutrons and protons (nucleons) comprising nuclei are the stable members of the baryon family of so-called elementary particles. A heavier member of the baryon family, the lambda particle, can be produced in very-high-energy reactions and has many of the characteristics of the nucleons. It differs, however, in having a different value for the quantum number that particle physicists have arbitrarily termed strangeness. In addition, it exists for only a quarter of a billionth of a second, although it must be emphasized that this is still a long time measured in terms of the "nuclear year," the time a photon takes to traverse a nucleon. In these units, the lambda exists for about 10^{13} nuclear years.

It was obvious from the outset that the existence of lambdas made possible an entirely new class of nuclei, the hypernuclei, in which one or more

Split-ring resonator, one of 18, will form the heart of the world's first superconducting heavy-ion linear accelerator, scheduled for completion at Argonne National Laboratory in 1978. The new accelerator is expected to provide heavy ions from Argonne's existing tandem Van de Graaff accelerator with energies sufficient to penetrate the heaviest target nuclei available.

Courtesy, Argonne National Laboratory, Argonne, Ill.

lambdas replaced nucleons. Also, since lambdas differ from nucleons, they can occupy filled shell-model orbitals and thus might be expected to yield quite new phenomena. Some years ago, Herman Feshbach of the Massachusetts Institute of Technology and his collaborators proposed that a hypernucleus might be made in a process that changed one nucleon in a nucleus into a lambda while leaving all else unchanged. Only during this past year has the relevant, very difficult experiment been carried out by Bogdan Povh and co-workers from the University of Heidelberg, West Germany, working at the CERN synchrotron. They scattered a beam of strange mesons (kaons) from nuclei; some of these kaons convert nuclear nucleons to lambdas, and the surplus momentum is carried off by a nonstrange pi meson (pion). From data obtained on targets of carbon, oxygen, and beryllium, John Schiffer and Harry Lipkin at the Argonne (Ill.) National Laboratory concluded that the West German group indeed had observed Feshbach's strangeness-exchange mechanism. With this production mechanism for hypernuclei now in hand, it is anticipated that their study will add an important new dimension to nuclear-science research.

Superatoms. While the search continues in many laboratories for new transuranic and super-transuranic nuclei of varying degrees of stability, attention during the past year has focused on a new class of entities, superatoms, originally proposed by Walter Greiner of Frankfurt, West Germany. It had long been recognized that normal electrodynamics would be expected to break down in an atom having a nucleus with a charge sufficiently large to bind its inner electrons by more than twice the electron rest mass. For a theoretical point nucleus this would require 137 protons, and for one of realistic finite dimensions, about 164 protons. Clearly there is no possibility of producing stable "normal" nuclei having this many protons because their mutual electrostatic repulsion would lead to instantaneous disruption. Greiner noted, however, that because the size of atoms is typically 100,000 times that of nuclei, if two lead nuclei, each having 82 protons, could be brought close together in a collision, the atomic electrons would be quite unable to distinguish the pair from a single supernucleus having 164 protons. The question was whether the pair stayed together long enough for atomic shells to form about them and for measurements to be made.

Although no accelerator currently in operation can give lead nuclei adequate energy to perform the required experiment, Jack Greenberg and his collaborators at Yale University have shown, by careful examination of the X-rays produced when electrons move from one atomic shell to another, that two nickel nuclei (28 protons each) brought close together in a collision function as an effective supernucleus about which the atomic shells do form for measurable periods. In addition to a whole range of new phenomena of interest to atomic physicists, this series of measurements has given great credence to Greiner's initial contention that the lead-lead collisions could provide a crucial test of quantum electrodynamics—the most precise and reliable of all existing physical theories—in a totally new fashion. A Darmstadt-Yale collaborating group is currently preparing such an experiment for the first lead beams from the Darmstadt UNILAC. The initial experiments will search for the positrons that should be emitted when an inner lead electron shell having one electron missing plunges into the Dirac antimatter sea of negative-energy electrons during the collision. This will be a truly fundamental process in that it will ionize, or change the charge of, the total universe by one electronic charge.

—D. Allan Bromley

Solid-state physics

Research in solid-state physics during the year was dominated by interest in an unconventional source of intense electromagnetic radiation whose unique properties made it an exceptionally useful scientific implement.

Synchrotron radiation. In recent months the scientific community has heartily welcomed what promises to be an experimental tool of considerable importance to solid-state physics and related fields. Called synchrotron radiation because it was first observed in 1947 as a by-product of synchrotron accelerators, only in the past few years has it become available as an intense continuum of electromagnetic radiation, from the infrared region well into the X-ray region of the spectrum. The continuum spectral distribution from the Sun is similar, except that it peaks in the visible and drops exponentially at shorter wavelengths (higher photon energies), until at wavelengths less than about 2500 Å it becomes insignificant compared with synchrotron radiation. By contrast, synchrotron radiation sources provide a smooth continuum of radiation over a photon energy range extending 10,000 times further in energy than the output of the Sun.

Synchrotron radiation is emitted from electrons traveling at relativistic velocities (near the speed of light) in circular paths. The maximum energy limit of this continuum of radiation increases as the fourth power of the electron energy; hence, electron energies in the range of billions of electron

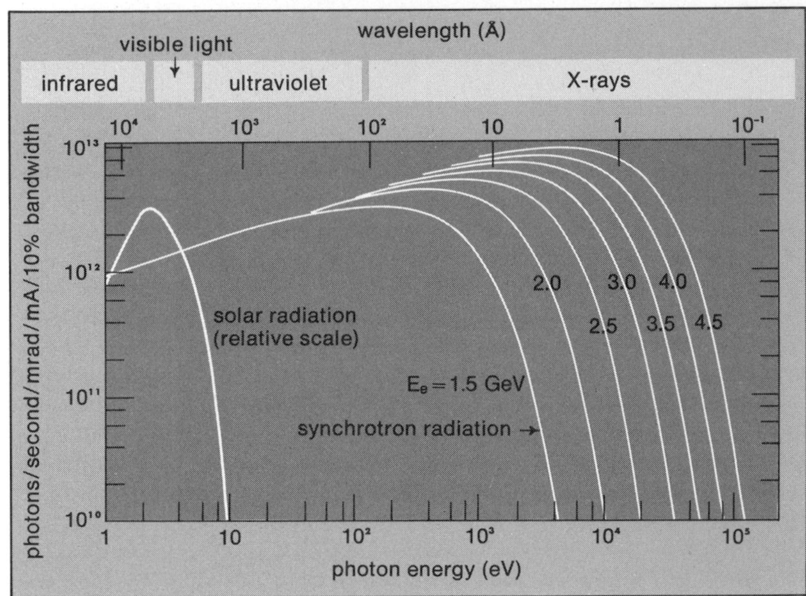

Comparison of the spectral distribution of solar radiation and synchrotron radiation obtained from the SPEAR positron-electron storage ring at Stanford University. E_e is the energy of the stored electrons for each of the seven synchrotron-radiation distribution curves.

volts (GeV) are necessary to obtain a continuum extending well into the hard X-ray region. Despite the importance of synchrotron radiation to solid-state physics and related fields, the development of synchrotron sources has remained dependent upon the development of accelerators and storage rings for high-energy physics. Its application to atomic and solid-state physics was begun in the 1950s, but only the recent construction of storage rings with electrons and positrons in the GeV range has provided it with sufficient intensity and stability to be practical for wide use. Two rings are presently operating at electron energies up to about 4 GeV. One, with the acronym SPEAR, is located at the Stanford (Calif.) Linear Accelerator Center (SLAC), and the other, called DORIS, is at the high-energy facility at Hamburg, West Germany. Experimental results have encouraged expansion of the synchrotron facility at SLAC. In addition, the U.S. National Science Foundation and the Energy Research and Development Administration have expressed interest in the development of a facility dedicated solely to the production and use of synchrotron radiation.

Synchrotron radiation is valuable because it allows many important experiments to be done that were either impossible or impractical using conventional sources of radiation. In its first year of operation, the Stanford Synchrotron Radiation Project (SSRP) at SPEAR emphasized three fields of research which illustrate this approach: X-ray absorption experiments directed toward obtaining information about the environment surrounding specific atoms in complex solids or molecules; X-ray diffraction studies of biologically important crystals; and photoemission studies of solids, surfaces, and surface reactions.

The potential of the first type of experiment, known as EXAFS (extended X-ray absorption fine structure), was first recognized in the 1930s; until the development of SSRP, however, the experiment was impractical. The nature of the experiment consists of examining the fine structure in the absorption of X-rays within about 1,000 electron volts (1 KeV) of the threshold of absorption from a specific atom in a solid or molecule. The absorptions threshold, which is typically in the KeV region, is set by the energy necessary to free an electron from one of the tightly bound atomic core states. As this excited electron moves away from its parent atom, it scatters from the surrounding atoms, and the probability for the absorption of an X-ray is modulated by this scattering. Since the modulation depends upon the relative location of the surrounding atoms, the fine structure in the absorption spectra can be translated into atomic positions. This modulation, however, is relatively weak and hard to detect. Conventional X-ray sources have strong characteristic emission lines but rather weak continuum radiation. Further, the continuum radiation usually has its own superimposed structure. Together these factors make it very difficult to detect fine structure extending over hundreds of electron volts in energy. On the other hand synchrotron radiation provides a smooth continuum of radiation of high intensity. As a result excellent spectra have been obtained at SSRP in half an hour. One interesting application of EXAFS has made it possible to measure the shift in position of an iron atom in the hemoglobin molecule when an oxygen atom is attached or released. Such information is essential to developing a better fundamental understanding of the absorption and transport of oxygen by blood.

Synchrotron radiation also has proved itself useful in X-ray diffraction studies of biologically important materials which, because of their complexity and small size, are difficult to study with conventional sources. Typical materials of interest are muscle, nerve, retina, and protein crystals. Because synchrotron radiation is emitted in a very narrow cone tangent to the electron path, it aids greatly in providing sharply focused, tunable monochromatic radiation. Due to the smooth continuum of radiation, optimum use can be made of the difference in X-ray scattering for photons just above and below the X-ray absorption edge of a given atom, simplifying the location of specific elements in the complex crystal. Also of fundamental importance was the discovery, by a Stanford University group working at SPEAR, that the use of an intense monochromatic X-ray beam effectively eliminated the effects of radiation damage to the sample, which had limited the information obtainable with conventional X-ray sources.

A third area of application has been in the study of surfaces and surface (including catalytic) reactions through the use of photoemission. In the photoemission process, an electron is excited by absorption of a photon in the solid. A certain fraction of the excited electrons escape from the solid. By measuring the distribution in energy of these electrons, much can be learned about the electronic structure of the solid. Such measurements are generally referred to as photoemission electron spectroscopy. Once again, the importance of synchrotron radiation lies in its ability to provide monochromatic radiation tunable over a large wavelength range. This is important for two reasons: First, the depth of escape of the photoelectrons (and thus the depth examined by the experiment) depends upon the energy of the emerging

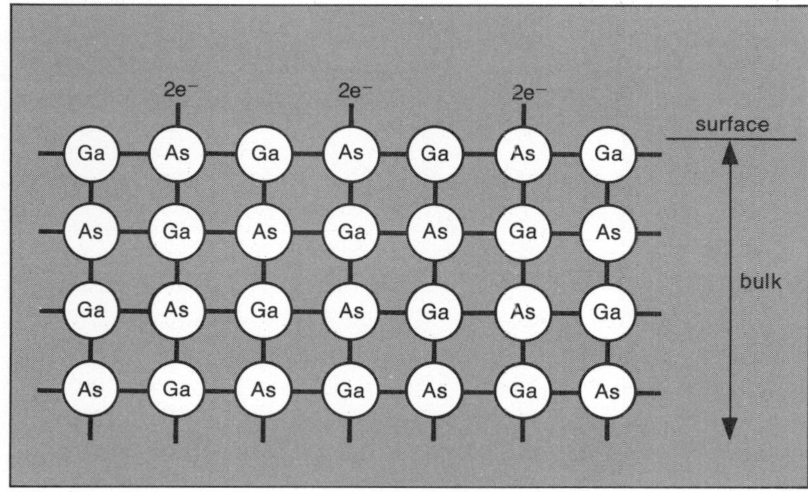

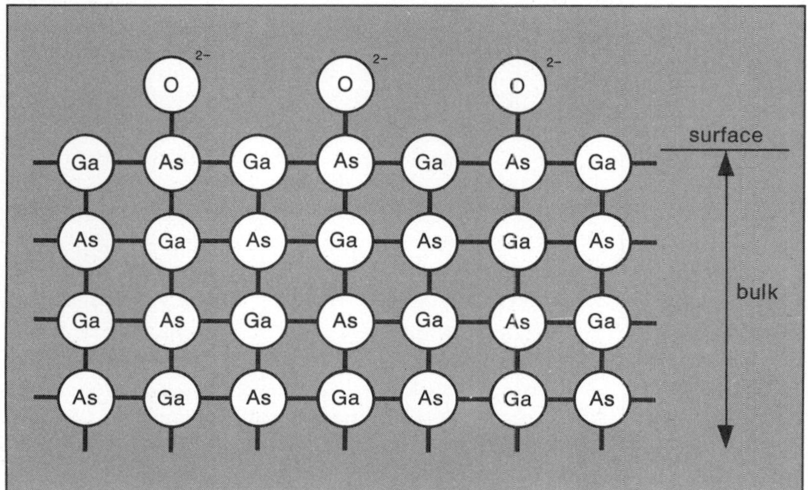

(Top) Schematic representation depicts a theoretical model for the electronic structure of a clean (unoxidized) surface of gallium arsenide (GaAs) crystal containing equal numbers of gallium and arsenic atoms. Note that unbonded electrons available for chemical reactions are found on arsenic surface atoms but not on gallium. As a result, oxygen atoms combine preferentially with arsenic surface atoms (bottom). A reversal of GaAs chemistry in bulk, where gallium is the electron donor and arsenic the acceptor, this model of GaAs surface chemistry was confirmed by experiments employing the unique qualities of synchrotron radiation.

Courtesy, C. Jeffries, University of California, Berkeley

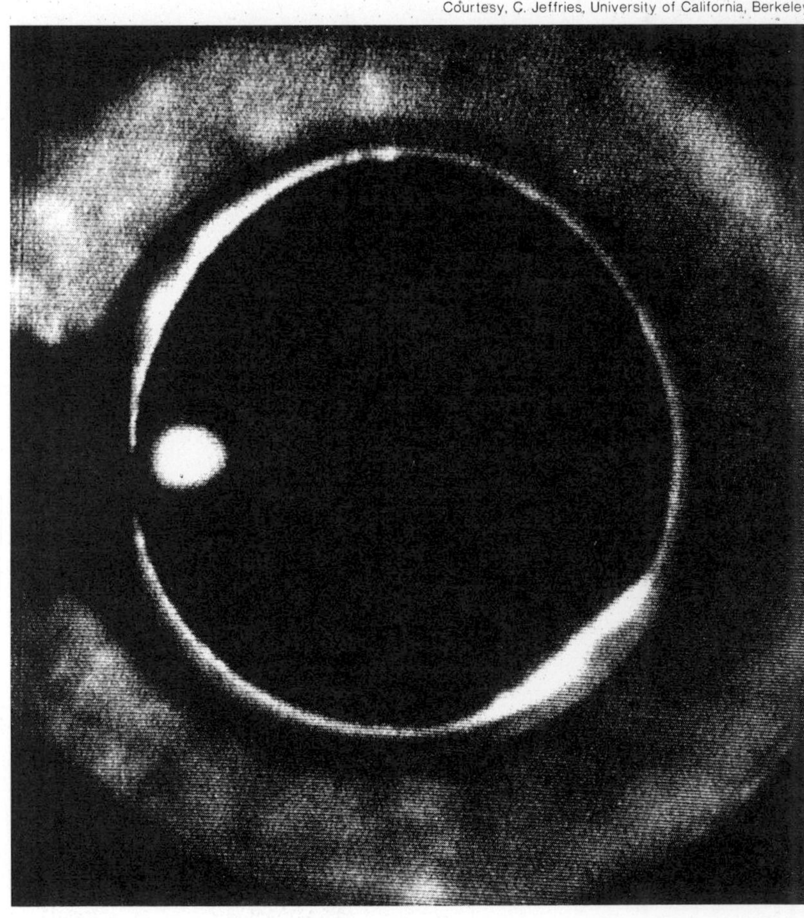

Luminescence from an electron-hole droplet stabilized in a germanium disk four millimeters in diameter is observable as a bright spot near the left edge of the disk. The light circular outline is luminescent light scattered from the edge of the disk.

photoelectron. In particular, it has been known for some time that, if a tunable monochromatic source was available, at some energy between 10 and 200 eV the escape depth could be reduced to the outer one or two atomic layers of the solid. This is, of course, the region of interest for surface studies. Second, synchrotron radiation is sufficiently energetic to excite electrons from core as well as valence energy levels. This is important for surface reactions, because a change in the valence electron distribution due to the chemical bonding of a foreign atom to the surface will result in a shift in energy of the core levels.

Surface science and surface chemistry. Although surfaces play a dominant role in many fields of solid-state physics, they are much less well understood than the bulk of solids. New experimental tools, including that of synchrotron radiation, and theoretical approaches have recently raised expectations that surfaces can be studied and defined with the precision already achieved in the bulk.

Recent advances in the study of gallium arsenide (GaAs) will serve to illustrate the knowledge being gained and the application of synchrotron sources to surface studies. GaAs is of importance because of its potential for semiconductor devices. It has intrinsic superiority over silicon (Si) for fabrication of integrated circuits; however, such use is prevented by the inability to terminate GaAs (*i.e.*, form a protective surface layer) in such a way that its favorable electronic properties are not compromised. Silicon is used in integrated devices because it can be successfully terminated by oxidation to form a protective SiO_2 layer.

Experimental and theoretical work over the last two years had resulted in a model for the surface of GaAs that implied some type of reversal from normal expectations in the surface chemistry of the crystal. It had been known that, in rupturing the crystal to form the surface, covalent bonds are broken, and "dangling" unbonded electrons are left at the surface, but an experiment had been needed that would directly probe the bonding of oxygen to the GaAs surface. Recently such an experiment was performed by experimenters at Stanford University, utilizing the unique qualities of the SSRP and a special monochromator to tune the synchrotron radiation. The group exploited the techniques of photoemission electron spectroscopy to examine the electrons in certain core states of surface gallium and arsenic atoms when the surface was both clean and oxidized. This in turn yielded information about the valence electron distribution of the surface atoms and, thus, about the bonding of oxygen to the crystal surface. The experiment confirmed that the unbonded electrons at the surface become localized on the surface arsenic atoms, to which the oxygen then preferentially bonds. This is a direct reversal of GaAs chemistry in bulk, where gallium is the electron donor and arsenic the electron acceptor. Such information is expected to bear importantly on efforts to adapt silicon semiconductor technology to GaAs.

Electron-hole droplets in semiconductors. A semiconductor requires a minimum energy (the band-gap energy) to excite an electron from its valence band, which is normally filled by electrons, into its conduction band, which is normally empty. Such excitation produces not only free electrons in the conduction band but also positively charged "holes" due to the empty states created in the valence band. In the late 1960s, Soviet physicists established that, at sufficiently low temperatures (below about 20° K), the electrons and holes in germanium would condense into phase-separated droplets that had many properties characteristic of liquid metals. These droplets, which represent the first example of a quantum liquid of constant density in a periodic crystal lattice, provide a fascinating field of fundamental study.

To date, droplet formation has been clearly observed in both germanium and silicon. The electron density in the droplets is usually approximately 2×10^{17}/cu cm and the radius of the drop less than ten micrometers (μm). The lifetime of the droplets, after the source of excitation energy is removed, is set by the rate at which the electrons return from the conduction band into the vacant valence band states and is normally 40 microseconds.

A group at the University of California, Berkeley, recently discovered that, because of the peculiarities of the quantum states of the germanium conduction band, the application of stress to the germanium crystal would reduce electron density in the droplets to about 7×10^{16}/cu cm and thus allow the formation of larger droplets (radius about 100 μm) with much longer lifetimes. The ability to vary droplet size and, particularly, to make large droplets has been very important in developing new methods for studying such properties of the droplets as microwave absorption and luminescence due to photon emission as the electrons return to the vacant states in the valence band.

—William E. Spicer

Psychology

This discussion surveys some of the highlights of the past year, concentrating on certain broad trends in the fields of experimental and applied psychology and salient events in the professional areas of the discipline.

Experimental psychology. A general trend of the past decade has been the challenge to traditional assumptions, objectives, and methodologies of experimental psychology. Substantively, this challenge has been and is being most strongly represented by the markedly decreased emphasis upon strictly behavioral (such as stimulus-response) research and the increased interest in and research upon all sorts of cognitive functions (such as the fantasy research described below). Methodologically, there has been considerable questioning of standards, including the so-called paradigm of the experimental method itself.

Reactions to this challenge have been varied. There has been considerable accommodation to these questions and recognition of much merit in them. As one veteran experimentalist put it in reviewing the 1975 (fourth) edition of the classic *Theories of Learning* by Ernest Hilgard and Gordon Bower, the book demonstrates "that the new psychology of learning is a coherent enterprise in a way that it has never been before, that its reach is now beyond the laboratory and that it is in touch with problems of significance, that it has traded off a little rigor for a great deal of relevance to naturally occurring behavior, and that the current mood in the area is one of high excitement." The following paragraphs provide a selective description of a small number of the interesting and potentially significant experimental researches reported in 1975.

Reflecting the renewed interest in consciousness, Jerome Singer's presidential address to the Eastern Psychological Association in April 1974 reviewed his and others' recent systematic research on thought processes. Among the many relationships between thinking and behavioral functions explored in this research was the finding of reduced eye movement in subjects actively engaged in fantasy; the suggestion that active daydreaming can serve some arousal function by keeping subjects more responsive to external signals during long, monotonous vigilance tasks; and the possibility that the direction of eye shifts correlates with the type of internal thought (fantasy or verbal) that is being processed. Singer offered the opinion that many of the subtle manifestations of thought (such as "altered states of consciousness" and "transcendental experiences") can be experimentally attacked as forms of information processing

Adapted from "Ultradian Rhythms: The 90-Minute Clock Inside Us" by Peretz Laire and Daniel F. Kripke, "Psychology Today," p. 56, April 1975

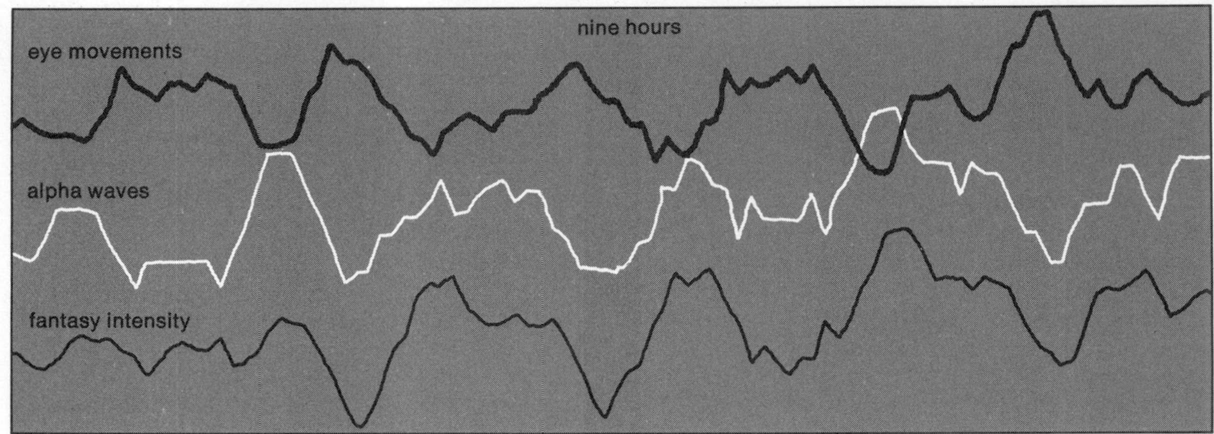

The thoughts of a person over nine hours, recorded every five minutes, reveal peaks of intensity for daydreams and alpha-wave brain activity and low points for eye movements every 90–100 minutes.

and not simply left to be "explained" by means of unanalyzed nonscientific terminology, as is customary for the various forms of Oriental meditation that have become so popular.

James Olds was one of the two discoverers, about a quarter of a century ago, of the so-called reward centers in the limbic system of the brain (the areas which, when stimulated electrically by means of permanently inserted microelectrodes, apparently produce such a pleasurable reaction that a preceding response, such as bar pressing by a rat, can be made to persist indefinitely, to the point of physical exhaustion and even death). Olds recently worked on certain basic neurophysiological factors in simple learning (conditioning). In his preliminary report on this research, he suggested that approximately the first ten milliseconds of stimulus input are crucial for learning or for the establishing of new associative relationships. After that period the message seems to be scrambled by the incredibly complex neurophysiological interrelationships that are activated, even in the simplest of situations.

Another recent piece of neuropsychological research raised questions about the generally accepted proposition that the hypothalamus is the primary controller of eating. H. P. Zeigler showed that a pigeon will stop eating immediately after experimental lesions are produced in its trigeminal nerve system, which mediates touch, temperature, and pain sensitivity in the mouth and face. Lesions in the trigeminal nerve of rats were found to have equivalent effects on eating by those animals. The fact that rats whose lateral hypothalamus is lesioned also eat less can be explained as a function of damage done to the trigeminal nerve, which passes through the hypothalamus.

Zeigler suggested that oral sensations have two major functions: to coordinate feeding and to serve to arouse appetite and motivate eating. He further suggested that differential susceptibility to the pleasurable sensation of food in the mouth and oversensitivity to food-related external cues may lead to obesity.

Daily and monthly internal cycles have been recognized for some time. Only recently, however, have certain more subtle 90-minute cycles been noted. Their identification was based on intensive, computerized analyses of such behavioral phenomena as daydreams, measured under both highly controlled laboratory and normal everyday conditions (the latter by means of cassette recorders into which volunteer subjects dictated their thoughts every ten minutes when a buzzer sounded). These results extended an earlier experimentation which noted 90-to-100-minute cycles of the rapid eye movement (REM) stage of sleep, in which REMs are accompanied by dreaming. The results also suggested the potential significance in our everyday lives of such short-term cyclic swings in mood, alertness, and appetite for eating and sexual activity.

Perhaps the most highly publicized social-psychological research of recent years has been Stanley Milgram's dramatic demonstration of the extent to which normal individuals will consistently obey an experimenter's instruction to administer apparent shock to an innocent victim. During the year Thomas Moriarty extended Milgram's results in a series of experiments that seem to indicate that normal individuals are surprisingly unwilling to defend even their own rights in social situations. These studies followed up the research on noninvolvement that was stimulated by the notorious Kitty Genovese murder case of 1964 (when many persons in nearby apartments who heard her screams did nothing to help, not even bothering to call police).

Moriarty used a wide variety of situations, ranging from submitting male college students to outrageously loud rock music while working to requesting callers in a public telephone booth to empty their pockets to show that they had not taken a ring allegedly left by a previous user of the booth. With remarkable consistency, approximately 80% of the subjects were found simply to accept the situation, doing nothing to protect their rights. Apparently under some conditions, people are not much more inclined to stand up for their own rights than they are to help others, again presumably because of their desire to avoid a "hassle."

At the ceremonies that took place when he received the International Kittay Award, Harry Harlow delivered a lecture entitled "Aggression and Love," which contained some interesting and timely accounts of his latest research. He reported on some of the effects in primates of lack of exposure to maternal and peer affection. In addition to an extremely high ("nearly suicidal") level of self-aggression, a battered-child syndrome was prominent in the behavior of such isolated female monkeys when they themselves became mothers. Their babies were either entirely ignored or were violently abused and even killed. Surprisingly, however, the loyalty of the abused infants to their mothers remained steadfast. Moreover, Harlow found that such loyalty was a solid foundation on which rehabilitation of the disturbed mothers could be instituted; the unremitting efforts at physical contact made by the abused infant monkeys sometimes induced their affectionless mothers to begin social activities in play.

Applied research. Recent research on infants has shown that the newborn are more sensitive to both fetal and neonatal stimulation than had formerly been supposed. In some new research on premature babies, marked behavioral adjustments were reported in infants provided with a special 30-day treatment consisting of systematic stroking and rocking. Significant gains were reported in weight, neurological development, and mental function. Moreover, mothers who provided the special treatment were pleased by the social interactions that occurred, which included smiling, eye contact, and vocalization.

Experimental investigations of crowding in rodents have generally showed it to have harmful

Courtesy, Wisconsin Primate Center

Young monkeys cling to cloth-covered surrogate mother and leave only for food provided by a wire surrogate. At the age of three, monkeys isolated from birth were considerably more aggressive and violent than those with real or surrogate mothers.

behavioral and biological effects on the animals. During the past year considerably more limited investigations on crowding in humans began to appear. Although definitive conclusions cannot yet be drawn, it appears that people are less negatively affected by crowding than are animals. The most recent studies on humans seem to indicate that sex differences have an influence on reactions to crowding. In one set of studies, for example, small groups of people were placed in various sizes of rooms for up to four hours of continuous discussion and work on mental tasks. In all-male groupings, as the room size became smaller the subjects were found to become more verbally aggressive and more competitive and to profess less liking for their co-workers. Exactly the opposite effects were found in all-female groupings. No reliable differences could be determined for mixed-sex groupings.

In another set of studies, sex differences were again reported. Men sitting in a large room for short periods of time tended to rate themselves and others more highly, and to look at each other more directly. Women showed these results when in a small room.

Professional affairs. In 1974 the American Psychological Association (APA) formed its own political-social advocacy organization, the Association for the Advancement of Psychology (AAP). Its primary purpose was to offer a moderate alternative to the vigorous political efforts of the independent Council for the Advancement of Psychological Professions and Sciences (CAPPS), which focused on such clinical problems as inclusion of psychological practice in various health insurance programs. Although the early prognosis for harmonious reconciliation of the two competing groups did not appear favorable, temperate heads seemed to have prevailed because during the summer of 1975 they agreed to merge. The name and formal structure are to be those of AAP, with former CAPPS leaders incorporated into the governing structure. Thus a potentially highly divisive split within psychology was averted, and all members of APA were urged to support AAP.

Of the more than 40,000 members of the APA, more than one-third are clinical psychologists. Thus it is hardly surprising that a clinician should be elected to the presidency of the national organization. In 1975, however, such an election did set an interesting precedent. For the first time this honor went to a clinician who was without a university or other institutional affiliation but who was strictly in private practice (Theodore Blau, of Tampa, Fla.). Whether this election is a harbinger of future similar events remains to be seen, but the political power of the practicing professionals within APA has been clearly demonstrated.

A landmark case relating to civil rights of mental patients may well have been decided in June 1975. A unanimous U.S. Supreme Court decision held for a 67-year-old ex-carpenter, Kenneth Donaldson, who had been detained against his will for 15 years in a Florida state mental hospital. The essence of the ruling was that "the mere presence of mental illness does not disqualify a person from preferring his home to the comforts of an institution . . . [if he] is capable of surviving safely in freedom by himself or with the help of willing and responsible family members or friends." The

Kenneth Donaldson, held in a state mental hospital against his will for 15 years, holds copy of a U.S. Supreme Court decision stating that the mentally ill cannot be confined involuntarily if they are not dangerous and can live safely in freedom.

Wide World

long-range significance of this decision is that it strongly supports the community-health movement as compared with traditional institutional treatment, and that it further weakens the so-called medical model (based on an analogy with physical illness) for behavior disorders.

An important shift in medical emphasis appeared in the 1975 version of "Forward Plan for Health," a long-range blueprint from the U.S. Department of Health, Education, and Welfare. The report details the shift in lethal health hazards from strictly physical pathological conditions (such as pneumonia, influenza, tuberculosis) to more environmentally determined factors (heart disease, cancer, stroke for older persons, and automobile accidents, homicide, and suicide for younger persons). It concludes that "A distinctive feature of these conditions is that most of them are caused by factors (*e.g.,* the environment and individual behavior) that are not susceptible to direct medical solution."

These views parallel even more environmentally based ones expressed in a 1974 Canadian document, entitled "A New Perspective on the Health of Canadians," by Canadian government official Marc Lalonde. He concludes that "unless the environment is changed and the self-imposed risks are reduced, the death rates will not be significantly improved."

These views converged with a growing consensus among responsible scientists and practitioners that environmental and cultural factors are the primary determinants of individual health. The behavioral significance of such views is obvious. Although there is yet no clear-cut blueprint for effective treatment, it may be expected that potentially helpful proposals will be easier to develop in the newly emerging climate of agreement about primary causal factors concerning individual health and welfare.

—Melvin H. Marx

Space exploration

Major developments in the exploration of space during the past year included the soft landings of Soviet probes on Venus and the launching of the U.S. Viking probes to Mars. A comparatively quiet year in manned flight was marked by a continuation of work on the reusable space shuttle and the space laboratory and by a Soviet mission of record duration. In July 1975 space crewmen from the U.S. and the Soviet Union achieved the first international docking of two spacecraft. For a full discussion of this accomplishment, see *1976 Yearbook of Science and the Future* Year in Review: SPACE EXPLORATION.

Manned flight

Among the highlights in manned space flight during the past year the development of the reusable space shuttle system proceeded in the United States, while in Europe the European Space Agency completed preliminary design of a space laboratory that could be carried into orbit and returned to Earth in the space shuttle orbiter payload bay. The Soviet Union continued its manned space research with its longest manned flight to date, a

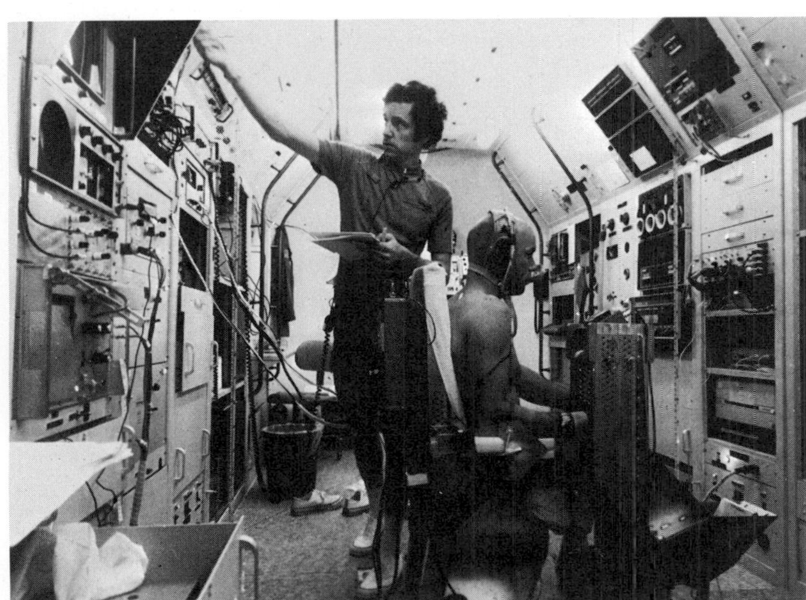

Courtesy, NASA

In spacelab simulator of U.S. space shuttle, scientist-astronaut Story Musgrave (seated) operates equipment that will provide data on skeletal muscle function during space flight. Cardiophysiologist Charles Sawin monitors the task.

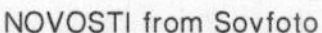
NOVOSTI from Sovfoto

Soviet manned spacecraft Soyuz 18 and its rocket booster are rolled out for launch at Tyuratam Space Center in 1975.

visit of more than 60 days to the orbiting Salyut space station.

U.S. manned space program. The U.S. continued the development of the space shuttle system. This new spacecraft will be unique in design and operation. Unlike previous U.S. and Soviet spacecraft, the space shuttle orbiter will be reusable and will resemble an aircraft. It will land on a runway rather than utilizing a parachute system for descent and landing. The orbiter will be approximately 122 ft long, the size of a DC-9 jet airliner. It will be launched with a rocket booster system into Earth orbit and will be able to fly missions of up to 30 days. The orbiter will have a cargo bay 15 ft in diameter and 60 ft long, which will be used to transport unmanned satellites into orbit or, possibly, to bring satellites back to Earth for repair or modifications. The cargo bay also will be used to carry scientific instruments that would be operated by the shuttle crew while in orbit. Additionally, a space laboratory can be carried in the cargo bay to provide a large facility in space for medical, engineering, and science experiments.

The shuttle will be flown by a pilot and co-pilot and can carry five additional crewmen. In the last year, major components for the orbiter were delivered, and the first spacecraft was being assembled at Palmdale, Calif. The initial flight tests were scheduled to be conducted in 1977 and will primarily demonstrate the approach and landing capability of the orbiter. In these tests, the orbiter will be transported by a Boeing 747 jumbo jetliner to a high altitude where the two will separate; a two-man crew will then fly the orbiter and land it on a conventional runway. Upon completion of this testing, the orbiter is to be transported to Cape Canaveral, Fla., for operational research tests. The shuttle is scheduled for operational use in 1979, and in the 1980s it will be used to transport men and equipment into space for scientific studies and to exploit the uses of space for the benefit of mankind.

In November the National Aeronautics and Space Administration (NASA) awarded contracts to two research teams, managed by scientists from the Johnson Space Center and the Marshall Space Flight Center, for "conceptual studies" of large permanent U.S. space stations. For the first stations, NASA envisioned facilities that would be a few hundred feet long at most and that would accommodate 6–12 persons. An extensive discussion of space stations can be found in the feature article COLONIES IN SPACE.

European Space Agency. A conglomerate of nations organized the European Space Agency to develop a capability for space exploration. In the manned spaceflight field, the agency assumed the responsibility to develop and furnish the Spacelab to be flown in the space shuttle program. The European nations are funding all costs for the Spacelab. This manned laboratory will be flight-tested in the early 1980s and will be used extensively in future manned space flight programs.

Spacelab was discussed at a meeting of British space scientists at London in December. They agreed that it would be especially useful for undertaking studies in the high-energy region of the spectrum (cosmic rays, gamma-rays, and X-rays). The most immediate astronomical benefits were likely to be photographs of the planets taken by a telescope aboard Spacelab. According to one scientist a one-meter telescope in Earth orbit would produce pictures of Jupiter with a resolution of about 250 km (155 mi) and pictures of Uranus as good as the best ground-based photographs of Jupiter. In comparison with space probes, Spacelab would have the advantage of being able to provide many pictures over a considerable length of time.

Soviet manned flight. In 1975 the Soviet Union successfully launched one manned mission and suffered a failure in the launch of a second. On April 5 the U.S.S.R. failed in the launch of a Soyuz spacecraft. The purpose of the mission was a second visit to the Salyut 4 space station. Because the Soviets do not officially designate mission numbers until a flight is successfully launched, the Soyuz 18A mission was not so named by the U.S.S.R. space team. The Soyuz 18A mission was aborted several minutes after launch when a failure occurred in the separation device between the second-stage and third-stage propulsion systems. At the time the mission was ended the spacecraft was near orbital altitude, and as a result of the failure it traveled in a suborbital trajectory to land in western Siberia some 1,000 mi from the launch site. The crewmen for this mission were Vasily Lazarev and Oleg Makarov, both veterans of a previous space mission in Soyuz 12. Both were recovered in good condition several hours after landing.

On May 24, 1975, the Soyuz 18 spacecraft was launched into Earth orbit with a two-man crew commanded by Pyotr Ilyich Klimuk with Vitaly Sevastyanov as flight engineer. The cosmonauts maneuvered their spacecraft into the same orbital plane as the Salyut 4 and docked with the space station. During the first few days they transferred equipment from the Soyuz spacecraft, activated the space station and its systems, and performed housekeeping tasks to make the space station habitable. The crewmen then conducted scientific studies on the Salyut for 63 days. On July 26, 1975, they separated the Soyuz spacecraft from the space station and proceeded with necessary maneuvers to reenter the Earth's atmosphere and land. The cosmonauts returned in excellent physical condition from the longest Soviet space mission to date.

—Richard S. Johnston

Space probes

Four of the most scientifically sophisticated probes ever developed were launched in 1975. The U.S.S.R. sent Veneras 9 and 10 toward Venus and the U.S. sent Vikings 1 and 2 to Mars.

Probing Venus. Venera was a two-part probe, one of which was designed to land upon the planet while the other stayed in orbit about it. Together the two units weighed about 10,200 lb. Thus, they became the heaviest probes yet to be launched by the Soviet Union. The launch vehicle for the craft was a three-stage Proton.

Venera 9 was launched from the Soviet space center at Tyuratam in Kazakhstan on June 8, 1975. Its voyage to Venus was apparently uneventful in that there were no major equipment or instrument failures. En route, instruments in the orbiter section measured the interplanetary magnetic fields and the solar wind. The probe entered a highly elliptical orbit around Venus in late October. On October 20 the descent vehicle was detached from the orbiter and two days later landed on the planet.

After leaving the sister ship, the lander was protected by a hemispherical heat shield that screened its instruments from the aerodynamic heating initially encountered in the thick and turbulent atmosphere of Venus. When it first encountered the atmosphere, the lander was traveling at a rate of 35,104 ft per sec; however, aerodynamic braking reduced the velocity to 820 ft per sec, at which point a parachute deployed and further reduced its rate of descent until the probe reached an altitude about 30 mi above the surface. From there, speed was checked by a ring surrounding the lander that also functioned as a reflector for its antenna. By the time it touched down, the lander was falling at a rate of only 25 ft per sec. Impact was minimized by a special crushable ring about the base.

During the descent, which took 1¼ hours, instruments in the lander analyzed the Venusian atmosphere, finding that it contained only 0.01% water vapor. They also studied the gaseous medium for its chemical constituents, searching for such compounds as sulfuric acid, a chemical that would suggest volcanic activity on the planet.

Once on the surface, Venera 9 had but a short time in which to accomplish its tasks. While the lander had been designed to withstand atmospheric pressures of 298 tons and temperatures up to 3,632° F, it was destined to operate for only 53 minutes. During that period, however, it took a panoramic picture and a detailed close-up view of its immediate vicinity. The panorama showed the horizon at a distance between 650–950 ft. The close-up included details of the base of the lander and showed rocks 12–16 in. across.

The pictures provided surprises to scientists, who had expected that because of the thick and dusty atmosphere of Venus, little light would filter down to the surface. Thus, Soviet scientists provided lamps on the lander. They were not needed, however, and there was more than enough light to

(Top) Sovfoto, (bottom) Pictorial Parade

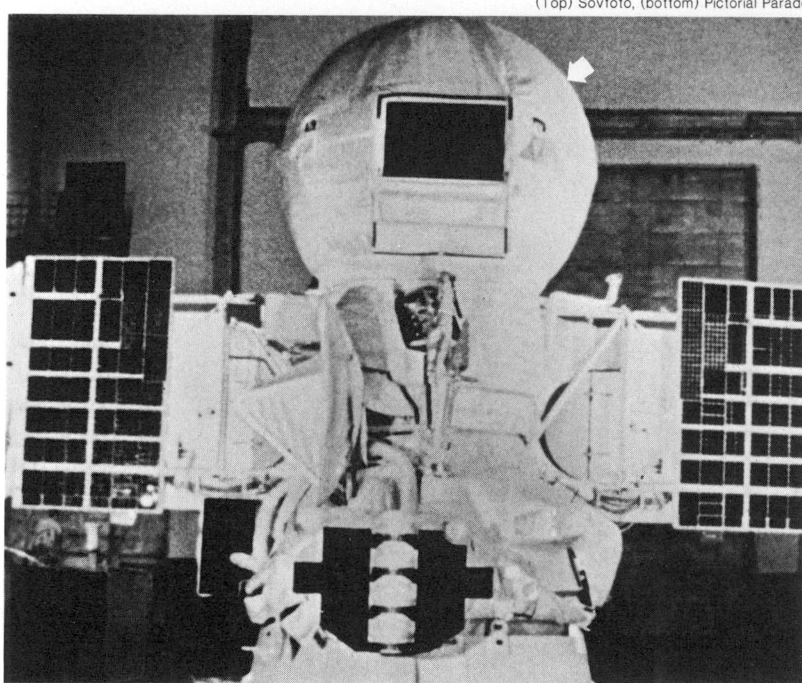

Prelaunch checkout of Venera 9 (right) reveals flight configuration of the Soviet space probe. The arrow points to the spherical heat shield that encased the lander. Photograph taken from the lander (below) shows the rocky surface of Venus; the curved section at the bottom is part of the vehicle.

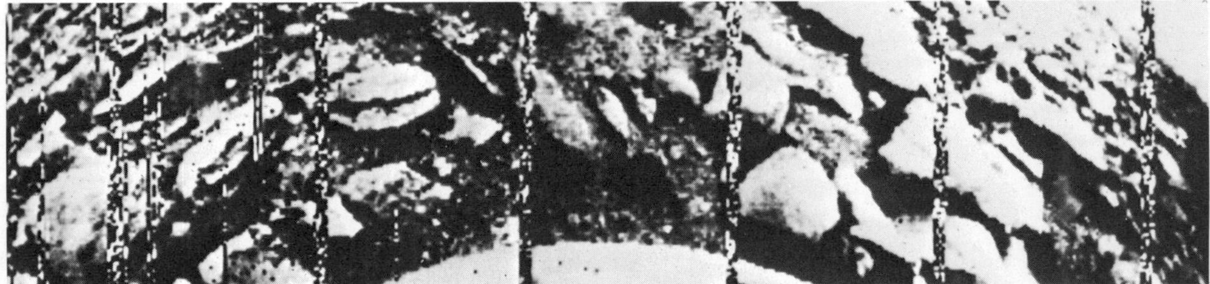

show that the rocks, in the close-up view, were angular and sharp-sided rather than worn smooth, as would be expected from wind erosion. The panoramic view showed that the horizon was convex rather than concave, as many scientists had predicted because of the light conditions and the refracting properties of the atmosphere.

Instruments aboard the lander also transmitted environmental information to the orbiter overhead. They showed that the atmospheric pressure at the landing site was approximately 1,350 psi and that the temperature was 905° F, hot enough so that had metals such as zinc, tin, or lead been present they would have existed in the liquid state. A wind velocity of only 8 mph was also registered.

Venera 10, using the same entry and landing technique, touched down on Venus on October 25 at a point about 1,365 mi away from Venera 9 and at a much lower altitude on the surface of the planet. As did its predecessor, the lander took a panoramic view of its landing site and then a close-up picture. Electronic data to constitute the pictures were telemetered to the orbiting unit, which, in turn, relayed it to tracking stations in the U.S.S.R. Some views of the rocks shown were closer to what had been expected but not seen in the pictures from Venera 9. They were rounded and smoothed through an erosive process. U.S. geologists, who had examined the photographs, estimated that the wind velocity in the area of Venera 10 would have about the same effect as a sea wave pounding on the coast of the Earth during rough weather with winds of 50 mph.

The Venera 10 lander operated on the surface for 65 minutes. During that period, it reported a local temperature of 869° F and an atmospheric pressure of 1,362 psi, both values of which are in good agreement with those measured by Venera 9. The wind velocity measured at the Venera 10 site was for all purposes the same as that at the Venera 9 location.

After examining the pictures returned by the two Venera probes, Arnold Selivanov, who had helped develop the cameras aboard the two spacecraft said, "We did not expect to get pictures of such good quality. . . .It was as bright on Venus as on a cloudy day in June in Moscow. . . . Even details of the surface, which were in deep perspective—at a distance of 50 to 100 meters—were seen clearly in both panoramas."

In analyzing the pictures from Venera 9, Soviet paleontologist Mikhail Marov said, "This is documental evidence that in the scale of evolution, the planet Venus should be placed with the young, still living planets. The sharp edges of the stones, which are characteristic of the panorama from Venera 9, clearly speak in favor of tectonic activity on the planet."

The two Venera orbiters were instrumented to study the Venusian atmosphere. Having finished their task of relaying data from the landers, the two craft continued to circle the planet and photograph the top of the atmosphere from an average distance of 932 mi. Eventually, Soviet scientists hoped to gather enough such pictures to have a photograph of the entire atmosphere.

Among the instruments aboard the orbiters were spectrometers developed by French scientists to measure temperatures in the upper atmosphere. The French were also planning a special balloon sonde to be released in the Venusian atmosphere during a future Venera launch. As 1975 ended, U.S. space scientists from the National Aeronautics and Space Administration (NASA) met with their Soviet colleagues to study the data returned from the two Venera probes to determine if they revealed anything that would necessitate changes in the U.S. Pioneer probe to be launched to Venus in 1978.

Vikings to Mars. The ambitious Viking program culminated in the launching of two craft toward Mars after a series of frustrating delays. Viking 1 was scheduled for launch on August 11; however, a malfunction in the Titan IIIE booster vehicle and a subsequent fault in the probe's battery charger forced a delay until August 20. On that date Viking 2 was launched but designated Viking 1. The two-part spacecraft consisted of a 5,125-lb orbiter and a 1,470-lb lander. A week after the launching, a course correction was made by firing the Viking's main engine. The course thus ensured that the probe would enter an orbit around Mars on June 19, 1976.

Prior to the delay in launching Viking 1, it was hoped that its lander would touch down on the planet on July 4, 1976. The delay, however, made a date of July 7 more desirable.

Viking 2 also had problems. Originally scheduled for launch on September 1, it suffered a failure in the orbiter's critical radio antenna. With the problem overcome, the probe was launched on September 9. Its rendezvous with Mars was thus set for Aug. 7, 1976, and a landing was to take place on September 4.

The lander on both Vikings was instrumented to perform eight investigations on the surface and one while descending through the planet's tenuous atmosphere. Shortly after separation from the orbiter, instruments in the lander were to begin monitoring the atmosphere by investigating its charged particles. They would continue to do so until the lander was about 70 mi above the surface. The atmosphere itself was to be sampled to determine its constituent gases, density, pressure, and temperature.

Once on the surface, the lander's two cameras were to take pictures of the landing site as well as panoramic views that would include the horizon. Because the two cameras are 39 in. apart, stereoscopic pictures can be taken. The pictures were to be sent to the orbiter for relay to the Earth, but it was also possible to bypass the orbiter and transmit the pictures directly.

A sophisticated biological laboratory in the lander was to search for the presence of bacteria in the Martian soil. Samples taken from the surface would be fed into a hopper that would send them to three different instruments designed to search for such forms of life through detection of products of their metabolism. The laboratory was to be maintained at a temperature of 46° to 63° F despite the fact that internal temperatures of the lander structure were expected to vary between −103° and + 95° F.

One of the biological experiments was designed to use radioactive carbon dioxide to see if it is fixed by bacteria in soil samples that have been incubated for five days. Gases from the incubation chamber were to be processed through a special trap to detect the presence of the radioactive components in organic compounds.

Another instrument was designed to test for evidence of metabolic activity in a soil sample moistened with a dilute aqueous solution of simple organic compounds such as alanine, glycolic acid, and glycine; all of these were to be tagged with radioactive carbon. Soil samples would be incubated in the dark for a period of 11 days. The detection of radioactive carbon dioxide or other gases would be indicative of a metabolic process by a microscopic form of life.

A third experiment, using no radioactively labeled materials, was to measure changes in an atmosphere (carbon dioxide and inert gases) surrounding incubated Martian soil samples in the presence of water with and without nutrients. Each sample was to be incubated for periods as long as 12 days and the atmosphere was to be periodically

(Top) Courtesy, NASA; (bottom) adapted from "The Road to Mars" by Jonathan Eberhart, "Science News," vol. 108, no. 5, p. 77, August 2, 1975

Viking lander (right) is basically a hexagonal prism 18 in. high with side panels alternately 43 and 22 in. wide. Among its experiments on the surface of Mars were those to determine the assimilation of carbon in an incubated soil sample (below left), the release from a nutrient-enriched soil sample of radioactively labeled gaseous carbon products (center), and the changes in an enclosed atmosphere surrounding a sample (right).

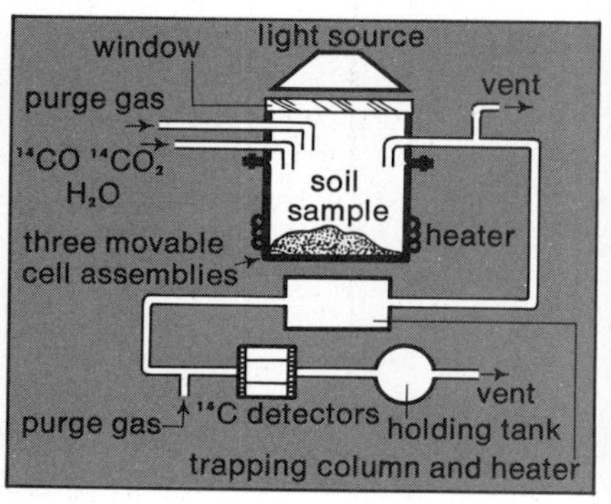

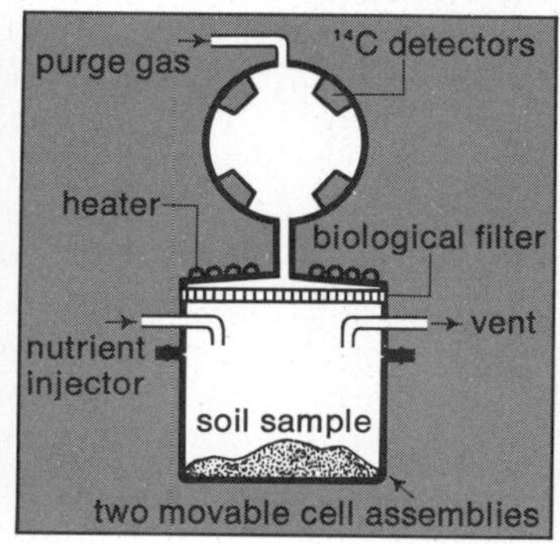

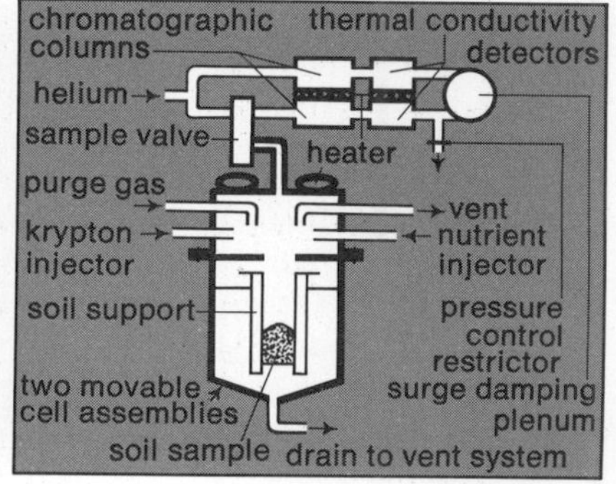

drawn off and analyzed in a gas chromatograph for changes in composition, including the presence of traces of molecular hydrogen, nitrogen, oxygen, and methane.

Other instruments aboard the lander were to search for organic compounds in the surface layer of Mars and attempt to identify them as being produced by living systems. An X-ray fluorescence spectrometer would seek to identify inorganic compounds in the surface material.

Additionally, instruments were to measure wind speed and direction, temperature, and atmospheric pressure 20 times a day. Miniature seismometers would monitor the planet for "Marsquakes." Using pictures from the lander's cameras, scientists planned to study the impression in the soil caused by the impact of the legs of the craft with it. They thus hoped to derive information about the mechanical properties of the soil. The soil was also to be tested by two permanent magnets to see if any of its constituents have magnetic properties. The radio signals passing between orbiter, lander, and Earth were to be analyzed in order to provide valuable information on the gravity field, figure, spin-axis orientation, and surface density of Mars; on pressure, temperature, and electron profiles in the planet's atmosphere; and on the characteristics of the interplanetary space between Mars and Earth.

While the lander performed these tasks, the orbiter was to be equally occupied with experiments of its own in addition to relaying data from the lander to Earth. Housed in the orbiter are two cameras. As Viking 1 approaches the planet, it will photograph Mars in three colors. Unlike the conditions that caused the fuzzy pictures returned by Mariner 9 in 1971, the weather on Mars is expected to be clear for the spacecraft's initial photographic sessions. After achieving orbit, the cameras will then photograph the chosen landing sites to make sure that they are in shape for reception of the lander. Once the lander is on the surface, the orbiter cameras will photograph those portions of the planet beneath its ground track.

The cameras of Viking 2 were to do the same thing, with one exception. After its lander has touched down, the orbiter will make a plane change so that it can photograph the polar regions of Mars. An infrared grating spectrometer aboard the orbiter was designed to monitor the planet's atmosphere for water vapor by detecting solar energy that has been reflected from the surface through the Martian atmosphere. Data so derived would be fed into special computers on Earth to make contour maps of water vapor abundance and pressure. Another infrared instrument was to measure the surface temperatures of the areas over which the orbiter passes.

In commenting on what can be gained from the Viking program, Richard Young, vice-chairman of the Viking Science Steering Group and one of 70 scientists involved in planning experiments for the probe, said, "It is an enormously profitable venture, from the point of view of understanding Earth. We will be able to compare what has been happening on Earth for the last four billion years, while it interacted with life, with another planet where there may be no life at all or where life should have evolved but didn't. Viking is not simply going to Mars to look for life. Viking is going to Mars to study the planet, to develop an understanding of the environment of Mars as best we can with the limited payload that we have so that we can see what's happening to another planet."

The Pioneers. Pioneer 11 continued on its way to a rendezvous with Saturn after swinging past Jupiter at a distance of only 26,000 mi. It is scheduled to reach the planet in September 1979. In mid-December 1975 Pioneer-Saturn, as it was renamed, successfully completed a hazardous maneuver. Controllers at the Ames Research Center commanded the probe to fire its engine and increase its velocity by 67 mph. In order to accomplish this they had to cut off all communications with the spacecraft for nearly two hours while maneuvers were being performed 287 million mi away in space. The short burst of its rocket engine ensured that Pioneer would fly between the rings of the planet and the planet itself or that it would come in under the rings and then pass upward outside of them. Several less harrowing maneuvers were expected to be necessary over the next three years, depending upon how well the initial trajectory change was made.

On Dec. 16, 1975, a precursor of Pioneer 11 celebrated its tenth year of sending back data from an orbit around the Sun. Pioneer 6, launched in 1965 with an estimated operational lifetime of six months, had by the end of 1975 circled the Sun 12 times and traveled more than 6 billion mi. Its data concerning conditions of the Sun are regularly used in weather predictions by the U.S. Federal Aviation Administration, commercial airlines, communications companies, the U.S. Department of Defense, and organizations concerned with surveying and navigation.

Probing the future. It was clear by the end of 1975 that future probes to Mars by both the U.S. and the U.S.S.R. would largely hinge on the outcome of the Viking missions. Even so, NASA was basing its future planetary probe program on a "mission model" utilizing the still-to-be-developed space shuttle and an undefined "interim upper stage," to be developed by the U.S. Air Force. Without the upper stage, interplanetary probes and geosynchronous Earth-orbiting satellites cannot be launched using the shuttle. A reusable "upper stage" probably will not be realized until after 1985.

While no funds had been allocated for such a project, NASA went ahead with studies for a probe that would orbit the Moon at its polar regions. It would be launched in 1980 by a Delta space carrier vehicle. The two-part probe would consist of a 200-lb radio-relay orbiter and a 900-lb scientific satellite. Placed in a circular orbit at about 60 mi above the Moon, the scientific orbiter would pass over 90% of its surface. Instruments would identify the composition of the lunar soil, measure the gravitational effects of the concentrations of dense material beneath its surface (known as "mascons"), determine the rate of heat flow from the Moon, investigate its magnetic properties, and photograph the surface. The cost of the project was estimated at $75 million, a sum that NASA hoped might be reduced by the financial and scientific participation of the U.K.

Taking the overview of the future, Herbert Friedman, director of the E. O. Hulburt Center for Space Research of the U.S. Naval Research Laboratory, said at the end of 1975, "Where might we stand in 1985, if we follow through in carrying out our presently considered programs? Reconnaissance will be extended from 5 to 19 astronomical units, from Jupiter to Saturn. Saturn's rings will be viewed in detail and Uranus in close-up resolution, along with several of its satellites. Composition, cloud patterns, magnetic fields, and other basic parameters will be established. Jupiter's magnetosphere may be understood to a degree comparable to Earth's magnetosphere. Captured satellites will be distinguished from primary objects formed together with a mother planet. All the information will contribute to a far greater depth of understanding of the origin of the solar system. Present speculation will seem pallid by comparison with the revealed truth."

—Mitchell R. Sharpe

Transportation

From the point of view of transport technologists, the past year could be characterized as one of compromise and accommodation. Faced with serious economic problems and growing fuel supply/cost pressures, policymakers were forced to ease many of their rules and regulations in safety and environmental areas. This gave those engaged in research more time to comply with tough guidelines.

Highway transport. Compromise was particularly noticeable in the highway transport area. The demands imposed by U.S. government agencies on auto and truck manufacturers to meet new standards for brakes, noise, and pollution had become so difficult that manufacturers were making strong protests. One maker of heavy-duty trucks, Chrysler Corp., announced that it was getting out of this phase of its business because the standards were too complex to permit profitable operations.

The U.S. Environmental Protection Agency (EPA), recognizing the difficulties faced by manufacturers, granted a one-year delay in its deadline for meeting new automobile emission standards for 1977 model cars. Environmentalists agreed to accept the delay in order to concentrate on blocking further attempts to delay the deadline. The EPA later delayed a deadline for its proposals for new sulfur-emission standards for autos, and it stated that its studies showed that cars with catalytic converters produced little more sulfuric acid than did cars without them.

Cube-treaded, low-pressure tires are designed to provide a comfortable ride and good traction on rugged ground or smooth roads.

The long, heated dispute over the adverse effect of lead in auto gasoline took another turn, as the du Pont Co. patented an emission-control catalyst that it claimed can accept high-octane, leaded fuel. This would allow a return to more efficient, high-compression engines. Du Pont also claimed that its new catalyst has such qualities as great durability, thermal stability, and resistance to lead, sulfur, and phosphorus.

Another example of retrenchment of government safety rules was a one-year suspension granted by the U.S. National Highway Traffic Safety Administration (NHTSA), until Jan. 1, 1977, of its requirement for installation of anti-skid brakes on transit and intercity buses. The requirement, which had gone into effect early in 1975, proved to be premature, as the performance of the new brakes in actual operations was erratic and thus necessitated more tests and development work. Truck lines that had been required to install similar anti-skid brakes made similar complaints about their performance; the NHTSA rejected the pleas of the truckers for a delay of the requirement but indicated that some modification of the rule may be forthcoming. Court actions were pending at the year's end.

Another U.S. government agency, the Bureau of Motor Carrier Safety, announced new rules that limited truck tire pressures strictly to those branded on side walls and that also limited truck loads to equivalent tire load ratings. When it issued the rules, the bureau said they were needed to stop the unsafe operating practices that it had documented during a 14-month study, and it stated further that it would not make any exceptions. Yet, a few months later, the bureau amended its rules, dropping maximum pressure restrictions and giving carriers until April 1, 1976, to install specified front-axle tires and until Oct. 1, 1976, to install specified rear-axle and trailer tires.

Urban mass transit. With full recognition of the new demands for greater safety and a cleaner environment, a new era of urban transit buses came another step closer. Three new 40-ft urban prototype buses developed by three different bus manufacturers (AM General Corp., General Motors, and Rohr Industries Inc.) under the Urban Mass Trans-

Courtesy, Washington Metropolitan Area Transit Authority

Metro train pulls into the Gallery Place subway station in Washington, D.C. After seven years of construction, the first subway line of Washington's Metro system opened in 1976.

portation Administration's Transbus program were given extensive operating tests in different U.S. cities. They are 25% wider than conventional city buses and have special safety, economy, and anti-pollution features. The purpose of the UMTA program is to produce the best performance specification for a new city bus.

General Motors unveiled what it called "the nation's first completely new mass production transit bus design since 1959." The 40-ft-long, 47-seat bus, the Rapid Transit Series, was to be built by welding together eight modules. It featured new suspension and transmission systems, plus a unique "kneeling" ability to permit easy boarding and exiting. GM said that the bus, which represents a $50 million investment in research and development, would be in full-scale production by late 1976.

One area where the potential of technological advances has been highly touted, and even classified as "futuristic," is that of a tracked system for so-called personal rapid transit (PRT). Such a system utilizes minibus vehicles that operate automatically, usually on a concrete roadway, by means of computer control. The troubled PRT system to link campuses of West Virginia University at Morgantown was in danger of outright discontinuation, despite nearly $60 million spent on it by UMTA. A recent development called for the university to accept ownership of the existing 45-vehicle system, with UMTA providing another $5 million for its operation.

UMTA continued to believe in the value of PRT, awarding three seven-month, $500,000 contracts to Boeing Vertol Co., Otis Elevator Co., and Rohr Industries in a competitive program to design a high-performance system that utilizes 12-seat vehicles. These programs may face delays, however, because the Office of Technology Assessment, an agency that advises and reports directly to the U.S. Congress, released a report urging UMTA to downgrade research and development on long-range advanced technology, such as PRT, and concentrate on practical, near-term projects. The agency cited a $95 million outlay over nine years by UMTA for PRT projects that had very little success and compared that with a total outlay of $125 million for 15 simple, successful private systems, such as intra-airport passenger shuttle systems at the Tampa International and Seattle–Tacoma International airports.

UMTA announced that its State-of-the-Art Cars (SOAC), built by St. Louis Car Co. to test innovative urban rail transit equipment, completed revenue tests in five cities. UMTA claimed enthusiastic rider acceptance and said commercial service with such cars would begin soon on Philadelphia's new Lindenwold Line. The two-car SOAC is capable of 80 mph and adjusts for different platform heights.

Aviation. Late in 1975 very strong pressure was put on U.S. leaders to allow the French-British supersonic air transport (SST) Concorde to operate out of John F. Kennedy International Airport in New York City and Dulles International Airport near Washington, D.C. Strong opposition from many congressional members, environmentalists, and the EPA forced U.S. Secretary of Transportation William T. Coleman, Jr., to promise an open hearing to hear both sides and then decide in early 1976 whether to permit such service. Not only did he have to consider such factors as noise, pollution, and safety but also U.S. relations with France and Great Britain.

Conflicting studies gave backing to both proponents and opponents of the SST. The U.S. Department of Transportation (DOT) released a three-year study that concluded that the 16 Concorde and 14 Soviet Tu-144 SSTs scheduled for service in 1976 would cause minimal damage to the upper atmosphere ozone layer, which protects the Earth from the Sun's ultraviolet rays. Shortly thereafter, another study by the U.S. National Academy of Sciences concluded that weakening of the ozone layer protection by a fleet of 100 SSTs would cause an estimated 1.4% increase in skin cancer and 300–400 SSTs would cause a 20% increase, approximately.

A study by the U.S. Federal Aviation Administration (FAA) showed that noise made by the Concorde on takeoff is twice that of a B-707 and four times that of a DC-10, which the EPA said would make landings "undesirable" at New York and "increasingly questionable" at Washington. Also, another government advisory agency, the Council on Environmental Quality, said that the Concorde poses a serious skin cancer threat and is far noisier than indicated by the earlier DOT environmental impact statement.

Secretary Coleman subsequently authorized a 16-month test of Concorde service to Kennedy (four daily round trips) and to Dulles (two daily round trips), at fares 17% above normal first-class rates. Further studies were to be made of the aircraft's noise and environmental impact; night takeoffs were prohibited and all flying over land would have to be at subsonic speeds. New York State quickly passed a law to block the Concorde's use of Kennedy, raising the legal issue of federal *vs.* state jurisdiction. Judicial and legislative challenges could also delay the aircraft's use of Dulles, a federal airport.

The uncertainty about the Concorde in the U.S. did not stop its entry into scheduled air transport service in early 1976. Interest, at least initially, was great, and the Paris–Rio de Janeiro and London–Bahrain flights both began on January 21. Load factors quickly dropped after the first few flights, however, and were unofficially reported as below 50%. Fares on these routes included a 15–20% surcharge over regular first-class rates, although competing airlines using subsonic jets were pressing the International Air Transport Association to

Air traffic controllers at one of the world's busiest airports, O'Hare International in Chicago.

UPI Compix

Wide World

The YC-15, a short-takeoff-and-landing (STOL) cargo transport, was built by McDonnell Douglas Corp. to compete with a similar Boeing Co. plane for a U.S. Air Force contract.

approve surcharges of at least 25%. There is no doubt that the Concorde, with a maximum capacity of only 144 and a cost of more than $60 million, cannot operate profitably at fares close to subsonic ones. How much subsidy will be necessary will depend to a large extent on whether the new plane can serve the high-density North Atlantic market.

The race to start the "world's first scheduled SST service" was won by the Soviet Union when in late December 1975 it began twice-weekly flights of its 120–140-passenger Tu-144 on internal flights between Moscow and Alma-Ata. The Tu-144, however, was to carry only cargo and mail over the 2,050-mi route until mid-1976 because of what Soviet authorities called "some unresolved question." Both the Concorde and Tu-144 are similar in appearance, with adjustable noses and delta wings; both have ranges of nearly 4,000 mi at cruising speeds close to 1,400 mph.

The U.S. Air Force (USAF) gave short-takeoff-and-landing (STOL) technology a big boost through its awards of about $100 million each to the McDonnell Douglas Corp. and Boeing Co. to compete in developing an advanced version with medium-range capability. The new STOL must be able to transport all of the U.S. Army's battlefield vehicles. The winner could capture a lucrative order for 300 USAF aircraft, and both could be much closer to developing a commercial STOL. McDonnell Douglas made the first flight with its version, the YC-15, which is a four-engine jet designed for fuel economy and high-lift capability and for use on 2,000-ft semiprepared runways. Boeing's YC-14 is a two-engine jet, which was slated for its first flight in mid-1976.

The newly introduced European-built A-300B Airbus provided regular London–Paris service, and operators claimed it was 98–99% reliable. As of 1976 no U.S. airlines had bought any of the craft, but South African Airways picked it over U.S. widebody jets for its domestic service, ordering four models for a total of $87 million, excluding parts and equipment. Although the Airbus was originally designed for short- and medium-distance routes, the completed plane had a 1,400-mi range. Lufthansa, which resisted purchase of the new craft initially, ordered three for a total of $68 million and took options on nine more. Its version can carry 153 passengers and 10 tons of cargo a distance of 1,612 mi. Air France began Airbus service between New York and Guadeloupe and Martinique in the Caribbean with a model that seats 262.

Similar remodeling of U.S. widebody aircraft took place, with Boeing making its first flight of its B-747SP, a "special performance" version designed for extra-long range and a flexible capacity for passengers and/or cargo. The new version is 47 ft shorter than the regular B-747, and it proved its long-haul capability on a 6,018-mi, New York-to-Tokyo, nonstop flight against peak winds. This model seats 254 in the economy section and 26 in first class.

Citing the excellent cost and service performance of its first all-cargo B-747, Air France ordered a second one, with front and side loading, for delivery in late 1976. Also, Lufthansa, successful innovators in use of the all-cargo B-747, ordered a combination passenger/cargo version of the transport for $35 million for late 1976 delivery. It would carry 239 persons and 6 freight containers (8 x 20 ft), or 169 persons and 12 containers. Boeing also won support from the British government to use Rolls-Royce engines with the B-747, which resulted in a $179 million order for four of the Boeing planes by British Airways. Boeing was thus able to sell B-747s with General Electric, Pratt & Whitney, or Rolls-Royce engines.

Air safety rules, like those for motor vehicles, are sometimes instituted faster than justified by the current state of technology. An example was an order by the FAA that all large turbine planes used by commercial air carriers must be installed with a ground proximity warning system to enable pilots to both see and hear when their aircraft drop below minimum flight and landing levels. Because of equipment shortages, the FAA gave airlines a six-month extension of its deadline for installation; then, citing an "unacceptable high number of false alarms," extended the deadline again, this time to Sept. 1, 1976.

Pipelines. Alyeska Pipeline Service Co. reported that its 798-mi, 48-in.-diameter trans-Alaskan crude oil pipeline was about half finished at the end of 1975, with completion of the entire project expected in time for mid-1977 shipments. Total cost of the project, including the pipeline, its nine pump stations, and the huge tanker and storage terminal at Valdez, rose to $6.5 billion.

What could be classified as a historic voyage was the partially successful convoy of 47 large oceangoing barges carrying 160,000 tons of vital supplies and equipment, worth more than $540 million, for construction of the trans-Alaskan line. Some of the barges were hauling housing modules as tall as a nine-story building. Gambling that a narrow channel between Point Barrow and Prudhoe Bay would be open sometime during the late summer weeks, the huge flotilla sailed out of Puget Sound on July 5 on its 3,300-mi voyage around Alaska to Prudhoe Bay. Unfortunately, the channel failed to stay open as expected, and only ten shallow-draft barges were able to get through. As a result, 22 of the barges had to return part of the way to have their loads transferred to trains, trucks, and aircraft for the rest of the journey. The remaining 15 barges, which contained the most vital equipment and supplies for meeting construction schedules, and also modules unsuited for other transport modes, were later moved to Prudhoe Bay—with the aid of aerial reconnaissance planes and U.S. Coast Guard icebreakers—through 6-in. ice, 20-knot winds, and 0° F cold.

Plans for building new transcontinental U.S. pipelines for shipments of Alaskan crude oil were announced. Standard Oil Co. of Ohio said that its line would run between the Los Angeles area and western Texas, a distance of about 1,050 mi, and would consist of a converted 700-mi natural-gas line and 350 mi of new lines. Six companies, including Burlington Northern Inc., announced plans to build a $1 billion, 1,500-mi, 36-in.-diameter line from Seattle, Wash., to Clearbrook, Minn. Construction was expected to take two years.

Proponents of coal slurry pipelines in the U.S., blocked in efforts to win state eminent domain rights to cross railroad tracks, stepped up their efforts to have a law passed by Congress to give

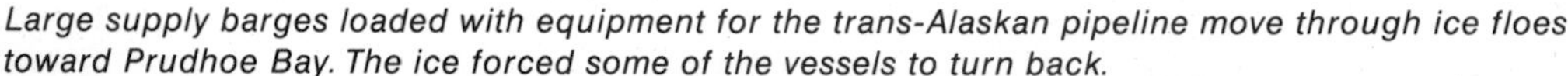

Large supply barges loaded with equipment for the trans-Alaskan pipeline move through ice floes toward Prudhoe Bay. The ice forced some of the vessels to turn back.

Wide World

Courtesy, Amtrak, Washington, D.C.

Amtrak's French-built turbotrain began twice-daily service between Chicago and Milwaukee, Wis.

them such rights at the federal level. A new Slurry Transport Association was formed in Dallas, Texas, as a means of "exchanging technical information and encouraging the further development of this relatively new method of transporting minerals and other materials." Coal slurry pipeline advocates, particularly Bechtel Corp., noted solutions to major operating problems through technological advances. These improvements assured better uniformity in slurry mixture, prevented corrosion caused by entry of air into lines, increased the size of coal chunks, permitted two-way movements of mixtures, and moved the slurry through the pipe without going through a pump. If eminent domain rights are obtained, several long-distance coal slurry lines are expected to be built quickly, as plans have been developed and the financing is available because of the certainty of long-term contracts with utilities.

The Marcona Corp. announced that it is building, as a joint venture with a Brazilian firm, S. A. Mineracao da Trindade, a 250-mi, 20-in.-diameter slurry pipeline to move iron ore concentrate in slurry form from a major new mine about 180 mi north of Rio de Janeiro to ocean shipping facilities. Initial output is expected to be about 7 million metric tons per year.

Railroads. The U.S. Department of Transportation proposed a $1.4 billion program to improve rail passenger service along the "Northeast Corridor" (Boston-New York-Philadelphia-Baltimore-Washington) by rehabilitating and upgrading the right-of-way to permit train speeds up to 105 mph. Congress went further by approving legislation to spend $2.4 billion for upgrading the track and facilities to permit train speeds of up to 150 mph, although most knowledgeable persons agreed that it would take two or more times that much money. U.S. Pres. Gerald Ford threatened to veto the legislation, and the bill was being held up at year's end to work out a compromise.

The quasi-government agency, Amtrak, operators of most U.S. rail passenger service, began twice-daily service between Milwaukee, Wis., and Chicago with new French-built turbotrains. Amtrak was also looking at a British passenger turbotrain, the Advanced Passenger Train, which is capable of reaching speeds of 155 mph and features a unique suspension system that enables the train to negotiate curves at speeds up to 50% higher than do conventional trains, without passenger discomfort. Amtrak, in addition, planned to test a lightweight, fast-accelerating Swedish locomotive that features a wheel-slip early warning device.

GATX, formerly General American Transportation Corp., received its first order for its innovative Tank Train, for thirteen 23,150-gal tank cars. Developed over three years, the train has a hose system that links all the cars, thus permitting loading and unloading at a single connection. GATX claimed that the system can load and unload up to 40 cars six times faster than conventional methods and does not need expensive terminal facilities. Savings of about 12% in transport costs thus are made possible. The Tank Train made its debut in Houston, Texas, where executives in the petroleum and chemical industries viewed its continuous-flow loading and unloading.

The Canadian National Railway reported operating 55-car, unit trains (those hauling a single commodity in shuttle-type service) for shipment of fuel oil on a 375-mi trip between a refinery at Saint Romuald, Que., and Bath, Ont. The oil is kept warm during the journey by fiberglass insulation of the 22,500-gal cars, which unload themselves while moving slowly over a 450-ft-long, between-the-rails trough; this is done by means of valves that are controlled manually.

Vessel designed to carry liquefied natural gas (LNG) is completed in Quincy, Mass. The first LNG ship in the U.S.-flag ocean fleet was launched in 1975.

Courtesy, General Dynamics

The future of unit train service appeared bright as a result of success of the Illinois Central Gulf Railroad's experimental six-hour Chicago–St. Louis piggyback service, which used 15-car trains, one locomotive, and a reduced crew. The railroad began the service in August 1975 with one train and by early 1976 had three in service.

To assure damage-free movement of new automobiles in transit, General Motors said that it was shipping Cadillacs to Oakland, Calif., and Miami, Fla., from 11 on-site railheads in computer-controlled, mechanically loaded Stac-Pac containers, each holding three cars. General Motors planned to expand such vandalproof shipments to ten additional cities.

ACF Industries, Inc. unveiled a new car that uses compressed air blasts to loosen thick, hard-to-unload items such as resin, synthetic rubber, and paint pigment for self-bottom unloading. Production of the $30,000 "Cannonaide" car began in the fall, with tests scheduled in moving asbestos.

Water transport. Advances in navigation and water carrier communications took several steps forward during the year. Moore McCormack Bulk Transport launched the first of three product carrier tankers, all to be built with relatively shallow 35-ft drafts for serving numerous draft-restricted ports on the U.S. Atlantic and Gulf coasts. The new tanker is also the first commercially owned U.S. vessel equipped with a Maritime Administration-sponsored experimental automated satellite navigation system and a ship-to-shore computer link through a satellite communication system.

A jointly funded U.S. government-industry research project was developed during the year to the point where a one-year test was scheduled to begin in September 1976. The project is a $1.3 million prototype communications system for barge lines operating on the lower Mississippi River, the cost being shared by the Maritime Administration and 15 carriers. The first phase of the project, called the Inland Waterways Communications System (IWCS), requires construction of eight VHF-FM stations along the Mississippi between Cairo, Ill., and Baton Rouge, La. The communications link is to be provided through VHF radio equipment installed aboard 25 river towboats. If the system contributes to safer, more dependable, and more economical commercial water transportation, as expected, it probably will be expanded to include all U.S. navigable waterways.

The first liquefied natural gas (LNG) carrier in the U.S.-flag ocean fleet was launched during the year at the shipyard of General Dynamics Corp. in Quincy, Mass. It was the first of eight LNG ships being built for operation by Burmah Charter as part of the Eastco Gas Project. The first three were scheduled to haul LNG from Algeria to U.S. East Coast ports. The ships are 936 ft in length and can carry 125,000 cu m of LNG at −265° F in tanks, or cargo spheres, that are designed to prevent any leakage because the gas, in liquefied form, is only 1/600th its volume as a gas. Upon arrival at U.S. ports, the liquid will be regasified and transported by pipeline to distributors or consumers.

Another type of relatively new merchant vessel, the large roll-on/roll-off container ship, took over a larger role in the U.S. merchant fleet. The first of four 684-ft ships was near completion at Bath (Maine) Iron Works for States Steamship Co. for use in the trans-Pacific trades. The ships were to have angled 24-ft-wide ramps in the stern, onboard cargo-handling equipment, and internal ramps connecting their five decks.

Another roll-on/roll-off vessel, called the "world's largest trailership" by its owner, Totem Ocean Trailer Express, went into service during the year between Seattle, Wash., and Anchorage, Alaska. The 790-ft ship operated at 24 knots, matching the top speeds of any other large merchant ships, and could carry 390 forty-foot trailers and 126 automobiles or other vehicles.

The great potential for shipping in the Alaskan and Arctic areas encouraged many technological advances. One was a joint $35 million project by the Canadian government and private industry to build a new 28,000-DWT (deadweight ton) bulk carrier with a special icebreaking bow that would permit operations through two-ft-thick ice at a constant three-knot speed. To be built specifically for Arctic service, it also featured a hull air-bubbling system to reduce hull friction when moving through ice and a controllable-pitch propeller to increase thrust at low speeds.

The Canadian government awarded a contract for study of a new, heavy-duty icebreaker, possibly to be powered by nuclear energy. Plans called for a 650-ft ship displacing about 33,000 tons, with a hull virtually double-skinned and possibly heated at the sides. The power plant, whether nuclear, gas turbine, or diesel-electric, must be capable of developing about 90,000 hp. Total cost of the ship was estimated to be between $130 million and $150 million.

In port developments, new facilities for handling container ships were built at Singapore and Hong Kong. The twin ports of Dairen and Port Arthur in China were enlarged to accommodate 100,000-ton tankers, and a new two-berth terminal eventually to be capable of handling 500,000-ton tankers was begun at Le Havre, France.

—Frank A. Smith

U.S. science policy

The arcane titles of basic research proposals for federal funding have always been vulnerable to attack and ridicule because of their apparent remoteness from everyday concerns. Why, then, don't scientists head off potential criticism by devising more general, more "relevant" titles? The question raises a classical issue concerning the uneasy relationship between science and politics, especially as it existed in 1975–76.

The productivity of basic research depends in large measure on the easy transfer of scientific knowledge. In a few rapidly advancing fields, such as molecular biology, where a handful of individuals have broken through to new ground, the transfer of knowledge may be almost entirely by personal contact and a handful of specialized journals. In most well-developed fields, however, scientific information is stored in large, computerized systems and retrieved by other investigators on the basis of research titles. In a large information system the usefulness of a title lies in its precision and specificity. Consequently, a generalized, politically useful title would have the least value to the scientific enterprise.

A still larger question is raised by this dilemma, one that has concerned scientists and government since the beginning of massive federal funding of academic research in the 1950s: How is it possible to safeguard the essential independence of the scientific community to chart its own course while at the same time ensuring that public funds are used wisely? In 1975 this question, coupled with a sudden and unexpected furor over a curriculum development program funded by the National Science Foundation (NSF), combined to present an extraordinarily serious threat to one of the mainstays of the process by which the federal government allocates funds to basic research, the peer-review system.

The MACOS controversy. The peer-review system is based on the fact that the granting agencies of the government cannot and perhaps should not support more than 20 to 50% of the proposals they receive. In some agencies, the choice is made by a small number of administrators according to the mission of the agency. In the largest agencies, however, selection is carried out with the aid of panels of outside scientists, usually in related but not competing fields. The judgments of these peer groups may be only advisory, or they may be binding on the agency to the extent that the peer-review panels will be asked to place all proposals submitted in rank order, with the understanding that the granting agency will attempt to fund all the proposals at the top of the list.

Courtesy, Committee on Science and Technology, and A. Dev O'Neill, Photographer to the Congress of the United States

Courtesy, Educational Development Center and Curriculum Development Associates

U.S. House of Representatives Committee on Science and Technology (left) criticized MACOS project for its film of Netsilik Eskimos (right) engaging in such practices as female infanticide.

The attack on the peer-review system began, as do many major controversies, in an apparently unrelated set of circumstances. In the 1960s a group of scholars at the Education Development Center, Inc., in Newton, Mass., began to develop a curriculum for fifth- and sixth-graders designed to show that one's ideas about the acceptability of human behavior are largely the result of one's own life experiences.

To expose the children to alternative life-styles, the producers of the new curriculum prepared a set of films, including one of the Netsilik Eskimos. The Netsilik, as it turned out, indulge in certain practices not socially acceptable in the constituencies of many congressmen—*e.g.,* putting female infants and the elderly to death and organized sexual promiscuity. Entitled "Man: A Course of Study," or MACOS, as it came to be known, the project was supported by the National Science Foundation and won a number of awards for its chief designer, Jerome Bruner. About 1,700 schools in 47 states used the MACOS materials in place of traditional social study courses.

The first attack on MACOS came from Rep. John Conlan (Rep., Ariz.) in March 1975, in testimony before the House Committee on Science and Technology. In addition to charging that the course was introducing repugnant concepts to young children, he revealed that the curriculum had been rejected by a number of commercial publishers and had only then been turned over by the NSF to a small publishing house in Washington, D.C., under an arrangement whereby the publishers were operating under a partial federal subsidy in competition with other publishers.

Already suffering from earlier congressional attacks on the relevance of its basic-research program, the NSF reacted swiftly. Within days, Director H. Guyford Stever announced that all funding of the MACOS project would be temporarily halted and a thorough review made of the NSF effort in precollege science course development. But the fat was already in the fire. A Conlan amendment to the authorization bill for the NSF that would have prevented the use of any authorized funds for the development of new curriculum materials without specific congressional approval was barely defeated, 215–196. Two other amendments aimed at restricting the freedom of the foundation's support of curriculum development were defeated by varying margins.

While these floor challenges were being beaten back, a much more fundamental challenge to the independence of the peer-review system came from Rep. Robert Bauman (Rep., Md.). On April 9 Bauman offered an amendment that would require the NSF to submit grant proposals every 30 days to the appropriate committees in the House and Senate for review. During a 30-day period following, either chamber could veto any proposal by majority vote. If no action was taken within the period, the grants would automatically be approved and the foundation would be permitted to undertake the usual review procedures. The amendment passed the House, 212–199.

The Senate version of the NSF authorization bill contained no provision for congressional approval of NSF grants. When the bill emerged from the House-Senate conference committee without the amendment, Bauman and a number of supporters fought vigorously to have it reinstated, but they lost finally, 321–79. Nevertheless, passage of the amendment in the House, despite the open opposition of several key members of the responsible committee, pointed up the fragility of the freedom and independence enjoyed by the agencies that dispense federal funds for purposes of research and development.

Peer review under fire. Still concerned about the need to exercise greater surveillance over federal expenditures in R and D, Bauman and his supporters sought firmer ground on which to fight. They found it in the peer-review system itself. Bauman had already begun to attack the basic premises of the system in his floor fight. He protested the fact that it operates out of sight and control of the Congress, that it is "secret" to the extent that names of the reviewers were not disclosed, and that the reviewers often reviewed proposals from each other's laboratories.

The peer-review system, which is used to evaluate not only research proposals but also fellowship applications and submissions to scientific journals, has traditionally protected the anonymity of the reviewers. The rejected applicant might be read a paraphrase of critical reviews, but it was held—in Stever's words—that preservation of the secrecy of verbatim contents of reviews and their authors' anonymity was "vital to the effective operation of the peer-review system." In a letter to the chairman of the House Committee on Science and Technology, Stever pointed out that the confidentiality of reviews had recently been upheld by the courts. The National Science Board, which sets policy for the NSF, had affirmed the confidentiality of both reports and reviewers in 1962, 1972, and again in May 1975.

But neither the National Science Board nor the courts could protect the foundation from attack by an outraged Congress. In examining a second NSF-supported curriculum development program, Conlan found that it had "shocking and objectionable sex and behavior modification content." What was worse, from the viewpoint of the NSF, he charged that the remarks of one of the reviewers had been so edited by the foundation staff as to include all the favorable comments and none of the unfavorable ones.

When Conlan demanded to see the verbatim contents of the original reviews, he was turned down on the basis of current NSF policy. Shortly thereafter, however, Stever received letters from friends in both the House and Senate, broadly supportive of the foundation but asking for documentation on the circumstances Conlan had cited. The NSF was clearly in a dilemma. If it provided the information requested, it would be violating an implied guarantee of confidentiality to the reviewers. If it did not, it would open itself to possible retaliatory action on the part of Congress.

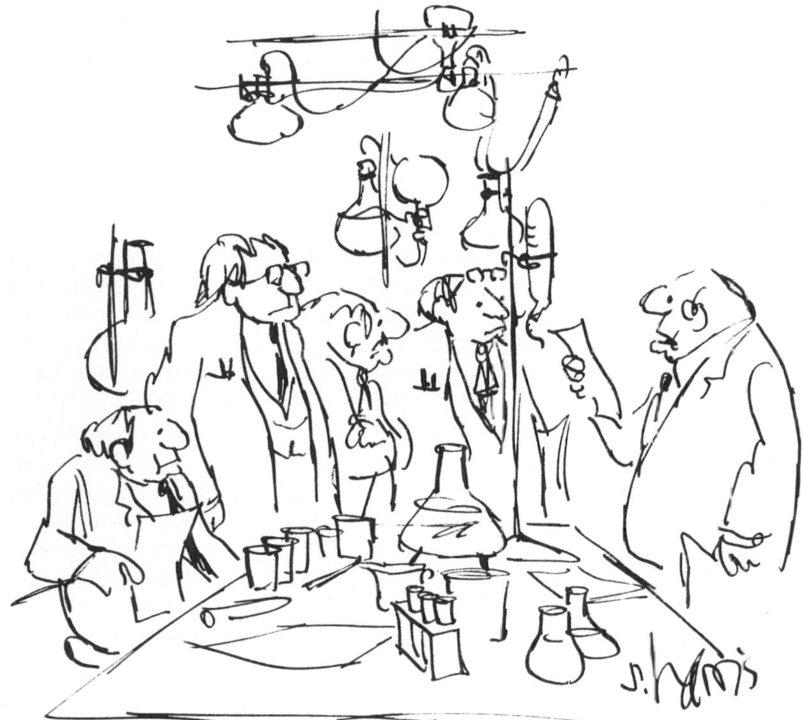

"Due to a tightening of the budget, we are forced to curtail our overtime and weekend schedule, and request that all major breakthroughs be achieved as early in the week as possible."

The NSF responded by falling back to what it considered a more defensible position. The National Science Board, which in May had unanimously reaffirmed the confidentiality of reviewers' comments, "reaffirmed" in June "its belief that the review process should be conducted as openly as possible . . . consistent with the effective evaluation of proposals." Specifically, it agreed to publish annually a list of all reviewers used by the foundation and to yield verbatim copies of reviews to the principal author of each proposal. The board said it would consider further the question of whether to reveal the identity of the authors of the specific comments.

Pressure from Congress continued—and from some powerful quarters. Rep. George H. Mahon (Dem., Tex.), the chairman of the House Committee on Appropriations, threatened in May to cut millions of dollars from the NSF budget if he found "damn fool" projects receiving NSF funds. Providing some indication of the kind of response being generated by unfavorable press comments on individual proposals, he declared himself to be "sick and tired of responding to correspondence from citizens who are blaming Congress for some of the idiotic things done by a few unstable people in the executive branch." The committee underlined his unhappiness by recommending to the House that it slice $35 million from the NSF basic-research budget.

Now the message was coming in loud and clear. After some reconsideration of the situation in the foundation, Stever wrote to the chairman of the House Subcommittee on Science, Research, and Technology, which has general oversight over NSF matters, that while the identity of individual reviewers would still be privileged, it could be revealed to appropriate chairmen of congressional committees. The foundation, Stever wrote, "has no intention of withholding such information when required by the Congress in performance of its investigatory or legislative functions."

The quarrel over the confidentiality of the peer-review system tended to drown out more fundamental questions that were being raised about the system within the scientific community itself. In his annual report to the members, Philip Handler, president of the National Academy of Sciences, defended the system against its congressional foes but added that "after two decades of operation, the peer review system warrants an intense examination to assure ourselves and our fellow citizens that it does indeed function as one might hope."

On November 21 the NSF announced a survey, in cooperation with the House Committee on Science and Technology, to determine how its various publics felt about the methods used to review and select programs for funding. Questionnaires were sent to a sample of 3,000 investigators, evenly divided between reviewers and those whose previous proposals had been accepted or rejected. Recipients were asked to comment generally on the effectiveness and equitability of proposal reviews and to forward their comments to the General Accounting Office. The GAO was to report on the survey in 1976. Meanwhile, the National Institutes of Health (NIH) decided not to wait for a costly nudge from Congress and therefore launched an independent survey of its own granting mechanisms.

Science and the citizen. As an integral part of its study, NIH scheduled three public hearings for February 1976—in Chicago, San Francisco, and at the NIH campus—to encourage scientists and lay groups to bring their comments to the survey committee. In choosing to go out to the nation, NIH was following a pattern of development that was recent insofar as the scientific community was concerned but that had already spread with remarkable rapidity.

The scientific community has always jealously guarded its freedom from political intrusion. Although—in the last few decades, at least—it reached out to inform the public of its successful endeavors, it did not often encourage any feedback from the public. More recently, however, it became only too apparent that a growing fraction of the public business of science was devoted to providing technical judgments on national policy issues of great moment—the widespread use of persistent pesticides, funding of the supersonic transport, depletion of the ozone layer, to name only a few. Clearly, the satisfactory resolution of science-based policy issues required a close working relationship between the scientific community, the policymakers, and the public.

Scientists and policymakers cautiously developed a modus vivendi, but it was evident that neither the soundest scientific judgment nor the wisest political decision would prevail if no response was made to the alarms and apprehensions of the public. The existence of energetic and highly articulate public-interest groups, often operating on minimal budgets, guaranteed that anything less than a completely reassuring response would not be acceptable. Furthermore, these groups found willing allies among the media, many of whom were only too happy to cast themselves as adversaries to the scientific establishment.

In an effort to adjust to this change in their environment, the major scientific institutions reached out beyond their public relations offices to develop public discussions about major issues. The

Courtesy, American Association for the Advancement of Science

Courtesy, ERDA, Aiken, South Carolina

Minnesota state legislators attend a workshop on the disposal of radioactive waste materials. At the right, a tank to contain such wastes is under construction near Aiken, South Carolina.

National Academy of Sciences organized a series of public forums on such questions as energy needs, human experimentation, and safety of pharmaceuticals and artificial sweeteners. The American Association for the Advancement of Science, with funding from the NSF, organized symposia to discuss regional aspects of national issues in energy and the environment.

Finally, in 1975, the scientific establishment opened its doors almost completely. Encouraged by a Federal Advisory Committee Act, government science agencies announced in advance the meetings of their external advisory committees and permitted attendance by the press and the public. A Freedom of Information Act made available all kinds of policy documents, including preliminary drafts and working papers, that previously had been withheld from the public.

It appeared that even more was called for. When a group of molecular biologists organized an international meeting in late February 1975 at Asilomar State Park in Pacific Grove, Calif., to discuss the possible hazards of genetic engineering, they permitted full reporting of the three-and-a-half-day meeting by 16 members of the press. Nevertheless, the organizers were later criticized by Sen. Edward M. Kennedy (Dem., Mass.) because no public-interest groups were represented. Even though the scientists in attendance had previously undertaken a voluntary moratorium on certain types of genetic research, Senator Kennedy was not satisfied because "scientists alone decided to impose the moratorium and scientists alone decided to lift it."

Since Kennedy had long been considered one of science's best friends in Washington, his rap on the knuckles of the Asilomar conference bewildered its organizers. In their view, they had shown unprecedented public responsibility by openly asking, as scientists, whether or not a productive line of scientific inquiry should be pursued. They were not making policy, as Kennedy had implied, but were calling the attention of policymakers to the possibility of a public hazard that otherwise would have escaped notice. Furthermore, all the recommendations of the Asilomar conference were transmittted directly to a committee of the NIH. The committee, in turn, would recommend policy to the NIH director, who would have the ultimate responsibility for establishing policy through the grant mechanism. Subsequent meetings of the NIH advisory committee, held under the provisions of the Federal Advisory Committee Act, made clear that there would be no lack of opportunity for public-interest representatives to participate in its deliberations.

Although the questions Kennedy raised about the Asilomar proceedings were washed away by later events, their implications were not. Shivers of apprehension ran through the ranks of molecular biologists as they envisaged the possibility of being required to seek the approval of panels of laymen before proceeding with new fields of research. No completely satisfactory formula for coupling scientific judgment with accountability to the public emerged, but one possible model was proposed by Stanley Cohen, a molecular biologist from Stanford University and one of the organizers of the Asilomar conference. "Radioisotope use," he noted, "is subject to regulations designed to ensure the safety of laboratory personnel and the general public, and there is public involvement in the enforcement of these safety procedures. However, the merit or lack of merit of specific experiments that employ radioactive materials is entirely a scientific judgment that is determined by the peer-review system."

Kennedy's interest in increasing citizen participation in science-policy questions emerged in more tangible fashion when he managed the Senate bill authorizing fiscal 1976 funds for the NSF. A provision of the act called on the foundation to come back to Congress with a program that would (1) improve public understanding of public-policy issues involving science; (2) facilitate the participation of experienced scientists and engineers in public activities aimed at the resolution of public-policy issues having significant scientific and technological aspects; and (3) enable nonprofit citizens' public-interest groups to acquire technical expertise that would assist them in dealing with the scientific and technological aspects of public-policy issues.

Since many science-based public interest groups lacked the funds to retain highly trained scientific advisers and many public-spirited scientists lacked the means to participate in policy issues on a national scale, there was considerable enthusiasm for the bill. Almost all the enthusiasm, however, was generated outside the NSF. Within, there was considerable unease as the foundation staff prepared to deal with groups they described as being "not our usual constituency."

Nevertheless, the NSF was credited with making a good-faith effort to develop better relations with its new "constituency." It held seven public hearings in as many cities in late 1975 to generate ideas for the "Science for Citizens Program." The National Science Board announced plans to hold a series of public meetings in 1976 for similar purposes, but these appeared to be aimed more at groups already within the scientific community than at the general public.

Some facts and figures. In December 1975 the National Science Board released the second in a series of "science indicators." They showed that the proportion of gross national product spent for R and D was declining steadily in the U.S. and growing substantially in the U.S.S.R., West Germany, and Japan. In the most recent year for which comparable figures were available, the fraction of GNP devoted to R and D was 2.4% in the U.S., 3.1% for the U.S.S.R., 2.4% in West Germany, and 1.9% for Japan. The number of scientists and engineers engaged in R and D per 10,000 population declined in the U.S. after 1969 but continued to grow in all other countries studied.

The U.S. was the largest producer of scientific literature in every field except chemistry and mathematics, and its research was cited more often than that of any other country. However, the proportion of major technological innovations originating in the U.S. declined from a high of 80% in the late 1950s to some 55–60% after the mid-1960s. West Germany and Japan increased their shares.

National expenditures for R and D in the U.S. increased steadily in current dollars between 1960 and 1974, but in terms of constant dollars there was no growth between 1968 and 1974. The fraction of R and D funding provided by industry rose from 33% in 1965 to 43% in 1974, while the government's share dropped from 65 to 53% in the same period. The proportion of R and D funds allocated to different types of activities remained nearly constant after 1965—13% to basic research, 23% to applied research, and 64% to development. In constant dollars, federal support of basic research declined 13% between 1968 and 1974 (down 25% in the physical sciences), but the government's share of basic-research support rose from nearly 60% in 1960 to almost 70% in 1974.

Academic basic research was concentrated in the life sciences (51% of all expenditures), engineering (12%), physical sciences (13%), social sciences (8%), and environmental sciences (7%). Universities were not only the largest producers of scientific literature (75%) but had supplanted industry as the most prolific contributors to technological advance. The total number of scientists and engineers engaged in R and D was 530,000 in 1974, down by more than 30,000 from the 1969 high; 68% were employed in industry, 13% in the academic sector, and 12% in the federal government. The proportion of science and engineering graduate students receiving federal support declined from 42% in 1967 to 25% in 1974.

The belief that science and technology had changed life for the better was held by 75% of the public in 1974, compared with 70% in 1972; in 1974, only 5% saw the change as being for the worse, compared with 8% in 1972. Similarly, the public continued to feel that, in the long run, science and technology would solve many of our major problems, but the fraction expecting most problems to be so solved declined from 30% in 1972 to 23% in 1974.

—Howard J. Lewis

Scientists of the Year

Honors and awards

The following is a selective list of recent awards and prizes in the areas of science and technology.

Architecture and civil engineering

Brown Medal. The Franklin Institute of Philadelphia named the Public Buildings Service of the U.S. General Services Administration as recipient of its 1975 Frank P. Brown Medal. The citation called attention to the agency's development of "the systems approach for designing, specifying, and constructing buildings" and its utilization of "the complete system using precoordinated, subsystems components," notably in the erection of three large federal office buildings.

Reynolds Award. Gustav Peichl, an Austrian architect also widely known as the caricaturist "Ironimus," was given the 1975 R. S. Reynolds Memorial Award for his design of the Austrian Broadcasting Corporation's radio and television center in Salzburg. In awarding the $25,000 prize, the jury praised Peichl's strikingly artistic use of aluminum, his solutions of technical problems, and the congenial blending of his design with existing structures. In particular, it commended the way in which the concept for the multipurpose, highly complex facility had been "matured by the design process."

Astronomy

Eddington Medal. The Royal Astronomical Society in London chose Stephen W. Hawking of Cambridge University and Roger Penrose of Oxford University as 1975 recipients of Eddington medals. The two conducted joint research that suggested the past existence of singularities (such areas as black holes, where traditional concepts of space and time are void) that are associated with very high densities.

Gould Prize. The National Academy of Sciences bestowed its 1975 Benjamin Apthorp Gould Prize on Lodewijk Woltjer of the European Southern Observatory in Chile. He was cited for distinguished astronomical research, especially regarding his work on the Crab Nebula and the galactic magnetic fields.

Magellanic and Vanderlinden Prizes. Ralph A. Alpher of the General Electric Research and Devel-

Ralph Alpher

Courtesy, General Electric Research and Development Center

Robert Herman

Courtesy, General Motors Technical Center, Michigan

opment Center and Robert Herman of General Motors Research Laboratories were named recipients of two separate honors for work done in the late 1940s at Johns Hopkins University, Baltimore, Md. Their prediction of 3° K blackbody radiation was confirmed for the first time at Bell Laboratories in 1965. The two awards were the 1975 Magellanic Premium of the American Philosophical Society and the Georges Vanderlinden Prize of the Belgian Royal Academy of Sciences, Letters, and Fine Arts.

Michelson Medal. In bestowing its 1975 Albert A. Michelson Medal on Irwin I. Shapiro of the Massachusetts Institute of Technology, the Franklin Institute of Philadelphia cited his pioneering role in developing radar astronomy and his application of radio and radar techniques to the investigation of fundamental problems in planetology, cosmology, and general relativity.

Pierce Prize. Each year the American Astronomical Society gives special recognition to a scientist not yet 35 years old for outstanding work in instrumentation or observational research. Recipient of the 1975 Newton Lacy Pierce Prize was Eric E. Becklin, a specialist in infrared astronomy at the California Institute of Technology.

Robertson Lectureship. Approximately every three years a scientist of international distinction is chosen by the National Academy of Sciences for the H. P. Robertson Memorial Lectureship. The 1975 honor, together with an honorarium of $2,500, went to Manfred H. Rees of the Laboratory for Atmospheric and Space Physics in Boulder, Colo. Rees has been mainly concerned with high-energy astrophysics.

Chemistry

Cope Award. The biennial Arthur C. Cope Award, administered by the American Chemical Society, is bestowed for outstanding achievement in the field of organic chemistry. The 1976 gold medal and an honorarium of $10,000 were given to Elias J. Corey of Harvard University. In addition, a $10,000 grant for research in organic chemistry is given to the winner, who may assign it to any university or nonprofit institution of his choice.

Garvan Medal. The American Chemical Society presented a gold Garvan Medal and an honorarium of $2,000 to Isabella L. Karle in 1976. Her research is conducted at the Laboratory for the Structure of Matter, part of the U.S. Naval Research Laboratory in Washington, D.C.

Gibbs Medal. Herman F. Mark, dean emeritus of the Polytechnic Institute of New York, was named recipient of the 1975 Willard Gibbs Medal by the Chicago section of the American Chemical Society. Mark was associated with the first X-ray analyses of natural polymers and with early applications of radioisotopes and photochemical techniques to polymer research. With others, he is also given credit for elucidating the relationships between the structures of polymers and their mechanical properties and thermal behavior.

John Cornforth

Wide World

Vladimir Prelog

UPI Compix

Langmuir Award. To recognize and encourage outstanding interdisciplinary research in chemistry and physics, the General Electric Foundation annually sponsors the Irving Langmuir Award, which carries with it an honorarium of $5,000. The American Chemical Society, which administered the award, selected John S. Waugh of the Massachusetts Institute of Technology for the 1976 honor.

Mendeleyev Medal. The most significant award in chemistry conferred in the Soviet Union is jointly sponsored by the nation's Academy of Sciences and the D. I. Mendeleyev All-Union Chemical Society in Moscow. The 1975 Gold Mendeleyev Medal was awarded to Vitalii I. Goldanskii, an authority on chemical reactions at low and very low temperatures. He heads the laboratory of nuclear and radiation chemistry at the Institute of Chemical Physics and is a professor at the Moscow Institute for Physical Engineering.

Nobel Prize. The Nobel Prize for Chemistry was shared in 1975 by John W. Cornforth and Vladimir Prelog. Both are experts in stereochemistry, which deals with the three-dimensional arrangement of atoms within molecules. The Royal Swedish Academy of Sciences, which presented the $143,000 honorarium, proclaimed their research "of fundamental importance to an understanding of biological processes." In essence, they demonstrated that the atomic configuration of a given molecule determines its behavior and properties. Such knowledge has far-reaching implications, notably in the fields of medicine, agriculture, and industry.

Cornforth, a native Australian, obtained his doctorate at Oxford University (1941), then worked with Sir Robert Robinson, who won a Nobel prize in 1947 for his research on the alkaloids. Prelog was born in Sarajevo (now in Yugoslavia) and was educated at the Technical University in Prague. In 1941 he moved to Switzerland, where he joined the Federal Institute of Technology.

Perkin Medal. Lewis H. Sarett, president of Merck Sharp & Dohme Research Laboratories, was named recipient of the 1976 Perkin Medal by the American section of the British Society of Chemical Industry. The gold medal acknowledges outstanding accomplishments in the field of applied chemistry.

Priestley Medal. In 1975 the American Chemical Society presented its highest award, the Priestley Medal, to George S. Hammond of the University of California at Santa Cruz. As a photochemist, Hammond's main interest has been the behavior of short-lived, energetic chemical species created by the action of light energy on larger, more stable molecules. In educational circles, the "Hammond curriculum" has also attracted considerable attention. The three-year program disregards the traditional categories of organic, inorganic, and physical chemistry in favor of an integrated approach, which stresses principles that unify all branches of chemistry.

Earth sciences

Bucher Medal. Lynn R. Sykes, head seismologist at the Lamont-Doherty Geological Observatory in New York, was awarded the 1975 Walter H. Bucher Medal of the American Geophysical Union for "original contributions to the basic knowledge of the earth's crust." Sykes, an authority on plate tectonics, headed a group of scientists who in 1973 made the first accurate prediction of an impending U.S. earthquake.

Chree Medal and Prize. The British Institute of Physics presented its 1975 Charles Chree Medal and Prize to Raymond Hide, head of the geophysical fluid-dynamics laboratory of the U.K. Meteorological Office. Hide was selected for his work on the hydrodynamics of rotating fluids, which included applications of his theories to motions in the major planets.

Day Medal. Since 1948 the Arthur L. Day Medal has been awarded annually for outstanding contributions to geological knowledge acquired through the application of physics and chemistry. In 1975 the Geological Society of America presented the award to Allan V. Cox of Stanford University. His research on the time scale of geomagnetic reversals helped advance theories of plate tectonics.

Pecora Award. Two U.S. government agencies, the Department of the Interior and the National Aeronautics and Space Administration (NASA), jointly sponsor the William T. Pecora Award for outstanding work that contributes to an understanding of the Earth by means of remote sensing. In 1975 one award was given to William Nordberg of the NASA Goddard Space Flight Center in Maryland and one to Carlos Brockmann, director of the Landsat-Bolivia Project. The award citation called attention to "the contributions they have made to man's understanding of his natural resources and environment and their relevancy to international betterment."

Penrose Medal. In 1927 the Geological Society of America established the Penrose Medal to honor geologists who have made significant contributions to the geological sciences. In 1975 the gold medal was presented to Francis J. Pettijohn, professor emeritus of Johns Hopkins University, for his outstanding work in sedimentary mineralogy, paleocurrent analysis, basin analysis, and Precambrian geology.

Courtesy, University of Chicago, photograph, L. E. Saunders

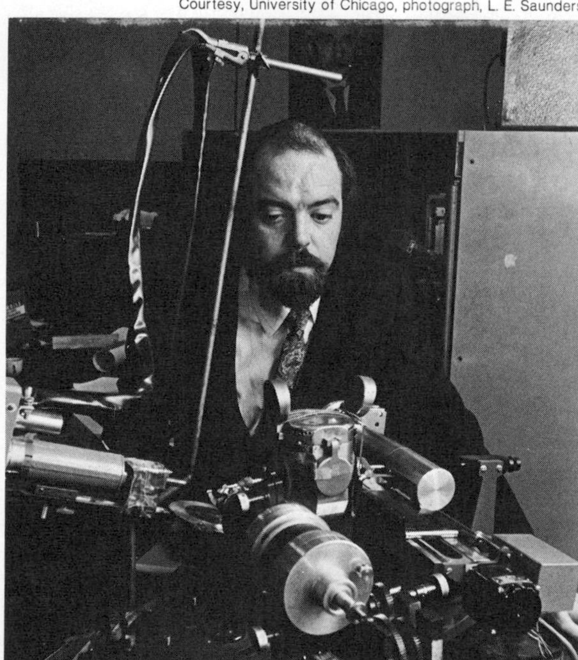

Paul Moore

Rosenstiel Award. The American Association for the Advancement of Science presented the 1975 Rosenstiel Award in Oceanographic Science to Kenneth O. Emery of the Woods Hole Oceanographic Institution in Massachusetts. Emery was cited "for outstanding studies and syntheses of the continental margins and marginal seas of the world," for exceptional contributions to national and international forums, and for his unparalleled record in training marine geologists.

Rossby Medal. Hans A. Panofsky, Evan Pugh Research Professor of Atmospheric Sciences at Pennsylvania State University, received the 1976 Carl-Gustaf Rossby Research Medal. In presenting the award, the American Meteorological Society recalled Panofsky's "many fundamental contributions to knowledge of the statistical properties of turbulence throughout the atmosphere."

Second Half Century Award. Roger M. Lhermitte of the University of Miami received the 1976 Second Half Century Award. The American Meteorological Society cited his "many significant contributions to the advancement of radar meteorology through the use of Doppler radar in probing the atmosphere, particularly for the dual-Doppler method, which has provided new insight into the internal motions of convective storms."

Senior U.S. Scientist Award. The Alexander von Humboldt Foundation of West Germany bestowed its Senior U.S. Scientist Award on Paul B. Moore of the University of Chicago. Besides receiving an honorarium of DM 30,000 (about $12,000), Moore also accepted an invitation to spend six months in West Germany doing further research. Most recently he has been investigating structural hierarchies in crystals, applying graphical enumeration theorems to classify known structures and to determine as yet unknown arrangements.

Sverdrup Medal. The American Meteorological Society awarded its 1976 Sverdrup Gold Medal to Robert W. Stewart of British Columbia, Canada, "for outstanding leadership in experimental and theoretical research in problems of the air-sea interface, and the adjacent turbulent boundary layers of the atmosphere and ocean. He has brought the high standards of measurement and analysis of the laboratory to field studies and has pioneered measurements of surface waves and turbulence in the upper ocean."

Energy

Glazebrook Medal and Prize. Each year the British Institute of Physics awards a silver medal and £250 to a prominent scientist. In 1975 Walter C. Marshall was honored for his outstanding direction of the research and development unit of the United Kingdom Atomic Energy Authority.

Hertz Prize. The first triennial Heinrich Hertz Prize, consisting of a gold plaque and DM 30,000

Alvin Weinberg

Courtesy, Oak Ridge Associated Universities, Oak Ridge, Tenn.

(about $12,000), was awarded in 1975 to Alvin M. Weinberg, director of the Oak Ridge (Tenn.) National Laboratory from 1955 to 1974. The award, sponsored by the Badenwerk-Aktiengesellschaft (an electricity company), was given to Weinberg for his contributions to the science and technology of nuclear energy.

Environment

Audubon Medal. The National Audubon Society presented its 1975 medal to Maurice F. Strong for "his unmatched achievement in creating a working structure to protect the global environment" while serving as executive director of the UN Environment Program.

Pollution Control Award. The American Chemical Society, which administers a $3,000 prize sponsored by the Monsanto Co. of St. Louis, Mo., announced that Thurston E. Larson was chosen to receive the Pollution Control Award for 1976. Larson's long association with the Illinois State Water Survey began in 1932.

Wildlife Conservation Prize. Salim A. Ali, an ornithologist of international renown, was honored in 1976 with the second annual J. Paul Getty Wildlife Conservation Prize. He received the $50,000 prize for lifelong dedication to the preservation of bird life on the Indian subcontinent and for tireless efforts to create "an environment for conservation in India." His writings include a ten-volume work (with S. Dilon Riplet) entitled *Handbook of the Birds of India and Pakistan* (1972).

Food and agriculture

Mullard Award. The Royal Society of London presented the 1975 Mullard Award to J. Bingham for successfully breeding new varieties of winter wheat while working with Great Britain's Agricultural Research Council. The award is given for scientific achievements that directly contribute to national prosperity in the United Kingdom.

Washburn Award. Each year since 1964 the Museum of Science in Boston has presented a gold medal and $5,000 to an individual "who has made an outstanding contribution toward public understanding of science, appreciation of its fascination, and the vital role it plays in all our lives." The 1975 Bradford Washburn Award was given to Jean Mayer for his lifelong service to the cause of better world nutrition. Among other responsibilities, Mayer directs the UN task force on child nutrition and is professor of nutrition at Harvard University's School of Public Health.

Wilson Award. Sterling Wortman, vice-president of the Rockefeller Foundation and a self-styled "corn breeder," was chosen in 1975 to receive the

Courtesy, United Nations

Maurice Strong

first $10,000 Joseph C. Wilson Award for achievement in international affairs. In administering the award for the Xerox Corp., the Rochester (N.Y.) Association for the United Nations particularly cited Wortman's dynamic role in establishing a network of agricultural research centers throughout the world.

Information sciences

Ballantine Medal. Three scientists were named recipients of the Franklin Institute's Stuart Ballantine Medal in 1975. Bernard C. DeLoach, Jr., of Bell Laboratories was selected "for his theoretical contributions and successful development of practical, solid-state, Impatt microwave oscillators." Martin M. Atalla, president of Atalla Technovations in California, and Dawon Kahng of Bell Laboratories were chosen "for their contributions to semiconductor silicon-silicon dioxide technology, and for the development of the MOS insulated gate, field-effect transistor."

Certificate of Merit. In 1975 seven scientists were presented with a Certificate of Merit by the Franklin Institute of Philadelphia. Frederic P. Heiman of Mars Money Systems, Steven R. Hofstein of Princeton Electronic Products, and Frank M. Wanlass of LSI Systems were cited "for their timely and significant contributions to the theory, experiments, and circuit applications of metal-oxide semiconductor devices, which resulted in materially advancing the practical realization of commercial MOS devices and systems."

Bernard DeLoach, Jr.

Courtesy, Bell Laboratories

Bruce E. Deal of Fairchild Camera and Instrument Corp., Andrew S. Grove of Intel Corp., Chih-tang Sah of the University of Illinois at Urbana, and Edward H. Snow of Reticon Corp. were honored "for the discovery and elucidation of the deleterious effects of alkali metal impurities in the silicon-oxide layer of MOS devices, and for the development of contamination-free processing leading to the practical commercial production of stable MOS field-effect devices."

Edison Medal. Sidney Darlington, who served on the technical staff of Bell Laboratories for more than 40 years, was awarded the 1975 Edison Medal by the Institute of Electrical and Electronics Engineers (IEEE). His main interests have included network theory, radar systems, and electronic circuits.

Founders Medal. In 1975 the IEEE selected John G. Brainerd as recipient of its annual Founders Medal. Brainerd headed the project that produced the first large-scale digital electronic general-purpose computer and directed the school of electrical engineering at the University of Pennsylvania for nearly half a century.

Founders Medal. The National Academy of Engineering presented its annual Founders Medal to James B. Fisk in 1975. An educator, a research scientist, and an executive at Bell Laboratories, Fisk was honored for "his leadership in the advancement of communications technology for the benefit of society, particularly his contributions to the development of the pulsed cavity magnetron, millimeter waveguide communication techniques, and many other improved telephone systems."

Potts Medal. The Franklin Institute of Philadelphia awarded its 1975 Howard N. Potts Medal to LeGrand G. Van Uitert of Bell Laboratories for demonstrating "outstanding ability in discovering and developing ferrites for microwave communication, garnets for magnetic bubbles and lasers and niobates for nonlinear optical devices."

Sarnoff Award. The Institute of Electrical and Electronics Engineers presented its 1975 David Sarnoff Award to Bernard C. DeLoach, Jr., of Bell Laboratories for his work on the impact avalanche and transit time device. DeLoach, who received a $1,000 honorarium, is currently working to develop solid-state lasers.

Zworykin Award. Jack S. Kilby of Dallas, Texas, was given the 1975 Vladimir K. Zworykin Award by the National Academy of Engineering "for outstanding achievement in the field of electronics applied in the service of mankind, with particular note of the pervasive influence of his contributions

Courtesy, University of Pennsylvania, photograph, Phillips Studio

Mildred Cohn

to the development of monolithic integrated electronic circuits." His work at Texas Instruments made the development of electronic calculators and wristwatches possible.

Life sciences

Cresson Medal. Mildred Cohn of the University of Pennsylvania received an Elliott Cresson Medal in 1975 for her work in molecular biology. At the presentation ceremonies, the Franklin Institute recalled Cohn's "pioneering work in the application of nuclear magnetic resonance techniques to elucidate the structure and function of enzymatic complexes."

Gairdner Awards. Each year the Gairdner Foundation in Ontario, Canada, selects a number of scientists as worthy of special recognition for scientific work that has significantly helped to conquer disease or relieve human suffering. In 1975 awards in the area of life sciences were presented to Ernest Beutler of Duarte, Calif., Baruch S. Blumberg of Philadelphia, and Hugh E. Huxley of Cambridge, England. Beutler was honored "for elucidating the biochemical and genetic basis of the hemolytic anemias related to glucose-6-dehydrogenase deficiency, and other studies on human erythrocytes." Blumberg was cited for "his discovery of the Australia antigen in 1963 and its association with hepatitis in 1967." Huxley received his award for research on the molecular basis of muscle contraction.

Gold Medal. The Entomological Society of Canada named Geoffrey George Edgar Scudder, an English-born zoologist and entomologist, recipient of the society's 1975 gold medal. Scudder has had an enduring interest in the Lygaeidae family of insects.

Gregory Medal. During the 13th Congress of the Pacific Science Association, the Bishop Museum in Honolulu presented its 1975 Herbert E. Gregory Medal to J. Linsley Gressitt, a noted zoologist. The quadrennial award may be given for distinguished Pacific area research in anthropology, botany, geology, or zoology.

Heineken Prize. The Royal Netherlands Academy of Arts and Sciences announced that L. L. M. Van Deenen of the University of Utrecht had been selected to receive the 1976 Heineken Prize. Van Deenen has done important research on the structure of biomembranes and on the mechanism of action of phospholipases. The prize, sponsored by Heineken Breweries, is worth 100,000 Dutch florins (about $37,500).

Horwitz Prize. Two biochemists at the Karolinska Institutet in Sweden shared the $25,000 Louisa Gross Horwitz Prize in 1975. Sune Bergström and Bengt Samuelsson determined the structures of the prostaglandins and then synthesized them chemically. Their work has important applications in the understanding and treatment of high blood pressure, ulcers, fevers, and pregnancy.

Lichtwitz Prize. Felix Bronner of the University of Connecticut received the 1975 Andre Lichtwitz Prize for his work on calcium transport. The prize of 7,000 francs (about $1,555), is awarded annually

Courtesy, Princeton University, Princeton, N.J.

Bruce Alberts

Courtesy, Harvard University, Cambridge, Mass.

Bruce Chalmers

by the National Institute of Health and Medical Research in Paris.

U.S. Steel Award. The National Academy of Sciences, which administers the annual U.S. Steel Foundation Award in Molecular Biology, bestowed the 1975 honor on Bruce M. Alberts of Princeton University. His achievements include the isolation of proteins required for DNA replication and for genetic recombination.

Weiland Prize. Michael S. Brown and Joseph L. Goldstein of the University of Texas Health Science Center in Dallas were presented with the 1975 Heinrich Weiland Prize. The Munich administrators of the DM 15,000 (about $6,000) award called attention to the recipients' outstanding work in lipid metabolism, notably their discovery of the biochemical genetic defect responsible for familial hypercholesterolemia.

Materials sciences

Bingham Medal. Alan N. Gent, a specialist on rubber and allied materials at the University of Akron (Ohio), received the 1975 Bingham Medal from the Society of Rheology. His research has included studies on elasticity, on strain-induced crystallization, and on the mechanisms of cohesive and adhesive fracture.

Gold Medal. The first annual Gold Medal presented by Acta Metallurgica, Inc. was awarded in 1975 to Bruce Chalmers, longtime Gordon McKay Professor of Metallurgy at Harvard University. The award, which is supported by Pergamon Press, acknowledged the many contributions Chalmers has made to physical metallurgy as a teacher and as the author of six books and more than 130 scientific articles.

Mathematics

Cole Prize. The American Mathematical Society awarded a 1975 Frank Nelson Cole Prize in Algebra to both Hyman Bass and to Daniel G. Quillen. Bass was chosen for his paper "Unitary Algebraic K-Theory," Quillen for "Higher Algebraic K-Theories"; both papers were published during 1973 in *Springer Lecture Notes*.

Steele Prize. During 1975 the American Mathematical Society presented LeRoy P. Steele prizes to H. Blaine Lawson and to George W. Mackey for outstanding papers. Lawson's "Foliations" was published in the *Bulletin of the American Mathematical Society* (No. 80, 1974); Mackey's "Ergodic Theory and Its Significance for Statistical Mechanics and Probability Theory" appeared in *Advances in Mathematics* (no. 12, 1974).

Wiener Prize. The second Norbert Wiener Prize in Applied Mathematics was awarded in 1975 to Peter D. Lax, who was cited "for his work on numerical and theoretical aspects of partial differential equations and on scattering theory." The quinquennial award, endowed by the department of mathematics of the Massachusetts Institute of Technology, is jointly administered by the American Mathematical Society and the Society for Industrial and Applied Mathematics.

Courtesy, Bell Laboratories

Hendrik Bode

Mechanical engineering

Longstreth Medal. "For his pioneering development and invention of the Direct Energy Balance Control System for automatically balancing boiler inputs to power demands, particularly as applied to modern once-through steam generators," Edward S. Bristol received the 1975 Edward Longstreth Medal from the Franklin Institute. Before his retirement, Bristol was employed by the Leeds and Northrup Co. of Pennsylvania.

Oldenburger Medal. The American Society of Mechanical Engineers presented its 1975 Rufus Oldenburger Medal to Hendrik W. Bode for his contributions to automatic control. During his 41 years with Bell Laboratories, especially during and after World War II, Bode made many contributions of great importance to military defense systems. After leaving Bell in 1967 he joined the faculty of Harvard University, where he is emeritus Gordon McKay Professor of Systems Engineering.

Medical sciences

Aebersold Award. Gordon L. Brownell, who is associated with the Massachusetts Institute of Technology, with the Harvard Medical School, and with the Massachusetts General Hospital, received the 1975 Paul C. Aebersold Award from the Society of Nuclear Medicine. Brownell developed a positron scanning system used to detect brain tumors, and in the field of gamma-ray dosimetry he originated the concept of the absorbed fraction in addition to the use of Monte Carlo techniques for calculations.

Banting Medal. The American Diabetes Association awarded the 1975 Banting Medal to Roger H. Unger, a professor at the University of Texas Southwestern Medical School and chief of medical research at the Veterans Administration Hospital in Dallas. Unger has compiled evidence indicating that the pancreatic hormone glucagon affects the balance of blood sugar within the body.

Barclay Medal. The Council of the British Institute of Radiology awarded its 1976 Barclay Medal to Roy Astley of the Children's Hospital in Birmingham, England. Astley, who specializes in pediatrics, was honored especially for work related to the cardiovascular and gastrointestinal systems of small children.

Gairdner Awards. The Gairdner Foundation of Ontario, Canada, awarded international prizes to three medical scientists in 1975. Henri G. Hers of Brussels, Belgium, was cited "for fundamental discoveries related to glycogen metabolism and its disorders, disorders of fructose metabolism and the lysosomal basis of several inborn errors of metabolism." John D. Keith of Toronto was honored for "his many and important contributions to our understanding of the natural history of congenital heart disease, which have formed the basis for the modern treatment of congenital heart malformations." William T. Mustard, also of Toronto, received an award for "his outstanding contributions in the area of cardiovascular surgery, in particular the achievement of the ingenious operation for the transposition of the great vessels, which operation now bears his name."

Lasker Awards. The $10,000 Albert Lasker Clinical Medical Research Award for 1975 was shared by Godfrey N. Hounsfield, a British electronics engineer, and William Oldendorf, a neurologist at the University of California, Los Angeles. Hounsfield invented a computerized X-ray brain scanner that has revolutionized the diagnosis of brain disorders. Its highly detailed pictures of the brain are widely regarded as the most significant advance in radiology since the discovery of X-rays. Oldendorf is credited with having suggested the idea of an X-ray brain scanner a decade earlier, but he was unable to persuade manufacturers of X-ray equipment to undertake the project.

Two Albert Lasker Basic Medical Research Awards, each worth $10,000, were also awarded in 1975. The first was shared by Roger C. L. Guillemin of the Salk Institute for Biological Studies in La Jolla, Calif., and by Andrew V. Schally of the Veterans Administration Hospital in New Orleans. Each helped to explain more fully the functions of that part of the brain called the hypothalamus. The

London Daily Express from Pictorial Parade

Renato Dulbecco

second Basic Medical Research Award was presented to Frank Dixon of the Scripps Clinic and Research Foundation in California and to Henry G. Kunkel of Rockefeller University. Their pioneering research contributed significantly to an understanding of the role played by immune complexes in such diseases as glomerulonephritis and rheumatoid arthritis.

Four scientists at the Merck Sharpe & Dohme Research Laboratories in New Jersey shared a special 1975 Lasker Award for creating the thiazide diuretics, a new class of drugs used to control high blood pressure and swellings associated with heart disease, with renal dysfunction, and with liver disease. This special Lasker award, which carried with it an honorarium of $10,000, was shared equally by John E. Baer, Karl H. Beyer, Jr., Frederick C. Novello, and James M. Sprague.

Nobel Prize. The 1975 Nobel Prize for Physiology or Medicine was shared by Renato Dulbecco, Howard Temin, and David Baltimore for "discoveries concerning the interaction between tumor viruses and the genetic material of the cell." In administering the award, the Nobel committee at Sweden's Karolinska Institutet noted that even though viruses have not yet been shown to be involved in the formation of malignant tumors in humans, it appears likely that research will eventually prove this to be so. The committee added that the "technology to study such a possible relationship is available today, and the conceptual foundation for an examination of this problem has been provided by the discoveries made by [these] Nobel Prize winners."

Dulbecco was born in Italy and educated at the University of Turin. He became a U.S. citizen before commencing research in London at the Imperial Cancer Research Fund Laboratory. Temin worked with Dulbecco in the late 1950s at the California Institute of Technology, his alma mater, and became a fellow of the U.S. Public Health Service. In 1969 he was appointed professor of oncology at the University of Wisconsin. Baltimore studied at the Rockefeller Institute (now Rockefeller University) in New York City and in 1968 joined the Massachusetts Institute of Technology.

Sandoz Prize. The British Pharmacological Society, which each year administers a £250 prize sponsored by Sandoz Ltd., bestowed the 1975 award on John Hughes of the University of Aberdeen, Scotland. His main research has centered on the effects of drugs on the transmission of nerve impulses.

Scientific Achievement Award. Rosalyn S. Yalow, senior medical investigator at the Bronx Veterans Administration Hospital, was given the 1975 Scientific Achievement Award of the American Medical Association. Yalow shared the honor with her longtime associate, the late Solomon A. Berson. Together they developed radioimmunoassay of plasma insulin, a technique for measuring minute quantities of insulin in the blood.

Optical engineering

Duddell Medal. The British Institute of Physics awarded the 1975 Duddell Medal and Prize to Ernst Ruska of West Germany for his work on elec-

Wide World

Howard Temin

Wide World

David Baltimore

tron microscopes and for his contributions to electron optics. The biennial award was established in 1923 to honor William du Bois Duddell, British inventor of the electromagnetic oscillograph.

Wood Prize. The first annual R. W. Wood Prize, sponsored by the Xerox Foundation and administered by the Optical Society of America, was awarded in 1975 to Emmett N. Leith of the University of Michigan and to Juris Upatnieks of the Environmental Research Institute of Michigan. The two received a $1,000 honorarium for their contributions to holography, in particular for recognizing and demonstrating the improvement in signal-to-noise to be gained by off-axis holography.

Physics

Boltzmann Prize. Kenneth G. Wilson of Cornell University, Ithaca, N.Y., was awarded the 1975 Boltzmann Prize of the International Union of Pure and Applied Science for his research in the field of statistical mechanics.

Born Medal and Prize. The Max Born Medal and Prize, jointly sponsored by the German Physical Society and the British Institute of Physics, is given alternately to German and British physicists. Trevor S. Moss of the Royal Aircraft Establishment, Farnborough, England, received the 1975 award for his achievements in the field of photoconductivity and for research on the optical properties of semiconductors.

Boys Prize. The 1975 Charles Vernon Boys Prize was given to Richard A. Stradling of Oxford University. In presenting the award, the British Institute of Physics cited Stradling's work on semiconducting and semimetallic solids, particularly in the areas of quantum-transport phenomena and magneto-phonon spectroscopy.

Cresson Medal. The Franklin Institute of Philadelphia presented a 1975 Elliott Cresson Medal to M. James Lighthill of the University of Cambridge "for his development of the concept of the acoustic quadrupole theory describing the generation of aerodynamic noise in fluid flow; for his aeroacoustic theory in the description of sound generated by a turbulent jet; and for his brilliant mathematical analysis of the whole mechanism of aerodynamic noise."

Europhysics Prize. The European Physical Society awarded the first Hewlett-Packard-Europhysics Prize in solid-state physics to four scientists for their discovery of the condensation of excitons. Sharing the 1975 award, which is worth 20,000 Swiss francs (about $7,700), were Victor S. Bagaev, Leonid V. Keldysh, and Jaroslaw E. Pokrovsky of the Lebedev Physical Institute in Moscow, and Michel Voos of the École normale supérieure in Paris.

Franklin Medal. The Franklin Institute of Philadelphia named John Bardeen of the University of Illinois recipient of the 1975 Franklin Medal. Bardeen, a Nobel laureate, was selected "for his comprehensive insight into the physical principles of electrical conductivity of solids; for his guidance of the thinking of scientists and engineers in the developing technology of semiconductor devices;

Wide World

Aage Bohr (left) and Ben Mottelson

Keystone

L. James Rainwater

and for his key role in developing the detailed theory of superconductivity."

Guthrie Medal and Prize. David Tabor of the Cavendish Laboratory in England was awarded the 1975 Guthrie Medal and a prize of £250 for his contributions to the study of surfaces. The award is presented annually by the British Institute of Physics.

Heinemann Prize. The West German Göttingen Academy of Science awarded a prize worth $12,000 to Philip W. Anderson of Bell Laboratories and Princeton University. He received the 1975 Dannie Heinemann Prize for his theoretical contributions to the physics of condensed materials, in particular for his elucidation of the phenomenon called "Anderson localization"—the inability of an electron to jump from site to site in a crystal.

Herzfeld Medal. In 1975 the Catholic University of America awarded the Karl F. Herzfeld Medal to John A. Wheeler of Princeton University. His special interests include the mechanisms of nuclear fission, the design of thermonuclear devices and plutonium-producing reactors, neutron stars, and black holes.

Maxwell Medal and Prize. Anthony J. Leggett of the University of Sussex received the 1975 Maxwell Medal and Prize from the British Institute of Physics. Leggett was honored "for his contributions to the theory of the behavior of condensed matter at very low temperatures, in particular, the elucidation of the transition in liquid ^{3}He."

Nobel Prize. The Royal Swedish Academy of Sciences awarded the 1975 Nobel Prize for Physics to Aage N. Bohr, Ben R. Mottelson, and L. James Rainwater for discovering "the connection between collective motion and particle motion in atomic nuclei and the development of the theory of the structure of the atomic nucleus based on this connection." This research, undertaken nearly a quarter-century ago, paved the way for nuclear fusion.

Bohr, whose father (Niels Bohr) became a Nobel laureate in physics in 1922, is director of the Niels Bohr Institute in Copenhagen. Mottelson, who was born in Chicago and educated at Harvard University, became a naturalized Danish citizen after joining Bohr at the Bohr Institute. Rainwater received his degree in physics from the California Institute of Technology and then began a long association with New York's Columbia University, where he worked for a time with Bohr.

Panetti Award. The Italian National Academy of Sciences named Ronald S. Rivlin of Lehigh University in Pennsylvania winner of the 1975 Premio Modesto Panetti in mechanics. Rivlin, who was born in London and educated at the University of Cambridge, became the eighth recipient of the gold medal and prize since 1958.

Pioneer Award. Ivan A. Getting, president of the Aerospace Corp., was named recipient of the 1975 Pioneer Award by the Aerospace and Electronic Systems Society of the Institute of Electrical and Electronics Engineers. Getting, honored for his contributions to microwave radar and fire-control systems, was responsible for the development of automatic tracking radar, which was used extensively during World War II.

Wetherill Medal. Donald N. Langenberg of the University of Pennsylvania, William H. Parker of the University of California at Irvine, and Barry N. Taylor of the National Bureau of Standards were named recipients of 1975 Wetherill medals by the

Franklin Institute of Philadelphia. The three were cited for "their outstanding experiments with the a.c. Josephson effect in the determination of $2e/h$ to high accuracy; for their improvement in knowledge of the fundamental constants; and for establishing the usefulness of the a.c. Josephson junction as a voltage standard."

Young Medal and Prize. "For his contributions to laser physics and particularly developments of the tunable dye laser," Daniel J. Bradley, professor of applied optics at London's Imperial College of Science and Technology, was awarded the 1975 Thomas Young Medal and Prize by the British Institute of Physics.

Psychology

Applications in Psychology Award. Nathan H. Azrin of the Rehabilitation Institute of Southern Illinois University was given the 1975 Distinguished Contribution for Applications in Psychology Award by the American Psychological Foundation (APF). He was cited "for diligence and imagination in applying learning principles to ameliorate a variety of human problems. . . . In particular, he has demonstrated the efficacy of multifaceted treatment programs for enuretics, stutterers, alcoholics, and individuals with marital problems and nervous habits."

Education in Psychology Award. Two psychologists were honored in 1975 for having made exceptional contributions to their profession as teachers. Bernice L. Neugarten of the University of Chicago and Philip G. Zimbardo of Stanford University each received a Distinguished Contributions to Education in Psychology Award from the American Psychological Foundation.

Gold Medal. The 1975 Gold Medal Award of the APF was given to Neal E. Miller of the Rockefeller University, a "scientific innovator and master experimentalist." His wide-ranging interests include psychoanalysis, learning theory, psychopharmacology, physiological psychology, and biofeedback.

Professional Contribution Award. The fourth annual Distinguished Professional Contribution Award of the APF was presented to George W. Albee of the University of Vermont in 1975. The citation noted that Albee's "active role in plotting the direction and independence of professional psychology . . . freed from the domination of older professions and older models" and his positive contributions to mental health have greatly influenced modern psychology.

Scientific Contribution Award. Three psychologists were honored by the APF in 1975 with a Distinguished Scientific Contribution Award and an honorarium of $1,000. Donald E. Broadbent received special commendation for his book *Perception and Communication* (1958), "the first systematic treatment of the organism as an information-processing system, with a specific structure that could be investigated by experiment." Robert R. Sears of Stanford received the award "for pioneering work and continuing leadership in making the study of socialization part of the science of psychology." David Shakow of the National Institute of Mental Health was honored "for his seminal contributions to research and theory that have pioneered the development of an experimental psychopathology of schizophrenia."

Courtesy, Rockefeller University, New York City

Neal Miller

Transportation

Collier Trophy. David S. Lewis of the General Dynamics Corp. and a U.S. Air Force-industry team were named recipients of the Robert J. Collier Trophy for 1975. Together they produced the F-16 fighter, which the National Aeronautic Association (NAA) viewed as a great advance in design and technology. The lightweight, relatively low-cost plane has 2.3 times the combat radius of previous fighters, and, the selection committee noted, "its high thrust-to-weight ratio and low wing loading permit it to out-turn or out-maneuver any threat aircraft in the classic air combat arena."

RAS Medals. The 1975 Gold Medal of Britain's Royal Aeronautical Society was given to David

Keith-Lucas, an aerodynamicist and designer at the Cranfield Institute of Technology, England, for developing the Short SC1 VTOL (vertical takeoff and landing) experimental vehicle and for his contributions to higher education. The Silver Medal was presented to W. A. Mair of the University of Cambridge for his research on fluid motion and aerodynamics. The Bronze Medal went to R. L. Bickerdike of the Royal Aircraft Establishment for improving aerospace materials and for furthering understanding of the relationship between microstructure and strength.

Wright Brothers Trophy. Clarence L. ("Kelly") Johnson was presented with the 1975 Wright Brothers Memorial Trophy by the NAA "for his vital and enduring contributions over a period of 40 years to the design and development of military and commercial aircraft." During a long career with the Lockheed Aircraft Corp., Johnson was chief architect of the P-38, the Constellation and Super Constellation, the F-80, the Jet Star, and the high-altitude U-2 reconnaissance plane.

Science journalism

AAAS-Westinghouse Awards. Each year the American Association for the Advancement of Science administers science journalism awards sponsored by Westinghouse Electric Corp. A $1,000 prize is assigned to each of three categories: articles appearing in U.S. newspapers having a daily circulation exceeding 100,000; articles appearing in newspapers having more limited circulation; and articles appearing in magazines that reach the general public. The first of the 1975 awards went to David Perlman for articles published in the *San Francisco Chronicle*; four dealt with hybrid DNA molecules (February–March 1975), others with the Pioneer 11 mission to Jupiter (November–December 1975), and three with unrelated scientific topics. Elizabeth J. Maggio, writing for the *Arizona Daily Star* in Tucson, was selected for "Aerosols Threaten Ozone Layer" (Nov. 10, 1974), "Man's Knowledge of Rainmaking Still Cloudy" (Nov. 3, 1974), and "Clovis Man: He Lived and Hunted Here 11,000 Years Ago, Then Disappeared" (Oct. 6, 1974). The magazine award went to Paul Brodeur for "Annals of Chemistry: Inert," which appeared in *The New Yorker* (April 7, 1975).

AIP-U.S. Steel Award. Frederic Golden, an associate editor with *Time* magazine, was named recipient of the 1976 American Institute of Physics-U.S. Steel Foundation Science-Writing Award in Physics and Astronomy. He received an honorarium of $1,500 for "Forecast: Earthquake," which appeared in *Time* on Sept. 1, 1975.

Grady Award. The 1976 James T. Grady Award for Interpreting Chemistry for the Public, which consists of a gold medal and a $2,000 honorarium, was given to Gene Bylinsky, an associate editor of

Paul Brodeur

Linus Pauling and Pres. Gerald Ford

Courtesy, National Science Foundation

Fortune magazine. The American Chemical Society, which administers the award each year, cited 26 articles written by Bylinsky over the past ten years as examples of outstanding reporting on a wide variety of scientific topics.

Miscellaneous

National Medal of Science. The United States government's highest scientific award, the National Medal of Science, is presented annually by the president of the U.S. to persons who have done outstanding work in the physical, biological, mathematical, or engineering sciences. The 13 recipients of 1974 gold medals were: Nicolaas Bloembergen, a physicist at Harvard University; Britton Chance, a biophysicist at the University of Pennsylvania; Erwin Chargaff, a biochemist at Columbia University; Paul J. Flory, a chemist at Stanford University; William A. Fowler, a physicist at the California Institute of Technology; Kurt Gödel, a mathematician at the Institute for Advanced Study in New Jersey; Rudolph Kompfner, a professor of electronics at the University of Oxford; James Van Gundia Neel, a geneticist at the University of Michigan; Linus Pauling, a chemist at Stanford University; Ralph B. Peck, a civil engineer in Albuquerque, N.M.; Kenneth S. Pitzer, a chemist at the Berkeley campus of the University of California; James A. Shannon, a medical investigator at Rockefeller University; and Abel Wolman, emeritus professor of sanitary engineering at Johns Hopkins University.

Science Talent Awards. The 35th annual Science Talent Search, sponsored by the Westinghouse Educational Foundation and administered by Science Service, produced the following winners in 1976. The first-place award, a $10,000 scholarship, was given to Edward S. Phinney III of Amherst Regional High School in Massachusetts. His project involved the use of color filter photography to determine characteristics of certain clusters of stars. The second-place $8,000 scholarship went to Scott T. Cohen of the Bronx High School of Science in New York City for a project in mathematics centered on amicable numbers. Winner of the third-place $8,000 scholarship was Mark A. Gubrud of Wilde Lake High School in Columbia, Md., for his project in optics; he evaluated a method for recording holograms. Three other winners each received a $6,000 scholarship: William A. Schwartz of Massapequa, N.Y.; Joel Fajans of Teaneck, N.J.; and Diane H. Wooden of Rockville, Md. Four others received scholarships worth $4,000: Alice Chow of Queens, N.Y.; David J. Boothby of Acton, Mass.; George D. Yancopoulos of Queens, N.Y.; and Rebecca A. Kurth of Los Angeles, Calif.

Obituaries

The following persons, all of whom died in recent months, were widely noted for their scientific accomplishments.

Baer, Erich Eugen Ferdinand (March 8, 1901–Sept. 23, 1975), German-born chemist, received his Ph.D. from Friedrich Wilhelm University in Berlin (1927) and then undertook research at the University of Berlin (1927–32) and the University of Basel in Switzerland (1932–37). In 1937 he transferred to the University of Toronto in Canada. Baer's research centered on the preparation of organic compounds of phosphorus that participate in physiological processes vital to both plants and animals. Among these compounds were simple substances involved in the formation, storage, and utilization of glucose and other carbohydrates. Baer also studied much more complex materials, the phospholipics, some of which take part in the reactions that occur in the clotting of blood, while others make up membranes and other structural components of cells and nerve fibers.

Bronk, Detlev Wulf (Aug. 13, 1897–Nov. 17, 1975), U.S. scientist and educator, is generally considered to be the father of modern biophysics, the application of physics to living organisms. After obtaining his doctorate from the University of Michigan, Bronk taught physiology and biophysics and headed the Institute of Neurology at the University of Pennsylvania. He then became president of Johns Hopkins University (1949–53) and later was president of Rockefeller Institute (1953–68), a

Detlev Bronk

Courtesy, Rockefeller University, New York City

medical research facility that became Rockefeller University in 1965. During his tenure at Johns Hopkins, he introduced an innovative program that permitted qualified undergraduates to sign up for graduate courses, and he later transformed Rockefeller into a high-caliber institution that awards only a few select doctorates in science each year. His influence on research and scholarship was further extended as head of the U.S. National Academy of Sciences (1950–62) and as president of the American Association for the Advancement of Science.

Chamberlain, Richard Hall (May 25, 1915–Dec. 5, 1975), U.S. radiologist, spent the major portion of his professional career at the University of Pennsylvania, where he headed the department of radiology from 1961 until his retirement in 1975. Besides pioneering the use of radionuclides for both diagnosis and therapy, he made contributions to rotation and neutron therapy, photoscanning, image amplification, and ambient temperature film processing. Chamberlain, who obtained his medical degree (1939) from the University of Louisville in Kentucky, was also the U.S. delegate to the UN Scientific Committee on the Effects of Atomic Radiation and was a member of the expert advisory panel on radiation at the World Health Organization. In 1969 he was awarded a gold medal by the American College of Radiology.

Cook, Sir James Wilfred (Dec. 10, 1900–Oct. 21, 1975), British scientist, graduated in chemistry from University College, London, and after a period of lecturing and research on coal-tar derivatives was invited by Sir Ernest Kennaway to join the Royal Cancer Hospital in London. He later became regius professor of chemistry at both the University of London (1935–39) and the University of Glasgow (1939–54). While in London Cook demonstrated the relationship of coal-tar derivatives to carcinogenesis, a major step in cancer research, and in 1936 shared with Kennaway the prize of the International Union Against Cancer. In 1938 he was elected a fellow of the Royal Society. Cook began a second career in 1954 as principal of the University College of the South West, which later became the University of Exeter. Soon after his retirement in 1965 he became vice-chancellor of the University of East Africa, Kampala, Uganda (1966–70), and then chairman of the Academic Advisory Committee of the New University of Ulster at Coleraine. He was knighted in 1963.

Dobzhansky, Theodosius (Jan. 25, 1900–Dec. 18, 1975), Russian-born geneticist, won international renown for his work on evolution and genetics and their implications for modern man. After graduating from the State University in Kiev (1921), he taught for six years in Kiev and Leningrad before going to the U.S. (1927) as a Rockefeller Foundation fellow. From 1929 to 1940 Dobzhansky taught genetics at the California Institute of Technology and from 1940 to 1962 was professor of zoology at Columbia University in New York. His *Genetics and the Origin of Species* (1937) was the first substantial synthesis of Darwinian evolution and Mendelian genetics and established evolutionary genetics as an independent discipline. Other writings include *Mankind Evolving* (1962), which has had great impact on anthropologists, *The Biological Basis of Human Freedom* (1956), *The Biology of Ultimate Concern* (1967), and *Genetics of the Evolutionary Process* (1970).

Courtesy, Rockefeller University, New York City

Theodosius Dobzhansky

Doxiadis, Konstantinos Apostolos (May 14, 1913–June 30, 1975), Greek architect, was an urban planner whose vision of man-centered rather than machine-dominated cities affected the lives of millions of persons all over the world. The science of human settlement that he developed—called "ekistics"—relies on the contributions of architects, engineers, and sociologists. Doxiadis studied at the Athens and Berlin polytechnic schools, graduating in 1936 as engineer-architect. The following year he was appointed director of town planning for the Athens-Piraeus area. After World War II he was director general of the Ministry of Reconstruction, planning adviser to various companies in Australia, consulting engineer on housing projects in the Middle East, and a teacher at Harvard University and the Massachusetts Institute of Technology.

Hertz, Gustav Ludwig (July 22, 1887—Oct. 30, 1975), German physicist, shared the Nobel Prize for Physics with James Franck in 1925. After studying at the universities of Göttingen, Munich, and Berlin, he was appointed an assistant in physics at the University of Berlin (1913), where he began to work with Franck. Through experiments, the two discovered the laws governing the impact of an electron upon an atom, thereby confirming an atomic theory earlier proposed by Niels Bohr. In 1925 Hertz transferred to the University of Halle and in 1928 moved to the Technische Hochschule in Berlin. After Soviet forces entered Berlin in 1945, he was sent to the U.S.S.R., where he continued his research until 1954. He then returned to East Germany as professor of physics and director of the Physics Institute in Leipzig (1954–61).

Historia-Photo

Gustav Hertz

Hodge, Sir William Vallance Douglas (June 17, 1903—July 7, 1975), British mathematician, was keenly interested in algebraic geometry, differential geometry, and pure mathematics. The theory of harmonic integrals that he developed has been especially influential in modern geometry. Hodge spent most of his professional life at the University of Cambridge, where he was Lowndean Professor of Astronomy and Geometry from 1936 to 1970. He also served as master of Pembroke College from 1958 to 1970. During this time he revitalized geometry by developing its relationship with analysis and topology. Ever eager to stimulate progress in mathematics, he helped inaugurate the annual British Mathematical Colloquium and played a leading role in the formation of the International Mathematical Union in 1952. He also presided over the International Congress of Mathematicians in 1958 and injected new vitality into several professional organizations while serving as their president. Hodge was elected a fellow of the Royal Society at the age of 35, was awarded the society's Royal Medal in 1957, and was knighted in 1959.

Holford, William Graham Holford, 1st Baron (March 22, 1907—Oct. 17, 1975), British architect, was a highly respected urban planner whose influence extended beyond Britain to many other parts of the world. After graduating in architecture from the University of Liverpool, he studied in Rome and visited the U.S. before entering private practice. He was appointed to the chair of civic design at Liverpool in 1937 and was head (1948–70) of town planning at University College, London, succeeding Sir Patrick Abercrombie in both posts. In 1944 he became adviser to the new Ministry of Town and Country Planning and with the minister, Lewis (later Lord) Silkin, drafted the Town and Country Planning Act of 1947. Holford's most notable building was the Army Museum in London. He was also a consultant in Australia and South Africa and drew up plans for the precincts of St. Paul's Cathedral and Piccadilly Circus in London. Other responsibilities included the presidency of the Royal Institute of British Architects (1960–62) and membership in both the Royal Fine Arts Commission and the Historic Buildings Council. Among his many honors and awards were the gold medal of the Town Planning Institute (1961) and the royal gold medal for architecture (1963).

Jones, Sir (Bennett) Melvill (Jan. 28, 1887—Oct. 31, 1975), British aeronautical engineer, was largely responsible for accumulating data that led to fundamental improvements in monoplane design. He joined the National Physical Laboratory in 1910 and worked at Farnborough and Orford Ness on aeronautics and aerial gunnery during World War I. In 1919 he was elected to the new chair of aeronautical engineering at Cambridge University, where his research led to important results in reducing the risks of fatal stalling, in streamlining, and, through flow studies, in minimizing drag. During World War II, while with the Ministry of Aircraft Production in London, he resumed work on aerial gunnery, advancing toward the gyro gunsight. He was elected a fellow of the Royal Society in 1939 and was knighted in 1942.

Josi, Enrico (1885?—Sept. 1, 1975), Italian archaeologist, was professor of topography at the Pontifical Institute of Christian Archaeology in Rome for 45 years (1925–70). His international reputation was already well established when Pope

Pius XII commissioned him (1939) to assist with the excavations taking place beneath St. Peter's Basilica. In time Josi and his colleagues compiled a detailed report documenting their discovery of the tomb of St. Peter the Apostle. Many years earlier Josi discovered a network of catacombs (1920) and the cemetery of St. Panfilus (1921) in Rome. The cemetery, which was uniquely intact, has since become a prime source of information on early Christian burials.

Kay, George Marshall (Nov. 10, 1904—Sept. 3, 1975), U.S. geologist, made important contributions to Paleozoic stratigraphy and was an authority on geosynclines, large troughs in which sedimentary rocks accumulated during past geologic eras. His research on continental drift modified earlier geologic theories on the origin of North America and suggested that in time Japan would become part of the Asian mainland and Alaska part of Siberia. Kay, who was associated with Columbia University for nearly half a century, received the Penrose Medal of the Geological Society of America in 1971 and was the author of *North American Geosynclines* (1951) and *Stratigraphy and Life History* (1965).

Kouwenhoven, William Bennett (Jan. 13, 1886—Nov. 10, 1975), U.S. electrical and biomedical engineer, developed heart resuscitating mechanisms and techniques that have since been used to save countless lives. His defibrillator imparts electrical shocks through the chest wall to suppress uncoordinated contractions of the heart so that its normal rhythm can be restored. Kouwenhoven, who spent some 60 years on the faculty of Johns Hopkins University, was also responsible for the worldwide use of external heart massage, an effective nonsurgical method for reversing cardiac arrest by applying brisk, repeated pressure on the breastbone.

William Kouwenhoven

Courtesy, Johns Hopkins University, Baltimore, Md.

Mishima, Tokushichi (Feb. 24, 1893—Nov. 19, 1975), Japanese metallurgist, gained international recognition for his invention (1931) of M. K. steel, a strongly magnetic alloy containing nickel and aluminum; it is now generally preferred for permanent magnets. Mishima alloys are also widely used in the manufacture of automobiles, airplanes, and electronics equipment. After graduating from Tokyo University, Mishima taught for a time and then studied in Europe before returning to his alma mater as professor of metallurgy and alloy science (1938). Both before and after his retirement (1950), he received many honors at home and in other countries, including the Imperial Academy Award in 1938, and the Albert Sauveur Achievement Award of the American Society for Metals in 1957.

Okazaki, Reiji (Oct. 8, 1930—Aug. 1, 1975), Japanese molecular biologist, made the important discovery (1966) that new DNA strands are synthesized during the process of DNA duplication in short sections (Okazaki pieces) that only later join together to form an uninterrupted unit. His final goal, unattained, was to determine how the small pieces start to form, the direction of their synthesis, and how they eventually unite. Okazaki's initial discovery explained how two new strands of DNA could be synthesized by duplicating both components of a parental double helix. He was professor of molecular biology at Nagoya University, his alma mater.

Proudman, Joseph (Dec. 30, 1888—June 26, 1975), British oceanographer and mathematician, studied at Trinity College, Cambridge, and at the University of Liverpool before commencing a 41-year career at Liverpool in 1913. He was appointed professor of applied mathematics in 1919, and in 1933 was given the chair of oceanography, which he occupied until his retirement in 1954. Three years after Proudman helped prepare a report (1916) on the state of ocean tide research for the British Association, he helped found the Tidal Institute (now the Bidston Laboratory of the Institute of Oceanographic Sciences), and served as its director from 1919 to 1946. He was also president of the International Association of Physical Oceanography (1951–54). Among other honors, he was awarded the Hughes Medal of the Royal Society (1957) and the Agassiz Medal of the U.S. National Academy of Sciences (1946).

Sauer, Carl Ortwin (Dec. 24, 1889—July 18, 1975), U.S. geographer, obtained his Ph.D. (1915) from the University of Chicago, then taught at the University of Michigan (1915–23) before serving as chairman of the department of geography at the University of California in Berkeley (1923–54). For Sauer, geography was inseparable from human history inasmuch as the Earth, its resources and its environment, is profoundly affected by man. He thus found it quite natural to delve into anthropology, archaeology, and sociology as extensions of his geographical studies. From him students learned respect for the material world in which they live and the need to use it benignly. In his writings Sauer expressed a wide variety of views, including the contention that Irish monks visited America long before Leif Ericson and that the New World was widely settled some 40,000 years ago. Among the many honors he received was the 1957 Vega Medal of the Swedish Society of Anthropology and Geography.

Courtesy, University of California, Berkeley, photograph by Dennis Galloway

Carl Sauer

Sjöqvist, Erik (July 15, 1903—July 16, 1975), Swedish archaeologist, participated in archaeological expeditions to Greece (1927) and Cyprus (1927–31) before joining the staff of the Royal Library in Stockholm (1932–39). He later directed the Swedish Archaeological Institute in Rome (1940–48), taught briefly at Princeton University (1948–49), and acted as private secretary to the king of Sweden (1949–51). During the next 20 years he was professor of classical archaeology at Princeton (1951–71). A mountainous region in Sicily was the site of Sjöqvist's most successful excavation. In 1954 a Princeton team under his direction unearthed the ancient walled city of Morgantina, which is mentioned in the chronicles of Greek and Roman historians. From well-preserved ruins, Sjöqvist and his colleagues were able to reconstruct much of the city's civic, social, and commercial history.

Tatum, Edward Lawrie (Dec. 14, 1909—Nov. 5, 1975), U.S. biochemist, shared the 1958 Nobel Prize for Physiology or Medicine with American geneticists George Beadle and Joshua Lederberg. While serving as research assistant at Stanford University in California (1937–41), Tatum, working with Beadle, confirmed that all biochemical processes are resolvable into a series of chemical reactions and that, when a genetic mutation can be shown to affect a specific chemical reaction, the enzyme catalyzing that reaction will be altered or missing. They thus demonstrated that each gene in some way determines the structure of a specific enzyme, thereby creating the modern science of molecular genetics. At Yale University, where he served as associate professor of botany (1945–46) and professor of microbiology (1946–48), Tatum worked with Lederberg. Their research was largely responsible for modern-day emphasis on bacteria as the primary source of information concerning the genetic control of biochemical

Edward Tatum

Courtesy, Rockefeller University, New York City

processes in the cell. Tatum returned to Stanford (1948–57) before joining the faculty of New York's Rockefeller Institute (now Rockefeller University).

Thompson, Sir (John) Eric Sidney (Dec. 31, 1898 —Sept. 9, 1975), British archaeologist, was a leading authority on the ancient Maya civilization of Middle America. After World War I he worked as a *gaucho* on an Argentine cattle ranch before entering the University of Cambridge. He was an assistant curator at the Field Museum of Natural History in Chicago (1926–35) and played a major role in the British Museum's expedition to British Honduras under T. A. Joyce in 1927. One of his major accomplishments was a precise correlation of the complicated Maya calendar with Christian calendars, now known as the Thompson Correlation. He joined the Carnegie Institute of Washington in 1935, retired in 1958, and was knighted in 1975.

Thomson, Sir George Paget (May 3, 1892—Sept. 10, 1975), British physicist, shared the Nobel Prize for Physics in 1937 with Clinton J. Davisson for their independent discovery of electron diffraction. The son of Sir J. J. Thomson, who discovered the electron in 1897 and won the Nobel Prize in 1906, he attended Trinity College, Cambridge. In 1922 he became professor of natural philosophy at the University of Aberdeen, where he conducted experiments demonstrating that a beam of electrons is diffracted upon passage through a crystalline substance. This work confirmed the 1924 theoretical predictions of Louis de Broglie that electrons have wavelike properties as well as properties common to particles. Thomson was elected a fellow of the Royal Society in 1930 and was knighted in 1943.

Sir George Thomson

The Mansell Collection

Tomkins, Gordon Mayer (June 4, 1926—July 22, 1975), U.S. biochemist, was an outstanding research scientist who specialized in hormonal control mechanisms and in the molecular biology of mammalian cells. His findings have been especially valuable in the fields of endocrinology, genetics, cellular and developmental biology, and cancer research. After obtaining an M.D. degree from Harvard University (1949) and a Ph.D. in physiology from the University of California at Berkeley (1953), Tomkins spent 16 years, in various capacities, at the National Institutes of Health in Bethesda, Md. In 1969 he transferred to the University of California Medical Center in San Francisco, where he became professor of biochemistry.

Vinogradov, Aleksandr Pavlovich (Aug. 21, 1895 —Nov. 16, 1975), Soviet scientist, was internationally known for his research on the origin of ores and on the distribution of elements in bedrock and soils, and for his hypothesis on the formation and evolution of the Earth's oceans and crust. He also undertook geochemical studies of other planets and of the Moon and manifested considerable knowledge of atomic energy in the course of presenting two papers at the Geneva International Conference on the Peaceful Uses of Atomic Energy in 1955. Vinogradov was long associated with the Soviet Academy of Sciences. He began as an assistant in the biogeochemical laboratory and in 1947 was named first director of the V. I. Vernadsky Institute of Geochemistry and Analytical Chemistry. He later became a full member of the academy and a vice-president. For some 20 years he also headed the department of geochemistry at Moscow M. V. Lomonosov State University.

Weis-Fogh, Torkel (March 22, 1922—Nov. 13, 1975), Danish physiologist, obtained his doctoral degree from the University of Copenhagen (1952) and then became research assistant to August Krogh, recipient of the 1920 Nobel Prize for Physiology or Medicine. Using a technique involving electron probe X-ray microanalysis, Weis-Fogh became an authority on the desert locust, especially on its flight mechanism. Among other things, he discovered a new type of elastic protein at the wing base of locusts and contractile material in protozoa. After serving for eight years as head of the Zoophysiological Laboratory in Copenhagen (1958–66), he was elected professor of zoology at the University of Cambridge and a fellow of Christ's College.

Index

Index entries to feature and review articles in this and previous editions of the *Yearbook of Science and the Future* are set in boldface type, *e.g.*, **Astronomy.** Entries to other subjects are set in lightface type *e.g.*, Radiation. Additional information on any of these subjects is identified with a subheading and indented under the entry heading. The numbers following headings and subheadings indicate the year (boldface) of the edition and the page number (lightface) on which the information appears.

Astronomy 77–263; 76–264; 75–181
  Center for Short-Lived Phenomena 75–151
  climatology 76–184
  employment problems 76–414
  Formation of Stars, The 77–76
  holography use 77–377
  honors 77–413; 76–418; 75–247
  hydrologic research 77–296
  Molecules in Space 77–104

All entry headings, whether consisting of a single word or more, are treated for the purpose of alphabetization as single complete headings and are alphabetized letter by letter up to the punctuation. The abbreviation "il." indicates an illustration.

A

B

C

D

E

F

G

H

I

J

K

L

M

N

O

Q

R

S

T

U

V

Acknowledgments

6 Photographs and illustrations by (left to right, top to bottom) Ron Villani; Jane Burton from Bruce Coleman Inc.; © 1974 by the Association of Universities for Research in Astronomy, Inc., The Kitt Peak National Observatory; Stanley Rosenfeld; © 1975 by the Association of Universities for Research in Astronomy, Inc., The Kitt Peak National Observatory; courtesy, Milwaukee Public Museum

42 Photographs by (top and center) Stanley Rosenfeld; (bottom) George Silk © Time Inc. from Time-Life Picture Agency

43 Russ Kinne from Photo Researchers

79, 82 Illustrations by Dennis Magdich

98 Data collected by R. J. Allen, R. D. Ekers, J. P. Terlouw, photograph produced by T. Cram; The Westerbork Radio Observatory is operated by the Netherlands Foundation for Radio Astronomy with the financial support of the Netherlands Organization for the Advancement of Pure Research (Z.W.O.); The National Radio Astronomy Observatory is operated by Associated Universities Inc. under contract with the National Science Foundation

108 © 1974 by the Association of Universities for Research in Astronomy, Inc., The Cerro Tololo Interamerican Observatory

109 Illustrations by Dennis Magdich, adapted from (top to bottom) "Interstellar Ammonia" by M. Morris et al., "The Astrophysical Journal" (AJ), vol. 186, p. 509, Dec. 1, 1973, © The American Astronomical Society (AAS), all rights reserved; "Water Vapor: Observations of Galactic Sources" by M. R. Meeks et al., "Science," vol. 165, p. 181, July 11, 1969, © AAAS; "Discovery and CO Observations of a New Molecular Source Near M17" by Charles Lada, Dale F. Dickinson, and Hays Penfield, AJ, vol. 189, p. L37, April 1, 1974, © AAS; "Radio Detection of Interstellar Sulfur Dioxide" by L. E. Snyder et al., AJ, vol. 198, p. L81, June 1, 1975, © AAS; "Formaldehyde Absorption in Dark Nebulae" by P. Palmer et al., AJ, vol. 156, p. L148, June, 1969, © The University of Chicago

119 Micrographs courtesy, Ann Arbor Science Publishers, Inc. and Walter C. McCrone Associates Inc.: (top left and center, bottom center) W. C. McCrone, R. G. Draftz, and J. G. Delly, "The Particle Atlas," (others) John G. Delly, "The Particle Atlas," 2nd ed.

167–175 Illustrations by Dave Beckes

209 Illustrations adapted from "Grzimek's Animal Life Encyclopedia," vol. 1; Van Nostrand Reinhold Co.

237–248 Illustrations by Dave Beckes

250 Photographs, courtesy (top to bottom) Herman Hertzberger, photo by Aerophoto Schiphol; Eugene A. Shinn, U.S. Geological Survey; General Electric Research and Development Center; G. E. Mattingly, Princeton University

294 Adapted from "Aseismic Uplift in Southern California," R. O. Castle, J. P. Church, M. R. Elliott, "Science," vol. 192, pp. 251–253, fig. 1, April 16, 1976, © AAAS

302 Illustrations adapted from (top to bottom) James Humphrey, "Business Week," p. 41, March 8, 1976; Robert McAuley, "Business Week," p. 51, November 10, 1975; James Humphrey, "Business Week," p. 18C, February 16, 1976

335 Illustrations adapted from (left to right) "Anatomy of Seed Plants" by Katherine Esau, Copyright © 1960 by John Wiley & Sons, Inc.; "Relation Between Mesophyll Surface Area, Photosynthetic Rate, and Illumination Level During Development for Leaves of Plectranthus parviflorus Henckel" by Park S. Nobel, Lawrence J. Zaragoza and William K. Smith, "Plant Physiology" (1975) vol. 55, pp. 1067–1070

368–369 Illustrations by Dennis Magdich, adapted from (p. 368) "Medical Illustrations of Total Hip and Total Knee Replacement" by Robert J. Demarest; (p. 369) "Manual of Internal Fixation" by M. E. Müller et al., Springer-Verlag, New York